Operations Management Research and Cellular Manufacturing Systems:

Innovative Methods and Approaches

Vladimir Modrák
Technical University of Kosice, Slovakia

R. Sudhakara Pandian
Kalasalingam University, India

BUSINESS SCIENCE Reference

Managing Director: Lindsay Johnston
Book Production Manager: Sean Woznicki
Development Manager: Joel Gamon
Development Editor: Mike Killian
Acquisitions Editor: Erika Carter
Typesetters: Chris Shearer and Deanna Jo Zombro
Print Coordinator: Jamie Snavely
Cover Design: Nick Newcomer

Published in the United States of America by
Business Science Reference (an imprint of IGI Global)
701 E. Chocolate Avenue
Hershey PA 17033
Tel: 717-533-8845
Fax: 717-533-8661
E-mail: cust@igi-global.com
Web site: http://www.igi-global.com

Library of Congress Cataloging-in-Publication Data

Operations management research and cellular manufacturing systems: innovative methods and approaches / Vladimir Modrak and R. Sudhakara Pandian, editors.
p. cm.
Includes bibliographical references and index.
Summary: "This book presents advancements in the field of operations management, focusing specifically on topics related to layout design for manufacturing environments"--Provided by publisher.
ISBN 978-1-61350-047-7 (hardcover) -- ISBN 978-1-61350-048-4 (ebook) -- ISBN 978-1-61350-049-1 (print & perpetual access) 1. Production management. 2. Manufacturing cells. 3. Operations research. I. Modrak, Vladimir, 1957- II. Pandian, R. Sudhakara, 1977-
TS155.O594 2011
658.5--dc23

2011020006

British Cataloguing in Publication Data
A Cataloguing in Publication record for this book is available from the British Library.

Table of Contents

Preface.......... ix

Acknowledgment.......... xiv

Section 1
Methods and Trends in Manufacturing Cell Formation

Chapter 1
Developments in Modern Operations Management and Cellular Manufacturing.......... 1
Vladimír Modrák, Technical University of Kosice, Slovakia (Slovak Republic)
Pavol Semančo, Technical University of Kosice, Slovakia (Slovak Republic)

Chapter 2
Decision Support Framework for the Selection of a Layout Type 21
Jannes Slomp, University of Groningen, The Netherlands
Jos A.C. Bokhorst, University of Groningen, The Netherlands

Chapter 3
Comparison of Connected vs. Disconnected Cellular Systems: A Case Study 37
Gürsel A. Süer, Ohio University, USA
Royston Lobo, S.S. White Technologies Inc., USA

Chapter 4
Design of Manufacturing Cells Based on Graph Theory.......... 53
José Francisco Ferreira Ribeiro, University of São Paulo, Brazil

Chapter 5
Genetic vs. Hybrid Algorithm in Process of Cell Formation.......... 68
R. Sudhakra Pandian, Kalasalingam University, India
Pavol Semančo, Technical University of Kosice, Slovakia
Peter Knuth, Technical University of Kosice, Slovakia

Chapter 6
Design of Cellular Manufacturing System Using Non-Traditional Optimization Algorithms 99
P. Venkumar, Kalasalingam University, India
K. Chandra Sekar, Sardar Raja College of Engineering, India

Chapter 7
Similarity-Based Cluster Analysis for the Cell Formation Problem 140
Riccardo Manzini, University of Bologna, Italy
Riccardo Accorsi, University of Bologna, Italy
Marco Bortolini, University of Bologna, Italy

Chapter 8
An Estimation of Distribution Algorithm for Part Cell Formation Problem 164
Saber Ibrahim, University of Sfax, Tunisia
Bassem Jarboui, University of Sfax, Tunisia
Abdelwaheb Rebaï, University of Sfax, Tunisia

Chapter 9
Cellular or Functional Layout? 189
Abdessalem Jerbi, University of Sfax, Tunisia
Hédi Chtourou, University of Sfax, Tunisia

Section 2
Production Planning and Scheduling in Cellular Manufacturing Environment

Chapter 10
Cell Loading and Family Scheduling for Jobs with Individual Due Dates 208
Gürsel A. Süer, Ohio University, USA
Emre M. Mese, D.E. Foxx & Associates, Inc., USA

Chapter 11
Production Planning Models using Max-Plus Algebra 227
Arun N. Nambiar, California State University, USA
Aleksey Imaev, Ohio University, USA
Robert P. Judd, Ohio University, USA
Hector J. Carlo, University of Puerto Rico - Mayaguez, Puerto Rico

Chapter 12
Operator Assignment Decisions in a Highly Dynamic Cellular Environment 258
Gürsel A. Süer, Ohio University, USA
Omar Alhawari, Royal Hashemite Court, Jordan

Chapter 13
Alternative Heuristic Algorithm for Flow Shop Scheduling Problem 277
Vladimír Modrák, Technical University of Kosice, Slovakia
R. Sudhakra Pandian, Kalasalingam University, India
Pavol Semančo, Technical University of Kosice, Slovakia

Chapter 14
Optimization and Mathematical Programming to Design and Planning Issues in Cellular Manufacturing Systems under Uncertain Situations 298
Vahidreza Ghezavati, Islamic Azad University, Iran
Mohammad Saidi-Mehrabad, University of Science and Technology, Iran
Mohammad Saeed Jabal-Ameli, University of Science and Technology, Iran
Ahmad Makui, University of Science and Technology, Iran, Seyed Jafar Sadjadi University of Science and Technology, Iran

Chapter 15
Planning Process Families with PROGRES 317
Linda L. Zhang, IESEG School of Management, France

Section 3
Related Issues to Cellular Manufacturing Systems

Chapter 16
Lean Thinking Based Investment Planning at Design Stage of Cellular/Hybrid Manufacturing Systems 342
M. Bulent Durmusoglu, Istanbul Technical University, Turkey
Goksu Kaya. Istanbul Technical University, Turkey

Chapter 17
Performance Comparison of Cellular Manufacturing Configurations in Different Demand Profiles 366
Paolo Renna, University of Basilicata, Italy
Michele Ambrico, University of Basilicata, Italy

Chapter 18
Petri Net Model Based Design and Control of Robotic Manufacturing Cells 385
Gen'ichi Yasuda, Nagasaki Institute of Applied Science, Japan

Chapter 19
Equipment Replacement Decisions Models with the Context of Flexible Manufacturing Cells 401
Ioan Constantin Dima, Valahia University of Târgovişte, Romania
Janusz Grabara, Częstochowa University of Technology, Poland
Mária Nowicka-Skowron, Częstochowa University of Technology, Poland

Chapter 20
Multi-Modal Assembly-Support System for Cellular Manufacturing 412
Feng Duan, Nankai University, China
Jeffrey Too Chuan Tan, The University of Tokyo, Japan
Ryu Kato, The University of Electro-Communications, Japan
Chi Zhu, Maebashi Institute of Technology, Japan
Tamio Arai, The University of Tokyo, Japan

About the Contributors 428

Index 435

Preface

ABOUT THE SUBJECT

In the globalization era, the production environment of all countries comes to the stage of realizing the real prosperity. With the growth of markets towards globalization, all the firms need to deal with the challenges facing it. This has resulted in the materialization of automated industries with high performance of manufacturing systems. Traditional manufacturing systems are not able to satisfy these requirements. In the global market there is an increasing trend toward achieving a higher level of integration between designed and manufacturing functions in industries to make the operations more efficient and productive. Operations management needs to reflect on these challenges. "Cellular Manufacturing Systems" (CMS) is one among the emerging trends, which can be implemented without losing much of production run time, with low set up time, low work-in-process inventory (WIP), short manufacturing lead time, high machine utilization, and high quality of products.

Manufacturing systems traditionally fall into three categories of layouts: job shop production, batch production, and mass production. Obviously, a batch production presents the topical problem for layout designers and manufacturing mangers. Since, in batch production the parts move in batches from one process to another process, each part in a batch must wait for the remaining parts in its batch to complete processing before it moves to the next stage. This will lead to increased production time, high level of in-process inventory, high production cost, and low production rate.

Taking this into account, this book is providing further understanding the subject with more fruitful ideas to academic researchers and managers of organizations in the pipeline.

ORGANIZATION OF THE BOOK

This book is compilation of 20 contributions to the field of operations management, especially of advance topics related to the layout design for manufacturing environments and production planning and scheduling in cellular manufacturing environment. These 20 chapters are written by a group of 43 authors from prestigious universities and firms.

"*Operations Management Research and Cellular Manufacturing: Innovative Methods and Approaches*" is organized in three sections.

Section 1: "*Methods and Trends in Manufacturing Cell Formation*" presents selected problems in plant layout designing. Decision making process in selecting the plant layout design is considered to be one of critical steps in a development of cellular manufacturing systems. Among other chapters in this sections are those devoted to the development and comparison of optimization algorithms and techniques for cell formation problems.

Section 2: "*Production Planning and Scheduling in Cellular Manufacturing Environment*" offers some advanced tools and approaches in this domain. It is not by chance that classical theories of Scientific Management give the first consideration to production scheduling. Equally, the distributed scheduling for cellular manufacturing systems plays important role in achieving the effectiveness and success of cellular manufacturing.

Section 3: "*Related Issues to Cellular Manufacturing Systems*" covers a wider spectrum of viewpoints by specialists in their respective fields. In this section some aspects of flexible manufacturing cells and robotic manufacturing cells, apart from other objects of interest, are discussed. These forms of cellular manufacturing, in addition to other advantages, observe principles of agile manufacturing and thereby help to satisfy the growing requirements of customization.

The first section includes 9 chapters summarized below.

Chapter 1, "*Developments in Modern Operation Management and Cellular Manufacturing*" by Vladimír Modrák and Pavol Semančo, maps the major publications/citations in these fields and their evolving research utility over the decades. This survey traces modern concepts and tools of operations management and cellular manufacturing in a successive order. Finally, the relationships between concept or/and tools in both areas that are empirically considered as consequences or coincidences present an object of interest.

Chapter 2, "*Decision Support Framework for the Selection of a Layout Type*" by Jannes Slomp and Jos A. C. Bokhorst, presents a decision support framework based on the analytic hierarchy process approach for the selection of a manufacturing layout. The value of the framework is illustrated by means of a case application.

Chapter 3, "*Comparison of Connected vs. Disconnected Cellular Systems: A Case Study*" by Gürsel A. Süer and Royston Lobo, discusses differences between connected vs. disconnected cellular systems with respect to average flowtime and work-in-process inventory under make-to-order demand strategy. The study was performed in a medical device manufacturing company.

Chapter 4, "*Design of Manufacturing Cells Based on Graph Theory*" by José Francisco Ferreira Ribeiro, offers a comparative study between sequential heuristics, simulated annealing, tabu search and threshold algorithm for graph coloring and its application for solving the problem of the design of manufacturing cells in a job shop system production. The results obtained with these algorithms on several examples found in the literature are consistently equivalent with the best solution hitherto known in terms of numbers of inter-cell moves and dimensions of cells.

Chapter 5, "*Genetic vs. Hybrid Algorithm in Process of Cell Formation*" by R. Sudhakara Pandian, Pavol Semančo, and Peter Knuth, focuses on presentation of hybrid algorithm and genetic algorithm that are helpful in production flow analysis to solve the cell formation problem. The evaluation of hybrid and genetic algorithms are carried out against the K-means algorithm and C-linkage algorithm that are well known from the literature. The comparison uses performance measure and the total number of exceptional elements in the block-diagonal structure of machine-part incidence matrix using operational time as an input.

Chapter 6, "*Design of Cellular Manufacturing System Using Non-Traditional Optimization Algorithms*" by P. Venkumar, describes an experimental study based on the implementation and comparison of meta-heuristics for cell formation problems with an objective of minimizing exceptional elements.

The meta-heuristics were implemented on ten 16 X 30 sized benchmark problems. The final sections include the comparison of computational time for the compared algorithms and pertinent conclusions.

Chapter 7, "*Similarity-Based Cluster Analysis for the Cell Formation Problem*" by Riccardo Manzini, Riccardo Accorsi, and Marco Bortolini, describes an application of hierarchical clustering method for the cell formation based problem on the application of a threshold level of group similarity. The experimental analysis represents the first basis for the identification of the best setting of the cell formation problem. This chapter confirms the importance of this threshold cut value for the dendrogram when it is explained in percentile on the number of nodes.

Chapter 8, "*An Estimation of Distribution Algorithm for Part Cell Formation Problem*" by Saber Ibrahim, Bassem Jarboui, and Abdelwaheb Rebaï, presents a new heuristic algorithm for machine-part cell formation problem. The objective of this chapter is to identify part families and machine groups and consequently to form manufacturing cells with respect to minimizing the number of exceptional elements and maximizing the grouping efficacy. The proposed algorithm is based on a hybrid algorithm that combines a variable neighborhood search heuristic with the estimation of distribution algorithm.

Chapter 9, "*Cellular or Functional Layout?*" by Abdessalem Jerbi and Hédi Chtourou, essentially focuses on the development of an objective methodology framework to compare the cellular layout (CL) to the classical functional layout (FL). This methodology can be easily applied to any manufacturing context and provides trustworthy results with a minimum experimentation effort.

Section 2, "*Production Planning and Scheduling in Cellular Manufacturing Environment*" is composed is composed of the following six chapters.

Chapter 10, "*Cell Loading and Family Scheduling for Jobs with Individual Due Dates*" by Gürsel A. Süer and Emre M. Mese, introduces a cell loading and family scheduling in a cellular manufacturing environment. What separates this study from others is the presence of individual due dates for every job in a family. Authors in this chapter propose two different approaches to tackle this complex problem namely, mathematical modeling and genetic algorithms. An experiment is carried out using both approaches and later the results are compared and a sensitivity analysis is also performed with respect to due dates and setup times.

Chapter 11, "*Production Planning Models Using Max-Plus Algebra*" by Arun N. Nambiar, A. Imaev, R. P. Judd, and H. J. Carlo, presents a novel building block approach to developing models of manufacturing systems. The chapter develops a generic modelling block with three inputs and three outputs. It is shown that this structure can model any manufacturing system. It is also shown that the structure is hierarchical, that is, a set of blocks can be reduced to a single block with the same three inputs and three output structures. Finally, several numerical examples are given throughout the development of the theory.

Chapter 12, "*Operator Assignment Decisions in a Highly Dynamic Cellular Environment*" by Gürsel A. Süer and Omar Alhawari, discusses concepts such as learning and forgetting rates with the aim to show how operator skill level varies from time to time; thus, the assignment decision is affected. The objective of this chapter is to propose better mathematical models for operator assignment and also compare the performance of two major strategies, Max and Max-Min, in highly dynamic cellular environments.

Chapter 13, "*Alternative Heuristic Algorithm for Flow Shop Scheduling Problem*" by Vladimír Modrák, R. Sudhakra Pandian, and Pavol Semančo, describes an alternative heuristic algorithm that is assumed for a deterministic flow shop scheduling problem. The algorithm is addressed to an m-machine and n-job permutation flow shop scheduling problem for the objective of minimizing the make-span when idle time is allowed on machines. In order to compare the proposed algorithm against the benchmarked, for this purpose, selected heuristic techniques and genetic algorithm have been used. In a realistic situation, the proposed algorithm can be used as it is without any modification and come out with acceptable results.

Chapter 14, "*Optimization and Mathematical Programming to Design and Planning Issues in Cellular Manufacturing Systems under Uncertain Situations*" by Vahidreza Ghezavati, Mohammad Saidi-Mehrabad, Mohammad Saeed Jabal-Ameli, Seyed Jafar Sadjadi, and Ahmad Makui, introduces basic concepts about uncertainty themes associated with cellular manufacturing systems and brief literature survey for this type of problem. The chapter also discusses the characteristics of different mathematical models in the context of cellular manufacturing.

Chapter 15, "*Planning Process Families with PROGRES*" by Linda L. Zhang, develops a PROGRES-based approach to model: planning data, knowledge and planning reasoning. The PROGRES-based process family planning models are hierarchically organized. At the top level, a meta-model is defined to conceptualize process family planning in general. Based on this meta-model, generic models are defined for planning process families for specific product families. Finally, instance models are obtained by instantiating the generic models, representing production processes for given product family members. The proposed approach is illustrated with planning processes for a textile spindle family.

Section three, "Related Issues to Cellular Manufacturing Systems," includes chapters 16-20.

Chapter 16, "*Lean Thinking Based Investment Planning at Design Stage of Cellular/Hybrid Manufacturing System*" by M. Bulent Durmusoglu and Goksu Kaya, focuses on providing a methodology for lean thinking based investment planning from the perspective of cellular or hybrid manufacturing systems. Its first part provides a general explanation of why lean thinking is so beneficial for managing manufacturing processes. The purpose of the second part is to explore axiomatic design approach it provides an overall view of what to do. The third part presents the actual use of the methodology with implementation of hybrid system at a furniture factory.

Chapter 17, "*Performance Comparison of Cellular Manufacturing Configurations in Different Demand Profiles*" by Paolo Renna and Michele Ambrico, aims to compare different configurations of cellular models through the main performance. These configurations are fractal CMS and cellular manufacturing systems with remainder cells, compared to classical CMS used as a benchmark. A simulation environment based on Rockwell ARENA® has been developed to compare different configurations assuming a constant mix of demand and different congestion levels.

Chapter 18, "*Petri Net Model Based Design and Control of Robotic Manufacturing Cells*" by Gen'ichi Yasuda, describes the methods of modelling and control of discrete event robotic manufacturing cells using Petri nets. A conceptual Petri net model is transformed into the detailed Petri net model based on task specification. Subsequently, detailed Petri net model is decomposed into constituent local Petri net based on controller tasks. Finally, simulation and implementation of the control system for a robotic workcell are described.

Chapter 19, "*Equipment Replacement Decisions Models with the Context of Flexible Manufacturing Cells*" by Ioan Constantin Dima, Janusz Grabara, and Mária Nowicka-Skowron, presents selected econometric models that are intended to solve a multiple machine replacement problem in flexible manufacturing cells with several machines. Firstly, models for a simple case multiple machine replacement problems are presented. Thereafter, the more complicated case is considered where technological improvement is taken into account.

Chapter 20, "*Multi-Modal Assembly-Support System for Cellular Manufacturing*" by Feng Duan, Jeffrey Too Chuan Tan, Ryu Kato, and Tamio Arai, proposes a multi-modal assembly-support system (MASS) which aims to support operators from both information and physical aspects. To protect operators in MASS system, five main safety designs as both hardware and control levels are also discussed. With the information and physical support from the MASS system, the assembly complexity and burden to the assembly operators are reduced. To evaluate the effect of MASS, a group of operators were required to execute a cable harness task.

TARGET AUDIENCE

The book is intended to support the academicians and industrialists (teachers, doctoral scholars, decision makers in industry, and students educated in this field). It is also intended to support subjects of operations management.

Vladimir Modrák
Technical University of Kosice, Slovakia

R. Sudhakara Pandian
Kalasalingam University, India

Acknowledgment

It is indeed our pleasure and honor to acknowledge those who have helped to make this book possible.

We would first like to thank Ms. Jan Travers, Director of Intellectual Property and Contracts at IGI Global and Mr. Mike Killian, Assistant Development Editor at IGI Global for their professional support in editing this book.

We are grateful to all who decided to participate in this publication project as the members of Editorial Advisory Board, contributors and reviewers, respectively. Also we are thankful to most of the authors who were willing to simultaneously serve as reviewers for chapters written by other authors.

Our special thanks go to Professor Ján Buda, Emeritus Fellow of CIRP, the International Academy for Production Engineering, for his continuing support and encouragement leading to the publication of this edited book.

Last, but not least, we would like to express our sincere thanks to our families for their understanding and patience during the editing of the book.

Vladimir Modrák
Technical University of Kosice, Slovakia

R. Sudhakara Pandian
Kalasalingam University, India

Section 1
Methods and Trends in Manufacturing Cell Formation

Chapter 1
Developments in Modern Operations Management and Cellular Manufacturing

Vladimír Modrák
Technical University of Kosice, Slovakia (Slovak Republic)

Pavol Semančo
Technical University of Kosice, Slovakia (Slovak Republic)

ABSTRACT

Operations management as a knowledge domain appears to be gaining position as a respected and dynamic academic discipline that is undergoing constant development. Therefore, from time to time it is sensible to monitor and analyze its developments by summarizing new features into comprehensive ideas. To support this necessity, the major publications/citations in this field and their evolving research utility over the decades are identified in this chapter. Because the goal of this book is to present the advancements in the area of operations management research, especially of advanced topics related to the layout design for cellular manufacturing, the second part of this chapter is focused on developments in cellular manufacturing approaches and methods by mapping literature sources during the last decade. Finally, the relationships between concept or/and tools in both areas that are empirically considered as consequences or coincidences are identified.

INTRODUCTION

Although the overviews of detailed historical developments in each cognition domain are useful, this survey will discuss modern eras of operations management and cellular manufacturing in a successive order.

Operations management (often called production management) may be defined in different ways depending upon one's attitude or point of view. Since this discipline is a field of management, it focuses on carefully managing processes

DOI: 10.4018/978-1-61350-047-7.ch001

to produce and distribute products faster, better and more cheaply than competitors. Operations management (OM) practically concerns all the operations within the organization and the objectives of its activities focus on the efficiency and effectiveness of processes. The modern history of production and operations management was initiated in the 1950s by the extensive development of operations research tools such as waiting line theories, decision theories, mathematical programming, scheduling techniques and other theories. However, the material covered in higher education was quite fragmented without the umbrella of what is called production and operations management (POM). Subsequently, the first publications 'Analysis of Production Management' by Bowman and Fetter (1957) and 'Modern Production Management' by Elwood Buffa (1961) represented an important transition from industrial engineering to operations management. Operations management finally appears to be gaining a position as a respected academic discipline. Thus, this may be a good time to update the evolution of the field. To achieve this goal, the major publications/citations in this field and their evolving research utility over the decades will also be identified in this chapter. Subsequently, opportunities and challenges of a modern operations management that managers were facing during the last decade will be examined.

Because the goal of this book is to present the advancements in the area of operations management, especially advance topics related to the layout design for manufacturing environments, the second part of this chapter focuses on developments in cellular manufacturing approaches and methods. A large body of literature has attracted a number of researchers to present different reports on the state of the art at different points in time. Several researchers have reviewed the literature and categorized the different methods. Our intention in this chapter is to analyze production-oriented cell formation methods based on the review mapping literature sources from 2000 to 2010.

Finally, in this chapter, we will note the relationships between concept or/and tools in both areas that are empirically considered as consequences or coincidences.

OPERATIONS MANAGEMENT IN THE CONTEMPORARY ERA

The process of building operations management theory and the definition of its scope or area has been treated by a number of authors. As mentioned above, the modern era of POM is closely connected with the history of industrial engineering (IE). The development of the IE discipline has been greatly influenced by the impact of operations research (Turner et al. 1993). Operations research (OR) was originally aimed at solving difficult war-related problems through the use of mathematics and other scientific branches. The diffusion of new mathematical models, statistics and algorithms to aid decision-making had a dramatic impact on industrial engineering development. Major industrial companies established operations research groups to help solve their problems. In the 1960s, expectations from OR were extremely high, and as was commented by Luss and Rosenwein (1997), "over the years it often appeared that the mathematics of OR became the goal rather the means to support solving real problems." This caused OR groups in companies to be transferred to traditional organization units within companies. As a reaction to this disappointment Corbert and Van Wassenhove (1993) classified OR specialists into three classes: theoreticians, management consultants, who focus on using the available methods to solve practical problems, and the "in-between" specialists called operations engineers, who adapt and enhance methods and approaches in order to solve practical problems. The term "operations engineers" was formulated due to the lack of a better term and accordingly the group could also be referred to as operations managers and the field conducting applied research to help solve

Figure 1. Conceptual map of relations between OM, IE, AM and OR

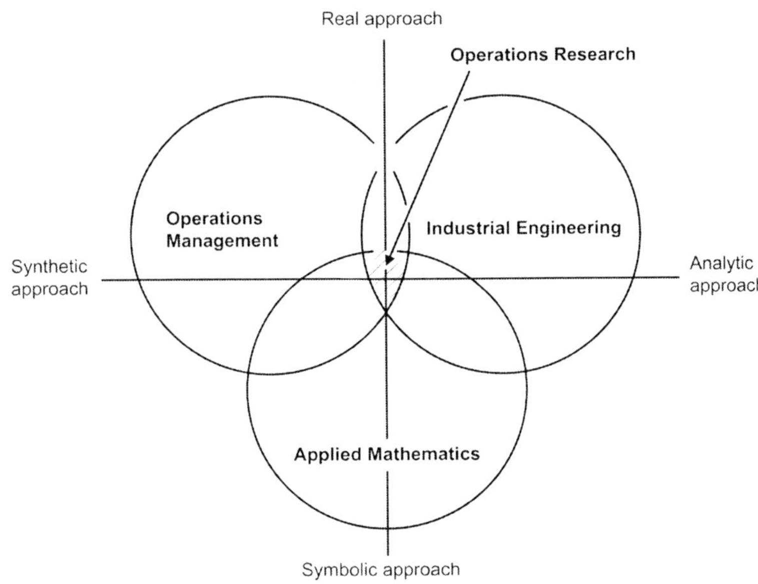

practical problems could be named production and operations management. In further developmental stages of OR, the term POM was consolidated and presented as concepts, methods and approaches related directly to productive systems and enhancing their management. Based on such a derivation of the mentioned disciplines it is obvious that IE, OR and OM have commonalities and similarities in their definitions. However, it is also important to specify the main differences among them. OM is a field of management, OR is a branch of applied mathematics (AM) and IE is an engineering discipline (Chase and Aquilano, 1989). In addition, according to Anderson (2002), OM and OR differ substantially, since "OM is managerially and activity oriented while OR is mainly technique and mathematically oriented involving modeling a situation or a problem and finding an optimal solution for it." Figure 1 illustrates a conceptual map of the compared disciplines with two axes: Symbolic/Real and Analytic/Synthetic.

As can be seen from the figure above, OM and OR follow two complementary routes that create a win-win scenario. Fuller and Martinec (2005), based on their analysis of the parallels between OM and OR, mentioned that both disciplines can be considered as "innovations" of the twentieth century.

Development Features of Operations Management

To discuss the important features of operations management, the reasonable action is to define and explain how the term can be understood from the viewpoint of the book's theme. For this purpose the following definition can be adopted: Operations management is concerned with the ways of achieving the most effective and efficient use of an organization's resources to produce goods and services needed by customers. It goes without saying that there are many other definitions that are more or less similar to the above definition. Although Chase and Aquilano (1989) precisely documented the historical development of OM starting with its real roots given by Taylor (1911), operations management is as old as industry itself (Bicheno and Elliot, 1997) and was articulated in the context of industrial production only after the 1960s (Baber, 1996; Landes, 1998). Because Chase and

Table 1. Historical summary of the history of modern OM (adopted from Chase and Aquilano, 1989)

Decade	Concept or Tool	Originator or Developer
1950s	Extensive development of OR tools of simulation, Queuing theory, Decision theory, Project scheduling techniques of PERT and CPM	David Georg Kendall (UK) Erich Leo Lehmann (USA)
1960s	Mathematical programming in industrial applications, Extensions of linear programming	Tjalling Koopmans (Netherlands) George Bernard Dantzig (USA)
1970s	Software packages for: Shop scheduling problems, Layout design, Forecast methods, Material requirement planning (MRP)	Joseph Orlicky (USA) Oliver Wight (USA) George W. Plossl (USA)
1980s	JIT, TQC and TQM, Factory automation, CIM, FMS, CAD/CAM Manufacture resource planning (MRP II)	Taiichi Ohno (Japan) W. Edwards Deming (USA) Armand V. Faigenbau (USA) Joseph M. Juran (USA) Mikell. P. Groover (USA)
1990s	The principles process innovations and business process reengineering, Logistics and Supply Chain Management, Optimized Production Technology (OPT), Theory of Constraints,	Thomas H. Davenport, (USA) Michael Martin Hammer (USA) James A. Champy (USA) Richard J. Schonberger (USA) Martin Christopher (UK) Eliyahu M. Goldratt (Israel) Eliyahu M. Goldratt (Israel)
2000s	Management of technology change, Disruptive Innovations and Organizational Change Operations strategy,	Clayton M. Christensen (USA) Wick Skinner (USA) Slack Nigel (UK)

Aquilano in their above-mentioned book mapped the history of the field of OM between circa 1910 and 1990, the following Table 1 gives an updated view of the historical development of operations management from the beginning of its modern era. The aim of this overview is to trace the key concepts and tools by decades from 1950 to 2010.

We are aware that the above specification of the latest developments in OM may not fully encompass all the decisive development directions, since operations management is a fair-sized and diversified field.

Opportunities and Challenges of Modern Operations Management

During the latest decennium operation managers had to react to unforeseen situations more frequently than they had needed to before. In this context they were facing new challenges of market globalization, information and communication technology advances as well as opportunities for organizational improvements and efficiencies.

One of the important challenges for the development of modern OM was the emergence of so-called global growth companies (GGCs). According to Jones (2005), during the mid-nineteenth century, thousands of European companies were formed exclusively to operate internationally with no prior domestic business. Those companies became momentous actors of economic globalization in international business. The emergence and influence of a new breed of high-growth global companies paying special attention to China and India was discussed by Kiggundu and Ji (2008). Concentrating on China and India as representatives of emerging economies explains the fact that these two countries were the best represented at the Dalian meeting in 2007 organized by the Centre for Global Growth Companies. By then,

Table 2. Selected challenges of modern operations management

Challenge	Description
Global Competition	Global market is increasingly complex and constantly changing. Products are traded internationally and components are sourced internationally. It requires a greater degree of international and cross-cultural communications, collaborations, and cooperation than at any time before. All companies have to think in global terms as regional companies are rapidly becoming a thing of past. (Steers and Nardon 2006).
Developments in strategic management approaches	Hambrick and Fredericson (2001) in their paper have talk about their uncertainty of whether that most organizations do actually have a strategy. According to them a meaningful strategy might consist of five elements, providing answers to following questions: Where will we be active? How will we get there? How will we win in the market-place? What will be our speed and sequence of moves? How will we obtain our returns? In reality, most strategic plans emphasize one or two of the elements without giving any consideration to the others.
Supply Chain Standardization and Integration	During the last decade has been proved the slogan that, much competition occurs between supply chains, not just between individual firms. This is due to the fact that company can't act as isolated entity, but as a part of supply chain integrated system.
Complex external environments	It is of crucial importance to understand how external environment impacts on organization. Therefore, companies are quite interested in knowing about macro environment situation representing the information on trends for demography, market geography, technologies energy demand growth, labor productivity growth, etc.. The environment in a global economy and its interactions with organizations is not only a complex phenomenon, but it is constantly changing in nature. Accordingly, any aspects of the environment can't be study as deterministic entities. By Kazmi (2008), "the organization and the environment are, in reality, more unpredictable, uncertain and non-linear". Therefore, for their study the complexity theory including chaos theory and their applications are applicable.

"firms in emerging economies and developing countries tend to have weaker systems of corporate governance than those in developed economies." In this connection, findings from differences between emerging economies and developed economies provide excellent opportunities for the study of corporate governance among global growth companies.

The aim of specifying general decisive and substantial challenges that managers have faced during the last 10 years comprises a substantial task, as it depends on different aspects. In an attempt to complete this task, in the Table 2 we depict some topical challenges that are related to the latest concepts and tools shown in Table 1.

In the continuing text, some of the main features of accented concepts and tools assigned to the last decade (shown in the table 1) will be illustrated with the aim of proving their topicality.

Management of Technology Change

According to Thomas and B. Grabot (2006), two main factors have dramatically changed the industrial context in the manufacturing area: specialization and technological changes that have recently occurred in the information technology area. Attention to that fact along with a large diffusion of innovations in industries during the twentieth century most likely evoked the emergence of the new managerial discipline of management of technology (MoT). The term itself was first introduced at the European Management Forum held in Davos in 1981. There are several definitions of MoT, which differ in their understanding of the very object of technology management in the sense of what needs to be managed. Drejer (2002), in this context, commented that "the discipline of MoT is characterized by a vast number of contributions emerging in a divergent manner rather than a convergent one." A succint definition of MoT has been formulated, for example, by Bueno et al. (1997), according to whom it is "the combination of competences allowing technological capabilities aiding the achievement of business objectives to be promoted and controlled." Although definitions of MoT are specific to a concrete target platform, the main object of interest in this chapter is its relevant context to business activities. One of the important roles of MoT is to promote innovation. It is especially topical for organizations that face a serious problem when technological changes are necessary in response to market signals. The internal conditions for implementing advanced technology for routine production are not always adequate for achieving this aim. Draft (2010), in this context, saw a problem with the organization of work. He argued that this problem can be solved only through innovative-oriented organization, which is typically associated with change and is considered the best for adapting to a changing environment. Therefore, programmes for the development of employees' creativity have become an important element of a cohesive corporate strategy.

Disruptive Innovations and Organizational Change

Presently, distinctions between disruptive technologies versus sustaining technologies are frequently discussed. According to the findings of Bowers and Christensen (1995), disruptive changes in technology had a significant impact on industries and many leading companies failed when they were confronted with them. Paradoxically, these failed firms were well-managed companies that invested aggressively in new technologies, carefully studied market requirements and opportunities and sharpened their competitive edges. Christensen (2002) proposed five principles of disruptive technologies in order to find a way to understand and harness this phenomenon. In his fourth principle he focuses on an organization's capabilities and disabilities, stating that "to succeed consistently, good managers have to be skilled not only just in choosing, training, and motivating the right people for the right job, but in choosing, building and preparing the right organization for the job as well." So, it is axiomatic that the phenomena of disruptive innovations and management of technology change are mutually reinforcing.

Operations Strategy

Admittedly, the operations or manufacturing strategy is considered as an inherent part of the long-term corporate strategy. Chase et al. (2004) offered with his sketch of a short history of operations strategy a broader insight into current operations strategy research and determined its role in contributing operations management functions to a firm's ability to achieve its competitive advantage in the marketplace. Since a firm's strategies are often changing and developing, it implies making sensible decisions that affect the business performances directly. In that context Swink and Way (1995) saw the position of manufacturing strategy as "the decisions and plans affecting resources and policies directly related to the sourcing, production and delivery of tangible products." Slack and Lewis's (2002) view of operations strategy is that it is not only a single decision, but the total pattern of the decisions that include the extent and ability of its capacity; delivery of products and services; approach to developing process technology, etc. The importance of operations strategy follows on from the fact that the long-term success of manufacturing firms depends on their ability to vary their operations quickly enough to fill the changing requirements of customers. The key factor that makes the operations function faster is called the manufacturing vision. For this reason, all principal world-class manufacturers have explicitly formulated a strategic manufacturing vision. Practically, it means that all the decisions related to system design, planning, control and supervision made by shop-floor managers are consistent with a corporate vision. On the other hand, world-class manufacturing ambition is not the only issue that matters. Therefore, it is not always optimal to adopt the most offensive manufacturing concepts that are inherent in world-class manufacturers. Accordingly, investing in improving marketing activities, product design or manufacturing operations can be as effective.

DEVELOPMENTS IN CELLULAR MANUFACTURING APPROACHES AND METHODS

Cellular manufacturing (CM), considered as an application of GT philosophy and its principle that focuses on the identification of similar parts to the benefit of a particular production, offers promising alternative solutions for manufacturing systems. CM can essentially be comprehended as a strategy that divides machines and parts into small groups or cells, where each cell can produce a family of parts completely. The manufacturing cells are basically composed of the heterogeneous machines to produce particular families' parts, which are allocated to these cells. For this purpose, various approaches and methods have been developed. Moreover, the CM approach benefits both the job and mass production. The main enhancement of cellular manufacturing implementation incorporates reductions in set-up time, throughput time and material handling and improved quality management.

One of the basic problems that has to be solved before implementing CM is the cell formation problem (CFP). The objective of the CF is to establish the family of parts and the group of machines for subsequent processes. The process of cell formation differs with respect to whether manufacturing cells have been created by rearranging existing facilities on the shop floor or whether new facilities are acquired for the cells. During the decades, a significant amount of research papers have been devoted to this problem. In this regard, several attempts at the classification of CF methods have been introduced. The classification of CF approaches was introduced for instance by Offodile et al. (1994) and Irani et al. (1999) In order to generalize previous classification frameworks, Table 3 shows a basic categorization of CF approaches that is in accord with the already-introduced contributions. The cell formation methods in Table 3 are accordingly divided into three basic groups. Each group has its own direction for part or machine identification.

In the last four decades, CMS research has mainly focused on production-oriented approaches. Therefore, the comprehensive reviews and taxonomy of studies that are devoted to CFP have been presented in previous research works. The following authors participated with their research studies in arranging all these CF methods into groups based on criteria like the minimization of inter-cell moves, machine utilization and others. Wemmerlöv and Hyer (1986) categorized more

Table 3. General classification of cell formation methods

Category	Brief description
Visual inspection methods	Visual inspection methods or eyeballing rely on the visual identification of the particular part families and machine groups.
Part classification and coding methods	PCA-based methods are oriented to design or shape feature. They attempt to group identical or similar design and manufacturing attributes into families. Therefore, they are ideal for reduction of product variety.
Production-oriented methods	The aim of these methods is to apply principles of line production to other types of production than mass production, even when the output is small and there is a large diversity of product. The PFA-based methods seek the optimal solution of cell formation in regard to objective and constraints.

than 70 papers into 4 representation groups. Subsequently, Selim et al. (1998) reviewed the literature (from 1963 to 1998) aimed at the cell formation problem, which is considered a fundamental issue in the CM environment. A comprehensive mathematical formulation of the CF problem has also been presented. The classification of reviewed papers based on multi-criteria cell design was employed by Mansouri et al. (2000), who applied the number of criteria as a measure for classification. They presented a review regarding the multi-criteria objective decision models that take into consideration the manufacturing cell formation problem.

Another view of the proposed taxonomy framework was introduced by Papaioannou and Wilson (2009). They also provided a review and comparison of 52 CF methods.

Production-Oriented Methods for CFP

Following the literature, the main scope within manufacturing cell formation methods focuses on production-orientated methods. In this section we present a modified classification framework for production-orientated CF methods and approaches based on the previous ones as shown in Figure 2.

In association with the proposed classification a brief review of the production-oriented CF methods in the following particular subsections is presented.

Descriptive Methods

One of the earliest descriptive approaches was developed by Burbidge, who began with the wave of CM systems. The proposed method by Burbidge (1971) is referred to as production flow analysis (PFA). The aim of PFA is to analyze the information from route cards to form cells. A manual method of Burbidge's PFA (1977) for CFP solutions is named nuclear synthesis.

Cluster Analysis Methods

Another group of approaches is presented as cluster analysis (CA). The objective of these methods is to group objects or entities or their attributes into clusters. Diverse techniques are applied in order

Figure 2. Classification of production-oriented methods for CF problem

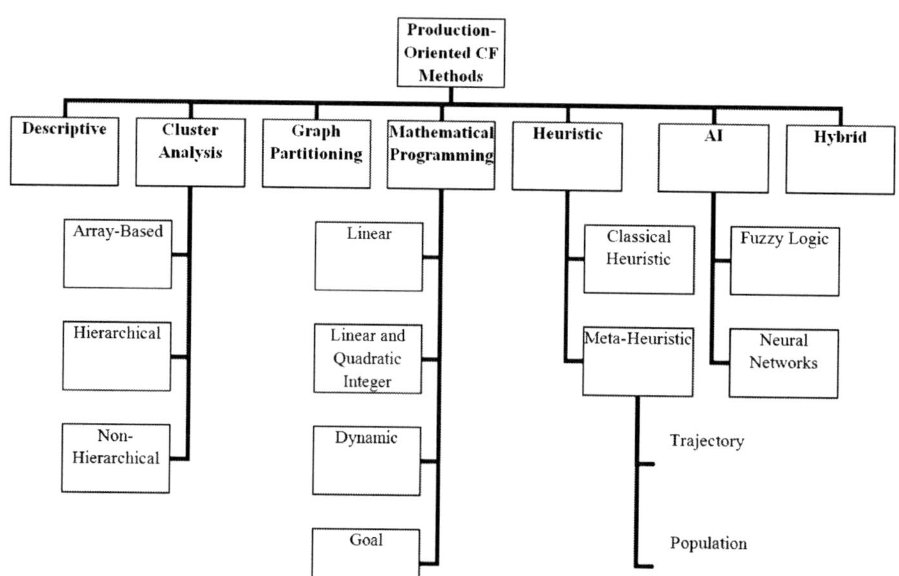

to create clusters. Those techniques can be further classified as array-based clustering methods, hierarchical clustering methods and non-hierarchical clustering methods within their own subsection. The first one rearranges the order of rows and columns to find a block-diagonal structure of a machine-part incidence matrix (MPIM). They are also known as array-based methods. McCormcik et al. (1972) are counted among the first researchers to have developed the cluster analysis method for the CM environment. They introduced the bond energy analysis (BEA) method. The other well-known CA-based methods are rank order clustering (ROC) by King (1980), direct clustering analysis (DCA) by Chan and Milner (1982) and modified rank order clustering (MODROC) by Chandrasekharan and Rajagopalan (1986). The block-diagonal structure (BDS) uses the minimization of the number of exceptional elements (EE) as its objective. EE represent inter-cell moves, which means in practice that they imply undesirable movement of parts to machines between the individual cells. An exceptional element basically means a bottleneck machine allocated to a cell while it is required in the other cells simultaneously, or a part in a family that requires the capabilities of machines allocated to other cells. The traditional MPIM is also represented as the binary (zero-one) matrix. The rows of the MPIM are machines and columns stand for parts. The entries in the matrix are '0's and '1's, which indicate whether a part needs to be a machine for production or not. The mathematical formulation to create a binary machine-part incidence matrix is defined as follows.

$$a_{ij} = \begin{cases} 1 & \textit{if part j visits machine i,} \\ 0 & \textit{otherwise.} \end{cases} \tag{1}$$

where 'i' is the machine index (i = 1, 2, 3,..., M), 'j' the part index (j = 1, 2, 3,..., P), M stands for the number of machines, and P the number of parts.

Figure 3 a presents the initial machine-part incidence matrix with a size of 9 machines and 17 parts that is formed by Equation (1). Figure 3 b shows the block-diagonal structure that includes one exceptional element.

The hierarchical methods are aimed at the separation of the MPIM data in several stages. In the first stage the MPIM data are grouped into a few broad cells. Subsequently the broad cells are partitioned into smaller groups until terminal groups are generated. The most frequent hierarchical methods applied to cell formation have been the three linkage methods. Single linkage (SL) was used by McAuley (1972), average linkage

Figure 3. Illustration of (a) binary MPIM with size of 9x17 and (b) block-diagonal structure with one EE

Part

Machine	1	2	3	4	5	6	7	8	9	10	11	12	13	14	15	16	17
1	0	0	0	1	0	0	0	0	1	0	0	0	0	0	1	1	0
2	0	1	1	0	1	0	0	1	0	0	1	0	0	1	0	0	1
3	0	0	0	0	0	1	1	0	0	1	0	1	1	0	0	0	0
4	1	0	0	1	0	1	0	0	1	0	0	0	0	0	1	1	0
5	0	0	0	0	0	1	1	0	0	1	0	1	1	0	0	0	0
6	0	1	1	0	1	0	0	1	0	0	1	0	0	1	0	0	1
7	1	0	0	1	0	0	0	0	1	0	0	0	0	0	1	1	0
8	0	0	0	0	0	1	1	0	0	1	0	1	1	0	0	0	0
9	0	1	1	0	1	0	0	1	0	0	1	0	0	1	0	0	1

a)

Part

Machine	1	4	9	15	16	20	2	3	5	8	11	14	17	6	7	10	12	13
1	**0**	1	1	1	1	1	0	0	0	0	0	0	0	0	0	0	0	0
4	1	1	1	1	1	1	0	0	0	0	0	0	0	**1**	0	0	0	0
7	1	1	1	1	1	1	0	0	0	0	0	0	0	0	0	0	0	0
2	0	0	0	0	0	0	1	1	1	1	1	1	1	0	0	0	0	0
6	0	0	0	0	0	0	1	1	1	1	1	1	1	0	0	0	0	0
9	0	0	0	0	0	0	1	1	1	1	1	1	1	0	0	0	0	0
3	0	0	0	0	0	0	0	0	0	0	0	0	0	1	1	1	1	1
5	0	0	0	0	0	0	0	0	0	0	0	0	0	1	1	1	1	1
8	0	0	0	0	0	0	0	0	0	0	0	0	0	1	1	1	1	1

b)

(AL) was used by Seifoddini and Wolfe (1986) and complete linkage (CL) was used by Mosier (1989). A representation of the hierarchical methods can be made by inverted tree structures also known as dendograms. The last of the cluster analysis methods, non-hierarchical methods, are iterative methods that need an initial partition of the data set. One of the well-known methods is ZODIAC, developed by Chandrasekharan and Rajagopalan (1987).

Graph Partitioning Methods

Graph partitioning methods consider machines or parts as nodes that are connected by arcs that represent the production flow between the machines. Graph methods enhance other methods like cluster analysis methods. Rajagopalan and Batra (1975) proposed the method that combines the use of a similarity coefficient and graph theory to solve the cell formation problem.

Mathematical Programming

Since the 1980s, a large number of research papers have been published in the field of mathematical programming with the aim of solving cell formation problems. Kusiak (1987), with his integer mathematical programming approach, was among the first of the authors to apply these methods to CFP. The formulation of mathematical programming (MP) can be employed to model CMS problems in a number of circumstances concerning a wide range of manufacturing data. The objective of MP is regularly maximization of the total number of part similarities in each cell, or minimization of inter-cell material handling costs. Most of the MP-oriented research papers are introduced and discussed by Selim et al. (1998). MP can be classified into four further groups with regard to their type of formulation: linear programming, linear and quadratic integer programming, dynamic programming and goal programming. Boctor (1991) dealt with the mathematical programming method. He proposed a linear formulation of the machine-part cell formation problem. Mathematical programming approaches belong to very time-consuming methods, which is why researchers have turned their attention to heuristic methods with their implementation in CMS.

Heuristic Methods

The heuristic methods are very fast in contrast to mathematical programming methods or others. It is generally known that the heuristic methods do not guarantee to find the optimal solution. However, if they are properly implemented and tuned up, the solution found will represent the optimum in most cases. They reach an optimal or pseudo-optimal solution in a reasonable amount of time. Heuristic methods start from a feasible solution then generate other random solutions, evaluate them and improve the effectiveness or goodness of the solution as time progresses. The presented classification framework in Figure 1 considers the further division of heuristics into classical heuristic methods and meta-heuristic methods, which incorporate evolutionary-based methods and population-based methods. There are numerous different heuristic approaches that are summarized in published review studies. Some of them are mentioned further.

Artificial Intelligence

Another significant group of methods aimed at CMS is introduced as artificial intelligence (AI) that is inspired by nature itself. Fuzzy logic and neural networks are the main approaches of this group. Kaparthi and Suresh (1992) proposed an application of neural networks to solve the cell formation problems. AI can be used to find patterns in manufacturing data in the CM environment. In most cases artificial intelligence approaches represent a robust and adaptive system. During the learning process they can perform a structure modification based on the information that flows through the network. Yang and Yang (2008) proposed a modified ART1 AI-based method to group data into machine-part cells. Guerrero et al. (2002) introduced the self-organizing neural network (SONN) approach, which solves the CF problem using a two-phase strategy. The first phase is dedicated to part-families formation and the second one assigns the machines to each part-family. Other contributors who have dealt with the artificial intelligence methods are shown in Table 2.

Hybrid Methods

The last group, frequently referred to as hybrid methods, solves the CF problem by combinations of two different methods. In this case they are based on combinations of cluster analysis methods, mathematical programming methods, heuristics and AI approaches. The hybrids take advantage of both methods by applying them to the cell formation problems they can solve efficiently. Based on the literature, in the last decades the hybrid methods have become very popular for the cell formation process in CMS. Caux (2000) proposed an approach that combines both the simulated annealing (SA) and branch-and-bound (BB) algorithms. The proposed approach (SABB) can simultaneously solve CF problems consisting of grouping machines into manufacturing cells and selecting one process plan for each part. Other hybrid methods are incorporated in Table 3.

Review of Modern Methodologies for the Cell Formation Problem

Based on the literature we can introduce an overall view of the evolution of production-oriented methods in CMS. This overall view corresponds with the previously presented classification of CF methods. It is divided into decades, starting with the 1960s, as shown in Figure 4.

Based on the Figure 4 we can propose another classification of CF methods according to progressivity as follows: classical optimization CF methods and modern CF methods. The first category includes descriptive methods, cluster analysis methods, graph partitioning methods, mathematical programming methods and some of the heuristic methods. The second category is formed by meta-heuristics, hybrid methods and AI approaches.

Figure 4. Overall view of CF methods evolution

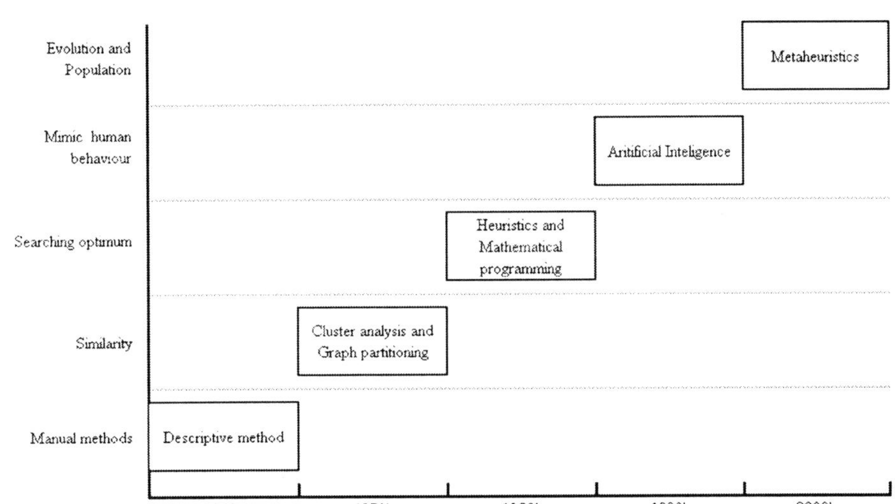

Table 4. Review of the modern cell formation approaches

	Source		CF	Categories							Performance			
No.	Author(s)	Year	method	1	2	3	4	5	6	7	A	B	C	D
1	Caux et al.	2000	Hy	·	•	□	□	•	□	•	(20x20)	·	✗	
2	Mak et al.	2000	GA	·	·	□	□	•	□	·	(40x100)	·	✗	
3	Lozano et al.	2001	Hy	·	•	□	□	·	•	•	(50x100)	•	✓	
4	Onwubolu and Mutingi	2001	GA	·	□	□	□	•	□	·	(20x45)	•	✓	4
5	Ravichandran and Rao	2001	Hy	·	•	□	□	·	•	•	(9x9)	·	✓	
6	Guerrero et al.	2002	NN	·	□	□	□	□	•	·	(40x24)	·	✓	1
7	Lozano et al.	2002	TS	·	·	□	·	•		·	(50x150)	·	✓	
8	Mukattash et al.	2002	H	·	·	□	□	•	□	·	(13x13)	·	✓	
9	Soleymanpour et al.	2002	Hy	·	·	·	•	·	•	•	(40x100)	•	✓	3
10	Logendran and Karim	2003	Hy	·	·	□	•	•	□	•	(15x30)	·	✗	
11	Park and Suresh	2003	Hy	·	·	□	□	·	•	•	(25x40)	·	✓	
12	Spiliopoulos and Sofi-anopoulou	2003	TS	·	·	□	□	•	□	·	(30x30)	•	✗	
13	Cao and Chen	2004	TS	·	·	·	·	•		·	(8x15)	•	✗	
14	Goncalves and Resende	2004	Hy	·	·	·	·	•	·	•	(40x100)	•	✓	6
15	Kim et al.	2004	H	·	·	□	□	•	□	·	(40x40)	·	✗	
16	Solimanpur et al.	2004	GA	·	·	□	□	•	□	·	(15x30)	•	✗	
17	Won and Lee	2004	Pm	·	·	·	•	·	·	·	(50x150)	·	✗	
18	Wu et al.	2004	TS	·	·	□	□	•	□	·	(18x5)	·	✗	
19	Albadawi et al.	2005	Hy	·	•	·	·	•	·	•	(16x43)	•	✓	4
20	Islier	2005	ACO	·	·	□	□	•	□	·	(40x100)	·	✓	3
21	Moghaddam et al.	2005	SA	·	·	·	·	•	·	·	(20x30)	·	✓	2
22	Prabhaharan et al.	2005	ACO	·	·	□	□	•	□	·	(40x100)	·	✓	
23	Venkumar and Haq	2005	NN	·	·	□	□	·	•	·	(40x100)	·	✓	
24	Andrés and Lozano	2006	PSO	·	·	□	□	•	□	·	(10x10)	•	✗	
25	Car and Mikac	2006	GA	·	·	·	·	•	·	·	(15x15)	•	✓	3
26	Defersha and Chen	2006	Hy	·	·	·	•	•	·	•	(30x90)	•	✗	
27	Nsakanda et al.	2006	H	·	·	□	□	•	□	·	(10000x25)	•	✗	
28	Foulds et al.	2006	TS	·	·	□	□	•	□	·	(5x7)	•	✗	
29	Lei and Wu	2006	TS	·	·	□	□	•	□	·	(15x30)	·	✓	
30	Nsakanda et al.	2006	Hy	·	·	·	·	•	·	•	(15x150)	•	✓	3
31	Torkul et al.	2006	NN	·	·	□	□	·	•	·	(18x25)	·	✓	
32	Venkumar and Haq	2006	NN	·	·	□	□	·	•	·	(40x100)	·	✓	
33	Won and Currie	2006	Pm	·	·	·	•	·	·	·	(41x30)	•	✓	1
34	James et al.	2007	Hy	·	·	·	·	•	·	•	(40x100)	•	✓	7
35	Saidi-Mehrabad and Safaei	2007	NN	·	·	□	□	·	•	·	(9x10)	•	✗	
36	Tavakkoli-Moghaddam et al.	2007	Hy	·	·	□	□	•	•	•	(10x10)	•	✗	
37	Wu et al.	2007	GA	·	·	□	□	•	□	·	NA	·	✗	

continues on following page

Table 4. Continued

No.	Source		CF method	Categories							Performance			
	Author(s)	Year		1	2	3	4	5	6	7	A	B	C	D
38	Boulif and Atif	2008	Hy	·	·	·	·	•	•	•	(20x24)	•	✓	2
39	Defersha and Chen	2008	GA	·	·	□	□	•	□	·	6x12)	•	✗	
40	Durán et al.	2008	SS	·	·	□	□	•	□	·	(12x12)	·	✗	
41	Kao and Li	2008	ACO	·	·	□	□	•	□	·	(50x150)	•	✓	
42	Megala and Rajendran	2008	ACO	·	·	□	□	•	□	·	(40x100)	·	✓	
43	Papaioannou and Wilson	2008	F	·	·	□	□	·	•	·	(9x9)	·	✗	
44	Safaei et al.	2008	F	·	·	□	□	·	•	·	(6x8)	•	✗	
45	Safaei et al.	2008	Hy	·	·	·	•	·	•	•	(6x8)	•	✓	1
46	Spiliopoulos and Sofi-anopoulou	2008	ACO	·	·	·	·	•	·	·	(37x53)	·	✓	
47	Tavakkoli-Moghaddam et al.	2008	SA	·	·	·	·	•	·	·	(17x30)	•	✗	
48	Wu et al.	2008	SA	·	·	·	·	•	·	·	(40x100)	•	✓	3
49	Yang and Yang	2008	NN	·	·	·	·	·	•	·	(46x105)	•	✗	
50	Ahi et al.	2009	Hy	·	•	·	·	·	•	•	(20x51)	•	✓	2
51	Bajestani et al.	2009	PSO	·	·	·	·	•	·	·	(9x10)	·	✓	
52	Bajestani et al.	2009	SS	·	·	·	·	•	·	·	(9x10)	•	✓	3
53	Mahdavi, et al.	2009	GA	·	·	·	·	•	·	·	(40x100)	•	✓	6
54	Oliveira et al.	2009	CA	·	•	·	·	·	·	·	(46x100)	•	✗	
55	Wu et al.	2009	Hy	·	·	·	·	•	·	•	(40x100)	•	✓	3
56	Wu et al.	2009	Hy	·	·	·	·	•	·	•	(50x120)	•	✓	8
57	Deljoo et al.	2010	GA	·	·	·	·	•	·	·	(10x10)	•	✗	
58	Li et al.	2010	ACO	·	·	·	·	•	·	·	(50x150)	•	✓	11
59	Naadimuthu et al.	2010	Hy	·	·	·	·	·	•	•	(6x5)	•	✗	
60	Neto and Filho	2010	Hy	·	·	·	·	•	·	•	(13x13)	·	✗	
61	Noktehdan et al.	2010	Hy	·	·	·	·	•	·	•	(37x53)	•	✓	2
62	Pailla et al.	2010	GA	·	·	·	·	•	·	·	(40x100)	•	✓	8
63	Wu et al.	2010	WFA	·	·	·	·	•	·	·	(50x100)	·	✓	2

Notes:
(I) Source list of contributors
(II) Acronym of CF approach
ACO = ant colony optimization CA = cluster analysis F = fuzzy logic GA = genetic algorithm H = classical heuristic.
Hy = hybrid NN = neural network Pm = p-median PSO = particle swarm optimisation SA = simulated annealing.
SS = search scatter TS = tabu search WFA = water flow-like algorithm.
(III) Major categories of production-oriented cell formation methods
1 = Descriptive methods 2 = Cluster analysis 3 = Graph partitioning 4 = Mathematical programming.
5 = Heuristic methods 6 = Artificial intelligence 7 = Hybrid methods.
(IV) Performance and computational results of the CF methods used
A = max size of data set used (machines x parts) B = provides performance measure for max size of data set used.
C = comparison to other existing methods ('✓' = yes and '✗' = no) D = number of other existing methods used for comparison NA = not available.

Figure 5. Frequency of individual categories for production-oriented CF methods

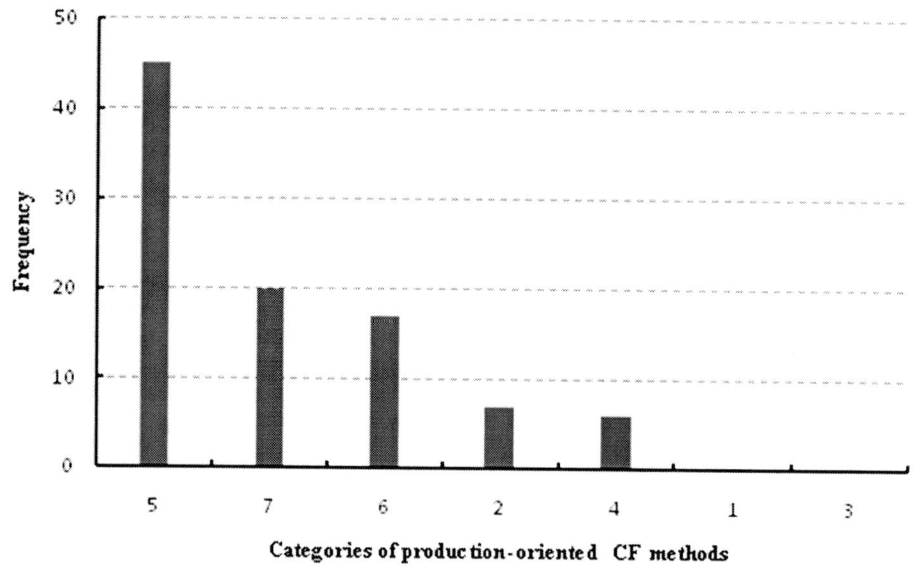

Figure 6. The percentage usage of meta-heuristic and heuristic classical methods for CFP

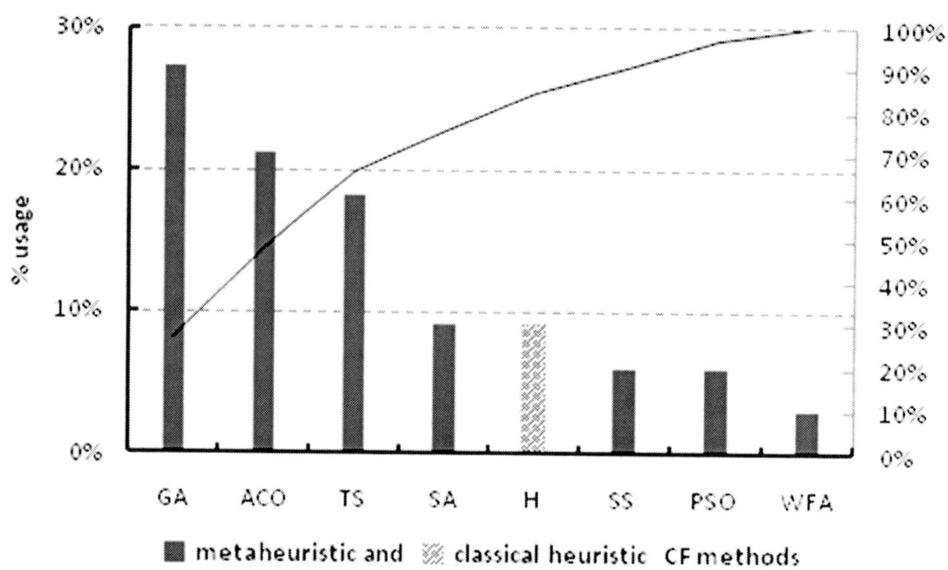

Because there are a number of existing studies mapping the time period from 1960 to 2000, our review presented in Table 4 focuses on the last decade. For the purpose of this review a classification based on descriptive approaches, cluster analysis approaches, graph partitioning approaches, mathematical programming approaches, heuristics, artificial intelligence and hybrid methodologies has been applied to categorize recent works. In addition, the methods are reviewed by key elements, such as the maximum size of the problems solved and the performance measures used.

From the survey depicted in the graph (see Figure 5) it is possible to identify the frequency of production-oriented cell formation methods.

The graph shows that the most frequent category of production-oriented cell formation methods is heuristics, including meta-heuristics (category # 5). Heuristics have been employed in CFP due to their promptness and acceptable solutions. Besides the heuristics approach there is another category that has attracted attention over the last few years. This category includes hybrid-based CF methods, which have increased significantly among the researchers in combinatorial optimization. Hybrids exploit the best from both methods combined in one technique. This provides more efficient behavior and higher flexibility. Based on the findings depicted in Figure 5, hybrids along with AI-based CF methods (categories # 7 and # 6) create the second level of importance. The next level of importance includes cluster analysis and mathematical programming methods (categories # 2 and # 4).

Even though the mathematical programming and the heuristic method for the CF problem were firstly introduced in about the same decade, the use and further development of mathematical programming methods in CFP were affected by computational limitations for large-scale problems and they were also time consuming.

The heuristic category has been chosen to compare the usage of particular CF methods due to the fact that it is the most frequent category of all. Based on Figure 6, the most utilized CF methods of the meta-heuristic methods in decreasing order are GA, ACO and TS. The cumulative percentage of the methods used is about 67%. The rest of the methods do not exceed the 10% bound, including the classical heuristic-based CF methods along with the simulated annealing method, scatter search method, particle swarm optimization (PSO) method and water flow-like algorithm (WFA). The order of particular methods in Figure 5 is obviously influenced by the chronology of their development, which has been provided in Table 4. From the mentioned review of the modern cell formation approaches in the last decades it is evident that the scatter search (SS) and water flow-like algorithms (WFA) have just recently received attention for CME.

CONCLUSION

The presented parallel survey on modern operations management and cell formation approaches independently maps the major development features in both POM and CM areas. As shown by Figure 1, CM research also represents a subset of POM. Then, the relationships between concept and/or tools in both areas can be empirically considered as consequences or coincidences. The causal relationship depends on the chronological order and assumes that X precedes Y otherwise Y would not occur. Accordingly, the surveys are based on taxonomies (classes or categories of items) and the chronological timing of the development changes.

Based on the comparison of the two areas it is possible to imply that mathematical programming is a mutual technique for POM and CM, even though this technique was dominantly applied in the given areas in different time periods. It shows that CM research partially applies generic methods or techniques of POM.

From another point of view it is viable to identify causal relationships between concepts and tools in the two areas. This relationship can be seen, for example, between the JIT concept in the 1980s and an expansion of the heuristic methods in the 1980s. The extensive acceptance of the just-in-time (JIT) philosophy by various industries supported the development of cell manufacturing in firms as it is an important element in the successful implementation of just in time. With the rapidly increasing demand for such solutions, the need for more effective cell formation methods was naturally growing. This demand is likely to have accelerated the development of cell formation methods based on heuristics and mathematical programming.

Moreover, the surveys of POM and CM show some topical development directions, such as the importance of operation strategy in POM and the dominance of meta-heuristic techniques in cell formation problems.

REFERENCES

Ahi, A., Aryanezhad, M. B., Ashtiani, B., & Makui, A. (2009). A novel approach to determine cell formation, intracellular machine layout and cell layout in the CMS problem based on TOPSIS method. *Computers & Operations Research*, *36*(5), 1478–1496. doi:10.1016/j.cor.2008.02.012

Albadawi, Z., Bashir, H. A., & Chen, M. (2005). A mathematical approach for the formation of manufacturing cells. *Computers & Industrial Engineering*, *48*(1), 3–21. doi:10.1016/j.cie.2004.06.008

Anderson, D., Sweeney, D., & Williams, T. (2002). *An introduction to management science: Quantitative approaches to decision making* (10th ed.). Cincinnati, OH: South-Western Publishing Company.

Baber, Z. (1996). *The science of empire*. Albany, NY: State University of New York Press.

Bajestani, M. A., Rabbani, M., Rahimi-Vahed, A. R., & Khoshkhou, G. B. (2009). A multi-objective scatter search for a dynamic cell formation problem. *Computers & Operations Research*, *36*(3), 777–794. doi:10.1016/j.cor.2007.10.026

Bicheno, J., & Elliott, B. B. (1997). *Operations management: An active learning approach*. Oxford, UK: Blackwell Publishing.

Boctor, F. F. (1991). A linear formulation of the machine-part cell formation problem. *International Journal of Production Research*, *29*(2), 343–356. doi:10.1080/00207549108930075

Boulif, M., & Atif, K. (2008). A new fuzzy genetic algorithm for the dynamic bi-objective cell formation problem considering passive and active strategies. *International Journal of Approximate Reasoning*, *47*(2), 141–165. doi:10.1016/j.ijar.2007.03.003

Bower, J. L., & Christensen, C. M. (1995). Disruptive technologies: Catching the wave. *Harvard Business Review*, *73*(1), 43–53.

Bowman, E. H., & Feter, R. B. (1957). *Analysis for production management*. Homewood, IL: R. D. Irwin, Inc.

Bueno, E., Morcillo, P., & Rodriguez Anton, J. M. (1997). Management of technology: Proposal for a diagnosis model. *The Journal of High Technology Management Research*, *8*, 63–87. doi:10.1016/S1047-8310(97)90014-6

Buffa, E. S. (1961). *Modern production management*. New York, NY: Wiley.

Burbidge, J. L. (1971). Production flow analysis. *Production Engineering*, *50*, 139–152. doi:10.1049/tpe.1971.0022

Burbidge, J. L. (1977). A manual method of production flow analysis. *Production Engineering*, *56*(1), 34–38. doi:10.1049/tpe.1977.0129

Cao, D., & Chen, M. (2004). Using penalty function and Tabu search to solve cell formation problems with fixed cell cost. *Computers & Operations Research*, *31*(1), 21–37. doi:10.1016/S0305-0548(02)00144-2

Car, Z., & Mikac, T. (2006). Evolutionary approach for solving cell-formation problem in cell manufacturing. *Advanced Engineering Informatics*, *20*(3), 227–232. doi:10.1016/j.aei.2006.01.005

Caux, C., Bruniaux, R., & Pierreval, H. (2000). Cell formation with alternative process plans and machine capacity constraints: A new combined approach. *International Journal of Production Economics*, *64*(1-3), 279–28. doi:10.1016/S0925-5273(99)00065-1

Chan, H. M., & Milner, D. A. (1982). Direct clustering algorithm for group formation in cellular manufacturing. *Journal of Manufacturing Systems*, *1*, 65–74. doi:10.1016/S0278-6125(82)80068-X

Chandrasekharan, M. P., & Rajagopalan, R. (1986). MODROC: An extension of rank order clustering for group technology. *International Journal of Production Research*, *24*, 1221–1233. doi:10.1080/00207548608919798

Chandrasekharan, M. P., & Rajagopalan, R. (1987). ZODIAC: An algorithm for concurrent formation of part-families and machine-ceils. *International Journal of Production Research*, *25*(6), 835–850. doi:10.1080/00207548708919880

Chase, R. B., & Aquilano, N. J. (1989). *Production and operation management: A life cycle approach* (5th ed.). Homewood, IL: Richard D. Irwin, Inc.

Chase, R. B., Aquilano, N. J., & Jacobs, F. R. (2004). *Operations management for competitive advantage* (10th ed.). Boston, MA: McGraw-Hill.

Christensen, C. M. (2002). *The innovator's dilemma: The revolutionary book that will change the way you do business*. New York, NY: Harper Business Essential.

Corbett, C. J., & Van Wassehove, L. N. (1993). The natural drift: What happened to operations research? *Operations Research*, *41*(4), 625–640. doi:10.1287/opre.41.4.625

Deljoo, V., Al-e-Hashem, S. M. J. M., Deljoo, F., & Aryanezhad, M. B. (2010). Using genetic algorithm to solve dynamic cell formation problem. *Applied Mathematical Modelling*, *34*(4), 1078–1092. doi:10.1016/j.apm.2009.07.019

Drejer, A. (2002). Towards a model for contingency of management of technology. *Technovation*, *22*(6), 363–370. doi:10.1016/S0166-4972(01)00029-3

Fuller, J. A., & Martinec, C. L. (2005). Operations research and operations management: From selective optimization to system optimization. *Journal of Business & Economics Research*, *3*(7), 11–15.

Gonçalves, J. F., & Resende, M. G. C. (2004). An evolutionary algorithm for manufacturing cell formation. *Computers & Industrial Engineering*, *47*(2-3), 247–273. doi:10.1016/j.cie.2004.07.003

Guerrero, F., Lozano, S., Smith, K. A., Canca, D., & Kwok, T. (2002). Manufacturing cell formation using a new self-organizing neural network. *Computers & Industrial Engineering*, *42*(2-4), 377–382. doi:10.1016/S0360-8352(02)00039-6

Hambrick, D. C., & Frederickson, J. W. (2001). Are you sure you have a strategy? *The Academy of Management Executive*, *15*(4). doi:10.5465/AME.2001.5897655

Irani, S. A., Subramanian, S., & Allam, Y. S. (1999). Introduction to cellular manufacturing systems. In Irani, S. A. (Ed.), *Handbook of cellular manufacturing systems*. USA: John Wiley & Sons. doi:10.1002/9780470172476.ch

James, T. L., Brown, E. C., & Keeling, K. B. (2007). A hybrid grouping genetic algorithm for the cell formation problem. *Computers & Operations Research*, *34*(7), 2059–2079. doi:10.1016/j.cor.2005.08.010

Jones, G. (2005). *Multinationals and global capitalism: From the nineteenth to the twenty-first century*. New York, NY: Oxford University Press.

Kaparthi, S., & Suresh, N. C. (1992). Machine-component cell formation in group technology: A neural network approach. *International Journal of Production Research*, *30*, 1353–1367. doi:10.1080/00207549208942961

Kazmi, A. (2008). *Strategic management and business policy, Tata*. New Delhi, India: McGraw Hill Publishing Company Ltd.

Kiggundu, M. N., & Ji, S. (2008). Global growth companies in emerging economies: New champions, new challenges. *Journal of Business and Behavioral Sciences*, *19*(1), 70–90.

King, J. R. (1980). Machine-component grouping in production flow analysis: An approach using rank order clustering algorithm. *International Journal of Production Research*, *18*, 213–232. doi:10.1080/00207548008919662

Kusiak, A. (1987). The generalized group technology concept. *International Journal of Production Research*, *25*, 561–569. doi:10.1080/00207548708919861

Landes, D. (1998). *The wealth and poverty of nations: Why some are so rich and others so poor*. New York, NY: W.W. Norton.

Li, X., Baki, M. F., & Aneja, Y. P. (2010). An ant colony optimization metaheuristic for machine–part cell formation problems. *Computers & Operations Research*, *37*(12), 2071–2081. doi:10.1016/j.cor.2010.02.007

Luss, H., & Rosenwein, M. B. (1997). Operations research applications: Opportunities and accomplishments. *European Journal of Operational Research*, *97*(2), 220–244. doi:10.1016/S0377-2217(96)00194-4

Mahdavi, I., Paydar, M. M., Solimanpur, M., & Heidarzade, A. (2009). Genetic algorithm approach for solving a cell formation problem in cellular manufacturing. *Expert Systems with Applications*, *36*(3), 6598–6604. doi:10.1016/j.eswa.2008.07.054

Mansouri, S. A., Husseini, S. M., & Newman, S. T. (2000). A review of the modern approaches to multi-criteria cell design. *International Journal of Production Research*, *38*, 1201–1218. doi:10.1080/002075400189095

McAuley, J. (1972). Machine grouping for efficient production. *Production Engineering*, *51*(2), 53–57. doi:10.1049/tpe.1972.0006

McCormick, W. T., Scweitzer, P. J., & White, T. W. (1972). Problem decomposition and data reorganization by a cluster technique. *Operations Research*, *20*, 993–1009. doi:10.1287/opre.20.5.993

Mosier, C. T. (1989). An experiment investigating the application of clustering procedures and similarity coefficients to the GT machine cell formation problem. *International Journal of Production Research*, *27*(10), 1811–1835. doi:10.1080/00207548908942656

Naadimuthu, G., Gultom, P., & Lee, E. S. (2010). Fuzzy clustering in cell formation with multiple attributes. *Computers & Mathematics with Applications (Oxford, England)*, *59*(9), 3137–3147. doi:10.1016/j.camwa.2010.02.038

Neto, A. R. P., & Filho, E. V. G. (2010). A simulation-based evolutionary multiobjective approach to manufacturing cell formation. *Computers & Industrial Engineering*, *59*(1), 64–74. doi:10.1016/j.cie.2010.02.017

Noktehdan, A., Karimi, B., & Kashan, A. H. (2010). A differential evolution algorithm for the manufacturing cell formation problem using group based operators. *Expert Systems with Applications: An International Journal*, *37*(7), 4822–4829. doi:10.1016/j.eswa.2009.12.033

Nsakanda, A. L., Diaby, M., & Price, W. L. (2006). Hybrid genetic approach for solving large-scale capacitated cell formation problems with multiple routings. *European Journal of Operational Research*, *171*(3), 1051–1070. doi:10.1016/j.ejor.2005.01.017

Offodile, O. F., Mehrez, A., & Grznar, J. (1994). Cellular manufacturing: A taxonomic review framework. *Journal of Manufacturing Systems*, *13*(3), 196–220. doi:10.1016/0278-6125(94)90005-1

Oliveira, S., Ribeiro, J. F. F., & Seok, S. C. (2009). A spectral clustering algorithm for manufacturing cell formation. *Computers & Industrial Engineering*, *57*(3), 1008–1014. doi:10.1016/j.cie.2009.04.008

Onwubolu, G. C., & Mutingi, M. (2001). A genetic algorithm approach to cellular manufacturing systems. *Computers & Industrial Engineering*, *39*(1-2), 125–144. doi:10.1016/S0360-8352(00)00074-7

Pailla, A., Trindade, A. R., Parada, V., & Ochi, L. S. (2010). A numerical comparison between simulated annealing and evolutionary approaches to the cell formation problem. *Expert Systems with Applications*, *37*(7), 5476–5483. doi:10.1016/j.eswa.2010.02.064

Rajagopalan, R., & Batra, J. (1975). Design of cellular production systems - A graph theoretic approach. *International Journal of Production Research*, *13*, 56–68. doi:10.1080/00207547508943029

Safaei, N., Saidi-Mehrabad, M., Tavakkoli-Moghaddam, R., & Sassani, F. (2008). A fuzzy programming approach for a cell formation problem with dynamic and uncertain conditions. *Fuzzy Sets and Systems*, *159*(2), 215–236. doi:10.1016/j.fss.2007.06.014

Seifoddini, H., & Wolfe, P. M. (1986). Application of the similarity coefficient method in group technology. *IIE Transactions*, *18*(3), 271–277. doi:10.1080/07408178608974704

Selim, H. M., Askin, R. G., & Vakharia, A. J. (1998). Cell formation in group technology: Review, evaluation and direction for future research. *Computers & Industrial Engineering*, *34*, 2–30. doi:10.1016/S0360-8352(97)00147-2

Slack, N., & Lewis, M. (2002). *Operations strategy*. London, UK: Financial-Times Prentice Hall.

Soleymanpour, M., Vrat, P., & Shankar, R. (2002). A transiently chaotic neural network approach to the design of cellular manufacturing. *International Journal of Production Research*, *40*(10), 2225–2244. doi:10.1080/00207540210122284

Steers, R. M., & Nardon, L. (2006). *Managing in the global economy*. Armonk, NY: M. E. Sharpe, Inc.

Swink, M., & Way, M. H. (1995). Manufacturing strategy: Propositions, current research, renewed directions. *International Journal of Operations & Production Management*, *15*(7), 4–26. doi:10.1108/01443579510090381

Tavakkoli-Moghaddam, R., Aryanezhad, M. B., Safaei, N., Vasei, M., & Azaron, A. (2007). A new approach for the cellular manufacturing problem in fuzzy dynamic conditions by a genetic algorithm. *Journal of Intelligent and Fuzzy Systems*, *8*(4), 363–376.

Tavakkoli-Moghaddam, R., Safaei, N., & Sassani, F. (2008). A new solution for a dynamic cell formation problem with alternative routing and machine costs using simulated annealing. *The Journal of the Operational Research Society*, *23*, 916–924.

Taylor, F. W. (1911). *The principles of scientific management*. New York, NY: Harper & Brothers.

Thomas, A., & Grabot, B. (2006). *Key technologies and strategies for creating sustainable manufacturing organizations*. IPROMS 2006 - Virtual International Conference on Intelligent Production Machines and Systems, Manufacturing Engineering Centre (MEC) Cardiff University Queen's Buildings.

Turner, W. C., Mize, J. H., Case, K. E., & Nazemetz, J. W. (1993). *Introduction to industrial and systems engineering*. Englewood Cliffs, NJ: Prentice Hall.

Wemmerlöv, U., & Hyer, N. L. (1986). Procedures for partfamily/machine group identification problem in cellular manufacturing. *Journal of Operations Management*, *6*, 125–145. doi:10.1016/0272-6963(86)90021-5

Won, Y., & Currie, K. R. (2006). An effective p-median model considering production factors in machine cell/part family formation. *Journal of Manufacturing Systems*, *25*(1), 58–64. doi:10.1016/S0278-6125(06)80033-6

Won, Y., & Lee, K. C. (2004). Modified p-median approach for efficient GT cell formation. *Computers & Industrial Engineering*, *46*(3), 495–510. doi:10.1016/j.cie.2004.01.010

Wu, T., Chang, C., & Chung, S. (2008). A simulated annealing algorithm for manufacturing cell formation problems. *Expert Systems with Applications: An International Journal*, *34*(3), 1609–1617. doi:10.1016/j.eswa.2007.01.012

Wu, T. H., Chang, C. C., & Yeh, J. Y. (2009). A hybrid heuristic algorithm adopting both Boltzmann function and mutation operator for manufacturing cell formation problems. *International Journal of Production Economics*, *120*(2), 669–688. doi:10.1016/j.ijpe.2009.04.015

Wu, T. H., Chung, S. H., & Chang, C. C. (2009). Hybrid simulated annealing algorithm with mutation operator to the cell formation problem with alternative process routings. *Expert Systems with Applications*, *36*(2), 3652–3661. doi:10.1016/j.eswa.2008.02.060

Wu, T. H., Chung, S. H., & Chang, C. C. (2010). A water flow-like algorithm for manufacturing cell formation problems. *European Journal of Operational Research*, *205*(2), 346–360. doi:10.1016/j.ejor.2010.01.020

Wu, T.-H., Low, C., & Wu, W.-T. (2004). A tabu search approach to the cell formation problem. *International Journal of Advanced Manufacturing Technology*, *23*, 916–924. doi:10.1007/s00170-003-1766-z

Wu, X., Chao-Hsien, C., Wang, Y., & Yan, W. (2007). A genetic algorithm for cellular manufacturing design and layout. *European Journal of Operational Research*, *181*, 156–167. doi:10.1016/j.ejor.2006.05.035

Yang, M. S., & Yang, J. H. (2008). Machine-part cell formation in group technology using a modified ART1 method. *European Journal of Operational Research*, *188*(1), 140–152. doi:10.1016/j.ejor.2007.03.047

Chapter 2
Decision Support Framework for the Selection of a Layout Type

Jannes Slomp
University of Groningen, The Netherlands

Jos A.C. Bokhorst
University of Groningen, The Netherlands

ABSTRACT

One of the most important design decisions in a firm is the choice for a manufacturing layout type. This chapter shows which aspects have to be taken into account and suggests a systematic method for the decision problem. The method can be seen as a decision support framework, which links the various aspects. The framework is based on the AHP (Analytic Hierarchy Process) approach. A case study, concerning a Dutch firm, illustrates the applicability of the framework in a practical instance.

INTRODUCTION

The choice for a manufacturing layout is a strategic issue and has a significant impact on the performance of the operations function of a company (Meijers and Stephens, 2004, Francis et al. 1992). A variety of manufacturing layout types may be applicable in a practical situation. Table 1 presents some alternative layout types for high-variety/low-volume situations. The most dominant layout type in practice is the process-oriented functional layout, where machines of the same type are located in the same area (Slomp et al., 1995). An important alternative is the socalled Cellular Layout type, where machines are grouped in cells and each cell is responsible for the complete manufacturing of a part family. This product-oriented layout type has gained substantial attention in literature and in practice (Wemmerlöv and Hyer, 1989, and Wemmerlöv and Johnson, 1997). Both types of manufacturing layout have their advantages and disadvantages. Several authors present alternative layout types to cope with the disadvantages of the functional and/or cellular layout type. Rosenblatt (1986) suggested a dynamic plant layout where cellular

DOI: 10.4018/978-1-61350-047-7.ch002

configurations periodically change depending on the demand in each period. Balankrishnan and Cheng (1998) present a review on the dynamic plant layout problem. Venkatadri et al. (1997) and Montreuil et al. (1999) propose a socalled fractal layout for job shop environments in order to gain the flow time advantages of Cellular Manufacturing and the flexibility of a functional layout. This type of layout is robust with respect to changes in demand and product mix. Another robust design, the socalled holographic or holonic layout, is proposed by Montreuil et al. (1993). Here individual machines, or machines types, are strategically distributed through the facility. Production orders are assigned to available machines which are located in the same area of the plant. A special case of the holonic layout is the socalled distributed layout (Benjaafar and Sheikhzdeh, 2000 and Benjaafar et al., 2002) where machine replicates are strategically distributed across physical space. Some researchers stress the need for a hybrid layout system which combines several layout types (e.g. Irani, 1993). Irani and Huang (2000) and Benjaafar et al. (2002) define a modular layout in which products have to be manufactured by one or more modules. Each module may have its own internal layout. A modular layout is an example of a hybrid layout. Wemmerlöv and Hyer (1989) show that many companies apply a hybrid layout.

This chapter presents a general decision support framework for the selection of a manufacturing layout type. Our focus lies on the selection of objectives, aspects and contributing elements for the selection problem. The framework applies the AHP (Analytic Hierarchy Process) approach (Saaty, 1980). This approach is useful for multi-criteria decisions where intuitive, qualitative and quantitative aspects play a role. The approach includes a hierarchical decomposition of the decision problem and a further decomposition of each decision level into pairwise comparisons of decision elements. Next, the "eigenvalue" method is used to estimate the relative weights of the decision elements. For a further explanation of the AHP method, we refer to Saaty (1980) or Zahedi (1986). As will be made clear in the remainder of this chapter, the AHP method offers several advantages in the layout type selection problem.

The next section will provide some further background to the selection problem. We will make clear that it is important to approach the layout type selection problem from a strategic viewpoint. In a subsequent section, we will discuss how layout types influence generic objectives of a company. We will then show how various aspects of layouts have an impact on manufacturing performance. The performance objectives and the various aspects are presented in the form of a decision hierarchy, according the AHP approach. After specifying the AHP approach, we present a case study to indicate the generic value of the defined decision hierarchy. The last section of this chapter is meant to reflect on the proposed selection methodology and to draw conclusions.

BACKGROUND

Literature on layout design problems falls into two major categories, algorithmic and procedural approaches (Yang and Kuo, 2003). Algorithmic approaches make use of simplified design constraints and objectives and can be used to generate layout alternatives efficiently (Meller and Gau, 1996). Algorithmic approaches are useful as a step in the design of a detailed layout. They assume the choice of a layout type. Procedural approaches may incorporate the choice of layout type and take care of both qualitative and quantitative objectives in the whole design process (Muther, 1973). A major disadvantage of a procedural approach is its dependence on the subjective judgement of one ore more experienced designers. Furthermore, procedural approaches divide the design problem in several steps which may lead to suboptimality. In order to overcome this suboptimality, designers may develop alternative layouts, based upon different layout types, and a well-working methodology

Table 1. Layout types and some major advantages and disadvantages

Type of Layout	Explanation	Major advantages	Major disadvantages
Process Layout or Functional Layout	Machines of the same type are located in the same area.	Routing Flexibility. Specialization in process type.	Complexity of coordination between departments.
Cellular Layout	Machines are grouped in cells and each cell is responsible for the complete manufacturing of a part family.	Short setup times because of the dedication of families to cells.	Sensitive for unbalance in the load of identical machines in different cells. Inflexible for the introduction of new products.
Dynamic Cellular Layout	A reconfigurable cellular layout.	Enables the cell layout to respond to product changes.	Costs of reallocating machines in case of product changes
Fractal Layout	Machines are grouped in various fractals, which are (more or less) identical cells able to produce all products.	Enables the cell layout to deal with changes in product mix.	Limited specialization of workers and machines.
Holonic Layout or Holographic Layout	Each machine (type) is an autonomous entity (holons) and is seemingly random (=random or based upon transition probabilities) located throughout the plant.	Provides efficient process routes for any production order. As orders arrive, routings are constructed by searching for compatibility between order requirements and machine availability, location, and capability.	Complexity of coordination between machine requirements of the various production orders
Distributed Layout or Scattered Layout. (Distributed or scattered layouts can be seen as special cases of the holonic layout)	Distributed or scattered layouts are those where machine replicates are strategically distributed across physical space.	Flexibility of assigning manufacturing orders to available machines which are located in the same area.	Limited specialization of workers and machines. Complexity of coordination.
Hybrid layout	Several layout types exist within one department	Fit between the various characteristics of the product types of a company and the various layout options.	Complexity of planning and control
Modular layout (Modular layouts can be seen as a special case of a hybrid layout)	Machines are clustered in modules. Each module has its own layout and is responsible for a number of operations to be performed on a product	Recognizes the layout needs of the various operations needed per product.	Complexity of the linkage of the various modules

may support the selection of the best layout. This chapter is devoted to the presentation of such a methodology. The methodology recognizes major differences between layout types.

Several authors propose methodologies to simultaneously cope with qualitative and quantitative objectives in the selection of a layout. Cambron and Evans (1991) applied Saaty's Analytic Hierarchy Process (AHP) to consider the problem's multiple objectives. The approach is illustrated by means of a problem involving the layout of a commercial printing and binding facility. Partove and Burton (1992) also propose AHP for layout selection. Yang and Kuo (2002) and Ertay et. al (2006) apply AHP and combine this with the data envelopment analysis (DEA) approach to solve the layout selection problem. Qualitative performance measures were weighted by AHP. DEA was then used to solve the multiple-objective layout problem. Yang and Kuo (2002) used a practical case study, an IC packaging company, to illustrate the efficiency and effectiveness of their methodology. Ertay et. al (2006) illustrate the applicability by means of a case study in which a choice has to be made between 17 alternative layouts for a company producing plastic profiles. Abdi and Labib (2003) apply AHP for the selection of a reconfigurable manufacturing system. Yang and Hung (2007) explore the use of multiple-attribute decision making (MADM) in the selection of an appropriate layout design. Two methods were proposed in solving the case study problem: the technique for order preference by similarity to ideal solution (TOPSIS) and fuzzy TOPSIS. The methodologies for layout selection, as presented in literature, focus on the methods to deal with conflicting quantitative and/or qualitative objectives. Limited attention is devoted to the selection of appropriate objectives and criteria. This chapter pays substantial attention to the choice of objectives and the specification of criteria.

Objectives and criteria in the layout literature are usually just a listing of some relevant elements. Researchers do not link the elements to the various strategic objectives of a company. The evaluation of layout alternatives is based on objectives such as (i) minimizing material handling costs, (ii) improving flexibility for arrangement and operation, (iii) utilising the available area most effectively, and (iv) minimising overall production time (Francis et al, 1992). Raman et. al (2009) distinguish three layout effectiveness factors—facilities layout flexibility (FLF), productive area utilisation (PAU) and closeness gap (CG). They claim that the measurement of these factors enables the decision-maker of a manufacturing enterprise to analyse a layout, based on which they can make decisions towards productivity improvement. They do not link the three factors to generic performance objectives of manufacturing companies. In this chapter we stress the importance of clarifying the link between layout decision criteria and the performance objectives of the company. This is needed to place the layout decision in its strategic context.

A major decision in many companies is the choice between a product-oriented and a process-oriented design philosophy. In a product-oriented design philosophy, machines and workers are grouped according to manufacturing needs of product types. A group of machines and workers is responsible for the complete manufacturing of (a set of) product types. The various groups in a product layout are relatively independent from each other. In a process-oriented design philosophy, machines and workers are grouped according the various functions needed to perform all product types. The functionally based groups are highly dependent on each other. Products flow from group to group. Most studies on the selection of a manufacturing layout implicitly assume a functional, or process-oriented, layout and are concerned with the allocation of the different functional groups. This chapter explicitly recognizes that companies may select a process-oriented or a product-oriented layout type, or a mix of both types. Table 1 gives an overview of possible layout types in a high-variety/low-volume environment. In our viewpoint, the type of layout has an important impact on the various performance objectives of a company.

The choice between a product- and a process-oriented layout is, many times, not an obvious decision. Case studies and survey articles (see e.g. Wemmerlöv and Hyer 1989, Burbidge et al. 1992) illustrate the enormous advantages of the introduction of a product-oriented, cellular manufacturing layout (CML). Other case studies indicate that several firms move from a cellular manufacturing layout towards a process-oriented, functional layout (FL) (see e.g. Slomp 1998, Molleman *et al.* 2002). Numerous simulation studies have been performed in order to compare the performance of a CML and a FL in various situations (for an overview, see Johnson and Wemmerlöv 1996, Argarwal and Sarkis, 1998, or Shambu et al., 1996). These studies indicate important factors, which have to play a role in the layout choice. Johnson and Wemmerlöv (1996), however, state that the simulation studies cannot assist practitioners in making specific choices between existing layouts and alternative cell systems. They indicate various mismatches between the model world and reality and suggest that decisions to change the existing layout should be made on a case-by-case basis for each potential cell application. A general framework, as presented in this chapter, may support managers in the selection of an appropriate layout type. The case study in this paper shows how the framework has supported the managers involved.

THE LINK BETWEEN STRATEGIC OBJECTIVES AND LAYOUT TYPE

An essential condition in the selection of a layout is the ability of decision makers to link the strengths and weaknesses of each layout alternative with the market demands, or performance objectives, with which the firm has to deal. Slack *et al.* (2001) distinguish five major performance objectives: price, quality, speed, flexibility, and dependability. The flexibility objective can be further split in product/service flexibility, mix flexibility and volume/demand flexibility. These objectives have to play an essential role in the selection of a manufacturing layout. We have added the objective "quality of work" to the set of objectives. This criterion is especially important in environments where labor is scarce. In the next subsections, we link the various performance objectives with aspects of the manufacturing layout type. We will refer to the functional (process-oriented) and cellular (product-oriented) layout type, as described in Table 1, in the discussion of the various aspects. Important words in the next subsections are written in italics. These words concern the aspects and elements which need to be dealt with in the layout selection problem. They are summarized in Table 2.

The Impact of Layout Type on the Price of a Product

From the perspective of a production manager, the price of a product has to be related to the manufacturing costs. A reduction of the manufacturing costs can be a reason to lower the price of products. Elements of the manufacturing costs are *equipment costs*, *personnel costs*, *material costs* and *inventory costs*. Several aspects of a

Table 2. Objectives, aspects and elements in the selection of a manufacturing layout

Objectives	aspects	elements
price	costs	• equipment costs • personnel costs • material costs • inventory costs
quality	• specialization of workers • advanced machinery • control loops	
speed	throughput time	• transport time • machining time • waiting time
flexibility • product/service • mix • volume/demand	• response • range	
dependability	• interchangeability of workers • interchangeability of equipment • control capacity	
quality of labor	• skill variety • task identity • task significance • autonomy • feedback	

manufacturing layout will have impact on these costs (see Figure 1). More *machines and tools* may be needed in a product-oriented cell layout in order to create independent, autonomous groups of workers and machines. A survey of 32 U.S. firms involved with cellular manufacturing, reported in Wemmerlöv and Hyer (1989), showed that new equipment and machine duplication was a major expense category for cell implementation. Specialization in a process layout may lead to a higher *production speed*, which may reduce equipment and personnel costs. On the other hand, *setup times* are usually lower in a product layout (see e.g. Flynn and Jacobs 1986, Wemmerlöv and Hyer 1989, Wemmerlöv and Johnson 1997) and reduce equipment and personnel costs in this type of layout. Furthermore, less *transport equipment* is usually required in a product layout because of the shorter transport distances. All these aspects have to be considered in order to estimate the impact on the equipment costs by the various types of layout. Personnel costs are, as mentioned above, related to the factors that have impact on equipment costs. Personnel costs in a product layout can also be lower than in a process layout because of a reduced need of *middle managers*. Farrington and Nazemetz (1998) indicate, by means of simulation studies, that a cellular system is easier to manage than a job shop. Empirical studies show the reduced need for indirect labor where firms convert from a functional layout to a cellular layout (Wemmerlöv and Hyer 1989, Burbidge 1992, Slomp *et al.* 1993). On the other hand, the *salaries* in a product-oriented layout may be higher, since more tasks are, probably, decentralized to the autonomous groups and workers need higher qualifications. These aspects of personnel costs should be taken into account when assessing the various layout alternatives. Material costs can be influenced by the layout through the effect of layout choice on *waste*. It is conceivable that workers in a product layout feel more responsibility for the reduction of the amount of waste. On the other hand, more advanced equipment and more specialized workers in a process layout may also reduce waste. Inventory costs can be lower in a product layout because of the smaller *lot sizes* that can be produced efficiently in this type of layout. This efficiency is due to the smaller setup times in a product layout. Further, *flow times* in a product layout are often lower than in a process layout and this reduces the work-in-process inventory. Reductions in throughput time and work-in-process inventory have been reported in surveys of plants that implemented cellular manufacturing (Wemmerlöv and Hyer 1989, Wemmerlöv and Johnson 1997).

The Impact of Layout Type on the Quality of a Product

The type of manufacturing layout can also influence the quality of products. In a process layout, workers are probably more *specialized* and will provide for a better product quality. In a product layout, experts are divided among the various groups and the best worker will not always be assigned to the most complex task. Furthermore, a process layout may apply more *advanced machinery*, which has a positive effect on the quality of products. On the other hand, the *control loops* in a product layout are short and may, in comparison to a process layout, have a positive effect on product quality.

The Impact of Layout Type on the Speed to Serve Customers

Speed concerns the time needed to fulfill the needs of internal or external customers. The throughput time of an average job is a measure for speed. This throughput time consists of *transport*, *machining* and *waiting times* (see Figure 2).

A product layout usually involves less *transport operations* for manufacturing jobs because of the proximity of the required machines. It may also be possible to produce in an *overlapping mode* (Shafer and Charnes 1993, Shafer and Meredith

Figure 1. Impact of layout factors on manufacturing costs

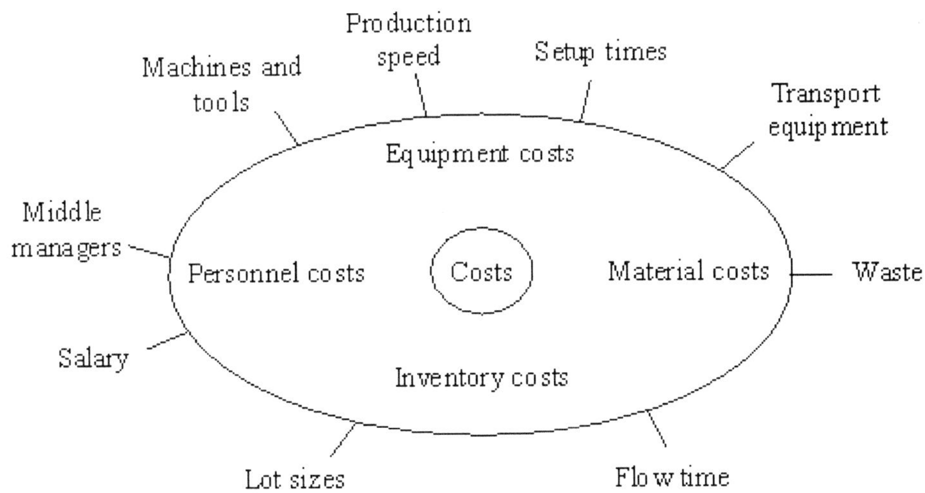

Figure 2. Impact of layout factors on speed

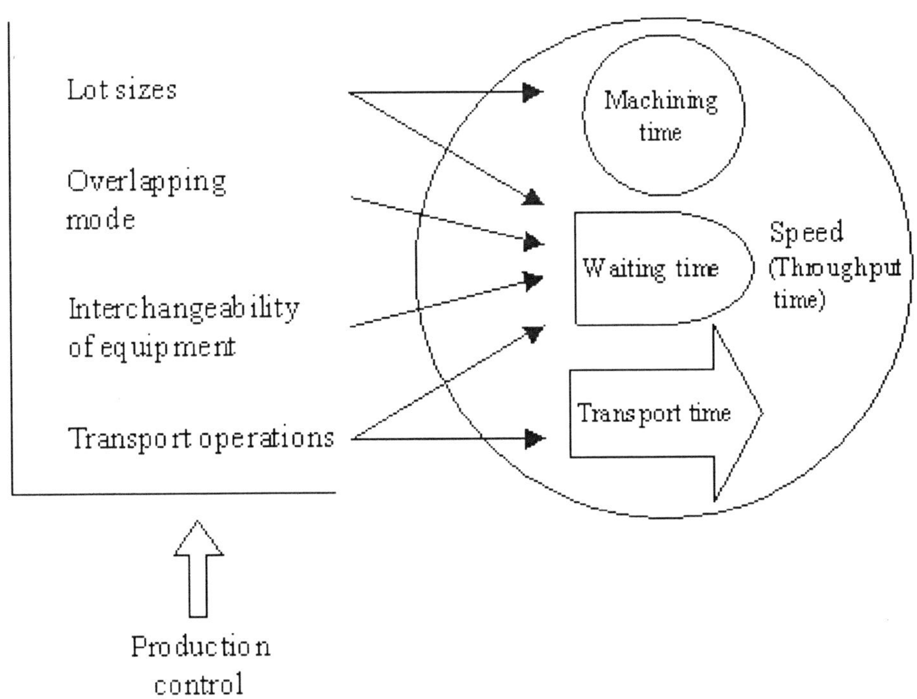

1993), which reduces the "lot size" waiting times. This is more easily realized in a product layout. Problematic for the waiting times in a product layout may be the lack of pooling synergy (Suresh and Meredith 1994). Identical machines are probably split over more than one group and the *interchangeability of equipment* is less than in a process layout. The throughput time will also be influenced by the *lot sizes*. Reduced setup time, which can be realized in a CM environment, may make smaller lot sizes acceptable. This may have a positive impact on the throughput time (Suresh, 1991). *Production control*, finally, plays an important role in the ability to realize short throughput times. An inflexible control system, for instance, may frustrate the production in an overlapping mode. When assessing the effect of different manufacturing layouts on speed, it is important to consider the requirements with respect to the production control system.

The Impact of Layout Type on the Flexibility to Serve Customer

As mentioned earlier, the flexibility objective can be further split in product/service flexibility, mix flexibility and volume/demand flexibility (Slack *et al.* 2001). The importance of these types of flexibility for a particular instance may differ substantially. Therefore, these types of flexibility have to be seen as different performance objectives. Flexibility, in general, can be defined as the ease (time, effort and/or money) by which changes can be realized. Two aspects determine the flexibility of a manufacturing layout: (i) *range* and (ii) *response*. "Range" refers to the scope of a layout and indicates the variety of situations that can be dealt with without a serious change of the production layout. "Response" indicates the speed by which the layout can be adapted to changing circumstances. Slack (1987) and Upton (1994) have observed that managers think along these lines with respect to the term flexibility.

Product flexibility indicates the ease by which new products can be introduced in a firm. This type of flexibility is higher in a product layout if the new product can be assigned to a single existing product group (quicker response). If a new product has impact on the design of the manufacturing cells, then a process layout is more stable and has more product flexibility (larger range).

Mix flexibility indicates the ease by which a firm can vary the mix of products. Important for the assessment of mix flexibility is insight in the effect of mix changes on the need of various manufacturing processes. A process layout is more range-flexible if the impact of mix changes on the need for manufacturing processes is limited. A product layout is range-flexible if work can be reallocated between the various groups. An important advantage of a product layout concerns the multi-functionality of workers: there are more capabilities at the work floor to deal with changes in the product mix.

Volume/demand flexibility concerns the ease by which the production volume can be increased or decreased. Temporary workers and the extension of working time are possibilities to increase the production volume. It is conceivable that autonomous teams in a product layout can respond more quickly to the need for additional capacity than functional groups in a process layout (i.e. response flexibility). The need for additional capacity is localized and only one group is involved in the need for more capacity for a particular product family. On the other hand, extension of capacity can be realized more easily in a process layout, because more workers with the same capabilities are eligible to work overtime (i.e. range flexibility). Also, temporary workers can probably best be integrated in a process layout.

The Impact of Layout Type on the Dependability to Deliver on Time

The performance objective "dependability" points to the importance of being dependable with respect to delivery times, the quality of the products, and such. To be dependable, it is important that the manufacturing activities can be buffered from all kinds of disturbances. Machine breakdowns and unexpected absenteeism of workers may complicate the dependability of a manufacturing system. *Interchangeability of machines* (or the ability to subcontract) and the *possibility to replace workers* (or increase the working times of some workers) indicate to what extent a manufacturing system can be reliable in various circumstances. The interchangeability and the possibility to replace workers are probably higher in a process layout because of the clustering of identical capacities. Another aspect of dependability concerns the ability to *control* the flow of products. It is likely that the throughput times in a product layout can be controlled better; the control responsibility can be decentralized to autonomous groups which are able to respond quickly to disturbances.

The Impact of Layout Type on the Quality of Work

Quality of work can be investigated in several ways. A well-known approach concerns the job characteristics model of Hackman and Oldham (1980). This model is used by Huber and Hyer (1985) and Shafer *et* al. (1995) to investigate human issues in cellular manufacturing. The job characteristics model distinguishes five task characteristics that have impact on quality of labor: (i) *skill variety*, (ii) *task identity*, (iii) *task significance*, (iv) *autonomy*, and v) *feedback*. Skill variety refers to the extent to which the work requires a variety of activities involving different skills and talents of the workers. Task identity concerns the extent to which the work enables the worker to complete a whole task from start to finish. Task significance relates to the impact of the work on other people within or outside of the organization. Autonomy indicates to what extent a worker has the freedom to plan, to organize, and to perform the tasks in his/her own way. Finally, feedback refers to the extent to which the worker receives information on the effectiveness of his/her performance.

A product layout likely supports a higher/better skill variety, more autonomy, and a better feedback mechanism: workers can perform a variety of tasks, they are responsible for the internal organization of the group, and they get a quick feedback on their activities. The task identity and task significance in a process layout is probably better: the tasks to be performed are clear for all workers and they will be respected because of their specialization.

DECISION SUPPORT FRAMEWORK

The previous section presented major performance objectives of a manufacturing system and indicated which layout-related aspects play an important role. Table 2 summarizes these aspects. As can be seen, the performance objectives price and speed consist of several elements. These elements together constitute the related performance indicators. We do not distinguish aspects for these two performance objectives.

The ultimate goal is to select the best layout out of a set of alternative layouts. Based upon the scheme of Table 2, the selection problem can be split in three sets of questions:

1. What are the relative scores of the various alternative layouts on the aspects mentioned in Table 2? This question involves a comparison of the alternative layouts with respect to the various aspects. Answering this question requires knowledge of operational issues on the work floor and the ability to assess the impact of an alternative layout on the aspects. The answer to this question determines value $\pi(i,j)$, see Table 3.
2. What is the relative importance of the various aspects for the performance objectives? Table 2 gives an overview of all the aspects. The answer to this question determines value $\pi(j,k)$. The sum of the elements, mentioned in column 3 of Table 2, forms an indication for the performance of respectively the price and speed objective.
3. What is the relative importance of the various performance objectives for the firm? This is basically a strategic question, which has

Table 3. Notation

$\pi(k)$ = relative importance of performance objective k for the firm, $\sum_k \pi(k) = 1$;
$\pi(j,k)$ = relative importance of aspects j on performance objective k, $\sum_j \pi(j,k) = 1$;
$\pi(i,j)$ = relative scores of the layout i on aspect j, $\sum_i \pi(i,j) = 1$;
$R(i)$ = relative performance of alternative *i*, $\sum_i R(i) = 1$.

to be answered by the management of the firm. It requires knowledge about customers and competitors. The answer to this question determines value $\pi(k)$.

The answers to these sets of questions enable the calculation of the relative performance of the alternatives:

$$R(i) = \sum_j \sum_k \pi(i,j)\ \pi(j,k)\ \pi(k) \qquad (1)$$

The three sets of questions and the way in which the relative performance of the alternatives are calculated can be seen as an example of using the weighted-score method (see e.g. Slack *et al.* 2001, p. 166). A major issue is the difficulty to "determine" the values of $\pi(k)$, $\pi(j,k)$, $\pi(i,j)$, and $R(i)$. It requires the ability to weight different types of issues.

The three sets of questions and the issue of weighting different types of issues fits in the AHP-approach of Saaty (1980). AHP forces the decision maker(s) to make all assessments explicit. The decomposition of the main problem in several smaller problems also enables an effective participation of employees in the decision problem, using their specific expertise and responsibilities. In the next section, we will illustrate the use of the AHP-approach for layout selection on hand of a practical instance.

CASE STUDY

The case study presented here concerns the sheet metal processing department of the firm Holec Algemene Toelevering B.V., a supplier of parts, tools, and services for the electro-mechanical industry. Before the layout study started, the sheet metal processing department consisted of four autonomous manufacturing cells with some exchangeability between the cells: (i) an automated flexible system for sheet metal working (<3 mm), (ii) numerical sheet metal working (>3mm), (iii) sheet metal construction processing (>5 mm), and (iv) conventional sheet metal processing. Basic processes to be performed in the cells are sawing, punching, cutting, tapping, squaring, welding, and bench work. The firm started to produce in manufacturing cells in 1987. This has led to significant improvements in manufacturing throughput time and efficiency. In the course of years, however, there were several reasons to move back to a more functional layout, such as the complexity and productivity of new equipment, the possibility of workers to operate more than one machine simultaneously, and the increased variety of part types. Other parts of the manufacturing facility of the firm were already transformed to a more process-oriented layout (see Molleman *et al.* 2002). A layout study at the sheet metal processing department started in 2000. Four alternative layouts were generated on the basis of a production flow analysis (Burbidge 1991,

Slomp 1998): (i) a group-technology-oriented alternative, (ii) a product-type-oriented alternative, (iii) a capability-oriented alternative, and (iv) a process-oriented alternative. These alternatives are schematically depicted in Figure 3.

In the group-technology-oriented alternative, the department is divided in two relatively autonomous cells with minimal intercell movements. Some part types can be produced in both groups, which simplifies the balancing of the workload. The product-type-oriented alternative consists of three manufacturing cells, each responsible for a particular type of product. Two cells are responsible for the production of repetitive part types, while one cell is mainly focused on the production of quick orders. An important advantage of this layout is that the production of repetitive part types is not disturbed by quick orders. This may simplify the production control. On the other hand, the cell that is responsible for the quick orders may face undesirable fluctuations of demand. The capability-oriented alternative consists of three manufacturing cells. Basic viewpoint of this alternative is that all assembly work (welding and bench work) needs to be performed in one manufacturing cell (C). The other cells (A and B) are autonomous cells, which have their own product-oriented capabilities. Therefore, intercell movements are minimal. The process-oriented alternative consists of two manufacturing cells. Sawing, punching and cutting is performed in cell A, while tapping, squaring, welding, and bench work is done in cell B. Each cell consists of small groups of identical machines.

The four alternatives are compared by means of the decision framework of Table 2 and by using the AHP methodology. We used the software package Expert Choice. Figure 4 presents the results of the comparisons of the four alternatives. The bars in Figure 4 indicate the relative importance of the various performance objectives. The four lines in the figure show the relative scores of each alternative on the performance objectives. The position of the alternatives at "total" shows the final judgment of the alternatives. As can be seen, the process-oriented layout is preferred

Figure 3. Layout type alternatives for the sheet metal processing

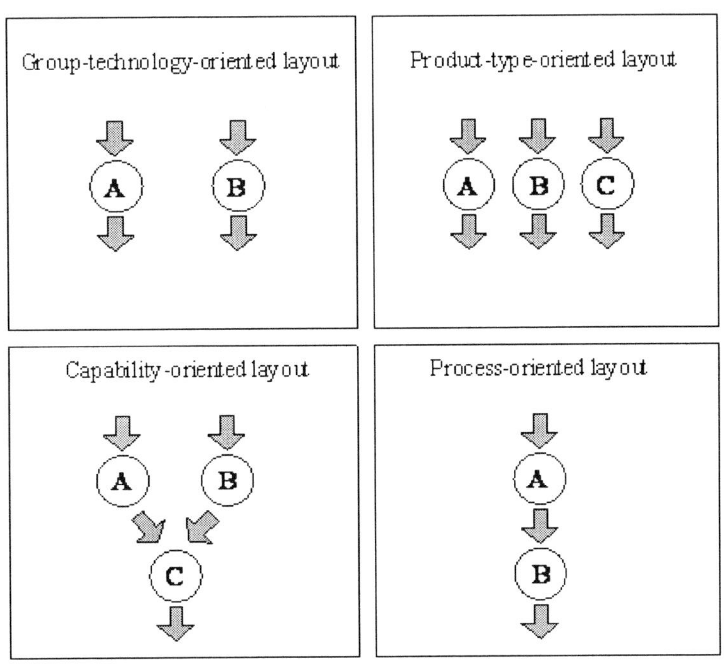

Figure 4. Scores of the four alternatives

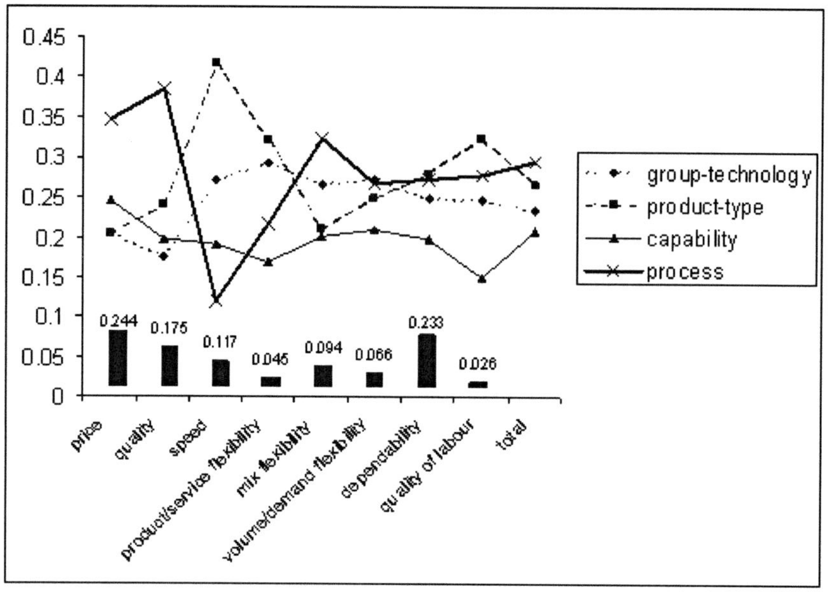

because of its positive effect on price, quality, mix flexibility and, to less extent, dependability. Especially price and quality are important reasons to select the process-oriented layout.

It is interesting to see the almost equal end scores of the product-type-oriented and the process-oriented layout, despite their completely different orientation. The product-type-oriented layout performs well with respect to the performance objective speed. In the assessment of the speed factors of Table 2, the managers of the firm assumed that the production control in the process-oriented layout is more complex and will perform worse than in the product-type-oriented layout. At that moment, the firm did not have a good registration system (bar-coding system) on the work floor that is connected with the production control system. A better shop floor control system, which was under study at the firm at the moment of deciding for a new manufacturing layout, would likely improve the score of the process-oriented layout on the performance objective speed. This kind of sensitivity analysis is also useful for the assessment of the scores on the performance objectives quality and price. In this particular case, the impact of shorter control loops on the quality of the products is assessed as being minimal. This assessment has a negative impact on the final score of the product-type-oriented layout. The software package Expert Choice supports sensitivity analysis. It appears that if short control loops do have a major impact on the quality of the products, the product-type-oriented layout performs better than the process-oriented alternative.

Based upon the results of the analysis, the firm changed the layout of the sheet-metal processing department into a process-oriented layout, see also Molleman *et al.* (2002). The systematic approach of the selection problem is seen at the firm as a major help to canalize the discussions about the required layout of manufacturing departments.

CONCLUSION AND REFLECTIONS

This chapter presents a systematic approach for the selection of a manufacturing layout type. The approach includes the use of the AHP-methodology. An important element of the approach is the construction of a decision hierarchy and the pairwise comparisons of decision elements. In this section we will first reflect on the AHP methodology and next we will draw conclusions on the use of AHP for the layout selection type problem.

As in all Multi-Criteria-Decision-Methods, AHP is sensitive for issues such as the specification of the selection problem, its decomposition, and the scales used for the pairwise comparisons (see Pöyhönen *et al.* 1997). The quality of the outcome of an AHP analysis is largely determined by the quality of the problem specification. For instance, adding aspects or regrouping decision elements may lead to different outcomes. A particular problem concerns the issue of "rank reversal". This means that the priority of alternatives may change if alternatives are removed from and/or other alternatives are added to the selection problem (see e.g. Belton and Gear 1983). The problem of rank reversal plays a role if almost identical alternatives are taken into consideration. Finally, the number of pairwise comparisons may be problematic and may lead to unreliable results. Employees who have to make the pairwise comparisons may get tired and lose the required concentration. Another issue, which has to be taken into consideration when applying the AHP approach, is the translation of verbal or graphical assessments in numerical figures. Pöyhönen *et al.* (1997) show that it is not advisable to mix different types of assessment within the levels of the AHP hierarchy.

This chapter has presented a decision support framework based on the AHP approach for the selection of a manufacturing layout. The value of the framework is illustrated by means of a case application. Important advantages of using the AHP approach are (1) the ability to decompose the complex decision problem in smaller problems, (2) the possibility of an efficient and effective employee participation, and (3) the detailed assessment of the selected layout alternative, which helps to define further improvement actions. These advantages of using the AHP approach are also illustrated by Abdi and Labid (2003).

Interesting point in the case study, as presented in this chapter, is that opposite alternatives do have the best scores. This illustrates the group technology debate, as it takes place in practice. Both alternatives appear to be acceptable and have their pros and cons. The proposed approach has the advantage that it gives insight in whether the two alternatives do have similar scores. A debate about the differences on the scores of the various aspects will help the decision process and the acceptability of the final decision.

The systematic approach as presented in this chapter is developed around 1999 (see also Slomp *et al.* 1999a, b) and is applied in several practical situations, mostly master projects of Industrial Engineering students. The evaluation criteria (see Table 2) and the use of AHP has proven to be a robust framework for the selection of a layout in many situations.

REFERENCES

Abdi, M. R., & Labib, A. W. (2003). A design strategy for reconfigurable manufacturing systems (RMSs) using analytical hierarchical process (AHP): A case study. *International Journal of Production Research*, *41*(10), 2273–2299. doi:10.1080/0020754031000077266

Agarwal, A., & Sarkis, J. (1998). A review and analysis of comparative performance studies on functional and cellular manufacturing layouts. *Computers & Industrial Engineering*, *34*(1), 77–89. doi:10.1016/S0360-8352(97)00152-6

Balakrishnan, J., & Cheng, C. H. (1998). The dynamic plant layout problem—A state-of-the-art review. *Omega*, *26*(4), 507–521. doi:10.1016/S0305-0483(97)00078-9

Belton, V., & Gear, T. (1983). On a short-coming of Saaty's method of analytic hierarchies. *Omega, 11*(3), 228–230. doi:10.1016/0305-0483(83)90047-6

Benjaafar, S., Heragu, S. S., & Irani, S. A. (2002). Next generation factory layouts: Research challenges and recent progress. *Interfaces, 32*(6), 58–77. doi:10.1287/inte.32.6.58.6473

Benjaafar, S., & Sheikhzadeh, M. (2000). Design of flexible layouts. *IIE Transactions, 32*, 309–322. doi:10.1080/07408170008963909

Burbidge, J. L. (1992). Change to group technology: process organization is obsolete. *International Journal of Production Research, 30*(5), 1209–1219. doi:10.1080/00207549208942951

Burbidge, J. L., Partridge, J. T., & Aitchison, K. (1991). Planning group technology for Davy Morris using production flow analysis. *Production Planning and Control, 2*(1), 59–72. doi:10.1080/09537289108919331

Cambron, K. E., & Evans, G. W. (1991). Layout design using the analytic hierarchy process. *Computers & Industrial Engineering, 20*(2), 211–229. doi:10.1016/0360-8352(91)90026-3

Ertay, T., Ruan, D., & Tuzkaya, U. R. (2006). Integrating data envelopment analysis and analytic hierarchy for the facility layout design in manufacturing system. *Information Sciences, 176*, 237–262. doi:10.1016/j.ins.2004.12.001

Farrington, P. A., & Nazemetz, J. W. (1998). Evaluation of the performance domain of cellular and functional layouts. *Computers & Industrial Engineering, 34*(1), 91–101. doi:10.1016/S0360-8352(97)00153-8

Flynn, B. B., & Jacobs, F. R. (1986). A simulation comparison of group technology with traditional job shop manufacturing. *International Journal of Production Research, 24*(5), 1171–1192. doi:10.1080/00207548608919795

Francis, R. L., McGinis, L. F., & White, J. A. (1992). *Facility layout and location—An analytical approach*. Englewood Cliffs, NJ: Prentice Hall.

Hackman, J. R., & Oldham, G. R. (1980). *Work redesign*. Reading, MA: Addison-Wesley.

Huber, V. L., & Hyer, N. L. (1985). The human factor in cellular manufacturing. *Journal of Operations Management, 5*(2), 213–228. doi:10.1016/0272-6963(85)90008-7

Irani, S. A., Cavalier, T. M., & Cohen, P. H. (1993). Virtual manufacturing cells: Exploiting layout design and intercell flows for the machine sharing problem. *International Journal of Production Research, 31*(4), 791–810. doi:10.1080/00207549308956757

Irani, S. A., & Huang, H. (2000). Custom design of facility layouts for multi-product facilities using layout modules. *IEEE Transactions on Robotics and Automation, 16*, 259–267. doi:10.1109/70.850644

Johnson, D. J., & Wemmerlöv, U. (1996). On the relative performance of functional and cellular layouts – An analysis of the model-based comparative studies literature. *Production and Operations Management, 5*(4), 309–334. doi:10.1111/j.1937-5956.1996.tb00403.x

Labib, A. W., Williams, G. B., & O'Connor, R. F. (1998). An intelligent maintenance model (system): An application of the analytic hierarchy process and a fuzzy logic rule-based controller. *Operational Research Society, 49*, 745–757.

Meller, R. D., & Gau, Y. K. (1996). The facility layout problem: Recent and emerging trends, and perspectives. *Journal of Manufacturing Systems, 15*, 351–366. doi:10.1016/0278-6125(96)84198-7

Meyers, F. E., & Stephens, M. P. (2004). *Manufacturing facilities design and material handling* (3rd ed.). Prentice Hall.

Molleman, E., Slomp, J., & Rolefes, S. (2002). The evolution of a cellular manufacturing system - A longitudinal case study. *International Journal of Production Economics*, *75*(3), 305–322. doi:10.1016/S0925-5273(01)00139-6

Montreuil, B., LeFrancois, P., Marcotte, S., & Venkatadri, U. (1993). *Holographic layout of manufacturing systems operating in chaotic environments.* Technical Report 93-53, Document de Recherche GRGL, Faculte des Sciences de l'Administration, Universite Laval, Quebec.

Montreuil, B., Venkatadri, U., & Rardin, R. L. (1999). Fractal layout organization for job shop environments. *International Journal of Production Research*, *37*(3), 501–521. doi:10.1080/002075499191643

Muther, R. (1973). *Systematic layout planning* (2nd ed.). Boston, MA: Cahners Books.

Partove, F. Y., & Burton, J. (1992). An analytic hierarchy approach to facility layout. *Computers & Industrial Engineering*, *22*(4), 447–457. doi:10.1016/0360-8352(92)90020-K

Pöyhönen, M., Hämäläinen, R. P., & Salo, A. A. (1997). An experiment on the numerical modeling of verbal ratio statements. *Journal of Multi-Criteria Decision Analysis*, *6*, 1–10. doi:10.1002/(SICI)1099-1360(199701)6:1<1::AID-MCDA111>3.0.CO;2-W

Raman, D., Nagalingam, S. V., & Lin, G. C. I. (2009). Towards measuring the effectiveness of a facilities layout. *Robotics and Computer-integrated Manufacturing*, *25*, 191–203. doi:10.1016/j.rcim.2007.06.003

Rosenblatt, M. J. (1986). The dynamics of plant layout. *Management Science*, *32*(1), 76–86. doi:10.1287/mnsc.32.1.76

Saaty, T. L. (1980). *The analytic hierarchy process.* New York, NY: McGraw-Hill.

Shafer, S. M., & Charnes, J. M. (1993). Cellular versus functional layouts under a variety of shop operating conditions. *Decision Sciences*, *24*(3), 665–681. doi:10.1111/j.1540-5915.1993.tb01297.x

Shafer, S. M., & Meredith, J. R. (1993). An empirically-based simulation study of functional versus cellular layouts with operations overlapping. *International Journal of Operations & Production Management*, *13*(2), 47–62. doi:10.1108/01443579310025303

Shafer, S. M., Tepper, B. J., Meredith, J. R., & Marsh, R. (1995). Comparing the effects of cellular and functional manufacturing on employees' perceptions and attitudes. *Journal of Operations Management*, *12*, 63–74. doi:10.1016/0272-6963(94)00011-3

Shambu, G., Suresh, N. C., & Pegels, C. C. (1996). Performance evaluation of cellular manufacturing systems: A taxonomy and review of research. *International Journal of Operations & Production Management*, *16*(8), 81–103. doi:10.1108/01443579610125796

Slack, N. (1987). The flexibility of manufacturing systems. *International Journal of Operations & Production Management*, *7*(4), 35–45. doi:10.1108/eb054798

Slack, N., Chambers, S., & Johnston, R. (2001). *Operations management.* Harlow, MA: Financial Times/Prentice Hall.

Slomp, J. (1998). Design of manufacturing cells: PFA applications in Dutch industry. In Suresh, N. C., & Kay, J. M. (Eds.), *Group technology and cellular manufacturing: A state-of-the-art synthesis of research and practice* (pp. 153–168). Boston, MA: Kluwer Academic. doi:10.1007/978-1-4615-5467-7_9

Slomp, J., Molleman, E., & Gaalman, G. J. C. (1993). Production and operations management aspects of cellular manufacturing – a survey of users. In I. A. Pappas & I. P. Tatsiopoulos (Ed.), *Advances in production management systems* (pp. 553-560). North-Holland: Elsevier Science Publishers B.V.

Slomp, J., Molleman, E., & Suresh, N. C. (1995). Empirical investigation of the performance of cellular manufacturing, *Proceedings Management and New Production Systems, EUROMA conference*, (pp. 432-441).

Slomp, J., Vrolijk, H., & Bokhorst, J. (1999). [a]). Beoordeling van alternatieve productiestructuren (deel 1). [in Dutch]. *MB-dossier*, *65*(6), 6–11.

Slomp, J., Vrolijk, H., & Bokhorst, J. (1999). [b]). AHP als methode voor de keuze van een geschikte productiestructuur - Gestructureerde besluitvorming in de keuze tussen alternatieve structuren (deel 2). [in Dutch]. *MB-dossier*, *65*(7/8), 6–12.

Suresh, N. C. (1991). Partitioning work centers for group technology: insights from an analytical model. *Decision Sciences*, *22*(4), 772–791. doi:10.1111/j.1540-5915.1991.tb00364.x

Suresh, N. C., & Meredith, J. R. (1994). Coping with the loss of pooling synergy in cellular manufacturing systems. *Management Science*, *40*(4), 466–483. doi:10.1287/mnsc.40.4.466

Upton, D. M. (1994). The management of manufacturing flexibility. *California Management Review*, *36*(2), 72–89.

Venkatadri, U., Rardin, R. L., & Montreuil, B. (1997). A design methodology for fractal layout organization. *IIE Transactions*, *29*, 911–924. doi:10.1080/07408179708966411

Wemmerlöv, U., & Hyer, N. L. (1989). Cellular manufacturing in the U.S. industry: A survey of users. *International Journal of Production Research*, *27*(9), 1511–1530. doi:10.1080/00207548908942637

Wemmerlöv, U., & Johnson, D. J. (1997). Cellular manufacturing at 46 user plants: Implementation experiences and performance improvements. *International Journal of Production Research*, *35*(1), 29–49. doi:10.1080/002075497195966

Yang, T., & Hung, C.-C. (2007). Multiple-attribute decision making methods for plant layout design problem. *Robotics and Computer-integrated Manufacturing*, *23*, 126–137. doi:10.1016/j.rcim.2005.12.002

Yang, T., & Kuo, C. (1997). A hierarchical AHP/DEA methodology for the facilities layout design problem. *European Journal of Operational Research, 147*(2003), 128–136

Zahedi, F. (1986). The analytic hierarchy process – A survey of the method and its applications. *Interfaces*, *16*, 96–108. doi:10.1287/inte.16.4.96

Chapter 3
Comparison of Connected vs. Disconnected Cellular Systems:
A Case Study

Gürsel A. Süer
Ohio University, USA

Royston Lobo
S.S. White Technologies Inc., USA

ABSTRACT

In this chapter, two cellular manufacturing systems, namely connected cells and disconnected cells, have been studied, and their performance was compared with respect to average flowtime and work-in-process inventory under make-to-order demand strategy. The study was performed in a medical device manufacturing company considering their a) existing system b) variations from the existing system by considering different process routings. Simulation models for each of the systems and each of the options were developed in ARENA 7.0 simulation software. The data used to model each of these systems were obtained from the company based on a period of nineteen months. Considering the existing system, no dominance was established between connected cells vs. disconnected cells as mixed results were obtained for different families. On the other hand, when different process routings were used, connected system outperformed the disconnected system. It is suspected that one additional operation required in the disconnected system as well batching requirement at the end of packaging led to poor performance for the disconnected cells. Finally, increased routing flexibility improved the performance of the connected cells, whereas it had adverse effects in the disconnected cells configuration.

DOI: 10.4018/978-1-61350-047-7.ch003

INTRODUCTION

Cellular Manufacturing is a well known application of Group Technology (GT). Cellular Design typically involves determining appropriate part families and corresponding manufacturing cells. This can be done either by grouping parts into families and then forming machine cells based on the part families or machine cells are determined first and based on these machine cells the part families may be formed or lastly both these formations can take place simultaneously. In a cellular manufacturing system, there may be a manufacturing cell for each part family or some of the manufacturing cells can process more than one part family based on the flexibility of the cells. The factors affecting the formation of cells can differ under various circumstances, some of them are volume of work to be performed by the machine cell, variations in routing sequences of the part families, processing times, etc.

A manufacturing system in which the goods or products are manufactured only after customer orders are received is called a make-to-order system. This type of system helps reduce inventory levels since no finished goods inventory is kept on hand.

In this chapter, two types of cellular layouts are analyzed, namely connected cells (single-stage cellular system) and disconnected cells (multi-stage cellular system) and their performance is compared under various circumstances for a make-to-order company. This problem has been observed in a medical device manufacturing company. The management was interested in such a comparison to finalize the cellular design. It was also important to research the impact of flexibility within each system for different combinations of family routings. A similar situation of connected vs. disconnected cellular design was also observed in a shoe manufacturing company, and in a jewelry manufacturing company. Authors believe that this problem has not been addressed in the literature before even though it has been observed in more than one company and therefore worthy to study.

BACKGROUND

The connected cells represent a continuous flow where the products enter the cells in the manufacturing area, complete the machining operations and exit through the corresponding assembly and packaging area after completion of the assembly and packaging operations. In other words, the output of a cell in the manufacturing area becomes the input to the corresponding cell in the assembly and packaging area. The biggest advantage of connected cells is that material flow is smoother and hence flowtime is expected to be shorter. This is also expected to result in lower WIP inventory. This paper focuses on a cellular manufacturing system similar to the system shown in Figure 1. There are three cells in the manufacturing area and three cells in the assembly and packaging area. In these cells, M1 through M3 represent the machines in the manufacturing area, A1, A2 and P1 through P3 represent the machines in the assembly and packaging area. The products essentially follow a unidirectional flow. The three cells in manufacturing area are similar since they have similar machines and all the products can be manufactured in any of the cells. However, the situation gets complicated in the assembly and packaging area. The three cells have restrictions in terms of the products that they can process. Therefore, deciding which manufacturing cell a product should be assigned is dictated by the packaging cell(s) it can be processed later on. This constraint makes the manufacturing system less flexible.

In the disconnected cell layout, the products enter the manufacturing area, complete the machining operations and exit this area. On exiting the manufacturing area, the products can go to more than one of the assembly and packaging cells. In other words, the output from the cells in

Figure 1. Connected cells

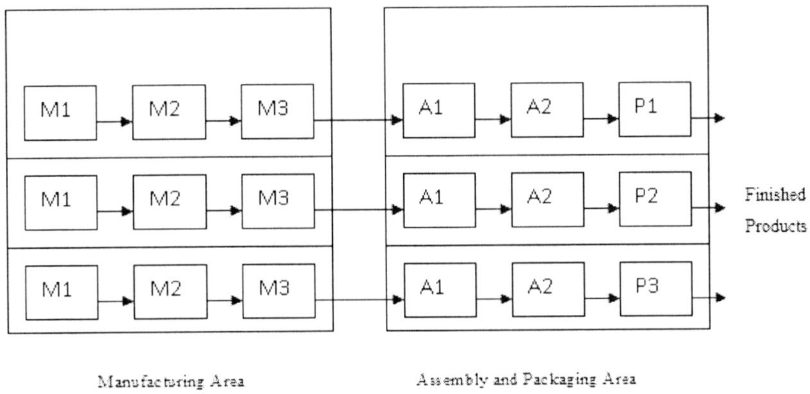

Figure 2. Disconnected cells with partial flexibility

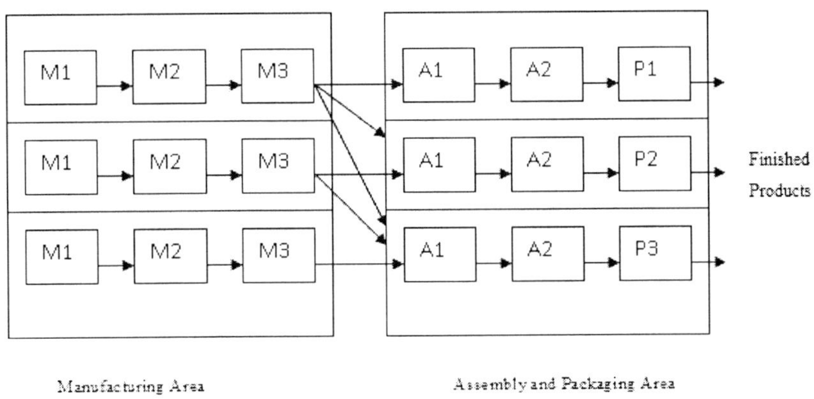

the manufacturing area can become an input for some of the cells in the assembly and packaging area (partially flexible disconnected cells) or all of them (completely flexible disconnected cells). Figure 2 shows a partially flexible disconnected cells case where the parts from cell 1 in the manufacturing area can go to any of the cells in the assembly and packaging area. Parts from cell 2 can only go to cell 2, and cell 3 of the assembly and packaging area. Parts from cell 3 of the manufacturing area can only go to cell 3 of the assembly and packaging area. The disconnected system design allows more flexibility. On the other hand, due to interruptions in the flow, some delays may occur which may eventually lead to higher flowtimes and WIP inventory levels.

LITERATURE REVIEW

A group of researchers compared the performance of cellular layout with process layout. Flynn and Jacobs (1987) developed a simulation model using SLAM for an actual shop to compare the performance of group technology layout against process layout. Morris and Tersine (1990) developed simulation models for a process layout and a cellular layout using SIMAN. The two performance measures used were throughput time and work-in-process inventory (WIP). Yazici (2005) developed a simulation model using Promodel based on data collected from a screen-printing company to ascertain the influence of volume, product mix, routing and labor flexibilities in the presence of fluctuating demand. A comparison between one-cell, two-cell configurations versus a job shop is made to determine the shortest delivery and highest utilization. Agarwal and Sarkis (1998) reviewed the conflicting results from the literature in regard to superiority of cellular layout vs. functional layout. They attempted to identify and compile the existing studies and understand conflicting findings. Johnson and Wemmerlov (1996) analyzed twenty-four model-based studies and concluded that the results of these work cannot assist practitioners in making choices between existing layouts and alternative cell systems. Shafer and Charnes (1993) studied cellular manufacturing under a variety of operating conditions. Queueing theoretic and simulation models of cellular and functional layouts are developed for various shop operating environments to investigate several factors believed to influence the benefits associated with a cellular manufacturing layout.

Another group of researchers focused on analyzing cellular systems. Selen and Ashayeri (2001) used a simulation approach to identify improvements in the average daily output through management of buffer sizes, reduced repair time, and cycle time in an automotive company. Albino and Garavelli (1998) simulated a cellular manufacturing system using Matlab to study the effects of resource dependability and routing flexibilities on the performance of the system. Based on the simulation results, the authors concluded that as resource dependability decreases, flexible routings for part families can increase productivity. On the contrary, from an economic standpoint they concluded that benefits will greatly reduce from an increase routing flexibility cost and resource dependability. Caprihan and Wadhwa (1997) studied the impact of fluctuating levels of routing flexibility on the performance of a Flexible Manufacturing System (FMS). Based on results obtained, the authors concluded that there is an optimal flexibility level beyond which the system performance tends to decline. Also, increase in routing flexibility when made available with an associated cost seldom tends to be beneficial. Suer, Huang, and Maddisetty (2009) discussed layered cellular design to deal with demand variability. They proposed a methodology to design a cellular system that consisted of dedicated cells, shared cells and remainder cell.

Other researchers studied make-to-order and make-to-stock production strategies. Among them, DeCroix and Arreola-Risa (1998) studied the optimality of a Make-to- Order (MTO) versus a Make-to-Stock (MTS) policy for a manufacturing set up producing various heterogeneous products facing random demands. Federgruen and Katalan (1999) investigated a hybrid system comprising of a MTO and a MTS systems and presented a host of alternatives to prioritize the production of the MTO and MTS items. Van Donk (2000) used the concept of decoupling point (DP) to develop a frame in order to help managers in the food processing industries to decide which of their products should be MTO and which ones should be MTS. Gupta and Benjaafar (2004) presented a hybrid strategy which is a combination of MTO and MTS modes of production. Nandi and Rogers (2003) simulated a manufacturing system to study its behavior in a make to order environment under a control policy involving an order release component and an order acceptance/rejection component.

Authors are not aware of any other study that focuses on comparing the performance of connected cells with disconnected cells and therefore we believe this is an important contribution to the literature.

DESCRIPTION OF THE SYSTEM STUDIED: THE CASE STUDY

This section describes the medical device manufacturing company where the experimentation was carried out. The products essentially follow a unidirectional flow. The manufacturing process is mainly divided into two areas, namely fabrication and packaging. Each area consists of three cells and cells are not identical. The one piece-flow strategy is adapted in all cells. The company has well defined families which are determined based on packaging requirements. Furthermore, the cells have been already formed. The average flowtime and the work-in-process inventory are the performance measures used to evaluate the performance of connected cells and disconnected cells.

Product Families

The products are grouped under three families: Family 1 (F1), Family 2 (F2), and Family 3 (F3). The finished products are vials consisting of blood sugar strips and each vial essentially contains 25 strips. The number of products in families 1, 2 and 3 are 11, 21 and 4, respectively.

The families that are described were already formed by the manufacturer based on the number of vials (subfamilies) included in the box. Family 1 requires only one subassembly (S), one box (B1), one label (L), and one Insert for instructions (I); family 2 (F2) requires 2 subassemblies, one box (B2), one label and one insert, and family 3 (F3) requires 4 subassemblies, one box (B3), one label and one insert to become finished product as shown in Table 1. Obviously, this family classification is strictly from manufacturing perspective and marketing department uses its own family definition based on product function related characteristics. The family definition has been made based on limitations of packaging machines. Not all packaging machines can insert 4 vials into a box. This seemingly simple issue becomes an obstacle in assigning products to packaging cells and furthermore becomes a restriction in assigning products to even manufacturing cells in connected cellular design.

Table 1. Product structures of families

	Components					
Family	**S**	**L**	**I**	**B1**	**B2**	**B3**
F1	1	1	1	1		
F2	2	1	1		1	
F3	4	1	1			1

Fabrication Cells

The fabrication area is where the subassemblies are manufactured. This area contains three cells which manufacture a single common subassembly and hence all three families can be manufactured in any of the three cells. The fabrication area has a conveyor system which transfers the products from one machine to another based on one-piece flow principle.

Operations in Fabrication Cells

There are three operations associated with the fabrication area:

- Lamination
- Slicing and Bottling
- Capping

The machines used for operation 1 in all three cells are similar and work under the same velocities (120 vials/min) but the number of machines within each cell varies. Operation 2 has machines that process 17 vials/min and 40 vials/min. Similarly, operation 3 has machines that process 78 vials/min and 123 vials/min. Table 2 shows the distribution of machines and velocities among the three cells.

Table 2. Number of machines and their production rates in fabrication cells

	Op. 1	Op.2		Op. 3		Output of Bottleneck (vials/min)
		Type I	Type II	Type I	Type II	
Production Rate (vials/min)	120	17	40	78	123	
Cell 1	1	2	2	0	1	114
Cell 2	1	4	0	1	0	68
Cell 3	2	3	2	0	2	131

Table 3. Feasibility matrix of families and packaging cells

Family	Packaging Cell 1	Packaging Cell 2	Packaging Cell 3
F1	X	X	
F2	X	X	X
F3	X		X

Packaging Cells

The packaging area also has a conveyor system similar to the fabrication area which transfers products within packaging cells and also from the fabrication cells to the packaging cells. In the packaging area, the subassemblies produced in the fabrication area are used to produce the various finished products. The packaging cell 1 is semiautomatic while cells 2 and 3 are automatic. This difference in the types of machines results in constraints that do not allow the packaging of certain products in certain cells. There are a total of 36 finished products which differ in the quantity of vials they contain, the type of raw material the vials are made of, and the destination of the country to where they are shipped. The original cell feasibility matrix for the families is given in Table 3 and the restrictions are due to constraints in the packaging of the vials.

Operations in Packaging Cells

There are five operations performed in packaging area and each operation requires one machine. The operations are described as follows:

- Feeding (This operation is only performed in the case of disconnected cells)
- Labeling
- Assembly (Automatic in cells 2 and 3, semi-automatic in cell 1)
- Sealing
- Bar Coding

Table 4 shows the production rates of the machines in all cells.

ALTERNATE DESIGNS CONSIDERED

In this section, the current product-cell feasibility restrictions are discussed for both connected and disconnected cellular systems.

Connected Cells

In this system, cells are set up such that the packaging cells form an extension or continuation of the respective fabrication cells. In other words, the output of a cell in fabrication area becomes the input for the corresponding packaging cell. Hence, it is referred to as a connected system. The connected system for the current product-

Table 4. Production rates for assembly-packaging machines in vials/minute

Cell	Family	Operation				
		4	5	6	7	8
Cell 1	Family 1	160	135	80	150	150
	Family 2	160	135	80	150	150
	Family 3	160	135	80	150	150
Cell 2	Family 1	160	135	100	150	150
	Family 2	160	135	180	150	150
	Family 3	NA	NA	NA	NA	NA
Cell 3	Family 1	NA	NA	NA	NA	NA
	Family 2	160	135	150	150	150
	Family 3	160	135	280	150	150

Figure 3. Cell routing of families for the connected system

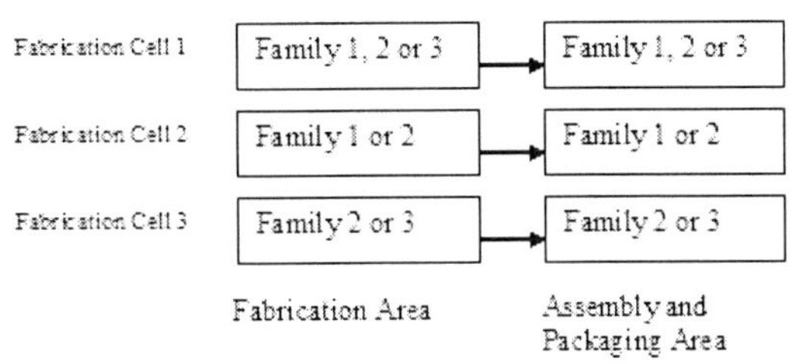

cell feasibility is shown in Figure 3. The output of family 1, family 2, and family 3 is essentially based on the bottleneck or the slowest machine in each cell of the fabrication or the packaging area and they are shown in Table 5.

Disconnected Cells

In this case, the output of a cell in the fabrication area can become an input for more than one cell in the packaging area depending upon the constraints in the packaging area. This can be considered to be a partially flexible disconnected cells type of system. The cell routing for each family is shown in Figure 4. In this figure, solid lines indicate that all the products processed in that particular fabri-

Table 5. Output rates for cells in the connected system

Cell #	Family #	Output Rate of the Bottleneck Machine in Fabrication Area (vials/min)	Output Rate of the Bottleneck Machine/Operator in Packaging Area (vials/min)	Output Rate (vials/min)
Cell 1	Family 1	114	80	80
	Family 2		80	80
	Family 3		80	80
Cell 2	Family 1	68	100	68
	Family 2		135	
Cell 3	Family 2	131	135	131
	Family 3		135	

Figure 4. Cell routing of families for disconnected system

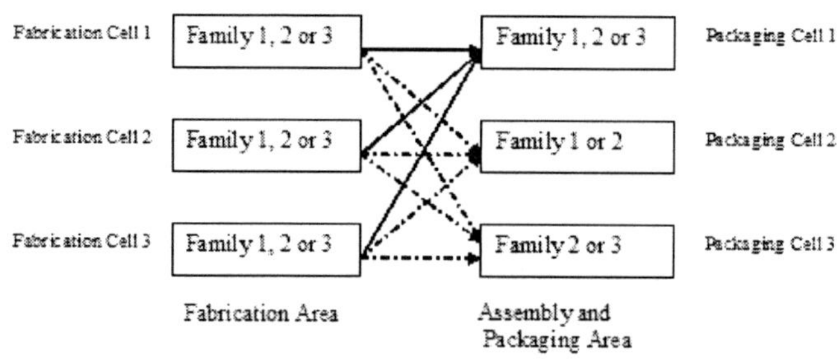

cation cell can be processed in the assembly and packaging cell that they are connected to. On the other hand, the dashed lines show that only some of the products processed in the fabrication cell can be processed in the corresponding assembly and packaging cell. This provides a greater amount of flexibility with respect to the routing of the parts in the cellular system. The output rates of family 1, family 2, and family 3 depend on the fabrication-packaging cell combination and they are determined by the slowest machine as shown in Table 6.

Cases Considered

The experimentation discussed in this chapter can be grouped in the following sections:

- Original Family-Cell Feasibility Matrix Production orders are based on customer orders.
- Various Family-Cell Feasibility Options Seven different family-cell feasibility options have been considered as given in Table 7. In this case too, production orders are based on customer orders.

METHODOLOGY USED

This section describes the methodology used to develop the different simulation models in Arena 7.0.

Input Data Analysis

Input data such as customer order distributions, their respective inter-arrival times, processing times, and routings were all obtained based on the data provided by the company. The data provided was basically the total sales volume in vials for each part belonging to one of the three families for a period of nineteen months. Table 8 shows the customer order sizes and the inter-arrival time distributions for each product.

Simulation Models

The models were run 24 hours a day which basically represented 3 shifts round the clock. Setup times and material handling times were negligible. Preemption was not allowed due to material control restrictions by FDA. Vials move based on one-piece flow between machines. The simulation models are discussed for different cases separately in the following paragraphs.

Case 1: Connected Cells: After the entities are created, they are routed to cells 1, 2 or 3 based on the type of family they belong to. The entities

Table 6. Output rate of each routing combination for the disconnected system

Family #	Fabrication Area Cell (Output of the Bottleneck Machine in vials/min)	Packaging Area Cell (Output of the Bottleneck Machine in vials/min)	Output Rate of Routing Combination (vials/min)
Family 1	Cell 1 (114)	Cell 1 (80)	80
	Cell 1 (114)	Cell 2 (100)	100
	Cell 2 (68)	Cell 1 (80)	68
	Cell 2 (68)	Cell 2 (100)	68
	Cell 3 (131)	Cell 1 (80)	80
	Cell 3 (131)	Cell 2 (100)	100
Family 2	Cell 1 (114)	Cell 1 (80)	80
	Cell 1 (114)	Cell 2 (135)	114
	Cell 1 (114)	Cell 3 (135)	114
	Cell 2 (68)	Cell 1 (80)	68
	Cell 2 (68)	Cell 2 (135)	68
	Cell 2 (68)	Cell 3 (135)	68
	Cell 3 (131)	Cell 1 (80)	80
	Cell 3 (131)	Cell 2 (135)	131
Family 3	Cell 1 (114)	Cell 1 (80)	80
	Cell 1 (114)	Cell 3 (135)	114
	Cell 2 (68)	Cell 1 (80)	68
	Cell 2 (68)	Cell 3 (135)	68
	Cell 3 (131)	Cell 1 (80)	80
	Cell 3 (131)	Cell 3 (135)	131

enter the fabrication area as a batch equivalent to the customer order size. Once a batch of entities enters the cell they are split and there is a one-piece flow in the cell. Entities belonging to a family go to one of its feasible cells based on the shorter queue length among 2nd operation. This is done because the second operation in each cell has been identified as the bottleneck operation based on trial runs conducted. In cell 1 and cell 3, the entities undergo operation 1 and go to operation 2 where there are two types of machines namely the slow (Type I) and fast (Type II) machines available for processing. The entities are routed to either type of machine based on a percentage which was decided after a number of simulation runs in order to minimize the queue lengths and hence the waiting time. In cell 1, 30% of the entities were routed to the Type I machine and the rest were routed to the Type II machine. In cell 3, 40% of the entities were routed to the Type I machine and the rest were routed to the Type II machine.

Each of the entities leaving the fabrication cells enters the corresponding packaging cells. For example, entities from cell 1 in the fabrication area will enter cell 1 of the packaging area. The entities entering the packaging area undergo processing through operation 4. In the fifth operation, the vials are grouped based on the type of family they belong to. Family 1 consists of only 1 vial, family 2 consists of 2 vials and family 3 consists of 4 vials. Thus, the vials that are batched in Arena after operation 5 are processed in operations 6, 7 and 8 where they are boxed, sealed and coded. In the final batching, the vials are batched together in a

Table 7. Different family-cell feasibility options

Cellular System	Cell Type	Cell	Options						
			O1	O2	O3	O4	O5	O6	O7
Connected Cells	Fab. Cells	C1	1	1,2,3	1,2	1,2,3	1	1,2	1,2
		C2	2	1,2,3	2,3	1,2	2	2,3	2,3
		C3	3	1,2,3	1,3	2,3	3	1,3	1,3
	Pack. Cells	C1	1	1,2,3	1,2	1,2,3	1	1,2	1,2
		C2	2	1,2,3	2,3	1,2	2	2,3	2,3
		C3	3	1,2,3	1,3	2,3	3	1,3	1,3
Disconnected Cells	Fab. Cells	C1	1	1,2,3	1,2	1,2,3	1	1,2,3	1,2
		C2	2	1,2,3	2,3	1,2	2	1,2,3	2,3
		C3	3	1,2,3	1,3	2,3	3	1,2,3	1,3
	Pack. Cells	C1	1,2,3	1,2,3	1,2,3	1,2,3	1	1,2	1,2
		C2	1,2,3	1,2,3	1,2,3	1,2,3	2	2,3	2,3
		C3	1,2,3	1,2,3	1,2,3	1,2,3	3	1,3	1,3

box based on the final customer order sizes. The final batch sizes are the same as the input batch sizes. There is a waiting time associated since the entities might have to wait till the required batch size is reached and only then get disposed.

The warm up time for the model was determined to be 2000 hours based on steady state analysis. The simulation was run for 2500 hours after the end of the warm-up period.

Case 1: Disconnected Cells: The entities enter the fabrication area in batches as explained for the connected system. The batches of entities in disconnected system are routed differently as compared to the connected system. Here, the batches of entities are routed to cell 1, cell 2, or cell 3 of the fabrication area based on the shortest queue length of the bottleneck operation which is operation 2 as explained earlier. The flexibility of routing the families to any of the cells in this type of system is the only major difference between the connected and disconnected systems in the fabrication area. The processing times of the machines and the sequence of operations for the entities for both systems are the same. Since the flow is disconnected in this system, the entities are batched again to the same customer order sizes at the end of the fabrication area.

The batches of entities entering the packaging area are routed to specific packaging cells based on shortest queue length as shown earlier in Table 4. These batches are then split and the entities follow a one-piece flow. Also, there is an extra feeding operation at the start of the packaging cells in order to accommodate the transfer of entities from fabrication to packaging. The method in which the entities are transferred from fabrication to packaging and the extra feeding operation is the only major difference between the connected and disconnected systems in the packaging area. The processing times of the machines and the sequence of operations for the entities for both systems are the same.

Case 2: It is very similar to case 1 except that the routings for products are varied as given in Table 7. In this table, Option 5 (O5) is the least flexible arrangement where each cell can process only one product family for both connected and disconnected cells. Option 2 (O2) is the most flexible arrangement with three cells capable of running all three product families both in connected cells and disconnected cells. The remain-

Table 8. Inter-arrival time and customer order size distributions for products

Family #	Product #	Inter-arrival Time Distribution	Customer Order Size Distribution
Family 1	1	0.999 + WEIB(0.115, 0.54)	1.09 + LOGN(1.56, 1.06)
	2	0.999 + WEIB(0.0448, 0.512)	TRIA(18, 23.7, 52)
	3	1.11 + EXPO(1.87)	9 + WEIB(7.66, 1.27)
	4	2 + LOGN(3.19, 3.68)	2 + 17 * BETA(0.387, 0.651)
	5	4 + LOGN(5.05, 14)	207 + LOGN(86.5, 139)
	6	UNIF(0, 26)	TRIA(6, 12.5, 71)
	7	-0.001 + 26 * BETA(0.564, 0.304)	UNIF(9, 80)
	8	TRIA(0, 6.9, 23)	EXPO(25.3)
	9	NORM(13.7, 7.49)	NORM(108, 30.8)
	10	6 + WEIB(3.78, 0.738)	TRIA(98, 120, 187)
	11	UNIF(0, 26)	UNIF(14, 34)
Family 2	12	0.999 + WEIB(0.0126, 0.405)	5 + WEIB(7.51, 0.678)
	13	1 + LOGN(0.99, 2.62)	2 + 11 * BETA(0.412, 0.527)
	14	1.24 + EXPO(1.46)	30 + 26 * BETA(0.643, 1.08)
	15	EXPO(7.06)	2 + 34 * BETA(0.321, 0.519)
	16	0.999 + WEIB(0.0313, 0.503)	NORM(149, 57.1)
	17	0.999 + WEIB(0.195, 1.12)	NORM(23, 14.2)
	18	TRIA(0, 11.2, 25)	101 * BETA(0.822, 0.714)
	19	26 * BETA(0.649, 0.42)	EXPO(154)
	20	EXPO(7.4)	UNIF(0, 90)
	21	UNIF(0, 26)	TRIA(0, 231, 330)
	22	28 * BETA(1.11, 0.547)	TRIA(0, 224, 325)
	23	27 * BETA(0.679, 0.429)	EXPO(119)
	24	28 * BETA(0.468, 0.255)	TRIA(425, 1.05e+003, 2.5e+003)
	25	1.16 + LOGN(2.48, 1.76)	NORM(867, 534)
	26	EXPO(7.03)	NORM(68, 32.8)
	27	TRIA(0, 4.44, 25)	EXPO(13.8)
	28	9 + 17 * BETA(0.559, 0.0833)	24 * BETA(0.67, 0.969)
	29	28 * BETA(0.466, 0.301)	NORM(420, 168)
	30	28 * BETA(0.932, 0.479)	NORM(267, 110)
	31	2 + 26 * BETA(0.314, 0.458)	TRIA(0, 274, 381)
	32	UNIF(0, 26)	TRIA(0, 297, 368)
Family 3	33	0.999 + WEIB(0.0117, 0.424)	TRIA(843, 1.19e+003, 2e+003)
	34	1.33 + 1.96 * BETA(0.3, 0.636)	WEIB(6.83, 0.613)
	35	1 + LOGN(5.23, 7.03)	37 + LOGN(147, 1.51e+003)
	36	4 + 22 * BETA(0.305, 0.197)	TRIA(0, 543, 591)

ing options vary in flexibility between O5 and O2. In Option 1, the system is highly inflexible in connected cells whereas it is very flexible in packaging cells of disconnected arrangement (three product families for each cell). In options 3, 4, 6 and 7, each product family can be run at least in two cells. In option 3, packaging cells of disconnected arrangement is more flexible (once again three product families for each cell). In option 4, a little bit more flexibility is added to both connected and disconnected cells (cell 1 can run three families). In option 6, more flexibility is now added to fabrication cells of disconnected system (three product families for each cell). In option 7, each family can be run in two cells. However, models for options 1 and 5 didn't stabilize and therefore they were not included in comparisons.

Production order quantities for products 33 and 36 were both reduced by 40% and 50%, respectively to fit into existing capacity for case 1. Validation and verification are an inherent part of any computer simulation analysis. Models were verified and validated before statistical analysis was performed for all scenarios.

RESULTS OBTAINED

The results obtained from simulation analysis for average flowtime and average work-in-process inventory are summarized in Tables 9 and 10, respectively. The results are based on 100 replications. The statistical analysis was conducted using the statistical functions available in Excel. A t-test assuming unequal variances for two samples was conducted for a 95% confidence interval for each family under each system. Table 11 displays the comparison for each family with respect to flowtimes and work-in-process between connected and disconnected systems. Table 12 displays comparisons for the families for the same perfor-

Table 9. Average flowtime results for all cases

Cases and Options	Connected Cells Configuration			Disconnected Cells Configuration		
	F1	F2	F3	F1	F2	F3
C1	42.66	50.52	87.53	31.19	54.39	71.61
C2-02	31.08	45.98	66.61	32.55	51.62	73.79
C2-03	24.91	39.84	67.06	27.24	46.93	83.48
C2-04	41.26	51.15	78.25	35.14	49.66	79.49
C2-06	Same as C2-03			31.88	51.17	73.80
C2-07	Same as C2-03			70.67	45.91	78.06

Table 10. Average work-in-process results for all cases

Cases and Options	Connected Cells Configuration			Disconnected Cells Configuration		
	F1	F2	F3	F1	F2	F3
C1	128.59	1403.77	1381.40	100.15	1622.52	1182.29
C2-02	90.00	1184.19	1052.06	99.70	1563.94	1267.13
C2-03	70.67	1046.90	1246.10	86.36	1425.42	1442.67
C2-04	126.10	1425.71	1269.42	111.27	1667.29	1409.31
C2-06	Same as C2-03			97.46	1555.34	1273.61
C2-07	Same as C2-03			80.34	1380.77	1532.79

Table 11. Connected vs. disconnected configuration for each family

Cases and Options	FLOWTIME			WIP		
	F1	F2	F3	F1	F2	F3
C1	S (D)	NS	NS	S (D)	S (C)	NS
C2 – O2	NS	S (C)	NS	S (C)	S (C)	S (C)
C2 – O3	S (C)	S (C)	S (C)	S (C)	S (C)	NS
C2 – O4	NS	NS	NS	NS	S (C)	NS
C2 – O6	S (C)	S (C)	NS	S (C)	S (C)	NS
C2 – O7	S (C)	S (C)	NS	S (C)	S (C)	NS

Table 12. Comparison between connected systems

Cases and Options	FLOWTIME			WIP		
	F1	F2	F3	F1	F2	F3
O2 VS O3	S (O2)	S (O2)	NS	S (O2)	S (O2)	NS
O2 VS O4	S (O2)	NS	NS	S (O2)	S (O2)	NS
O3 VS O4	S (O3)	S (O3)	NS	S (O3)	S (O3)	NS
C1 VS O2	S (O2)	NS	NS	S (O2)	S (O2)	NS
C1 VS O3, O6,O7	S (O3)	S (O3)	NS	S (O3)	S (O3)	NS
C1 VS O4	NS	NS	NS	NS	NS	NS

Table 13. Summary table of results for disconnected system: cases 1 and 2

Cases and Options	FLOWTIME			WIP		
	F1	F2	F3	F1	F2	F3
O2 VS O3	S (O3)	S (O3)	NS	S (O3)	S (O3)	NS
O2 VS O4	S (O4)	S (O4)	NS	S (O4)	S (O4)	NS
O2 VS O6	S (O6)	S (O6)	NS	S (O6)	S (O6)	NS
O2 VS O7	S (O7)	S (O7)	NS	S (O7)	S (O7)	NS
O3 VS O4	S (O3)	NS	NS	S (O3)	S (O3)	NS
O3 VS O6	S (O3)	S (O3)	NS	S (O3)	S (O3)	NS
O3 VS O7	NS	NS	NS	NS	NS	NS
O4 VS O6	NS	NS	NS	S (O6)	NS	NS
O4 VS O7	S (O7)	S (O7)	NS	S (O7)	S (O7)	NS
C1 VS O2	NS	NS	NS	NS	NS	NS
C1 VS O3	S (O3)	S (O3)	NS	S (O3)	S (O3)	S (C1)
C1 VS O4	NS	NS	NS	NS	NS	NS
C1 VS O6	NS	NS	NS	NS	NS	NS
C1 VS O7	S (O7)	S (O7)	NS	S (O7)	S (O7)	NS

mance measures but the comparisons are made between different connected systems from cases 1 and 2. Table 13 also displays comparisons for the families for the same performance measures but the comparisons are made between different disconnected systems from cases 1 and 2. Results are denoted as significant (S) or not significant (NS) based on the conclusions reached. Also whenever significant, better option was denoted in a parenthesis. The significance of the results was based on the p-value obtained from the T-test conducted for an alpha level of 0.05. As mentioned earlier, no results for options 1 and 5 were obtained as the system did not stabilize.

As observed in Table 11, for case 1, the flowtimes and work-in-process were observed to be different and the disconnected system had lower flowtimes and WIP for families F1 and F3 while the difference was significant for F1. On the other hand, WIP was significantly lower for F2 in the connected system. For case 2 with all the options considered, when there was a significant difference, this was always in favor of connected systems. For option 2, the flowtime for family 2 and the WIP for all three families for the connected system were significantly lower than those of in the disconnected system. For options 3, 6, and 7 which were the same for the connected system, the flowtimes and WIP for families 1 and 2 were significantly lower than the disconnected system. For option 4, the WIP for family 2 in the connected system was the only significant result. From Table 12, it can be observed that option 2 (O2) provided the best results when compared to rest of the options within the connected system with lower flowtimes and WIP followed by option 3 (O3). From Table 13, it can be observed that the flowtimes and WIP for options 3 and 7 (O3, O7) were consistently and significantly better when compared to the rest of the options in the disconnected cells configuration. Also, when these two options were compared against each other there was no significant difference observed for any of the families and performance measures. A comparison between models C1 and O2 did not yield any significant results either and were definitely less superior in performance when compared with the rest of the options.

CONCLUSION

In this chapter, the performance of connected and disconnected cellular systems was compared under make-to-order strategy in a real cellular setting. In the existing system (case 1), it was observed that no cellular manufacturing design dominated the other, i.e., mixed results were obtained as to which system did better for each family. The flowtime and work-in-process for family 1 for the disconnected system were lower. On the other hand, the WIP for family 2 in the connected system was lower. The other comparisons did not yield any significant results and hence dominance could not be established in terms of better cellular system.

In case 2, which is basically an extension of case 1, the impact of considering alternate cell routings for each part family was studied for both connected cells and disconnected cells. In most cases, connected cells outperformed disconnected cells with respect to both average flowtime and WIP, especially for family 1 and family 2. This leads to the conclusion that the connected system is the better system in this situation since family 1 and family 2 make up for 32 of the 36 products and comprise of about 85% by volume of the production orders in the system. The average flowtime and WIP conclusions are similar but not identical, i.e. there were incidents where flowtime was significantly better but not necessarily corresponding WIP and vice versa. If one wanted to choose the best connected cell configuration, that would be option 2. This is possibly due to option 2 having the highest flexibility among all options as each family could be routed to any of the fabrication and packaging cells. Options 3, 4 and case 1 followed in the order of performance leading to the conclusion that increase in routing flexibility of the families resulted in significantly lower flowtimes and WIP.

A similar comparison among all options developed for the disconnected system showed that options 3 and 7 performed better than the rest of the options. Option 3 had complete flexibility in the packaging area but limited flexibility in the fabrication area and option 7 had limited flexibilities in both the areas. Limited flexibility as applicable to these two options means that each family could go to at least two specified cells. On the other hand, option 2 was the worst performing system among the options for case 2 even though it had the highest flexibility. This can be attributed to the fact that routing decisions are made based on queue sizes only. Family 3 products have the highest processing times and it is possible that queues in all cells may contain products from family 3 thus leading to higher lead times for the parts that join that queue.

For case 1 and also option 2 from case 2, the disconnected system was modified to delete the extra feeding operation and the batching at the end of the fabrication area. This was done in order to determine the reason why the connected system performed better than the disconnected system in most of the comparisons made. The two modified simulation models were run and the results were statistically analyzed. In case 1, the flowtime for family 1 and the WIP for family 2 was significantly better for the disconnected system. In the original comparison, WIP and flowtime for family 1 in the disconnected system was better and the WIP for family 2 in the connected system was significantly better. The rest of the comparisons did not yield any significant results. For option 2, none of the comparisons yielded significant results as opposed to the original comparison when the connected system clearly performed better than the disconnected system. From these results it can be concluded that the extra operation and the extra batching increases the average WIP and flowtimes for each of the families and could be responsible for the disconnected system not performing as well as or better than the connected system.

REFERENCES

Agarwal, A., & Sarkis, J. (1998). A review and analysis of comparative performance studies on functional and cellular layouts. *Computers & Industrial Engineering*, *34*(1), 77–89. doi:10.1016/S0360-8352(97)00152-6

Albino, V., & Garavelli, C. A. (1998). Some effects of flexibility and dependability on cellular manufacturing system performance. *Computers & Industrial Engineering*, *35*(3-4), 491–494. doi:10.1016/S0360-8352(98)00141-7

Caprihan, R., & Wadhwa, S. (1997). Impact of routing flexibility on the performance of an FMS – A simulation study. *International Journal of Flexible Manufacturing Systems*, *9*, 273–278. doi:10.1023/A:1007917429815

DeCroix, G. A., & Arreola-Risa, A. (1998). Make-to-order versus make-to-stock in a production inventory system with general production times. *IIE Transactions*, *30*, 705–713. doi:10.1023/A:1007591722985

Federgruen, A., & Katalan, Z. (1999). The impact of adding a make-to-order item to a make-to-order production system. *Management Science*, *45*(7), 980–994. doi:10.1287/mnsc.45.7.980

Flynn, B. B., & Jacobs, F. R. (1987). A comparison of group technology and process layout using a model of an actual shop. *Decision Sciences*, *18*, 289–293.

Gupta, D., & Benjaafar, S. (2004). Make-to-order, make-to-stock, or delay product differentiation? A common framework for modeling and analysis. *IIE Transactions*, *36*, 529–546. doi:10.1080/07408170490438519

Johnson, J., & Wemmerlov, U. (1996). On the relative performance of functional and cellular layouts – An analysis of the model-based comparative studies literature. *Production and Operations Management*, *5*(4), 309–334. doi:10.1111/j.1937-5956.1996.tb00403.x

Morris, S. J., & Tersine, J. R. (1990). A simulation analysis of factors influencing the attractiveness of group technology cellular layouts. *Management Science*, *36*(12), 1567–1578. doi:10.1287/mnsc.36.12.1567

Nandi, A., & Rogers, P. (2003). Behavior of an order release mechanism in a make-to-order manufacturing system with selected order acceptance. *Proceedings of the 2003 Winter Simulation Conference.*

Selen, J. W., & Ashayeri, J. (2001). Manufacturing cell performance improvement: A simulation study. *Robotics and Computer-integrated Manufacturing*, *17*, 169–176. doi:10.1016/S0736-5845(00)00051-X

Shafer, S. M., & Charnes, J. M. (1993). Cellular versus functional layouts under a variety of shop operating conditions. *Decision Sciences*, *24*(3), 665–681. doi:10.1111/j.1540-5915.1993.tb01297.x

Süer, G. A., Huang, J., & Maddisetty, S. (2010). Design of dedicated, shared and remainder cells in a probabilistic environment. *International Journal of Production Research*, *48*(19), 5613–5646. doi:10.1080/00207540903117865

Van Donk, D. P. (2000). Make to stock or make to order: The decoupling point in the food processing industries. *International Journal of Production Economics*, *69*, 297–306. doi:10.1016/S0925-5273(00)00035-9

Yazici, J. H. (2005). Influence of flexibilities on manufacturing cells for faster delivery using simulation. *Journal of Manufacturing Technology*, *16*(8), 825–841. doi:10.1108/17410380510627843

Chapter 4
Design of Manufacturing Cells Based on Graph Theory

José Francisco Ferreira Ribeiro
University of São Paulo, Brazil

ABSTRACT

In this chapter a comparative study is presented between (I) sequential heuristics, (II) simulated annealing, (III) tabu search, and (IV) threshold algorithm for graph coloring and its application for solving the problem of the design of manufacturing cells in a job shop system production. The job shop production system has a very large proportion of all manufacturing activity. The principal concepts of manufacturing cells, graph theory, and heuristics are presented. The results obtained with these algorithms on several examples found in the literature are consistently equivalent with the best solution hitherto known in terms of numbers of inter-cell moves and dimensions of cells.

INTRODUCTION

Over 75% of all parts manufactured in the industry are produced in batches of 50 parts or less. Consequently, the production in batch and the production on demand constitute a considerable proportion of all manufacturing activity (Groover, 1987). Job shop is a production environment that produces parts in small batches. It's a production environment common in small and medium enterprises. The parts require different manufacturing operations and must be performed through various production departments and in different sequences (Oliveira, Ribeiro and Seok, 2009). Orders differ in the number of parts, design, processing times, setup times or urgency. The high demand for machinery and the different production sequences can cause long queues in the shop floor. The consequence is delivery times unreliable, whereas nowadays delivery times should be short and reliable (Ribeiro and Pradin, 1993). Clustering is the task of classifying a collection of objects,

DOI: 10.4018/978-1-61350-047-7.ch004

such as documents, parts, or machines, into natural categories. Clustering techniques are widely used in many areas such as machine learning, data mining, pattern recognition, image analysis and bioinformatics. The design of manufacturing cells or clustering consists in partitioning the set of parts to be manufactured in an industry into families and the available machines into groups or cells, so that each family is associated with one machine group, and vice-versa. Each family-group pair constitutes a manufacturing cell. This concept lies on grouping similar parts in families, proposing to produce them in cells that have specially selected machines to accomplish this. This procedure leads to greater automation, set up time reduction, standardization of the tools used and a reduction of manufacturing cycles (Ribeiro and Meguelati, 2002). A greater efficiency in management and manufacturing is expected, due to the decomposition of the production global system in sub-systems of reduced dimension. Workshops operating on this principle, called Group Technology (Burbidge, 1975), offer a reduction in unproductive manufacturing time, resulting in greater flexibility, just in time and productivity (Hyer and Wemmerlöw, 1989).

However, the design of manufacturing cells requires the solution of a complex mathematical problem: the block-diagonalization of an incidence matrix [parts × machines] corresponding to the global production system. This block-diagonalization uses as its optimization criterion the minimization of the number of elements in the matrix outside the diagonal blocks. These elements represent inter-cell moves and, in practice, imply undesirable movement of parts to machines in other cells that are not present in the cell to which the part is assigned.. That is the reason why, regarding the manufacturing cells, there is an attempt to minimize the number of inter cell moves, at the same time that a balance of workloads between the different cells projected is sought. This is treated as a combinatorial problem for which there is no polynomial time algorithm (Garey and Johnson, 1979), and the most common approach in the literature is to propose heuristic algorithms.

BACKGROUND

A large number of techniques have been used in recent years (Sing, 1993) to hit the block-diagonalization of the matrix, designing manufacturing cells and implementing Group Technology in the industries. For example:

a. Mathematical programming: Won (2000), Albadawi, Bashir, and Chen (2005), Panchalavarapu and Chankong (2005), Slomp, Chowdary, and Suresh (2005), Rajagopalan and Fonseca (2005), Adil and Ghosh (2005), Yin, Yasuda, and Hu (2005), Foulds, French, and Wilson (2006)
b. Branch and bound: Ramabhatta e Nagi (1998), Boulif and Atif (2006)
c. Fuzzy logic: Xu e Wang (1989), Chu and Hayya (1991)
d. Genetic algorithms: Suer, Vásquez, and Peña (1999), Zhao and Wu (2000), Dimopoulos and Mort (2001), Suer, Peña, and Vázquez (2003), Gonçalves and Resende (2004), Hicks (2004), Solimanpur, Vrat, and Shankar (2004), Rajagopalan and Fonseca (2005), Vin, De Lit, and Delchambre (2005), Gonçalves and Tiberti (2006), Jeon and Leep (2006)
e. Neural networks: Lozano, Canca, Guerrero, and Garcia (2001), Guerrero, Lozano, Smith, Canca, and Kwok (2002), Solimanpur, Vrat, and Shankar (2004), Pashkevich and Kazheunikau (2005)
f. Meta-heuristics—tabu search and simulated annealing: Caux, Bruniaux, and Pierreval (2000), Baykasoglu (2003), Spiliopoulos and Sofianopoulou (2003), Xambre and Vilarinho (2003), Cao and Chen (2005)
g. Data analysis: Ribeiro and Pradin (1993), Diallo, Pierreval, and Quilliot (2001), Rios,

Campbell, and Irani (2002), Ribeiro and Meguelati (2002), Ribeiro (2003), Oliveira, Ribeiro and Seok (2009)

h. Graph theory: Rajagopalan and Batra (1975), Askin and Chiu (1990), Rath, Das and Sahu (1995), Selim (2002), Ribeiro (2009)

The main different methods from the literature are approximate methods (not optimal), because the problem is NP-complete. The computational time is very high when we use exact methods to solve large-size examples.

GRAPH COLORING

A coloring of a graph G(V, A), with), with V a set of N nodes and A a set of M edges (Matula, Marble, and Isaacson, 1972, Christofides, 1975, Korman, 1979) is an assignment of any color belonging to set of colors C = {ci} for each node of V, where two nodes connected by an edge of A cannot have the same color.

I.e., a coloring of G is a function $f: V \rightarrow C \mid$ if $(i, j) \in A \Rightarrow f(i) \neq f(j)$. A k-coloring of G(V, A) is a coloring with k colors, i.e., a partition of V in k independent sets of nodes.

In this case, G is a k-coloring graph. The chromatic number $\lambda(G)$ of G is the smaller number of colors for which a k-coloring of G exists. The graph coloring problem is NP-complete (Garey and Johnson, 1979).

It is a high combinatorial problem: for example, a complete and exhaustive enumeration of all colorings consists in an $O(M.k^N)$ algorithm (Aho, Hopcroft and Ullman, 1983). But, some immediate results can be obtained of the definition, such as:

a. A graph G is bi-chromatic if and only if is bipartite
b. A complete sub-graph of G with t vertices Kt is t-chromatic
c. A graph G with two or more edges is at least 2-chromatic

The concepts of graph coloring, clique (a complete sub-graph of G) and independent set of nodes are close:

a. k colors are necessary to coloring k nodes of a clique with cardinality k, thus $\lambda(G)$ is greater than or equal to the cardinality of the greatest clique of G
b. Let us consider $S_1 \ldots S_k$ disjunctive subsets of S generated by the k-coloring; then $= \cup S_i = S$, $i = 1 \ldots k$, and each S_i is an independent set and each Si is an independent set in which any couple of edges is not connected.

GRAPH COLORING AND MANUFACTURING CELLS DESIGN

In a graph G(V, A) associated to a production system, V = {set of parts to be manufactured}. The dissimilarity indexes between parts are calculated, for example, based on the existing differences between their production sequences.

In this case, an edge connecting two nodes i and j exists if the dissimilarity dij is greater than or equal to a critical dissimilarity established a priori.

This critical dissimilarity can be modified in any instant with the objective of higher or lower number of edges of the graph and then, the number designed of cells.

Two nodes connected cannot have the same color, resulting in a coloring where the parts with different colors have dissimilarity indexes greater than or equal the fixed value for critical dissimilarity.

Parts with the same color will be assigned to the same family and manufactured in the same cell of machines.

The distribution of machines is carried out considering the number of operations carried out in each family: a machine is assigned to the family in which it will be most used.

Table 1. Matrix MPM-LOAD [parts × machines]

	Machine 1	Machine 2	Machine 3	Machine 4
Part 1	0	15	0	40
Part 2	25	0	60	0
Part 3	0	35	0	50
Part 4	90	0	70	0
Part 5	30	0	0	0

Example

From the initial data a matrix MPM-LOAD[parts × machines] is obtained (see Table 1), which provides the total time spent by each part transiting in the program of each machine. This matrix is called work load matrix and its coefficients are calculated as follows:

load[i,j] = unit[i] × ∑ duration[i, k]
k | program[i, j] = j

where:

unit[i] = unit number of parts[i] to manufacture.
duration[i, k] = duration of operation k on a part[i].
program[i, k] = type of machine used to perform operation k on a part[i].

Let the matrix MPM [parts × machines] (see Table 2), that informs the use ($MPM_{ij} = 1$) or not ($MPM_{ij} = 0$) of part i by the machine j. The matrix D [parts × parts] (see Table 3) gives the dissimilarity index between parts ("distances between parts") to be manufactured. Two manufacturing cells will be designed for this workshop. If the critical dissimilarity is equal to 1, G(V, A) is the graph shown in Figure 1, i.e., three-coloring, and not acceptable because the workshop must be partitioned in two cells. Then, the critical dissimilarity value is incremented to two and the two-coloring graph shown in Figure 2 is obtained. After defining the families of parts, machines are assigned to the most demanding families. The distribution of machines is carried out considering the number of operations carried out in each family: a machine is assigned to the family in which it will be most used. The matrix CELLS [parts × machines] in Table 4 presents the workshop partitioning in 2 manufacturing cells and this solution is very good: there are no inter-cell moves and the dimensions of cells are equal to [2 × 2] for cell 1 and [3 × 2] for cell 2.

DISSIMILARITY INDEXES BETWEEN PARTS

Similarity measures for parts or machines have a long history of use for manufacturing cells design problems. The first use of similarity measure for the manufacturing cells design problem was by McAuley (1972). Let:

x_{ij} = number of parts processed by machines mi and m_j (number of matches)
x_i = number of parts processed by machine m_i only

Table 2. Matrix MPM [parts × machines]

	Machine 1	Machine 2	Machine 3	Machine 4
Part 1	0	1	0	1
Part 2	1	0	1	0
Part 3	0	1	0	1
Part 4	1	0	1	0
Part 5	1	0	0	0

Table 3. Matrix D [parts × parts]

	Part 1	Part 2	Part 3	Part 4	Part 5
Part 1	—				
Part 2	4	—			
Part 3	0	4	—		
Part 4	4	0	4	—	
Part 5	3	1	3	1	—

Figure 1. Three-coloring graph

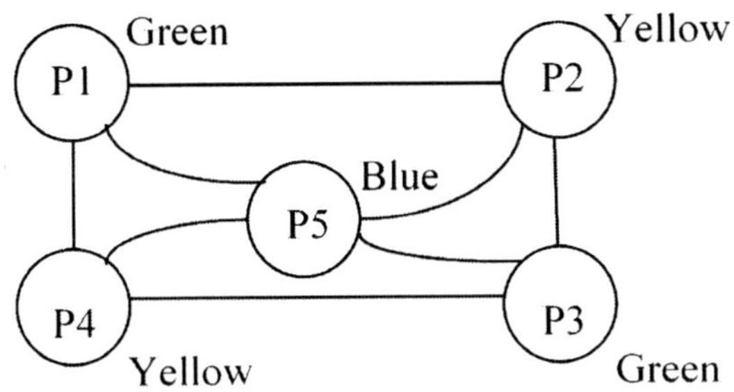

Figure 2. Two-coloring graph

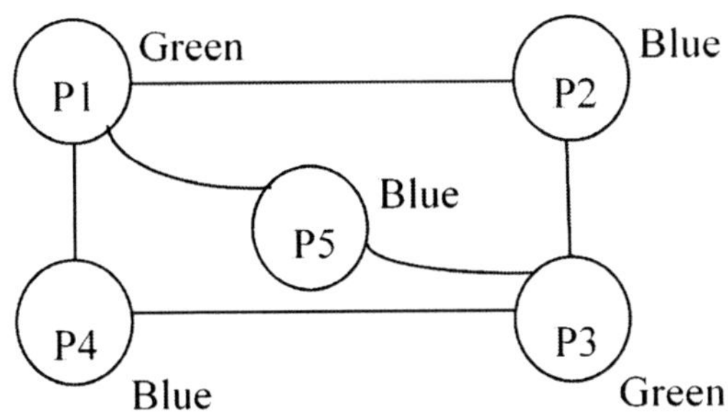

Table 4. Matrix CELLS [parts × machines]

	Machine 2	Machine 4	Machine 1	Machine 3
Part 1	1	1		
Part 3	1	1		
Part 2			1	1
Part 4			1	1
Part 5			1	

y_{ij} = number of parts that are not processed by either m_i or m_j (number of misses).

McAuley (1972) adopted a Jaccardian similarity measure in evaluating similarities of pairs of machines. This measure considers x_{ij} as a main factor for similarity and divides it by the number of parts which either machine m_i or m_j processes. After the use of the first similarity measure, many other different types of similarity measures have been introduced, for example: Bipartite, Kusiak, McAuley-Jaccard, BUB, Russel and Rao, Sorenson, Simple Matching, Ochiai (Oliveira, Ribeiro and Seok, 2008). Jaccardian similarity measures (Bipartite, McAuley-Jaccard, Sorenson and Ochiai) are adopted for many cell formation problems (Wei and Kern, 1989). Sorenson's measure is a simple modification of McAuley's measure, where xij has more weight (Romesburg, 1984). However, it is reported that these Jaccardian similarity measures are limited because they do not use the number of misses, yij (Moiser, 1989, Islam and Sarker, 2000). Selim and Abdel (2003) distinguished those Jaccardian similarity measures from non-Jaccardian measures which use yij (Kusiak, BUB, Russel and Rao, Simple Matching). Among them, Kusiak's measure is one of the simplest (Kusiak, 1987). As said before, they use not only xij but also yij in the similarities. Simple matching was first introduced in the field of medicine. This measure simply adds yij to both the top and bottom of McAuley's Jaccardian measure. The most successful non-Jarccardian measure is the BUB similarity measure. This measure has been successfully adopted by many clustering algorithms because their distribution of values are more normal and continuous (Baroni-Urban and Buser, 1976). Other notable non-Jarccardian measures are Russel and Rao's and Ochiai's in Romesburg (1984). Note that there are two big differences between the bipartite similarity matrices and other similarity coefficients usually used for manufacturing cells design problems. First, while we construct both similarity matrices for machines and parts, most array-based clustering algorithms use only a machine-machine similarity matrix to construct cells of closely related machines and then part families are constructed according to these machine cells. Second, diagonal entries of similarity coefficient matrices do not play any role, so all diagonal entries are considered 0. However, it is reasonable when the similarity of each machine or part with itself has bigger similarity than with others.

More details on the similarity measures can be found in Sarker (1996) and Sarker and Islam (1999) that are dedicated to analyzing most of the existing similarity measures in the literature based on a set of important properties developed by Baroni-Urban and Buser (1976).

In this chapter the dissimilarity indexes between parts i and j are calculated by quantifying the differences between their production sequences. The proposed method compares the production sequences of parts i and j utilizing the binary matrix MAT [parts × machines], where:

$$MAT_{ij} = \begin{cases} 1 & \text{if machine j is utilized to manufacture part i} \\ 0 & \text{if machine j is not utilized to manufacture part i} \end{cases}$$

Then, the dissimilarity index between parts i and k is computed as follows:

number of machines
$D[i, k] = \sum \boldsymbol{\mu}[MAT_{ij}, MAT_{kj}]$
j=1

where:

$$\mu[a_{ij}; a_{kj}] = \begin{cases} 1 & \text{if } MAT_{ij} \neq MAT_{ij} \\ 0 & \text{otherwise} \end{cases}$$

This procedure is based on the similarity computation used, for example, by Kusiak (1987), Wei and Kern (1989), Ribeiro and Meguelati (2002), Oliveira, Ribeiro and Seok (2008) and Ribeiro (2009).

ALGORITHMS

Sequential Heuristics

Let us consider a Table that defines an order for the N nodes of G. A sequential method S consists in coloring successively $V_1 \dots V_N$, assigning to these nodes the color of the minimum index not utilized by their neighbors. The choice of the node is imposed in each iteration and the choice of the color is trivial, corresponding to an algorithm sample and compact in $O(N^2)$.

But, this heuristic can be very detrimental, despite the existence of an optimal order. In fact, if $\lambda(G) = k$, then a partition of nodes exists in k independent set of nodes $S_1 \dots S_k$. Inserting in V the nodes in crescent order of the number of the independent set of nodes, this heuristic finds the optimal solution searching N! orders. An order by the nodes can be established, for example, by three manners as follows:

a. The simpler manner to implement S consists in utilizing the natural order of the numbers of nodes, i.e., $V_i = i$ for all node i. This heuristic is called FFS (First Fit Sequential).
b. By utilizing the crescent order of the degree of the nodes, Wesh and Powell (1967) propose the algorithm called LFS (Largest First Sequential).
c. An order is defined as follows: I) V_N is the node with minimum degree; II) For i = N-1 … 1, the node V_i is the node of minimum degree in the sub-graph generated by V – $\{V_N \dots V_{i+1}\}$. This order is the basis of the algorithm proposed by Matula, Marble, and Isaacson (1972) and it is known by SLS (Smallest Last Sequential).

These three heuristics obtain good and bad results at random. Then, the implementation of meta-heuristics simulated annealing and tabu search utilizes them for obtaining an initial solution.

Simulated Annealing

Kirkpatrick, Gelatt and Vecchi (1983) have created the simulated annealing meta-heuristic. This technique is based on the annealing of metals in metallurgy: a metal cooled very fast presents a lot of microscopic defeats, corresponding to a local minimum in optimization problems. If the cooling is slow, the atoms arrange its structure, the defeats missing and the metal has, then, an ordered structure, equivalent to the global minimum in optimization. Simulated annealing in combinatorial optimization has some similarity with thermodynamics. The energy of the system is represented by an arbitrary real number T, i.e., the temperature.

Simulated annealing begins by local search or heuristic procedure (FFS, LFS or SLS), because the method always begins from a feasible initial solution s, and the six steps bellow are activated:

a. Take at random a transformation s' instead to find the best or the first improved solution in the neighborhood
b. Build the resulting solution s' and calculate the variation of cost $\Delta f = f(s') - f(s)$
c. If $\Delta f \leq 0$ the cost is better. Make s = s'
d. If $\Delta f > 0$, the cost is worse. The penalty is higher when the temperature is low and Δf is big. An exponential function has the appropriate properties. A probability of acceptation $a = e^{-\Delta f/T}$ is calculated and, then a parameter p is taken at random in [0, 1]. If $p \leq a$, the transformation is accepted, even with degradation of the cost and the solution is modified: s = s'. Otherwise, the transformation is rejected and s is conserved for the next iteration
e. The convergence is assured by lowering T slowly at each iteration, for example, T=kT, k < 1, but close
f. The algorithm stops when T attempts a limit equal to a ε fixed close to 0. In the implementation, the neighborhood of the solutions is defined by lists, but not on the graph.

Tabu Search

Tabu search was created by Glover (1989, 1990) and has no stochastic character. For the same computational time, tabu search generally presents better results than simulated annealing. Three fundamental points constitute the technique:

a. In each iteration, the neighborhood V(s) of the actual solution s is completely examined and the best solution s' is chosen, even if the cost is higher
b. Tabu list T forgives the return to a solution recently obtained. This list stocks in a compact manner, the steps traversed by the algorithm. Then, the method seeks s' in V(s) – T
c. The best solution found is stocked, because unlike the simulated annealing, it is rarely the last one

The method stops after a maximum number of iterations or after a maximum number of iterations without improving the best solution or when V(s) – T = ∅. The problematic point is the capacity C of the tabu list T. Glover's research knows that C = 7 to 20 is sufficient to prevent cycles, for any dimension of the problem. T is utilized as a short term memory. In each iteration, the Cth transformation of T (the oldest) is substituted by the last transformation realized. In the implementation, T is simply generated as a data structure of type queue. Werra (1990) and Hertz (1991) propose tabu search to solve the sub-problem of the existence of the k-coloring, for k fixed a priori. The technique begins to assign k arbitrary colors to the N nodes. If the k-coloring is attempted, the coloring procedure is repeated for k-1 colors. Frequently, the assignment presents conflict, i.e., nodes connected with the same color. Then, the objective consists in minimizing the total number of conflicts, converting the problem of existence in an optimization problem on the number of conflicts. The neighborhood is formed by all assignments obtained by exchanging the color i of a node α to another color j (α must have at least one neighbor with color i). Tabu search forgives α to take its previous color during NT iterations, NT

= capacity of tabu list. In the iteration, the search seeks the best neighbor of the actual assignment, i.e., the neighbor that minimizes the number of conflicts. The search stops if the number of conflicts is equal to 0 (it means a k-coloring) or if the maximal number of searches in the neighborhood is over. The method is considered good for finding a coloring close to the optimal. It begins with a value of k corresponding to a superior limit, obtained, for example, by a heuristic procedure. If the tabu search solves the problem with k colors, the procedure is repeated for k-1.

The results described in the literature are very impressive, even for big graphs (10,000 nodes = 10,000 parts). The initial number of iterations of the search is lower (0 to 10 iterations), but increase quite a bit for the last values of k. This fact shows the hardiness of the existence problem when the procedure is close to the optimal solution.

Threshold Algorithm

Provided that the problem of partitioning an X collection of n objects with a dissimilarity of classes, in a fixed k number, so that the diameter d(P) among the classes is minimum. This corresponds to constructing a threshold G_s graph, a partial sub-graph of G(V<A), that should be k-colorable. To color the graphs of the vertices, a technique based on the algorithm by Guénoche (1993) was used. This algorithm enumerates all of the partitions in a fixed number of minimum diameter classes. It is based on a threshold graph in p colors, with each color defining a class. Many heuristics to approximate the diameter and the partitions of minimum diameter are enumerated only at the last stage. Let d(P) the diameter of the partition of V into p classes. A superior limit, s, for d(P) is heuristically determined. Then, the k-colorations for G_s are enumerated. As long as there is at least one possible k-coloration, the value s is decreased and a new iteration is processed. When the algorithm stops, the highest value obtained is equal to the remaining partition diameters. To enumerate all of the possible partitions of X, in minimum diameter K classes, this algorithm is applied, which is comprised of 3 stages and has an $O(k^{N-k})$ complexity. In the first stage a heuristic method is used to determine the maximum s, highest approximation to d(P), such that Gs will be k-colorable. The maximum s is determined by a method of dichotomic subdivisions of the variation interval of the dissimilarities in a sequential coloring method, in this case the saturation algorithm (called "Dsatur"). In this algorithm, at each iteration, the rate of saturation $DS_i(\alpha)$ is defined as the number of colors already employed by the neighbors of α. The procedure consists of:

a. coloring the most sizable vertex with the color 1
b. in the following stages, take the vertex free of maximum DS and color it with the least possible index color

In the second stage, once the maximum s has been set, all of the Gs colorings are enumerated in k colors. With the aid of "Dsatur", maximum if possible, a clique of G_s is obtained, where each node represents a class. Then, according the order of saturation, the other vertices are colored in all possible manners. If a vertex is near the colored vertices, these will not be able to be used to color them. Thus, all of the partitions in k classes of less than s diameter are obtained. In the third stage, a decreasing order of the edges of the dissimilarity values, starting from s, are considered. An edge can be inserted in the graph as long as there is a compatible partition, that is, if the edge connects different types of nodes.

Thus, each inserted edge eliminates some previously obtained partitions. The first edge that cannot be inserted, since there would be no more compatible partitions left, has equal value to the highest interclass dissimilarity value. The remaining partitions for a fixed number of classes are of minimum diameter.

RESULTS

Table 5 summarizes some tests carried out with the programs corresponding to the sequential heuristics, simulated annealing, tabu search and threshold techniques. These programs are written in MatLab and run on a microcomputer. Using examples found in the literature, a comparison is presented between the best solution known (BS) and the solutions obtained by sequential heuristics (SH), simulated annealing (SA), tabu search (TS) and threshold method (TM).

The result presented to the sequential heuristics is the best solution obtained by the three options described above. Parameters, such as temperature, ε, NT, etc., for running the programs corresponding to simulated annealing and tabu search algorithms were chosen differently for each example, by attempt, always with the objective to find the best solution. In the 1st column of Table 5, the example given is given; in the 2nd, the methods utilized for solving the example: BS (the method responsible by the best solution known), SH: sequential heuristics, SA: simulated annealing, TS: tabu search, TM: threshold method; in the 3rd, the number of cells; in the 4th, the number of parts and machines; in the 5th, the number of inter-cell moves; in the 6th, the dimensions of the cells obtained; in the 7th, the computational time (in seconds — Pentium 2.20 GHz, 1.99 GB RAM).

Table 5. Computational results

E	M	C	P × M	I	D	T
Ribeiro and Meguelati, 2002	BS	2	9 × 12	0	4×6,5×6	—
	SH			2	4×6,5×6	0.008
	SA			0	4×6,5×6	1.204
	TS			2	4×6,5×6	1.018
	TM			0	4×6,5×6	1.865
Ribeiro and Pradin, 1993	BS	3	30 16	16	5×3,10×6,15×7	—
	SH			24	5×4,10×6,15×6	1.987
	SA			16	5×3,10×6,15×7	8.133
	TS			24	5×4,10×6,15×6	6.098
	TM			16	5×3,10×6,15×7	19.334
Harhalakis, Nagi, and Proth,1990	BS	5	20 20	14	5×4,4×3,4×5,3×4,4×4	—
	SH			15	6×5,4×4,4×5,3×3,3×3	0.985
	AS			14	5×4,4×3,4×5,3×4,4×4	6.124
	TS			15	6×5,4×4,4×5,3×3,3×3	4.769
	TM			14	5×4,4×3,4×5,3×4,4×4	9.897
Harhalakis, Nagi, and Proth,1990	BS	4	20 20	11	7×7,6×5,4×5,3×3	—
	SH			11	7×7,6×5,4×5,3×3	0.824
	SA			11	7×7,6×5,4×5,3×3	5.655
	TS			11	7×7,6×5,4×5,3×3	4.086
	TM			11	7×7,6×5,4×5,3×3	8.897
Ribeiro and Pradin, 1993	BS	3	20 12	0	6×4,9×5,5×3	—
	SH			0	6×4,9×5,5×3	1.554
	SA			0	6×4,9×5,5×3	3.776
	TS			0	6×4,9×5,5×3	2.899
	TM			0	6×4,9×5,5×3	7.881
Kusiak, 1987	BS	2	5 ×4	0	2×2,3×2	—
	SH			0	2×2,3×2	0.005
	SA			0	2×2,3×2	0.402
	TS			0	2×2,3×2	0.678
	TM			0	2×2,3×2	0.813

CONCLUSION AND FUTURE RESEARCH DIRECTIONS

In this chapter, a comparison between sequential heuristics (SH), meta-heuristics simulated annealing (SA) and tabu search (TS) and threshold method (TM) developed for graph coloring with the objective of solving the cellular manufacturing design problem has been presented. The results obtained are either the best one known or already of a high standard for the simulated annealing, tabu search and threshold method. Threshold method and simulate annealing were better than tabu search and sequential heuristics in terms of the inter-cell movements or the dimensions of the solutions for several examples from the literature treated. Tables 6 and 7 present, respectively, the solutions obtained by sequential heuristics (SH) or tabu search (TM), and simulated annealing (SA) or threshold method (TM) for the workshop proposed by Ribeiro and Meguelati, 2002. The manufacturing cells designed by SA or TM present 0 inter-cell moves and by SH or TS 2 inter-cell moves. The computational time to find the best solution was equal to 0.006, 0.523, 0.779 and 0.989 seconds respectively for SH, TM, SA and TM.

The four algorithms take very little computational time. Thus, it is sure of obtaining a solution, which is either optimal or feasible within a very reasonable computational time. Nevertheless, for large examples generated at random, the meta-heuristic tabu search presented best results than

Table 6. Manufacturing cells obtained by SH and TS (2 inter-cell moves)

	M 1	M 2	M 4	M 6	M 8	M 11	M 3	M 5	M 7	M 9	M 10	M 12
Part 1	1	1		1	1	1						
Part 2	1	1	1	1	1							
Part 3	1		1	1	1							
Part 5		1	1		1	1					**1**	
Part 6	1	1	1		1	1						
Part 4							1		1	1	1	
Part 7								1		1	1	
Part 8							1	1	1			
Part 9	**1**						1	1		1		1

Table 7. Manufacturing cells obtained by SA and TM (0 inter-cell moves)

	M 3	M 5	M 7	M 9	M 10	M 11	M 1	M 2	M 4	M 6	M 8	M 12
Part 4	1		1	1	1							
Part 5	1	1		1	1	1						
Part 7		1			1	1						
Part 8	1	1	1									
Part 1							1	1		1	1	1
Part 2							1	1	1	1	1	
Part 3							1		1	1	1	
Part 6							1	1	1		1	1
Part 9							1	1	1		1	1

simulated annealing for the same computational time.

When the computational time is free, the results obtained generally are the same or threshold method is better.

The procedure of dissimilarity indexes computation is a very important step of the method. This parameter has a hard role in the partitioning of parts, because it measures the "distances" between parts. With this in mind, research on the computation of differences between parts was recently conducted in DMB implement factory for sugar cane located in Brazil, where 3,500 parts are manufactured by 101 machines. The objective of this study is to explore the specific characteristics of the industrial real case instances in order to obtain a high standard solution.

REFERENCES

Adil, G. K., & Ghosh, J. B. (2005). Forming group technology cells incrementally using grasp. *International Journal of Advanced Manufacturing Technology*, *26*(11), 1402–1408. doi:10.1007/s00170-004-2113-8

Aho, A., Hopcroft, J., & Ullman, J. (1983). *Data structure and algorithms*. Addison Wesley.

Albadawi, Z., Bashir, H. A., & Chen, M. (2005). A mathematical approach for the formation of manufacturing cells. *International Journal on Computers and Industrial Engineering*, *48*(1), 3–21. doi:10.1016/j.cie.2004.06.008

Askin, R. G., & Chiu, K. S. (1990). A graph partitioning procedure for machine assignment and cell formation in group technology. *International Journal of Production Research*, *28*(1), 1555–1572. doi:10.1080/00207549008942812

Baroni-Urban, C., & Buser, M. W. (1976). Similarity of binary data. *Zoology (Jena, Germany)*, *25*, 251–259.

Baykasoglu, A. (2003). Capability-based distributed layout approach for virtual manufacturing cells. *International Journal of Production Research*, *41*(11), 2597–2618. doi:10.1080/0020754031000087229

Boulif, M., & Atif, K. (2006). A new branch-and-bound-enhanced genetic algorithm for the manufacturing cell formation. *International Journal on Computers and Operations Research*, *33*(8), 2219–2245. doi:10.1016/j.cor.2005.02.005

Burbidge, J. L. (1975). *The introduction of group technology*. John Wiley.

Cao, D., & Chen, M. A. (2005). A robust cell formation approach for varying product demands. *International Journal of Production Research*, *43*(8), 1587–1605. doi:10.1080/00207540412331327754

Caux, C., Bruniaux, R., & Pierreval, H. (2000). Cell formation with alternative process plans and machine capacity constraints. *International Journal of Production Economics*, *64*(1), 279–284. doi:10.1016/S0925-5273(99)00065-1

Christofides, N. (1975). *Graph theory: An algorithmic approach*. Academic Press.

Chu, C. H., & Hayya, J. C. (1991). A fuzzy clustering approach to manufacturing cell formation. Proceedings of the 1991 International Industrial Engineering Conference, (pp. 495-500). Miami, USA.

Diallo, M., Pierreval, H., & Quilliot, A. (2001). Manufacturing cells design with flexible routing capability in presence of unreliable machines. *International Journal of Production Economics*, *74*(1), 175–182. doi:10.1016/S0925-5273(01)00124-4

Dimopoulos, C., & Mort, N. (2001). A hierarchical clustering methodology based on genetic programming for the solution of cell-formation problems. *International Journal of Production Research*, *39*(1), 1–19. doi:10.1080/00207540150208835

Foulds, L. R., French, A. P., & Wilson, J. M. (2006). The sustainable cell formation problem: Manufacturing cell creation with machine modification costs. *International Journal on Computers and Operations Research*, *33*(4), 1010–1032. doi:10.1016/j.cor.2004.09.001

Garey, M. R., & Johnson, D. S. (1979). *Computers and intractability: A guide to the theory of NP-completeness*. Freeman.

Glover, F. (1989). Tabu search – Part I. *ORSA Journal on Computing, 1*, 190-206.

Glover, F. (1990). Tabu search – Part II. *ORSA Journal on Computing, 2*, 4–32.

Gonçalves, E. V., & Tiberti, A. J. (2006). A genetic algorithm for the machine cell formation problem. *International Journal of Production Economics*, *102*(1), 1–21. doi:10.1016/j.ijpe.2004.12.029

Gonçalves, J. F., & Resende, M. G. C. (2004). An evolutionary algorithm for manufacturing cell formation. *International Journal on Computers and Industrial Engineering*, *47*(2), 247–273. doi:10.1016/j.cie.2004.07.003

Groover, M. P. (1987). *Automation production systems and computer integrated manufacturing*. Prentice Hall.

Guénoche, A. (1993). Enumeration des partitions de diamètre minimum. [in French]. *Discrete Mathematics*, *1*(11), 227–287.

Guerrero, F., Lozano, S., Smith, K. A., & Canca, D., 7 Kwok, T. (2002). Manufacturing cell formation using a new self-organizing neural network. *International Journal on Computers and Industrial Engineering*, *42*(2), 377–382. doi:10.1016/S0360-8352(02)00039-6

Harhalakis, G., Nagi, R., & Proth, J. M. (1990). An efficient heuristic in manufacturing cell formation for group technology applications. *International Journal of Production Research*, *28*(1), 185–198. doi:10.1080/00207549008942692

Hertz, A. (1991). Cosine: A new graph coloring algorithm. *Operations Research*, *10*, 411–415.

Hicks, C. (2004). A genetic algorithm tool for designing manufacturing facilities in the capital goods industry. *International Journal of Production Economics*, *90*(2), 199–211. doi:10.1016/S0925-5273(02)00467-X

Hyer, N. L., & Wemmerlöw, U. (1989). Group technology in the United States manufacturing industry. *International Journal of Production Research*, *27*(8), 1287–1304. doi:10.1080/00207548908942623

Islam, K. M., & Sarker, B. P. (2000). A similarity coefficient measure and machine-parts grouping in cellular manufacturing systems. *International Journal of Production Research*, *38*(3), 699–720. doi:10.1080/002075400189374

Jeon, G., & Leep, H. R. (2006). Forming part families by using genetic algorithm and designing machine cells under demand changes. *International Journal on Computers and Operations Research*, *33*(1), 263–283. doi:10.1016/j.cor.2005.03.033

Kirkpatrick, S., Gellat, C. D., & Vecchi, M. P. (1983). Simulated annealing. *Science*, *220*, 671–680. doi:10.1126/science.220.4598.671

Korman, S. M. (1979). The graph coloring problem. In *Combinatorial optimization*. John Wiley.

Kusiak, A. (1987). The generalized group technology concept. *International Journal of Production Research*, *26*(5), 561–569. doi:10.1080/00207548708919861

Lozano, S., Canca, D., Guerrero, F., & Garcia, J. M. (2001). Machine grouping using sequence-based similarity coefficients and neural networks. *Robotics and Computer-integrated Manufacturing*, *17*(5), 399–404. doi:10.1016/S0736-5845(01)00015-1

Matula, D. W., Marble, G., & Isaacson, J. D. (1972). Graph coloring algorithms. In *Graph theory and computing*. Academic Press.

McAuley, J. (1972). Machine grouping for efficient production. *Production Engineering*, *51*(2), 53–57. doi:10.1049/tpe.1972.0006

Moiser, C. T. (1989). An experiment investigating the application of clustering procedures and similarity coefficients to the group technology cell formation problem. *International Journal of Production Research*, *27*(10), 1811–1835. doi:10.1080/00207548908942656

Oliveira, S., Ribeiro, J. F. F., & Seok, S. C. (2008). A comparative study of similarity measures for manufacturing cell formation. *Journal of Manufacturing Systems*, *27*(1), 19–25. doi:10.1016/j.jmsy.2008.07.002

Oliveira, S., Ribeiro, J. F. F., & Seok, S. C. (2009). A spectral clustering algorithm for manufacturing cell formation. *International Journal on Computers and Industrial Engineering*, *57*(3), 1008–1014. doi:10.1016/j.cie.2009.04.008

Panchalavarapu, P. R., & Chankong, V. (2005). Design of cellular manufacturing systems with assembly considerations. *International Journal on Computers and Industrial Engineering*, *48*(3), 449–469. doi:10.1016/j.cie.2004.06.010

Pashkevich, A., & Kazheunikau, M. (2005). Neural network approach to trajectory synthesis for robotic manipulators. *Journal of Intelligent Manufacturing*, *16*(2), 173–187. doi:10.1007/s10845-004-5887-5

Rajagopalan, A., & Fonseca, D. J. (2005). Volume sensitivity analysis for manufacturing cells: A genetic algorithm. *Journal of Advanced Manufacturing Systems*, *4*(2), 167–183. doi:10.1142/S0219686705000667

Rajagopalan, R., & Batra, J. L. (1975). Design of cellular production systems: A graph-theoretic approach. *International Journal of Production Research*, *13*(6), 567–579. doi:10.1080/00207547508943029

Ramabhatta, V., & Nagi, R. (1998). An integrated formulation of manufacturing cell formation. *Operations Research*, *77*, 79–95.

Rath, S. K., Das, C., & Sahu, S. (1995). Graph colouring approach for cell formation problems. *Production Planning and Control*, *6*(5), 421–428. doi:10.1080/09537289508930299

Ribeiro, J. F. F. (2003). Selection and assignment of machines: A parallel approach. *Journal of Computing and Information Technology*, *11*(4), 263–270. doi:10.2498/cit.2003.04.01

Ribeiro, J. F. F. (2009). *Manufacturing cells formation based on graph theory.* Paper presented at 39th International Conference on Computers & Industrial Engineering, Paris, France.

Ribeiro, J. F. F., & Meguelati, S. (2002). Organização de um sistema de produção em células de fabricação. [in Portuguese]. *Journal Gestão e Produção*, *9*(1), 62–77.

Ribeiro, J. F. F., & Pradin, B. (1993). A methodology for cellular manufacturing design. *International Journal of Production Research*, *31*(1), 235–250. doi:10.1080/00207549308956723

Rios, M. C., Campbell, A. M., & Irani, S. A. (2002). An approach to the design of a manufacturing cell under economic considerations. *International Journal of Production Economics*, *78*(3), 223–237. doi:10.1016/S0925-5273(02)00102-0

Romesburg, C. (1984). *Cluster analysis for researchers*. Lifetime Learning Publications.

Sarker, B. P. (1996). The resemblance coefficients in group technology: a survey and comparative study of relational metrics. *International Journal on Computers and Industrial Engineering*, *30*(1), 103–116. doi:10.1016/0360-8352(95)00024-0

Sarker, B. P., & Islam, K. M. (1999). Relative performance of similarity and dissimilarity measures. *International Journal on Computers and Industrial Engineering*, *37*(7), 769–807. doi:10.1016/S0360-8352(00)00011-5

Selim, H. M. (2002). Manufacturing cell formation problem: a graph partitioning approach. *Industrial Management & Data Systems*, *102*(6), 341–352. doi:10.1108/02635570210432046

Selim, H. M., & Abdel, A. (2003). Formation of machine groups and part families: A modified SLC method and comparative study. *Integrated Manufacturing Systems*, *14*(2), 123–137. doi:10.1108/09576060310459429

Singh, N. (1993). Design of cellular manufacturing systems: An invited review. *European Journal of Operational Research*, *69*, 284–291. doi:10.1016/0377-2217(93)90016-G

Slomp, J., Chowdary, B. V., & Suresh, N. (2005). Design of virtual manufacturing cells: A mathematical programming approach. *Robotics and Computer-integrated Manufacturing*, *21*(3), 273–288. doi:10.1016/j.rcim.2004.11.001

Solimanpur, M., Vrat, P., & Shankar, R. (2004). A multi-objective genetic algorithm approach to the design of cellular manufacturing systems. *International Journal of Production Research*, *42*(7), 1419–1441. doi:10.1080/00207540310001638073

Spiliopoulos, K., & Sofianopoulou, S. (2003). Designing manufacturing cells: A staged approach and a tabu search algorithm. *International Journal of Production Research*, *41*(11), 2531–2546. doi:10.1080/0020754031000088156

Suer, G. A., Peña, Y., & Vázquez, R. (2003). Cellular design to minimize investment costs by using evolutionary programming. *International Journal on Manufacturing Technology and Management*, *5*(12), 118–132. doi:10.1504/IJMTM.2003.002536

Suer, G. A., Vásquez, R., & Peña, Y. (1999). Cellular design to minimize investment costs by using evolutionary programming. *Journal of Engineering Valuation and Cost Analysis*, *5*(1-2), 133–142.

Vin, E., De Lit, P., & Delchambre, A. (2005). A multiple-objective grouping genetic algorithm for the cell formation problem with alternative routings. *Journal of Intelligent Manufacturing*, *16*(2), 189–205. doi:10.1007/s10845-004-5888-4

Wei, J. C., & Kern, G. M. (1989). Commonality analysis: A linear cell clustering algorithm for group technology. *International Journal of Production Research*, *27*(12), 2053–2062. doi:10.1080/00207548908942674

Welsh, D. J. A., & Powell, M. B. (1967). An upper bound for the chromatic number of a graph and its application to time–tabling problems. *The Computer Journal*, *10*, 85–96. doi:10.1093/comjnl/10.1.85

Werra, D. (1990). Heuristics for graph coloring. *Computing*, *7*(Supplementum), 191–208.

Won, Y. (2000). Two–phase approach to group technology cell formation using efficient p–median formulations. *International Journal of Production Research*, *38*(7), 1601–1613. doi:10.1080/002075400188744

Xambre, A. R., & Vilarinho, P. M. (2003). A simulated annealing approach for manufacturing cell formation. *European Journal of Operational Research, 151*(2), 434–446. doi:10.1016/S0377-2217(02)00836-6

Xu, H., & Wang, H. P. (1989). Part family formation for group technology applications based on fuzzy mathematics. *International Journal of Production Research, 27*, 1637–1651. doi:10.1080/00207548908942644

Yin, Y., Yasuda, K., & Hu, L. (2005). Formation of manufacturing cells based on material flows. *International Journal of Advanced Manufacturing Technology, 38*(2), 385–395.

Zhao, C., & Wu, Z. (2000). A genetic algorithm for manufacturing cell formation with multiple routes and multiples objectives. *International Journal of Production Research, 38*(2), 385–395. doi:10.1080/002075400189473

Chapter 5
Genetic vs. Hybrid Algorithm in Process of Cell Formation

R. Sudhakra Pandian
Kalasalingam University, India

Pavol Semančo
Technical University of Kosice, Slovakia

Peter Knuth
Technical University of Kosice, Slovakia

ABSTRACT

The cell formation problem has met with a significant amount of attention in recent years by demonstrating great potential for productivity improvements in production environment. Therefore, the researchers have been developing various methods based on similarity coefficient (SC), graph theory approaches, neural networks (NN), and others with aim to automate the whole cell formation process. This chapter focuses on presentation of hybrid algorithm (HA) and genetic algorithm that are helpful in production flow analysis to solve the cell formation problem. The evaluation of hybrid and genetic algorithms are carried out against the K-means algorithm and C-linkage algorithm that are well known from the literature. The comparison uses performance measure and the total number of exceptional elements (EEs) in the block-diagonal structure of machine-part incidence matrix using operational time as an input. The final performance results are presented in the form of graphs.

INTRODUCTION

In recent years, the cell formation problem has received a significant amount of attention by demonstrating a great potential for productivity improvements of cellular manufacturing system (CMS), which groups machines with dissimilar function and workstation types, dedicated to family of similar components. The main problem in designing of cellular manufacturing system is the formation of part families and corresponding groups of machines. One of the methods for classification of cell formation is production flow analysis (PFA). The concept of PFA, proposed by Burbidge (1977) is probably the most well-known and most widely accepted. This method requires

DOI: 10.4018/978-1-61350-047-7.ch005

reliable and well-documented route sheets and is also time-consuming. The researchers have initiated development of various methods like similarity coefficient method, graph theoretic approaches, array based methods, etc. in this field with aim to automate the cell formation process.

The modelling of CMS through mathematical programming was started to incorporate more real life constraints on the problem. Later researchers began developing heuristics and meta-heuristics methods to explore the best optimal solutions for the Cell Formation (CF) problems. Since soft computing techniques nowadays expand their applications to various fields like telecommunications, networking, design and manufacturing, current research in CMS is being carried out using soft computing techniques.

Very few studies focus on CF considering production factors such as operational time, operational sequence, batch size etc. In this chapter some of the real-time production factors are considered. For this purpose, the zero-one binary machine part incidence matrix (MPIM) of CF problem is converted into real valued operational time data. The use of soft computing technique is found more suitable for such type of problems, because it is capable of producing reliable results.

One of the chapter objectives is introduction of heuristic and meta-heuristic approaches based on similarity coefficient for solving cell formation problem. Simultaneously some important production factors that cannot be ignored during cell formation are presented. Another objective is to propose suitable methodologies for cell formation considering real time production factors using hybrid and genetic algorithms. The evaluation of the particular algorithms will be carried out by use of modified grouping efficiency (MGE) as a measure of performance.

LITERATURE REVIEW

The concept of cellular manufacturing system (CMS) as the application of GT philosophy was originally proposed by Burbidge (1979) who defines it as "an approach to the organization of work in which the organizational units are relatively independent groups, each responsible for the production of a given family of products". Burbidge (1963) has developed a production flow analysis (PFA) approach that later has been incorporated by him in a manual method (Burbidge, 1977). PFA method has been evolved in many ways after that. The most significant contributions from the researchers are towards similarity coefficient methods (SCM) (McAuley, 1972), graph theory (Rajagopalan and Batra, 1975), mathematical programming (Arthanari and Dodge, 1981), meta-heuristics algorithms (Hendizadeh, et al., 2008), fuzzy set theory (Leem and Chen, 1996) and neural networks (Kulkarni and Kiang, 1995) that are used to solve cell formation (CF) problems. One of the key aspects of CMS is the formation of machine-part cells, in which parts and machines are assigned to distinct cells where the machine utilization within a cell is maximized and inter-cells movement of parts is minimized. The machine-part cell formation (MPCF) problem was formally defined by Burbidge (1971) with his work focused on heuristic approaches to solve the block-diagonal problem for a machine-part incidence matrix (MPIM). Many methods have been developed for the MPCF problem so far. However, the methods like SC methods (De Witte, 1980), rank order clustering (ROC) (King, 1980) and graph theory methods (Faber, & Carter, 1986) have been developed only to solve the machine grouping in the CF problem and grouping of parts into part families is done in the supplementary step of the procedure only. Later clustering methods such as the modified ROC (MODROC) (Chandrasekaran and Rajagopalan, 1986), ZODIAC (Chandrasekaran and Rajagopalan, 1987), MACE (Waghodekar and Sahu, 1984) and GRAFICS

(Srinivasan and Narendran, 1991) are reported for solving the cell formation problems. Subsequently many algorithms based on meta-heuristics approaches have also developed and adjusted for solving the CF problems. The simulated annealing (SA) (Kirkpatrick, et al., 1983) is one of the meta-heuristic methods for finding the global minimum of objective function that may possess several local minima. The inspiration for developing SA method has come from annealing in metallurgy, where it represents a technique involving heating and controlled cooling of a material to increase the size of its crystals and reduce their defects. Tabu search (TS) (Pierre, S. and Houeto, 2002) is also classified as meta-heuristic method. TS method uses the flexible structures memory where stores a potential solutions. Once a potential solution has been determined, it is marked as "taboo" so that the algorithm does not visit that possibility repeatedly. Genetic algorithm (GA) (Yasuda, et al., 2005) represents a search technique used in soft computing to find exact or approximate solutions and search problems. Once the genetic representation and the fitness function are defined, GA proceeds to initialize a population of solutions randomly, and then improves it through repetitive application of mutation, crossover, inversion and selection operators. The popular algorithms of this category include SA based algorithm proposed by Boctor (1991) and GA based algorithm proposed by Venugopal and Narendran (1992). Jayakrishnan Nair and Narendran (1998) proposed an algorithm called CASE which considers sequence of operations that a part undergoes through a number of machines. Fernando (2002), TS based algorithm proposed by Wu (2004). Meanwhile many researchers have proposed artificial neural network (ANN) based methodologies for solving the cell formation problems. There are many popular ANN models found in the literature (Kao and Moon, 1991), (Carpenter and Grossberg, 2002) and Kaparthi and Suresh, 1992) which are efficient in producing satisfactory solutions to these NP-hard problems. Kao et al. (1991) introduced back propagation neural network model for GT whereas Kaparthi and Suresh (1992) made an attempt to introduce adaptive resonance theory (ART1).

METHODOLOGY

In order to enhance cell formation process in the PFA analysis genetic and hybrid algorithms, have been compared. The proposed algorithms for specific problems are illustrated by the flow chart diagram with aim to visualize improvements of the algorithms.

Evaluation process of the proposed algorithms includes performance comparison with alternative modified ART1 algorithm, K-means clustering algorithm and C-linkage algorithm by use of modified grouping efficiency. Every algorithm is tested under the conditions of 24 data sets. The objective of the evaluation process is to determine optimal data set solution that has minimal inter-cell moves. Final results are visually presented in the graph form and discussed in the text.

GA-BASED ALGORITHM

As it can be seen from literature review, the CF problem focuses on the several objectives that are known from the literature like inter-cell and intra-cell moves, measure of performance and exceptional element. Unfortunately, all these objectives do not reflect exactly smooth flow of material leading to work-in-process (WIP) inventories reduction and productivity increase. For that reason a cell load variation must be considered.

Genetic Algorithm is suggested for cell formation with the objective to minimize both the total cell load variation and the exceptional elements. The cell load variation is calculated as the difference between workload on the machine and average load on the cell (Venugopal and Narendran, 1992), expressed as the objective function using equation (1).

$$Z = \sum_{i=1}^{m} \sum_{k=1}^{c} X_{ik} \sum_{j=1}^{n} \left(w_{ij} - m_{kj} \right) \quad (1)$$

$$m_{kj} = \frac{\sum_{i=1}^{m} X_{ik} \cdot w_{ij}}{\sum_{i=1}^{m} X_{ik}} \quad (2)$$

where:

$\sum_{i=1}^{m} X_{ik} \cdot w_{ij}$ is the total number of machines in cell 'k' that induced by part j and $\sum_{i=1}^{m} X_{ik}$ is the total number of machines in cell 'k'.

For a predefined number of cell k, the Z value is calculated using (1).

Z: Objective function $f(X,m,w) = Z$
m: number of machines. (i=1,2,3,…,m)
p: number of parts. (j=1,2,3,…,p)
c: number of cells. (k=1,2,3,…,c)
$[X_{ik}]$: machine cell (m x k) membership matrix where $X_{ik}=1$ if ith machine is in cell k, $X_{ik}=0$ otherwise.
$[w_{ij}]$: machine part (m x p) matrix in terms of workload on machine I induced by part 'j'.
$[m_{kj}]$: cell part (k x p) matrix of average cell load as in (2).

The solution of the problem is naturally represented in GA as a genome (or chromosome). The GA then creates a population of solutions and applies genetic operators such as mutation and crossover to evolve the solutions in order to find the best one(s). The three most important aspects of using GA are:

- Definition of the objective function.
- Definition and implementation of the genetic representation.
- Definition and implementation of the genetic operators.

GA is adopted to find out the machine clusters to form the cells. In GA a candidate solution is represented by sequence of genes named chromosome. A chromosome potential is called its fitness function, which is evaluated by the objective function. A set of selected chromosomes (population) is subjected to generations (number of iterations). In each generation crossover and mutation operators are performed to get new population.

GA Coding Scheme Representation

Representation form plays a key role in the development of GA based algorithm. The proposed GA uses coding method, where each solution is coded as a set of numbers. The sum of numbers will be equal to the sum of machines to be grouped. The position of a number denotes the machine number and the value of the machine cell number, where the particular machine belongs to.

Figure 1 refers to the illustrative chromosome of two-cell solution, where the machines 1, 3 and 5 belong to cell number two and the machines 2 and 4 belong to cell number one. After representation, the population of chromosomes in the range of 10 to 40 depending on the size of problem is

Figure 1. Decoded GA chromosome for five machines of two cells solution

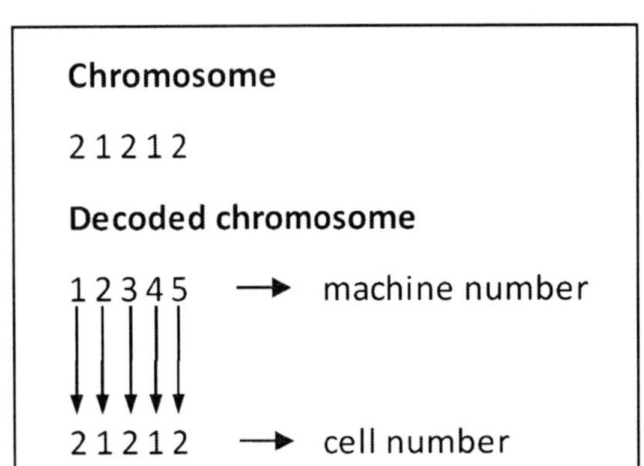

chosen. The initial solutions are generated randomly and the generated solution is subjected to generations (or iterations).

Reproduction

A fitness function value is computed for each string in the population and the objective is to find a string with the maximum fitness function value. Due to objective of minimizing both the total cell load variation and the exceptional elements, it is required to map it inversely and then maximize the resultant. Goldberg (1989) suggested a mapping function given as:

$$F(t)=Zmax-Z(t), \quad (3)$$

where $F(t)$ stands for fitness function of tth string and $Zmax$ is $max[Z(t)]$ of all strings (t). The advantage of this function is that the worst strings get zero fitness function value so that they are not going to be reproduced into the next generation.

Crossover and Mutation

With the GA coding scheme used in the proposed algorithm, the crossover operator is carried out with a probability known as crossover probability. In the crossover operation a pair of strings is selected randomly with a crossover probability. Crossover is exchange of a portion of strings at a point called crossover site (S). The genes (numbers) after the crossover site are swapped to produce the pair of offspring strings. Here partial mapped crossover given by Michalewicz (1996) is performed i.e., crossover site is selected and the genes of one string between the sites are swapped with genes of another string as shows Figure 2.

Also mutation is done randomly with some probability denoted as mutation probability. In this method inversion mutation is adopted where one gene is selected randomly, comes out from one cell and goes to another cell, while a machine from latter cell comes to the former cell as shown on Figure 2.

Elimination of Machine Cell with Single Machine

If a single machine is found in any cell, the following operations are carried out to merge the single machine cells with other cells:

1. the average workload of each part in the cell and the Euclidean distance between the cells are calculated,
2. the minimum Euclidean distance between cells is found out,
3. cells with a single machine are merged to the cells with minimum Euclidean distance.

Part Assignment

The following procedure given by Zolfaghari and Liang (2003) is used to assign parts into the machine cells. A machine cell which processes the part for a larger number of operations than any other machine cell is found out and the corresponding part is assigned into that cell. Ties are broken by choosing the machine cell which has the largest percentage of machines visited by the part. In the case of tie again the machine cell with the smallest identification number is

Figure 2. Genetic operators

Representation

12345 → Position of the machine number
21212 → Cell number (gene)

Crossover

Parent$_1$ 12 | **212** → Off spring 12 **112**
Parent$_2$ 12 | **112** → Off spring 12 **212**
S - (Cossover site)

Mutation

21**12**2 → 21**21**2

selected. Thus all the parts are assigned to all the cells which form part families using membership index given below.

$$P_{kj} = \frac{f_{kj}}{f_k} \cdot \frac{f_{kj}}{f_j} \cdot \frac{T_{kj}}{T_j}, \quad (4)$$

where:

P_{kj}: membership index of part 'j' belongs to cell 'k'.
f_{kj}: number of machines in cell 'k' required by part 'j'.
f_k: total number of machines in cell 'k'.
f_j: total number of machines required by part 'j'.
T_{kj}: processing time of part 'j' in cell 'k'.
T_j: total processing time required by part 'j'.

Sections and Particular Steps of Genetic Algorithm

The GA consists of the sections with several steps. The algorithm is presented by the flow chart diagram shown on the Figure 3. Subsequently, each section with its steps is described in detail. The proposed GA algorithm includes 5 subsections that are incorporated in the steps of main section of the given algorithm.

Section 1. Initialization

- Set the values of P_s, gen, P_c, P_m.
- Read the workload given in terms of processing time W_{ij} of part j on machine i.
- Create an initial population of size P_s and call it old population (P_{old}).
- Calculate the objective function using Equation (1).
- Sort string in the increasing order of objective function value.
- Set gen = 0.

Section 2. Reproduction

- Compute F(t) for P_{old}.
- Compute P_t of each string.
- Find the cumulative of P_t.
- Generate 'r_a' and select the string from P_{old} according to r and reproduce it in P_{new}.
- Repeat step 4 for P_s time.
- End.

Section 3. Crossover

- Generate 'r' if ($r<P_c$) go to step 2 else go to step 4.
- Select two strings t1 and t2 and swap genes between them by selecting crossover site S randomly.
- Repeat step 2 for $P_s/2$ times.
- End.

Section 4. Mutation

- Generate 'r_a'.

Figure 3. Flow chart diagram of GA algorithm

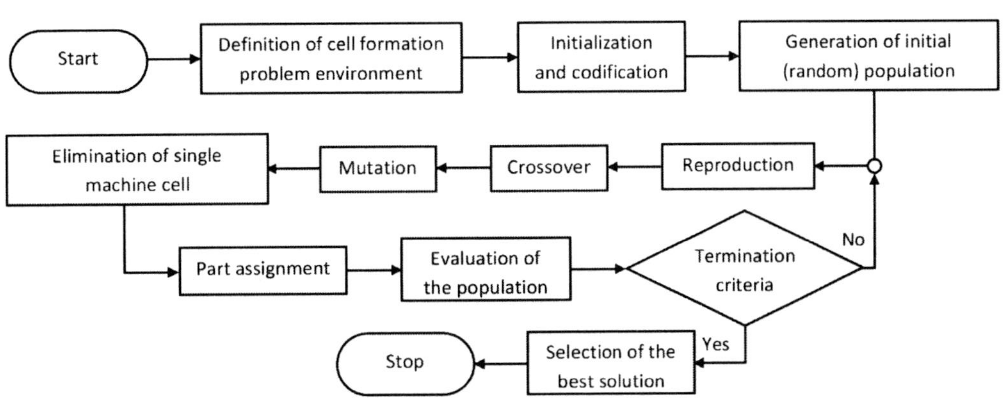

- If ($r_a < P_m$) go to step 3 else go to step 1.
- Select two machines randomly in t and interchange its positions.
- End.

Section 5. Part Assignment

- Find a machine cell that processes the part for a larger number of operations than any other machine cell and assign the part in that machine cell.
- If tie occurs, choose the machine cell that has the largest percentage of machines visited by the part and assign in that cell.
- If again tie occurs, select the machine cell with the smallest identification number and assign the part in that machine cell.
- End.

Main Section of the GA Algorithm

- Define cell formation problem, the number of cells c = k. (k = 2,3,…,m)
- Initialize the values and evaluate the objective function as given in section I.
- Do Reproduction as given in section II.
- Do Crossover as given in section III.
- Do Mutation as given in section IV.
- Do Part Assignment as given in section V.
- Increment counter.
- If (counter < gen) go to step 2 else step 11.
- Store the objective value in Z. Go to step 0. k=k+1.
- Print the best value of Z.
- Stop.

HYBRID ALGORITHM

Hybrid algorithm based on ART1 approach has natural formation of the machine cells, meaning that there is no constraint or objective function involved in the algorithm during clustering process. The basic purpose of ART1 approach is to develop a simple and efficient methodology to provide quick solutions for shop floor managers with least computational efforts.

The modified version of ART1 is adopted from the method proposed by Pao (1989) that accommodates analogue patterns (matrix with ratio level data) instead of binary form of input vectors (conventional MPIM) for machine cell formation problem.

The Modified ART1 Algorithm

Step 1. Initialize: Set nodes in the input layer equal to N (number of parts) and nodes in output layer equal to M (number of machines). Set vigilance threshold (φ).

Step 2. Initialize top-down connection weights as given in Equation (5)

$$wt_{ij}(0)=0, \quad (5)$$

where i is defined as 1,2,...,M and j *is defined as* 1,2,...,N.

Step 3. Let q =1. The first input vector X_1 (first row of the workload matrix) is presented to the input layer and assigned to the first cluster. Then, first node in the output layer is activated.

Step 4. The top-down connection weights for the present active node are set equal to the input vector.

Step 5. Let $q = q + 1$. Apply new input vector X_q. (input vectors are the rows of the workload matrix).

Step 6. Compute Euclidean distance between X_q and the exemplar stored in the top-down weights (wt_{ji}) for all active nodes i as given in the Equation (6). This distance function is used to calculate similarity between the stored pattern and the present input pattern. If the similarity value is less than or equal to ρ (vigilance threshold), the present input is categorized under the same cluster as that of stored pattern.

$$e_i = \sqrt{\sum_{j=1}^{N}\left(x_{qj} - wt_{ji}\right)^2}, \qquad (6)$$

Step 7. Perform vigilance test: Find out minimum Euclidean distance.

Step 8. If min $e_i \geq \varphi$ (threshold value), select output node for which Euclidean distance is minimum. If tie occurs, select the output node with lowest index number. Suppose output node k is selected then allocate the vector X_q to the node k (cell) and activate node k. Make increment to the number of machines in the active node k by one. If i e s for all active nodes are greater than ρ, then go to step 9.

Step 9. Start a new cell by activating a new output node.

Step 10. Update top-down weights of active node k using Equation (7). When a new vector is presented to the algorithm, its belongingness to existing nodes is judged by matching with respective top-down weights. The matching criterion is based on minimizing dissimilarity between existing exemplar stored as top-down weights and new input vector. Therefore, top-down weight updating principle warrants for storing combined information of previously stored exemplar and the present input pattern. Usually, higher weights are emphasized on stored exemplar than that of the new input vector. When a vector is selected (to be allocated to an output node), its top-down weights are updated using more information of the previously stored exemplar and a relatively less information of the input vector (pattern) as shown in Equation (7).

$$wt_{jk} = \left(\frac{n}{m} \cdot wt_{ji}\right) + \left(\frac{1}{m} \cdot x_{qj}\right), \qquad (7)$$

Step 11. Go to step 5 and repeat till all the rows are assigned in the output nodes (cells).

Step 12. Check for single machine cells. If a single machine is found in any cell, perform the following operations to merge the single machine cells into any other cells.

1. Determine average workload of each cell.
2. Calculate the Euclidean distance between the cells.
3. Merge a cell containing single machine with another in such a way that Euclidean distance between them is minimum.

Step 13. Assign parts to cells using the membership index given in Equation (4) and the maximum belongingness that can be calculated using Equation (8).

$$P_m = max(P_{kj}), \qquad (8)$$

where k is defined as 1,2,3,…,C. The value of P_m lies between 0 to 1 where $P_m = 1$ indicates that the part 'j' perfectly belongs to cell 'k'.

TESTING AND MEASUREMENT OF PROPOSED ALGORITHMS PERFORMANCE

There are several performance measures proposed by the researchers in last two decades. Each of them has its own advantages and drawbacks depending on the data considered for CF problem. However, no grouping efficiency can be considered for the generalized cell formation with maximum available information. Literature suggests that two popular measures the grouping efficiency and grouping efficacy are used to check the performance of block-diagonal structure generated by a cell formation technique.

Grouping Efficiency

Grouping efficiency is given by the Equation (9). Chandrasekharan and Rajagopalan (1986) defined grouping efficiency as a weighted average of two functions η_1 and η_2. It represents the very first performance measure in CF. The efficiency was proposed as a weighted average of two efficiencies and higher grouping efficiency will result in better grouping.

$$\eta = r \cdot \eta_1 + (1-r)\eta_2, \tag{9}$$

where:

$$\eta_1 = \frac{Number\ of\ 1's\ in\ the\ diagonal\ blocks}{Total\ number\ of\ elements\ in\ the\ diagonal\ blocks}, \tag{10}$$

$$\eta_2 = \frac{Number\ of\ 0's\ in\ the\ off-diagonal\ blocks}{Total\ number\ of\ elements\ in\ the\ off-diagonal\ blocks}, \tag{11}$$

r is a weighting factor that lies between zero to one ($0 <= r <= 1$) and its value is decided depending on the size of the matrix. Grouping efficiency considers two functions - packing density inside the cells (η1) and inter-cell moves (η2). Weighting factor is used to achieve a tradeoff between two functions depending on desirability of the decision maker. A higher value of η is supposed to indicate better clustering.

Grouping Efficacy

Kumar and Chandrasekharan (1990) have introduced grouping efficacy as a new performance measure, which has been proposed to overcome the drawbacks of grouping efficiency. High grouping efficacy will result as good CF.

$$\Gamma = \frac{1-\psi}{1-\phi}, \tag{12}$$

where:

$$\psi = \frac{Number\ of\ EEs}{Total\ number\ of\ operations}, \tag{13}$$

$$\phi = \frac{Number\ of\ voids\ (0's)\ in\ the\ diagonal\ blocks}{Total\ number\ of\ operations}, \tag{14}$$

Unlike grouping efficiency, grouping efficacy is not affected by the size of the matrix. However, both measures - grouping efficiency and grouping efficacy treat all operations equally and suitable only for the zero-one incidence matrix.

Modified Grouping Efficiency

Both the grouping efficiency and grouping efficacy treat all operations equally and they are suitable only for the binary (zero-one) incidence matrix. Therefore, a new measure for grouping efficiency termed as modified grouping efficiency denoted as MGE is presented to find out the performance of the CF method that deal with workload matrix due to consideration of voids (idle machines) inside the cells. For the cell formation problems using workload (operational time) information, the grouping efficiency has to be found out from the ratio of total workload inside the cells denoted as T_{pti}, and total workload of the matrix. When total workload is being calculated the number of voids presented inside the cells is taken into account and the proportionate value of voids with the number of elements presented inside the cells are calculated using the weighting factor to the voids ratio. The elements outside the cells represent exceptional elements, denoted as T_{pto}. The MGE is calculated using Equation (15).

$$MGE = \frac{T_{pti}}{T_{pto} + \sum_{k=1}^{C} T_{ptk} + \sum_{k=1}^{C} T_{ptk} \cdot \omega_v}, \tag{15}$$

Table 1. Results of the tested algorithms

Data Set No.	No. of Cells	K-means Algorithm		C-linkage Algorithm		Hybrid Algorithm		Genetic Algorithm	
		EE	MGE %	EE	MGE %	EE	MGE %	EE	MGE %
1	2	2	77.25	2	77.25	2	77.25	2	77.25
2	2	2	78.34	2	78.34	2	78.34	2	78.34
3	2	7	81.87	7	81.87	7	81.87	7	81.87
4	2	2	79.85	2	79.85	2	79.85	2	79.85
5	2	3	61.77	3	61.77	3	61.77	3	61.77
6	2	1	65.48	1	65.48	1	65.48	1	65.48
7	2	6	57.00	6	57.00	4	69.70	6	69.70
8	2	28	60.00	28	60.00	25	61.30	28	61.30
9	3	9	83.40	9	83.40	9	83.40	9	83.40
10	3	0	77.14	0	77.14	0	77.14	0	77.14
11	3	0	93.28	0	93.28	0	93.28	0	93.28
12	2	2	59.43	2	59.43	2	60.59	0	62.42
13	4	7	68.13	9	65.23	2	76.13	3	73.19
14	3	15	64.81	15	64.81	15	64.81	20	64.81
15	2	42	49.13	42	49.13	19	60.10	19	60.10
16	3	1	71.00	1	71.00	1	71.15	1	71.15
17	4	31	61.50	31	61.50	28	61.71	32	61.70
18	3	38	51.70	38	51.70	42	50.50	42	51.92
19	4	34	46.70	30	51.39	30	51.39	29	52.02
20	6	0	90.28	0	90.28	0	90.28	0	94.58
21	5	7	71.60	7	71.60	9	73.89	9	73.89
22	3	12	56.65	17	53.98	17	53.98	15	56.14
23	6	20	61.84	20	61.84	26	55.51	22	62.23
24	3	33	50.51	33	50.51	17	53.19	25	55.32

Figure 4. Comparison of Hybrid algorithm with (a) K-means algorithm and (b) C-Linkage algorithm

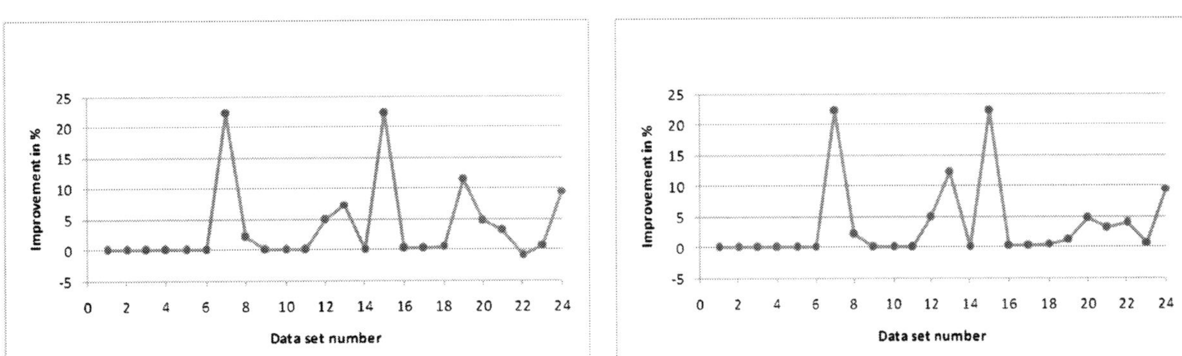

(a) (b)

Figure 5. Comparison of GA based algorithm with (a) K-means algorithm, (b) C-Linkage algorithm and (c) Hybrid algorithm

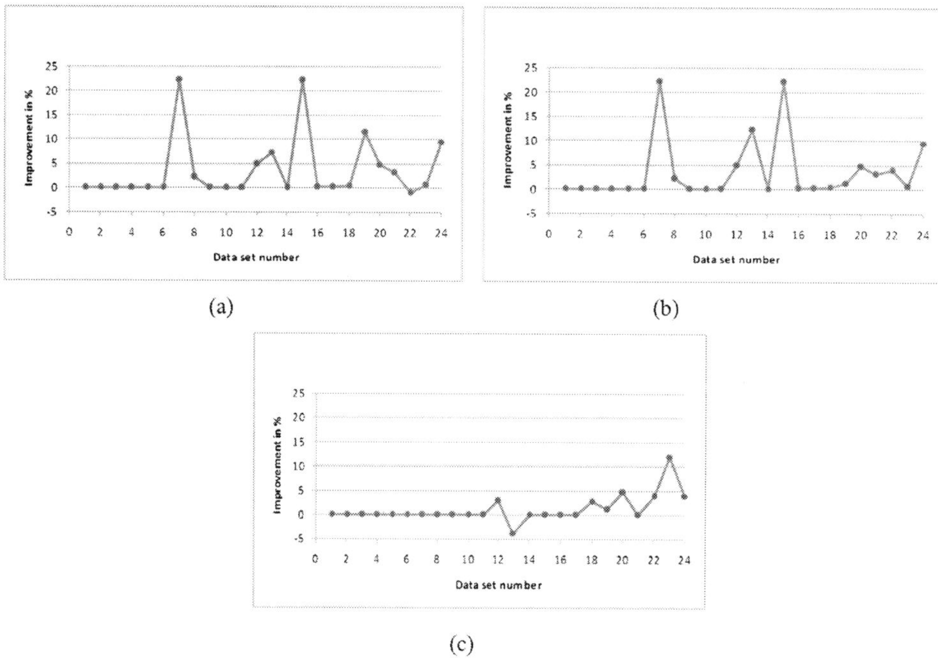

where:

$$\omega_v = \frac{Number\ of\ voids\ in\ the\ cell\ k}{Total\ number\ of\ elements\ in\ the\ cell\ k}, \quad (16)$$

RESULTS AND EVALUATION

In this chapter, an efficient algorithm based on genetic and hybrid algorithm has been proposed for cell formation problem considering operational time of the parts instead of conventional zero-one incidence matrix with the objective of minimizing total cell load variation. The algorithm is coded in C++ and run on Pentium IV PC, 2.4GHz processor.

The real valued matrix is produced by assigning random numbers in the range of 0.5 to 1 as uniformly distributed values by replacing the ones in the incidence matrix and zeros to remain in its same positions. Genetic and hybrid algorithm have been tested with 24 benchmark problems of varied sizes ranging from 5 x 7 to 30 x 50 from open literature. The results of the genetic and hybrid based algorithms are also compared with K-means clustering and C-linkage clustering algorithm given in Table 1.

Based on exhaustive experiments, the crossover (P_c) and mutation (P_m) probabilities are fixed to be 0.5 and 0.1 respectively. This probability can be varied depending upon the decision maker to tune the algorithm. The chromosome representation used in this study may result in the formation of an empty cell or violates some constraints. Particularly, crossover may result in the formation of a chromosome like 113331 when predefined number of cells is three. The above chromosome contains an empty cell where cell number 2 is missing. In such cases, the respective chromosomes are rejected. Crossover and mutation steps are repeated with other pairs of chromosomes till a useful chromosome is obtained.

Figure 4 and Figure 5 depict percentage differences of the hybrid and genetic algorithm against two traditional algorithms known from the open literature.

From Figure 4(a) it is evident that the hybrid algorithm has given better solution than K-means algorithm in 11 of 24 data sets. There is only one case (data set number 22) where hybrid algorithm has not performed well. The HA has similar even better solutions in comparison to C-linkage algorithm, as shown in Figure 4(b).

In the Figure 5(a), the best improvements are in the data set number 7, 15, 19 and 24 in comparison with K-means algorithm. A similar enhancement can be seen from Figure 5(b) where the genetic algorithm is compared to C-linkage algorithm. Figure 5(c) compares genetic algorithm with hybrid algorithm. The GA has outperformed hybrid algorithm.

DISCUSSION AND CONCLUSION

In this work the genetic and hybrid based algorithms with two traditional algorithms like C-linkage and K-means are used to solve the cell formation problem using the non-binary real valued work load data as an input matrix. The genetic algorithm and hybrid algorithm are tested with benchmark problems found in the literature and the results are compared with the traditional algorithms mainly K-means and C-linkage clustering algorithm. In addition to the commonly used measure of performance that is the number of exceptional elements, a newly developed performance measure namely modified grouping efficiency (MGE) is also applied to evaluate the efficiency of the GA and HA based algorithm. The genetic algorithm outperforms the HA and traditional techniques both in terms of exceptional elements and modified grouping efficiency. The GA based algorithm may be suitably modified and employ to solve the cell formation problem with other non binary real value data like machine capacity, production volume and product sequence.

REFERENCES

Arthanari, T. S., & Dodge, Y. (1981). *Mathematical programming in statistics*. New York, NY: Wiley.

Boctor, F. F. (1991). A linear formulation of the machine-part cell formation problem. *International Journal of Production Research*, *29*(2), 343–356. doi:10.1080/00207549108930075

Burbidge, J. L. (1963). Production flow analysis. *Production Engineering*, *42*, 742–752. doi:10.1049/tpe.1963.0114

Burbidge, J. L. (1971). Production flow analysis. *Production Engineering*, *50*, 139–152. doi:10.1049/tpe.1971.0022

Burbidge, J. L. (1977). A manual method of production flow Analysis. *Production Engineering*, *56*, 34–38. doi:10.1049/tpe.1977.0129

Burbidge, J. L. (1979). *Group technology in the engineering industry*. New York, NY: Wiley.

Carpenter, G. A., & Grossberg, S. (2002). *Adaptive resonance theory: A handbook of brain theory and neural networks*. Massachusetts, USA: MIT Press.

Chandrasekaran, M. P., & Rajagopalan, R. (1986). MODROC: An extension of rank order clustering for group technology. *International Journal of Production Research*, *24*(5), 1221–1233. doi:10.1080/00207548608919798

Chandrasekaran, M. P., & Rajagopalan, R. (1987). ZODIAC: An algorithm for concurrent formation of part families and machine cells. *International Journal of Production Research*, *25*(6), 835–850. doi:10.1080/00207548708919880

Cheung, Y. M. (2003). K-means: A new generalized K-means clustering algorithm. *Pattern Recognition Letters, 24*, 2883–2893. doi:10.1016/S0167-8655(03)00146-6

De Witte, J. (1980). The use of similarity coefficients in production flow analysis. *International Journal of Production Research, 18*, 503–514. doi:10.1080/00207548008919686

Defays, D. (1977). An efficient algorithm for a complete link method. *The Computer Journal, 20*(4), 364–366. doi:10.1093/comjnl/20.4.364

Faber, Z., & Carter, M. W. (1986). A new graph theory approach for forming machine cells in cellular production systems. In A. Kusiak (Ed.), *Flexible manufacturing systems: Methods and studies* (pp. 301-315). North- Holland: Elsevier Science Publishers B.V.

Fernando, J. M. (2002). *A hybrid genetic algorithm in cell formation problems. AT&T Labs Research Technical Report*. Florham Park, NJ: AT&T Labs Research.

Goldberg, D. E. (1989). *Genetic algorithms in search optimization, and machine learning*. Reading, MA: Addison-Wesley Press.

Hendizadeh, S. H., Faramarzi, H., Mansouri, S. A., Gupta, J. N. D., & El Mekkawy, T. (2008). Metaheuristics for scheduling a flowline manufacturing cell with sequence dependent family setup times. *International Journal of Production Economics, 111*(2), 593–605. doi:10.1016/j.ijpe.2007.02.031

Kao, Y., & Moon, Y. B. (1991). A unified group technology implementation using the backpropagation learning rule of neural networks. *Computers & Industrial Engineering, 20*(4), 425–437. doi:10.1016/0360-8352(91)90015-X

Kaparthi, S., & Suresh, N. C. (1992). Machine component cell formation in GT: A NN approach. *International Journal of Production Research, 30*, 1353–1367. doi:10.1080/00207549208942961

King, J. R. (1980). Machine/component grouping in production flow analysis: An approach using a rank order clustering algorithm. *International Journal of Production Research, 18*, 213–232. doi:10.1080/00207548008919662

Kirkpatrick, S., Gelatt, C. D., & Vecchi, M. P. (1983). Optimization by simulated annealing. *Science, 220*(4598), 671–680. doi:10.1126/science.220.4598.671

Kulkarni, U. R., & Kiang, M. Y. (1995). Dynamic grouping of parts in flexible manufacturing systems - A self-organizing neural networks approach. *European Journal of Operational Research, 84*, 192–212. doi:10.1016/0377-2217(94)00326-8

Kumar, S. C., & Chandrasekharan, M. P. (1990). Grouping efficacy: A quantitative criterion for goodness of block diagonal forms of binary matrices in group technology. *International Journal of Production Research, 28*(2), 233–243. doi:10.1080/00207549008942706

Leem, C. W., & Chen, J. J. G. (1996). Fuzzy-set-based machine-cell formation in cellular manufacturing. *Journal of Intelligent Manufacturing, 7*, 355–364. doi:10.1007/BF00123911

McAuley, J. (1972). Machine grouping for efficient production. *Production Engineering, 51*, 53–57. doi:10.1049/tpe.1972.0006

Michalewicz, Z. (1996). *Genetic algorithms + data structures = evolution programs*. New York, NY: Springer-Verlag.

Pao, Y. H. (1989). *Adaptive pattern recognition and neural networks*. New York, NY: Addison-Wesley Publishing Company, Inc.

Pierre, S., & Houeto, F. (2002). A tabu search approach for assigning cells to switches in cellular mobile networks. *Computer Communications, 25*, 464–477. doi:10.1016/S0140-3664(01)00371-1

Rajagopalan, R., & Batra, J. L. (1975). Design of cellular production system: A graph theoretic approach. *International Journal of Production Research*, *13*, 567–579. doi:10.1080/00207547508943029

Sinha, N. K., Gupta, M. M., & Zadeh, L. A. (2000). Outline of computational theory of perceptions based on computing with words. In *Soft computing & intelligent systems*. New York, NY: Academic Press.

Srinivasan, G., & Narendran, T. T. (1991). A non hierarchical clustering algorithm for group technology GRAFICS. *International Journal of Production Research*, *29*(3), 463–478. doi:10.1080/00207549108930083

Venugopal, V., & Narendran, T. T. (1992). Cell formation in manufacturing systems through simulated annealing. *European Journal of Operational Research*, *63*, 409–422. doi:10.1016/0377-2217(92)90158-6

Venugopal, V., & Narendran, T. T. (1992). A genetic algorithm approach to the machine component grouping problem with multiple objectives. *Computers & Industrial Engineering*, *22*, 469–480. doi:10.1016/0360-8352(92)90022-C

Waghodekar, P. H., & Sahu, S. (1984). Machine-component cell formation in group technology. MACE. *International Journal of Production Research*, *22*(6), 937–948. doi:10.1080/00207548408942513

Wu, L. W. (2004). A tabu search approach to the cell formation problem. *International Journal of Advanced Manufacturing Technology*, *23*, 916–924. doi:10.1007/s00170-003-1766-z

Yasuda, K., Hu, L., & Yin, Y. (2005). A grouping genetic algorithm for the multiobjective cell formation problem. *International Journal of Production Research*, *43*(4), 829–853. doi:10.1080/00207540512331311859

Zolfaghari, S., & Liang, M. (2003). A new genetic algorithm for the machine/part grouping problem involving processing times and lot sizes. *Computers & Industrial Engineering*, *45*, 713–731. doi:10.1016/j.cie.2003.09.003

APPENDIX

Table 2. Source and size of the data sets

Data Set No.	Source	Problem Size
1	King and Nakornchai (1982)	5x7
2	Waghodekar and Sahu (1984)	5x7
3	Seiffodini (1989)	5x18
4	Kusiak (1992)	6X8
5	Kusiak (1987)	7x11
6	Boctor (1991)	7x11
7	Seiffodini and Wolfe (1986)	8x12
8	Chandrasekaran et al (1986)	8x20
9	Chandrasekaran et al (1986)	8x20
10	Mosier et al. (1985)	10x10
11	Chan et al. (1982)	10x15
12	Askin et al. (1987)	14x23
13	Stanfel (1985)	14x24
14	Srinivasan et al. (1990)	16x30
15	Mosier et al. (1985)	20x20
16	Carrie (1973)	20x35
17	Boe et al. (1991)	20x35
18	Kumar et al. (1986)	23x20
19	Mccornick et al. (1972)	24x16
20	Chandrasekaran et al (1989)	24x40
21	Chandrasekaran et al (1989)	24x40
22	Kumar et al. (987)	30x41
23	Stanfel (1985)	30x50
24	Stanfel (1985)	30x50

Table 3. Data set No. 1

i/j	P1	P2	P3	P4	P5	P6	P7
M1	0	0.53	0	0.99	0.83	0.91	0
M2	0.82	0	0.83	0	0	0	0
M3	0.91	0	0.92	0	0	0.86	0.97
M4	0	0.79	0	0.56	0	0.88	0
M5	0.53	0	0	0	0.51	0	0.98

Table 4. Data set No. 2

i/j	P1	P2	P3	P4	P5	P6	P7
M1	0.53	0	0	0	0.99	0.83	0.91
M2	0	0.82	0.83	0.91	0.92	0	0
M3	0	0	0.86	0.97	0.79	0.56	0
M4	0.88	0.53	0.51	0.98	0	0	0
M5	0	0.83	0	0.71	0.58	0.54	0

Table 5. Data set No. 3

i/j	P1	P2	P3	P4	P5	P6	P7	P8	P9	P10	P11	P12	P13	P14	P15	P16	P17	P18
M1	0.53	0.99	0.83	0	0.91	0.82	0	0.83	0	0	0.91	0.92	0.86	0.97	0	0.79	0.56	0
M2	0.88	0	0.53	0.51	0	0.98	0.83	0.71	0	0.58	0.54	0.54	0.74	0	0.63	0	0	0.63
M3	0	0	0	0.53	0	0	0.69	0	0	0.63	0	0	0	0	0.68	0	0	0.51
M4	0.61	0.94	0.68	0	0.67	0.7	0	0.84	0	0	0.79	0.99	0.94	0.84	0	0.78	0.93	0
M5	0	0	0	0.73	0	0	0	0	0.98	0.92	0	0	0	0	0.92	0	0	0.7

Table 6. Data set No. 4

i/j	P1	P2	P3	P4	P5	P6	P7	P8
M1	0	0.53	0	0.99	0	0	0.83	0
M2	0.91	0.82	0.83	0	0.91	0.92	0.86	0.97
M3	0	0	0.79	0	0	0.56	0	0.88
M4	0	0	0	0.53	0	0	0.51	0
M5	0.98	0	0.83	0	0.71	0.58	0	0.54
M6	0	0	0	0.54	0	0	0.74	0

Table 7. Data set No. 5

i/j	P1	P2	P3	P4	P5	P6	P7	P8	P9	P10	P11
M1	0	0.53	0.99	0	0	0	0.83	0	0	0	0
M2	0.91	0	0	0	0.82	0	0	0	0	0	0.83
M3	0	0	0	0	0	0	0	0	0	0.91	0.92
M4	0.86	0	0.97	0	0	0.79	0	0	0	0	0
M5	0	0	0	0	0.56	0	0	0.88	0	0	0
M6	0.53	0	0	0.51	0	0	0	0.98	0.83	0.71	0
M7	0	0	0.58	0.54	0	0.54	0.74	0	0.63	0	0

Table 8. Data set No. 6

i/j	P1	P2	P3	P4	P5	P6	P7	P8	P9	P10	P11
M1	0.53	0.99	0	0	0	0.83	0	0	0	0	0
M2	0	0.91	0	0	0	0.82	0	0	0.83	0	0
M3	0.91	0	0.92	0	0	0	0.83	0	0	0	0.97
M4	0	0	0.79	0	0	0	0.56	0	0	0	0
M5	0	0	0.88	0.53	0	0	0	0	0	0	0.51
M6	0	0	0	0.98	0.83	0	0	0	0	0.71	0
M7	0	0	0	0	0.58	0	0	0.54	0	0.54	0

Table 9. Data set No. 7

i/j	P1	P2	P3	P4	P5	P6	P7	P8	P9	P10	P11	P12
M1	0.53	0.99	0.83	0.91	0	0	0	0	0	0	0	0
M2	0.82	0	0.83	0.91	0.92	0.86	0.97	0	0	0.79	0	0
M3	0	0	0.56	0.88	0.53	0.51	0.98	0.83	0.71	0	0	0
M4	0	0	0	0	0	0.58	0.54	0.54	0.74	0.63	0	0
M5	0	0	0	0	0	0	0.63	0.53	0.69	0.63	0	0
M6	0	0	0	0	0	0	0.68	0.51	0.61	0	0.94	0
M7	0	0	0	0	0	0	0	0	0	0	0.68	0.67
M8	0	0	0	0	0	0	0	0	0	0	0.7	0.84

Table 10. Data set No. 8

i/j	P1	P2	P3	P4	P5	P6	P7	P8	P9	P10	P11	P12	P13	P14	P15	P16	P17	P18	P19	P20
M1	0.53	0	0.99	0.83	0	0	0	0	0.91	0.82	0	0	0	0.83	0.91	0.92	0	0.86	0.97	0
M2	0	0.79	0.56	0.88	0	0.53	0.51	0	0.98	0	0.83	0	0	0	0	0	0	0.71	0	0.58
M3	0	0	0	0	0.54	0.54	0.74	0.63	0	0	0.63	0.54	0.69	0	0	0.63	0.68	0	0.51	0.61
M4	0	0	0.94	0.68	0	0	0.67	0.7	0.84	0.79	0	0	0.99	0.94	0.84	0	0.78	0.93	0.73	0.98
M5	0.92	0.92	0.7	0	0	0.89	0	0	0	0.52	0	0.52	0.54	0	0.77	0.76	0.96	0	0.6	0.61
M6	0.54	0.67	0	0	0.7	0	0.85	0.99	0	0.87	0.67	0.63	0	0.74	0.85	0.78	0	0.55	0.81	0
M7	0	0	0	0	0.63	0.97	0.54	0.52	0	0	0.85	0.55	0.99	0	0	0.93	0.94	0	0.8	0.68
M8	0.6	0.63	0.7	0.9	0.71	0	0	0.98	0.53	0.68	0	0	0.91	0.53	0	0.76	0.88	0	0	0

Table 11. Data set No. 9

i/j	P1	P2	P3	P4	P5	P6	P7	P8	P9	P10	P11	P12	P13	P14	P15	P16	P17	P18	P19	P20
M1	0	0.53	0.99	0	0	0	0	0.83	0.91	0	0.82	0	0.83	0.91	0	0.92	0.86	0	0.97	0
M2	0	0	0.79	0.56	0	0.88	0.53	0	0	0	0	0	0	0.51	0	0	0	0.98	0	0.83
M3	0	0.71	0	0	0	0	0	0.58	0.54	0	0.54	0	0.71	0.63	0	0.63	0.53	0	0.69	0
M4	0	0	0.63	0.67	0	0.51	0.61	0	0	0.94	0	0	0	0	0	0	0	0.68	0	0.67
M5	0.7	0	0	0	0.84	0.79	0	0	0	0.99	0	0.94	0	0	0.89	0	0.78	0	0	0
M6	0.93	0	0	0	0.73	0	0	0	0.98	0.92	0	0.92	0	0	0.7	0	0	0	0	0.89
M7	0	0	0.52	0.52	0	0.54	0.77	0	0	0	0.76	0.96	0	0	0	0	0	0.6	0	0.61
M8	0	0	0.54	0.67	0	0.7	0.85	0	0	0	0	0	0	0	0	0	0	0.99	0	0.87

Table 12. Data set No. 10

i/j	P1	P2	P3	P4	P5	P6	P7	P8	P9	P10
M1	0.53	0	0	0	0	0	0	0	0	0.99
M2	0	0	0.83	0.91	0	0	0	0.82	0	0
M3	0	0	0	0	0.83	0.91	0	0	0	0
M4	0.92	0	0	0	0	0	0	0	0	0
M5	0	0	0	0	0	0	0.86	0	0	0.97
M6	0.79	0	0	0	00	0	0.56	0	0	0.88
M7	0	0	0.53	0	0	0	0	0.51	0	0
M8	0	0	0	0	0	0.98	0	0	0.83	0
M9	0	0.71	0.58	0.54	0	0	0	0	0	0
M10	0	0.54	0.74	0.63	0	0	0	0.63	0	0

Table 13. Data set No. 11

i/j	P1	P2	P3	P4	P5	P6	P7	P8	P9	P10	P11	P12	P13	P14	P15
M1	0	0.53	0	0	0	0	0	0	0	0.99	0.83	0.91	0	0	0
M2	0	0	0.82	0	0.83	0	0	0.91	0	0	0	0	0.92	0	0.86
M3	0.97	0	0	0	0	0.79	0	0	0.56	0	0	0	0	0.88	0
M4	0.53	0	0	0.51	0	0	0	0	0.98	0	0	0	0	0.83	0
M5	0	0	0.71	0	0.58	0	0	0.54	0	0	0	0	0.54	0	0.74
M6	0.63	0	0	0.63	0	0.53	0	0	0.69	0	0	0	0	0.63	0
M7	0	0.68	0	0	0	0	0.51	0	0	0.61	0.94	0.67	0	0	0
M8	0	0	0.67	0	0.7	0	0	0.84	0	0	0	0	0.79	0	0.99
M9	0	0	0	0.94	0	0.84	0	0	0.78	0	0	0	0	0.93	0
M10	0	0.73	0	0	0	0	0.98	0	0	0.92	0.92	0.7	0	0	0

Table 14. Data set No. 12

i/j	P1	P2	P3	P4	P5	P6	P7	P8	P9	P10	P11	P12	P13	P14	P15	P16	P17	P18	P19	P20	P21	P22	P23
M1	0	0	0	0	0	0	0.53	0.99	0	0	0	0	0	0	0	0	0	0	0	0	0	0	0
M2	0	0	0	0.83	0.91	0	0	0	0	0	0	0	0	0	0	0	0	0	0	0	0	0	0
M3	0	0	0	0.82	0.83	0	0	0	0	0	0	0	0	0	0	0	0	0	0	0	0.91	0	0
M4	0	0.92	0.86	0	0	0	0	0	0	0	0	0	0	0	0	0	0.97	0	0.79	0.56	0	0.88	0
M5	0	0.53	0.51	0	0	0	0	0	0	0	0	0	0	0	0	0	0.98	0	0	0.83	0	0.71	0
M6	0.58	0	0	0	0	0	0	0	0	0.54	0.54	0.74	0	0.63	0.63	0.53	0	0	0	0	0	0	0
M7	0	0.69*	0.63	0	0	0	0	0.68	0	0	0	0	0	0	0	0	0.51	0	0	0.61	0	0	0
M8	0.94	0	0	0	0	0.68	0	0	0	0.67	0	0.7	0	0.84	0.79	0.99	0	0	0	0	0	0	0
M9	0	0	0	0	0	0.94	0	0	0	0.84	0	0.78	0.93	0	0.73	0	0	0	0	0	0	0	0
M10	0	0	0	0.96	0	0	0	0	0	0	0	0	0	0	0	0	0	0	0	0	0	0	0.92
M11	0	0	0	0.92	0.7	0	0	0	0	0	0	0	0	0	0	0	0	0	0	0	0.89	0	0
M12	0	0	0	0	0	0	0	0.52	0.52	0	0	0	0	0	0	0	0	0	0	0	0	0	0
M13	0	0	0	0	0	0	0.54	0.77	0.76	0	0	0	0	0	0	0	0	0.96	0	0	0	0.6	0
M14	0	0	0	0	0	0	0	0	0	0.61	0.54	0	0.67	0	0.7	0	0	0	0	0	0	0	0

Table 15. Data set No. 13

i/j	P1	P2	P3	P4	P5	P6	P7	P8	P9	P10	P11	P12	P13	P14	P15	P16	P17	P18	P19	P20	P21	P22	P23	P24
M1	0	0	0	0	0	0.53	0.99	0	0	0	0	0	0	0	0	0	0	0	0	0	0	0	0	0
M2	0	0	0.83	0.91	0	0	0	0	0	0	0	0	0	0	0	0	0	0	0	0	0	0	0	0
M3	0	0	0.82	0.83	0	0	0	0	0	0	0	0	0	0	0	0	0	0	0	0	0.91	0	0	0
M4	0.92	0.86	0	0	0	0	0	0	0	0	0	0	0	0	0	0	0.97	0	0.79	0.56	0	0	0.88	0
M5	0.53	0.51	0	0	0	0	0	0	0	0	0	0	0	0	0	0	0.98	0	0	0.83	0	0	0.71	0
M6	0	0	0	0	0	0	0	0	0.58	0.54	0.54	0.74	0	0.63	0.63	0.53	0	0	0	0	0	0	0	0
M7	0.69	0.63	0	0	0	0	0.68	0	0	0	0	0	0	0	0	0	0.51	0	0	0.61	0	0.94	0	0
M8	0	0	0	0	0.68	0	0	0	0.67	0.7	0	0.84	0	0.79	0.99	0.94	0	0	0	0	0	0	0	0
M9	0	0	0	0	0.84	0	0	0	0.78	0	0	0.93	0.73	0	0.96	0	0	0	0	0	0	0.92	0	0
M10	0	0	0.62	0	0	0	0	0	0	0	0	0	0	0	0	0	0	0	0	0	0	0	0	0.7
M11	0	0	0.89	0.52	0	0	0	0	0	0	0	0	0	0	0	0	0	0	0	0	0.52	0	0	0.54
M12	0	0	0	0	0	0	0.77	0.76	0	0	0	0	0	0	0	0	0	0	0	0	0	0	0	0
M13	0	0	0	0	0	0.96	0.6	0.61	0	0	0	0	0	0	0	0	0	0.54	0	0	0	0	0.67	0
M14	0	0	0	0	0	0	0	0	0.7	0	0.85	0	0.99	0	0.87	0	0	0	0	0	0	0	0	0

Figure 6. Data set No. 14

i/j	P1	P2	P3	P4	P5	P6	P7	P8	P9	P10	P11	P12	P13	P14	P15	P16	P17	P18	P19	P20	P21	P22	P23	P24	P25	P26	P27	P28	P29	P30
M1	0	0	0	0.53	0.99	0	0.83	0	0.91	0	0	0.82	0	0	0	0	0	0.83	0	0	0	0.91	0	0	0	0	0	0	0	0.92
M2	0.86	0	0	0.97	0	0	0	0	0	0.79	0	0	0	0	0	0.56	0	0.88	0	0.53	0	0	0	0	0	0	0	0	0	0
M3	0.51	0	0	0	0	0	0	0	0	0	0	0	0.98	0	0	0	0	0	0	0	0	0	0.83	0	0.71	0	0.58	0.54	0.54	0
M4	0	0.74	0	0.63	0	0	0.63	0	0.53	0	0	0	0	0	0	0	0	0.69	0	0	0	0.63	0	0	0.68	0	0	0	0	0.51
M5	0	0	0.61	0	0	0.94	0	0.68	0	0	0.67	0	0	0.7	0.84	0	0	0	0	0	0.79	0	0	0.99	0	0.94	0	0	0	0
M6	0	0	0	0	0	0	0	0	0	0	0	0	0	0	0	0	0	0	0	0	0	0	0.84	0	0.73	0	0.93	0	0.73	0
M7	0	0.98	0	0.92	0	0	0.92	0	0	0	0	0.7	0	0	0	0	0	0.89	0	0	0	0.52	0	0	0.52	0	0	0	0	0.54
M8	0	0.77	0	0.76	0	0	0.96	0	0.6	0	0	0.61	0	0	0	0	0	0.54	0	0	0	0.67	0	0	0	0	0	0.7	0	0
M9	0	0	0	0	0.85	0	0	0	0	0	0	0	0	0	0	0	0	0	0.99	0	0	0	0.87	0	0.67	0	0.63	0.74	0	0
M10	0	0	0	0	0	0.85	0	0.78	0	0	0.55	0	0	0.81	0.63	0	0	0.97	0	0	0.54	0	0	0	0	0.52	0	0	0	0
M11	0	0.85	0	0	0	0	0.55	0	0.99	0	0.93	0.94	0	0	0	0	0	0.8	0	0	0	0	0	0	0	0	0	0	0	0.68
M12	0	0.6	0	0.63	0	0	0.7	0	0	0	0	0.9	0	0	0	0	0	0.71	0.98	0.53	0	0.68	0	0	0	0	0	0	0	0
M13	0	0	0.91	0	0	0	0	0	0	0.53	0	0	0.76	0	0	0	0	0	0	0.88	0	0	0	0	0	0	0	0	0.79	0.52
M14	0	0	0	0	0	0.94	0	0.78	0	0	0.52	0	0	0.72	0.92	0	0.92	0	0	0	0.86	0	0	0	0	0	0.8	0	0	0
M15	0	0	0	0	0.67	0	0	0	0	0	0	0	0	0	0	0	0	0	0	0	0	0	0.53	0	0.69	0.59	0.54	0	0.66	0.87
M16	0	0	0	0	0	0.74	0.7	0.77	0	0	0	0	0	0.85	0	0	0	0	0	0	0	0	0	0.81	0	0.63	0.6	0	0	0

Table 16. Data set No. 15

i/j	P1	P2	P3	P4	P5	P6	P7	P8	P9	P10	P11	P12	P13	P14	P15	P16	P17	P18	P19	P20
M1	0.53	0	0	0	0	0	0.99	0	0	0	0	0	0	0	0	0	0	0.83	0	0.91
M2	0	0.82	0	0	0	0	0	0	0	0	0	0	0.83	0	0	0	0	0	0	0
M3	0	0.91	0.92	0	0.86	0	0	0	0	0	0	0	0	0	0.97	0	0	0	0	0.79
M4	0	0	0.56	0	0	0	0	0.88	0	0	0.53	0.51	0	0	0	0	0.98	0	0	0.83
M5	0	0.71	0	0	0	0.58	0	0	0	0	0.54	0	0	0	0.54	0.74	0.63	0	0	0
M6	0.63	0	0	0.53	0	0.69	0.63	0.68	0.51	0	0	0	0	0	0	0	0	0	0	0
M7	0	0	0	0	0	0	0.61	0	0	0	0	0	0	0	0.94	0	0.68	0.67	0	0
M8	0	0.7	0.84	0	0	0	0	0	0	0	0	0	0	0	0	0	0	0.79	0	0.99
M9	0.94	0	0	0.84	0	0	0.78	0.93	0.73	0	0	0	0	0	0.98	0	0	0	0	0
M10	0	0	0	0.92	0	0	0	0.92	0	0.7	0	0	0.89	0	0	0	0	0	0.52	0
M11	0.52	0	0.54	0	0	0	0	0	0	0.77	0	0.76	0.96	0	0	0.6	0	0	0	0
M12	0	0	0	0	0	0	0	0	0	0	0	0.61	0	0	0	0	0	0	0	0
M13	0.54	0	0.67	0	0.7	0	0	0	0	0.85	0.99	0	0.87	0.67	0	0	0	0	0	0
M14	0	0	0.63	0	0	0.74	0.85	0.78	0.55	0	0.81	0.63	0	0.97	0	0.54	0.52	0	0.85	0
M15	0	0	0	0	0.55	0	0	0.99	0	0.93	0.94	0	0	0	0	0	0.8	0	0	0
M16	0.68	0	0.6	0	0	0	0.63	0	0.7	0	0	0	0.9	0	0.71	0.98	0.53	0.68	0	0
M17	0	0.91	0	0	0	0	0	0	0	0	0	0.53	0.76	0.88	0	0	0	0	0	0
M18	0	0	0.79	0	0	0	0.52	0.94	0	0	0.78	0.52	0.72	0.92	0	0	0	0	0	0
M19	0	0	0	0	0	0	0.92	0	0.83	0.8	0.67	0	0	0.53	0	0	0	0	0	0.69
M20	0	0	0	0	0	0	0.59	0.54	0	0	0	0	0	0	0	0	0	0	0	0

Figure 7. Data set No. 16

i/j	P1	P2	P3	P4	P5	P6	P7	P8	P9	P10	P11	P12	P13	P14	P15	P16	P17	P18	P19	P20	P21	P22	P23	P24	P25	P26	P27	P28	P29	P30	P31	P32	P33	P34	P35
M1	0.53	0	0.9	0	0	0	0	0	0	0	0	0	0	0	0	0	0	0	0	0.83	0	0	0.91	0	0.82	0	0	0	0.83	0	0	0	0	0	0
M2	0	0.9	0	0	0	0	0.9	0	0	0.86	0	0.97	0.79	0	0	0	0	0.56	0	0	0	0	0	0.88	0	0	0.53	0	0	0	0.51	0	0	0	0
M3	0.98	0	0.8	0	0.7	0	0	0	0	0	0	0	0	0	0.58	0	0.54	0	0	0	0	0	0	0	0	0	0	0	0.54	0	0.74	0	0	0	0
M4	0	0.6	0	0	0	0	0.6	0	0	0	0	0.53	0.69	0	0	0	0	0	0	0	0	0	0	0.63	0	0	0.68	0	0	0	0	0	0	0	0
M5	0	0	0	0	0	0	0	0.5	0	0	0	0	0	0.61	0	0.94	0	0	0	0	0	0	0	0	0	0	0	0	0	0	0	0	0	0.68	0
M6	0	0	0	0	0	0	0	0.6	0	0	0	0	0	0.7	0	0.84	0	0	0.79	0	0	0.99	0	0	0	0.94	0	0	0	0	0	0	0	0.84	0
M7	0.78	0	0.9	0	0.7	0	0	0	0	0	0	0	0	0	0.98	0	0.92	0	0	0.92	0	0	0.7	0	0	0	0	0	0	0	0	0	0	0	0
M8	0.89	0	0.5	0	0.5	0	0	0	0	0	0	0	0	0	0.54	0	0.77	0	0	0.76	0	0	0.96	0	0.6	0	0	0	0.61	0	0	0	0	0	0
M9	0	0	0	0	0	0	0	0.5	0	0	0	0	0	0.67	0	0	0	0	0.7	0	0	0.85	0	0	0	0.99	0	0	0	0	0	0	0	0	0
M10	0	0	0	0	0	0	0	0.8	0	0	0	0	0	0.67	0	0.63	0	0	0.74	0	0	0.85	0	0	0	0.78	0	0	0	0	0	0	0	0	0
M11	0	0	0	0.5	0	0.81	0	0	0.6	0	0.97	0	0	0	0	0	0	0	0	0	0.54	0	0	0	0	0	0	0.52	0	0.85	0	0.55	0	0	0.99
M12	0	0	0	0.9	0	0.94	0	0	0.8	0	0.68	0	0	0	0	0	0	0	0	0	0.6	0	0	0	0	0	0	0	0	0	0	0	0.63	0	0
M13	0	0.7	0	0	0	0	0	0	0	0	0	0.9	0.71	0	0	0	0	0	0	0	0	0	0	0.98	0	0	0	0	0	0	0	0	0	0	0
M14	0	0.5	0	0	0	0	0.68	0	0	0.91	0	0.53	0.76	0	0	0	0	0.88	0	0	0	0	0	0.79	0	0	0.52	0	0	0	0.94	0	0	0	0
M15	0	0	0	0.7	0	0.5	0	0	0.7	0	0.92	0	0	0	0	0	0	0	0	0	0.92	0	0	0	0	0	0	0.86	0	0.8	0	0.67	0	0	0
M16	0	0	0	0.5	0	0.6	0	0	0.5	0	0.54	0	0	0	0	0	0	0	0	0	0.66	0	0	0	0	0	0	0.87	0	0.74	0	0.7	0	0	0
M17	0.77	0	0.8	0	0.8	0	0	0	0	0	0	0	0	0	0.63	0	0.6	0	0	0	0	0	0.96	0	0.53	0	0	0	0.9	0	0	0	0	0	0
M18	0	0.9	0	0	0	0	0	0	0	0.83	0	0.78	0.99	0	0	0	0	0.51	0	0	0	0	0	0.82	0	0	0	0	0	0	0.89	0	0	0	0
M19	0	0	0	0.6	0	0	0	0	0.5	0	0.97	0	0	0	0	0	0	0	0	0	0.88	0	0	0	0	0	0	0.8	0	0.72	0	0.7	0	0	0
M20	0	0	0	0	0	0	0	0.7	0	0	0	0	0	0.53	0	0	0	0	0.95	0	0	0	0	0	0	0.95	0	0	0	0	0	0	0	0	0

Figure 8. Data set No. 17

i/j	P1	P2	P3	P4	P5	P6	P7	P8	P9	P10	P11	P12	P13	P14	P15	P16	P17	P18	P19	P20	P21	P22	P23	P24	P25	P26	P27	P28	P29	P30	P31	P32	P33	P34	P35
M1	0.5	0	0	0	0	0.9	0	0	0	0	0.83	0	0	0.91	0.82	0	0	0.83	0	0.91	0	0	0.92	0.86	0.97	0	0	0	0	0.79	0	0.56	0	0.88	0.53
M2	0	0.5	0	0	0	0	0	0	0	0.98	0	0.83	0.71	0	0	0	0	0.58	0.54	0	0	0	0	0.54	0	0	0	0	0	0	0.74	0	0	0	0
M3	0.6	0	0.6	0	0.5	0	0	0	0	0	0	0	0	0	0.69	0	0	0	0	0	0	0	0	0	0	0	0	0	0.63	0	0.68	0	0	0	0
M4	0	0.5	0	0	0	0	0.6	0	0	0	0	0.94	0.68	0	0	0	0	0	0	0	0	0	0	0.67	0	0	0.7	0	0	0	0	0	0	0	0
M5	0	0	0	0	0.84	0	0	0	0	0	0	0	0	0.79	0	0	0.99	0	0	0	0	0	0.94	0	0	0.84	0	0	0	0	0	0	0	0	0
M6	0	0	0	0	0	0	0	0.7	0	0	0	0	0	0.93	0	0	0	0	0.73	0	0	0.98	0	0	0.92	0	0	0	0	0.92	0	0	0	0	0
M7	0.7	0	0.8	0	0.5	0	0.5	0	0	0	0	0.54	0	0	0.77	0	0.76	0	0.96	0.6	0	0.61	0.54	0.67	0	0.7	0	0	0.85	0.99	0.87	0.67	0	0	0
M8	0.6	0	0	0	0.7	0	0	0	0.85	0	0	0	0	0	0.78	0	0.55	0	0	0.81	0	0	0.63	0	0	0	0	0	0	0	0	0	0	0	0.97
M9	0	0	0	0	0	0	0	0.54	0	0	0	0	0	0.52	0	0	0	0	0.85	0	0	0	0.55	0	0	0	0	0	0	0	0	0.99	0	0	0.93
M10	0	0	0	0	0	0	0	0.94	0	0	0	0	0	0.8	0	0.68	0	0	0.6	0	0	0.63	0	0	0	0.7	0	0	0	0	0	0	0	0	0
M11	0	0	0	0.9	0	0.71	0	0	0.98	0	0.53	0	0	0	0	0	0	0	0	0	0.68	0	0	0	0	0	0	0	0	0	0	0	0.91	0	0
M12	0	0	0	0.5	0	0.76	0	0	0.88	0	0.79	0	0	0	0	0	0	0	0	0	0.52	0	0	0	0	0	0	0	0	0	0	0	0	0	0
M13	0	0.9	0	0	0	0	0	0	0	0	0	0.78	0.52	0	0	0	0	0	0	0	0	0	0	0.72	0	0	0	0	0	0	0	0	0	0	0
M14	0	0.9	0	0	0	0	0.92	0	0	0.86	0	0.8	0.67	0	0	0	0	0.53	0	0	0	0	0	0.69	0	0	0.59	0	0	0	0.54	0	0	0	0
M15	0	0	0	0.66	0	0.87	0	0	0.74	0	0.7	0	0	0	0	0	0	0	0	0	0.77	0	0	0	0	0	0	0.85	0	0.81	0	0	0	0	0
M16	0	0	0	0.63	0	0.6	0	0	0.96	0	0	0.53	0	0	0	0	0	0.9	0	0	0	0	0	0	0	0.92	0	0	0	0.83	0	0.78	0	0	0
M17	0.9	0	0.5	0	0.82	0	0	0	0	0	0	0	0	0	0.89	0	0.65	0	0	0.57	0	0	0	0	0.97	0	0	0	0	0	0	0	0	0	0.88
M18	0	0.8	0	0	0	0	0.72	0	0	0	0	0.7	0.78	0	0	0	0	0	0	0.53	0	0	0	0.95	0	0	0	0	0	0	0.95	0	0	0	0
M19	0	0	0	0.55	0	0	0	0	0.84	0	0.67	0	0	0	0	0	0	0	0	0	0.92	0	0	0	0	0	0	0.85	0	0.52	0	0.61	0	0	0
M20	0	0	0	0	0	0	0	0.55	0	0	0	0	0	0.62	0	0	0	0	0.8	0	0	0	0.81	0	0	0.98	0	0	0	0	0	0	0	0	0

Table 17. Data set No. 18

i/j	P1	P2	P3	P4	P5	P6	P7	P8	P9	P10	P11	P12	P13	P14	P15	P16	P17	P18	P19	P20
M1	0.53	0	0.99	0	0.83	0.91	0	0.82	0	0	0	0.83	0.91	0.92	0.86	0	0	0.97	0.79	0
M2	0	0	0.56	0	0	0.88	0	0	0.53	0.51	0	0.98	0	0	0	0	0	0	0	0
M3	0.83	0	0	0.71	0	0	0	0.58	0	0	0	0	0	0	0	0	0	0.54	0.54	0.74
M4	0	0	0	0	0.63	0	0	0	0	0	0	0	0.63	0	0	0	0	0	0	0
M5	0	0.53	0	0	0	0	0	0	0.69	0	0.63	0	0	0	0	0.67	0	0	0.51	0
M6	0	0	0	0	0	0	0	0	0	0.61	0	0	0	0	0	0	0.94	0	0	0
M7	0	0	0.68	0	0	0	0.67	0	0.7	0.84	0.79	0	0	0	0	0.99	0.94	0	0	0
M8	0	0	0	0	0	0	0.84	0	0	0.78	0	0	0	0	0	0	0	0	0	0
M9	0	0	0	0	0	0	0.93	0	0	0.73	0	0	0	0	0	0	0	0	0	0
M10	0.98	0	0	0	0.92	0.92	0	0	0.7	0	0	0.89	0.52	0	0	0	0	0.52	0	0
M11	0.54	0	0.77	0	0.76	0	0	0	0	0	0	0.96	0	0	0	0	0	0	0	0
M12	0	0	0	0.6	0	0	0	0.61	0	0	0	0	0	0.54	0.67	0	0	0	0.7	0.85
M13	0	0.99	0	0	0	0	0	0.87	0	0	0	0	0	0	0	0	0.67	0	0	0
M14	0	0.63	0	0	0	0	0	0	0	0	0	0	0	0	0	0.74	0	0	0	0
M15	0.85	0.78	0.55	0	0.81	0.63	0	0.97	0	0	0	0.54	0.52	0.85	0	0	0.55	0.99	0	0.93
M16	0	0	0	0	0	0	0	0.94	0	0.8	0	0	0	0.68	0	0	0	0	0	0.6
M17	0	0	0	0.63	0	0	0	0	0	0.7	0	0	0	0	0	0	0	0	0	0
M18	0	0	0.9	0	0	0	0	0	0.71	0	0	0.98	0.53	0	0.68	0.91	0.53	0	0	0
M19	0	0	0	0	0	0	0	0	0.76	0	0	0	0	0	0.88	0	0.79	0	0	0
M20	0	0	0	0	0	0	0	0	0	0	0	0.52	0.94	0	0	0.78	0	0	0	0
M21	0.52	0	0	0.72	0	0	0	0	0.92	0	0	0	0	0	0.92	0	0	0	0	0
M22	0.86	0	0.8	0.67	0	0.53	0	0.69	0.59	0.54	0	0	0	0.66	0.87	0	0	0	0	0.74
M23	0.7	0	0	0.77	0	0	0	0	0.85	0	0	0	0	0	0.81	0	0	0	0	0

Table 18. Data set No. 19

i/j	P1	P2	P3	P4	P5	P6	P7	P8	P9	P10	P11	P12	P13	P14	P15	P16
M1	0.53	0	0.99	0	0	0	0.83	0.91	0	0	0	0	0	0.82	0	0
M2	0	0	0	0	0	0.83	0	0	0	0.91	0	0.92	0.86	0	0	0
M3	0.97	0	0	0	0	0.79	0	0.56	0	0	0	0.88	0	0	0	0
M4	0.53	0	0.51	0	0	0	0	0	0	0	0	0	0	0.98	0	0
M5	0	0	0	0	0	0	0	0	0	0	0	0	0	0	0.83	0
M6	0.71	0	0.58	0	0	0	0	0	0	0	0	0	0	0.54	0.54	0
M7	0.74	0	0	0	0	0	0	0.63	0.63	0	0.53	0	0.69	0	0	0
M8	0.63	0	0	0	0	0	0	0	0	0	0	0	0.68	0	0	0
M9	0	0	0	0	0	0	0	0	0	0.51	0.61	0.94	0	0	0.68	0.67
M10	0.7	0.84	0	0	0	0	0	0.79	0	0.99	0	0	0	0	0	0
M11	0.94	0	0	0	0	0	0	0	0	0	0	0.84	0.78	0	0	0.93
M12	0	0	0	0	0	0	0	0	0	0	0	0	0	0.73	0	0
M13	0	0	0	0	0	0.98	0	0.92	0	0	0	0	0	0	0	0
M14	0	0	0	0	0	0	0	0	0	0.92	0.7	0.89	0	0.52	0	0.52
M15	0.54	0	0	0	0	0.77	0	0	0	0	0	0	0	0	0	0
M16	0	0	0.76	0.96	0	0	0	0	0	0	0	0	0	0	0	0
M17	0	0.6	0.61	0	0	0.54	0	0	0	0.67	0	0	0	0	0	0.7
M18	0.85	0	0	0	0	0.99	0	0	0	0	0	0	0	0	0	0
M19	0.87	0	0.67	0.63	0	0	0.74	0	0	0	0	0	0	0	0	0
M20	0	0	0	0	0	0	0	0	0	0.85	0.78	0.55	0	0	0.81	0.63
M21	0	0	0	0	0.97	0.54	0	0	0	0	0	0	0	0	0	0
M22	0.52	0.85	0	0	0	0	0	0.55	0	0.99	0	0	0	0	0	0
M23	0.93	0.94	0	0	0	0	0	0.8	0	0.68	0	0	0	0	0	0
M24	0	0	0	0	0	0	0.6	0	0.63	0.7	0.9	0.71	0	0.98	0	0.53

Figure 9. Data set No. 20

i/j	P1	P2	P3	P4	P5	P6	P7	P8	P9	P10	P11	P12	P13	P14	P15	P16	P17	P18	P19	P20
M1	0.5	0	0	0	0	0	0	0	1	0	0	0	0	0	0	0.8	0.9	0	0	0
M2	0	0	0	0	0	0	0	0	0	0.8	0	0	0.9	0.9	0	0	0	0	0	0
M3	0	0.6	0	0	0	0	0	0	0	0	0.9	0.5	0	0	0.5	0	0	0	0	0
M4	0	0	0	0	0	0	0	0.5	0	0	0	0	0	0	0	0	0	0	0.5	0
M5	0	0	0	0	0	0	0	0	0	0.6	0	0	0.7	0.5	0	0	0	0	0	0
M6	0	0	0	0.7	0.7	0	0	0	0	0	0	0	0	0	0	0	0	0.8	0	0
M7	0	0	0.8	0	0	0	0	0	0	0	0	0	0	0	0	0	0	0	0	0
M8	0	0	0	0.7	1	0	0	0	0	0	0	0	0	0	0	0	0	0.9	0	0
M9	0	0	0	0	0	0.5	0.5	0	0	0	0	0	0	0	0	0	0	0	0	0.5
M10	0	0	0	0	0	1	0.6	0	0	0	0	0	0	0	0	0	0	0	0	0.6
M11	0	0	0	0	0	0	0	0	0	0.7	0	0	0.9	1	0	0	0	0	0	0
M12	0	0	0	0.7	0.9	0	0	0	0	0	0	0	0	0	0	0	0	0.8	0	0
M13	1	0	0	0	0	0	0	0	0.5	0	0	0	0	0	0	0.5	0.9	0	0	0
M14	0	0	1	0	0	0	0	0	0	0	0	0	0	0	0	0	0	0	0	0
M15	0	0	0	0.8	0.7	0	0	0	0	0	0	0	0	0	0	0	0	0.6	0	0
M16	0	0	0	0	0	0	0	0.7	0	0	0	0	0	0	0	0	0	0	1	0
M17	0	0	0	0	0	0.9	0.8	0	0	0	0	0	0	0	0	0	0	0	0	0.5
M18	0	0	0	0.5	0.7	0	0	0	0	0	0	0	0	0	0	0	0	0.9	0	0
M19	0	0	0	0	0	0	0	0	0	0.7	0	0	0.5	0.7	0	0	0	0	0	0
M20	0	0.9	0	0	0	0	0	0	0	0	0.7	0.7	0	0	0.8	0	0	0	0	0
M21	1	0	0	0	0	0	0	0	0.5	0	0	0	0	0	0	0.9	0.9	0	0	0
M22	0.8	0	0	0	0	0	0	0	1	0	0	0	0	0	0	0.5	0.8	0	0	0
M23	0	0	0.7	0	0	0	0	0	0	0	0	0	0	0	0	0	0	0	0	0
M24	0	0	0.9	0	0	0	0	0	0	0	0	0	0	0	0	0	0	0	0	0

i/j	P21	P22	P23	P24	P25	P26	P27	P28	P29	P30	P31	P32	P33	P34	P35	P36	P37	P38	P39	P40
M1	0	0	0	0	0	0	0	0	0	0	0	0	0.8	0	0	0	0	0	0	0
M2	0	0.9	0	0	0	0	0	0	0	0	0	0	0	0	1	0.8	0	0	0	0
M3	0	0	1	0.8	0	0	0	0	0	0	0.7	0	0	0.6	0	0	0	0	0	0
M4	0.7	0	0	0	0	0	0	0.6	0	0	0	0	0	0	0	0	0.6	0.5	0.7	0
M5	0	0.6	0	0	0	0	0	0	0	0	0	0	0	0	0.9	0.7	0	0	0	0
M6	0	0	0	0	0	0.8	1	0	0	0.9	0	0	0	0	0	0	0	0	0	0
M7	0	0	0	0	0.8	0	0	0	0	0	0	0.9	0	0	0	0	0	0	0	0
M8	0	0	0	0	0	0.9	0.7	0	0	0.9	0	0	0	0	0	0	0	0	0	0
M9	0	0	0	0	0	0	0	0	0.8	0	0	0	0	0	0	0	0	0	0	0.8
M10	0	0	0	0	0	0	0	0	0.5	0	0	0	0	0	0	0	0	0	0	0.7
M11	0	0.9	0	0	0	0	0	0	0	0	0	0	0	0	0.7	0.6	0	0	0	0
M12	0	0	0	0	0	0.6	0.8	0	0	0.6	0	0	0	0	0	0	0	0	0	0
M13	0	0	0	0	0	0	0	0	0	0	0	0	0.6	0	0	0	0	0	0	0
M14	0	0	0	0	0.9	0	0	0	0	0	0	0.9	0	0	0	0	0	0	0	0
M15	0	0	0	0	0	0.6	0.7	0	0	0.9	0	0	0	0	0	0	0	0	0	0
M16	0.5	0	0	0	0	0	0	0.7	0	0	0	0	0	0	0	0	0.9	0.5	0.8	0
M17	0	0	0	0	0	0	0	0	0.9	0	0	0	0	0	0	0	0	0	0	0.8
M18	0	0	0	0	0	0.9	0.9	0	0	0.8	0	0	0	0	0	0	0	0	0	0
M19	0	0.6	0	0	0	0	0	0	0	0	0	0	0	0	0.5	0.7	0	0	0	0
M20	0	0	0.9	0.8	0	0	0	0	0	0	0.6	0	0	0.6	0	0	0	0	0	0
M21	0	0	0	0	0	0	0	0	0	0	0	0	0.8	0	0	0	0	0	0	0
M22	0	0	0	0	0	0	0	0	0	0	0	0	0.9	0	0	0	0	0	0	0
M23	0	0	0	0	0.6	0	0	0	0	0	0	1	0	0	0	0	0	0	0	0
M24	0	0	0	0	0.8	0	0	0	0	0	0	0.7	0	0	0	0	0	0	0	0

Figure 10. Data set No. 21

i/j	P1	P2	P3	P4	P5	P6	P7	P8	P9	P10	P11	P12	P13	P14	P15	P16	P17	P18	P19	P20
M1	0.5	0	0	0	0	0	0	0	1	0	0	0	0	0	0	0.8	0.9	0	0	0
M2	0	0	0	0	0	0	0	0	0	0.8	0	0	0.9	0.9	0	0	0	0	0	0
M3	0	0.8	0	0	0	0	0	0	0	0	0.6	0.9	0	0	0.5	0	0	0	0	0
M4	0	0	0	0	0	0	0	0.6	0	0	0	0	0	0	0	0	0	0	0.5	0
M5	0	0	0	0	0.5	0	0	0	0	0.7	0	0	0	0.6	0	0	0	0	0	0
M6	0	0	0	0.9	0.7	0	0	0	0	0	0	0	0	0	0	0	0	0.7	0	0
M7	0	0	1	0	0	0	0	0	0	0	0	0	0	0	0	0	0	0	0	0
M8	0	0	0	0.8	0.9	0	0	0	0	0	0	0	0	0	0	0	0	0.7	0	0
M9	0	0	0	0	0	0.7	0.9	0	0	0	0	0	0	0	0	0	0	0	0	0
M10	0	0	0	0	0	0.5	0.8	0	0	0	0	0	0	0	0	0	0	0	0	0.8
M11	0	0	0	0	0	0	0	0	0	0.6	0	0	0.5	0.7	0	0	0	0	0	0
M12	0	0	0	0.9	0.7	0	0	0	0	0	0	0	0	0	0	0	0	0.6	0	0.7
M13	0.8	0	0	0	0	0	0	0	0.6	0	0	0	0	0	0	1	0.5	0	0	0
M14	0	0	0.9	0	0	0	0	0	0	0	0	0	0	0	0	0	0	0	0	0
M15	0	0	0	0.9	0.9	0	0	0	0	0	0	0	0	0	0	0	0	0.8	0	0
M16	0	0	0	0	0	0	0	0.7	0	0	0	0	0	0	0	0	0	0	0.9	0
M17	0	0	0	0	0	0.5	0.8	0	0	0	0	0	0	0	0	0	0	0	0	0.9
M18	0	0	0	0.8	0.5	0	0	0	0	0	0	0	0	0	0	0	0	0.7	0	0
M19	0	0	0	0	0	0	0	0	0	0.8	0	0	0.7	0.5	0	0	0	0	0	0
M20	0	0.7	0	0	0	0	0	0	0	0	0.9	0.7	0	0	0.7	0	0	0	0	0
M21	0.6	0	0	0	0	0	0	0	1	0.5	0	0	0	0	0	0.9	0	0	0	0
M22	0.8	0	0	0	0	0	0	0	0.8	0	0	0	0	0	0	1	0.5	0	0	0
M23	0	0	0.9	0	0	0	0	0	0	0	0	0	0	0	0	0	0	0	0	0
M24	0	1	0.9	0	0	0	0	0	0	0	0	0	0	0	0	0	0	0	0	0

i/j	P21	P22	P23	P24	P25	P26	P27	P28	P29	P30	P31	P32	P33	P34	P35	P36	P37	P38	P39	P40
M1	0	0	0	0	0	0	0	0	0	0	0	0	0.8	0	0	0	0	0	0	0
M2	0	0.9	0	0	0	0	0	0	0	0	0	0	1	0	0	0	0	0	0	0
M3	0	0	0.5	1	0	0	0	0	0	0	0.8	0	0	0.7	0	0	0	0	0	0
M4	0.5	0	0	0	0	0	0	0.7	0	0	0	0	0	0	0	0	0	0.6	0.6	0
M5	0	0.7	0	0	0	0	0	0	0	0	0	0	0	0	0.5	0.6	0	0	0	0
M6	0	0	0	0	0	0.7	0.8	0	0	0.8	0	0	0	0	0	0	0	0	0	0
M7	0	0	0	0	0.9	0	0	0	0	0	0	0.8	0	0	0	0	0	0	0	0
M8	0	0	0	0	0	1	0.9	0	0	0.9	0	0	0	0	0	0	0	0	0	0
M9	0	0	0	0	0	0	0	0	0.5	0	0	0	0	0	0	0	0	0	0	0.5
M10	0	0	0	0	0	1	0	0	0	0	0	0	0	0	0	0	0	0	0	0.6
M11	0	0.7	0	0	0	0	0	0	0	0	0	0	0	0	0.9	1	0	0	0	0
M12	0	0	0	0.9	0	0	0.8	0	0	0.6	0	0	0	0	0	0	0	0	0	0
M13	0	0	0	0	0	0	0	0	0	0	0	0	0.5	0	0	0	0	0	0	0
M14	0	0	0	0	0.6	0	0	0	0	0	0	1	0	0	0	0	0	0	0	0
M15	0	0	0	0	0	0.7	0.6	0	0	0	0	0	0	0	0	0	0.6	0	0	0
M16	0.7	0	0	0	0	0	0	0	0	0	0	1	0	0	0	0	0.5	0.7	0.9	0
M17	0	0	0	0	0	0	0	0	0.8	0	0.5	0	0	0	0	0	0	0	0	0.9
M18	0	0	0	0	0	0.9	0.9	0	0	0.9	0	0	0	0	0	0	0	0	0	0
M19	0	0.7	0	0	0	0	0	0	0	0	0	0	0	0	0.6	0.5	0	0	0	0
M20	0	0	0.8	0.9	0	0	0	0	0	0	0.8	0	0	0.6	0	0	0	0	0	0
M21	0	0	0	0	0	0	0	0	0	0	0	0	0.9	0	0	0	0	0	0	0
M22	0	0	0	0	0	0	0	0	0	0	0	0	0.8	0	0	0	0	0	0	0
M23	0	0	0	0	0.7	0	0	0	0	0	0	0.6	0	0	0	0	0	0	0	0
M24	0	0	0	0	0	0	0	0	0	0	0	0.8	0	0	0	0	0	0	0	0

Figure 11. Data set No. 22

i/j	P1	P2	P3	P4	P5	P6	P7	P8	P9	P10	P11	P12	P13	P14	P15	P16	P17	P18	P19	P20	P21	P22	P23	P24	P25	P26	P27	P28	P29	P30	P31	P32	P33	P34	P35	P36	P37	P38	P39	P40	P41
M1	0	0	0	0	0	0	0	0	0	0.5	0	0	0	0	0	0	0	0	0	0	0	0	0	0	0	0	0	0	0	0	1	0.8	0.9	0	0	0	0	0	0.8	0	0.8
M2	0	0	0	0	0	0	0	0	0	0	0.9	0.9	0	0	0	0	0	0.9	0	0	0	0	0	0	0	0	0	0	0	0	0	1	0.8	0	0	0	0	0	0.6	0	0.9
M3	0	0	0	0	0	0	0	0	0	0	0	0.5	0	0	0	0	0	0	0	0	0	0	0.5	0	0	0	0	0	0	0	1	0.8	0	0	0	0	0	0	0	0.7	0
M4	0	0	0	0	0	0	0.6	0	0	0	0	0	0	0	0.5	0	0	0	0	0	0	0	0	0.5	0.7	0	0	0.6	0	0	0	0	0	0	0.6	0	0	0	0	0	0
M5	0	0	0	0	0.5	0	0	0	0	0	0	0	0	0	0	0	0.7	0	0	0	0	0	0	0	0	0	0	0	0	0	0	0	0	0.6	0	0.7	0	0	0	0	0
M6	0	0	0	0	0	0.5	0	0	0	0	0	0	0	0	0	0	0	0	0	0	0	0	0	0	0	0.6	0	0	0	0	0	0	0	0	0	0	0	0	0	0	0
M7	0	0	0	0	0	0	0	0	0	0	0	0	0	0	0	0.9	0	0	0	0	0	0	0	0	0	0	0.7	0	0	0	0	0	0	0.7	0	0.7	0	0	0	0	0
M8	0.8	0	0	0	0	0	0	0.8	1	0	0	0	0	0.9	0	0	0	0	0	0	0	0	0	0	0	0	0	0	0.8	0	0	0	0	0	0	0	0	0	0	0	0
M9	0	0	0.8	0	0	0	0	0	0	0	0	0	0.9	0	0	0	0	0	0	0	0.7	0	0	0	0	0	0	0	0	0	0	0	0	0	0	0	0	0	0	0	0
M10	0	1	0	0	0	0	0	0	0	0	0	0.9	0	0	0	0	0	0	0	0.9	0	0	0.7	0	0	0	0	0	0	0	0	0	0	0	0	0	0	0	0.9	0	0
M11	0	0	0	0	0	0	0	0	0	0.5	0	0	0	0	0	0	0	0	0	0	0	0	0	0	0	0	0	0	0	0	0	0	0.5	0	0	0	0	0	0.5	0	0.8
M12	0	0	0	0	0	0	0	0	0	0	0.8	1	0	0	0	0	0	0.6	0.6	0	0	0	0.5	0.7	0	0	0	0	0	0	0	0	0	0	0	0	0	0	0.7	0.9	0
M13	0	0	0	0	0	0	0	0	0	0	0	0	0	0	0	0	0	1	0	0	0	0	0	0	0.9	0	0	0	0	0	0	0	0	0	0	0	0	0.7	0	0	0
M14	0	0	0	0.6	0.7	0	0	0	0	0	0	0	0	0	0	0	0	0.9	0	0	0	0	0	0	0	0.8	0	0	0	0	0	0	0	0	0	0	0.6	0	0	0	0
M15	0	0	0	0	0	0	0	0	0	0	0	0	0	0	0	0	0.8	0	0	0	0	0	0	0	0	0	0	0	0	0	0	0	0	0	0	0	0.6	0	0	0	0
M16	0	0	0	0	0	1	0	0	0	0	0	0	0	0	0	0	0	0	0	0	0	0	0	0	0	0	0	0.5	0	0	0	0	0	0	0	0	0	0	0	0	0
M17	0	0	0	0	0	0	0.5	0	0	0	0	0	0	0	0	0.9	0	0	0	0	0	0	0	0	0	0	0.6	0	0	0	0	0	0	0	0	1	0	0	0	0	0
M18	0	0	0	0	0	0	0	0	0	0	0	0	0	0.9	0	0.9	0	0	0	0	0	0	0	0	0	0	0.8	0	0	0	0	0	0	0.7	0	0	0	0	0	0	0
M19	0.6	0	0.6	0	0	0	0	0	0	0	0	0	0	0	0	0	0	0	0	0	0.7	0.9	0	0	0	0	0	0	0	0.7	0	0	0	0	0	0	0	0	0	0	0
M20	1	0	0.5	0	0	0	0	0	0	0	0	0	0	0	0	0	0	0	0	0	0.7	0.9	0	0	0	0	0	0	0	0	0	0	0	0	0	0	0	0	0	0	0
M21	0	0	0	0	0	0	0	0	0	0.5	0	0.8	0	0	0	0	0	0	0	0	0	0	0	0	0	0	0	0	0	0	0.9	0.8	0.5	0	0	0	0	0	0.9	0.8	0
M22	0	0	0	0	0	0	0	0	0	0.5	0	0.7	0	0	0	0	0	0	0	0	0	0	0.9	0	0	0	0	0	0	0	0.9	0.9	0	0	0	0	0	0	0.8	0.7	0
M23	0	0.5	0	0	0	0	0	0	0	0	0	0.7	0	0	0	0	0	0	0	0	0	0	0.6	0	0	0	0	0	0	0	0.5	0	0	0	0	0	0	0	0.7	0	0
M24	0	0	0	0	0	0	0	0	0	0	0	0	0	0	0	0	0	0	0	0	0	0	0	0	0	0	0	0	0	0	0	0	0	0	0	0	0	0.9	0	0	0
M25	0	0	0	0.7	0	0	0	0	0	0	0	0	0	0	0	0	0	0	0	0	0	0	0	0	0	0	0	0	0	0	0	0	0	0	0	0	0	0	0	0	0
M26	0	0	0	0	0	0	0	0	0	0	0	0	0	0	0	0.7	0	0	0	0	0	0	0	0	0	0	0.8	0	0	0	0	0	0	0.9	0	0.8	0	0	0	0	0
M27	0	0	0	0	0	0	0	0.6	0	0	0	0	0	0	0.6	0	0	0	0	0	0	0	0	0	0	0	0	0	0	0	0	0	0	0	0	0	0	0	0	0	0
M28	0	0	0	0	0	0	0	1	0	0	0	0	0	0	0	0	0	0	0	0	0	0	0	0	0	0	0	0	0.5	0	0	0	0	0	0.9	0	0	0	0	0	0
M29	0.9	0	0.8	0	0	0	0	0	0.8	0	0	0	1	0	0	0	0	0	0	0	0.5	0	0	0	0	0	0	0	0	0.8	0	0	0	0	0	0	0	0	0	0	0
M30	0.9	0	0.7	0	0	0	0	0	0	0	0	0	0	0	0	0	0	0	0	0	0.6	1	0	0	0	0	0	0	0	0	0	0	0	0	0	0	0	0	0	0	0

Figure 12. Data set No. 23

i/j	P1	P2	P3	P4	P5	P6	P7	P8	P9	P10	P11	P12	P13	P14	P15	P16	P17	P18	P19	P20	P21	P22	P23	P24	P25	P26	P27	P28	P29	P30	P31	P32	P33	P34	P35	P36	P37	P38	P39	P40	P41	P42	P43	P44	P45	P46	P47	P48	P49	P50
M1	0.5	0	1	0.8	0	0.9	0	0	0	0	0	0	0	0	0.8	0	0	0	0	0	0	0	0	0	0	0	0	0	0	0	0	0	0	0	0	0	0	0	0	0	0	0	0	0	0	0	0	0	0	0
M2	0	0.8	0	0.6	0	0.9	0	0	0.5	0	0	0	0	0	0	0	0	0	0	0	0	0	0	0	0	0	0	0	0	0	0	0	0	0	0	0	0	0	0	0	0	0	0	0	0	0	0	0	0	0
M3	0.5	0	0	0	1	0	0	0	0	0	0	0	0	0	0.8	0	0	0	0	0	0	0	0	0	0	0	0	0	0	0	0	0	0	0	0	0	0	0	0	0	0	0	0	0	0	0	0	0	0	0
M4	0.7	0	0	0	0.5	0	0	0.7	0.6	0	0	0	0	0	0	0	0	0	0	0	0	0	0	0	0	0	0	0	0	0	0	0	0	0	0	0	0	0	0	0	0	0	0	0	0	0	0	0	0	0
M5	0	0	0.6	0	0	0	0.5	0	0	0	0.7	0.6	0.7	0.5	0	0	0	0	0	0	0	0	0	0	0	0	0	0	0	0	0	0	0	0	0	0	0	0	0	0	0	0	0	0	0	0	0	0	0	0
M6	0	0.6	0	0	0	0	0.9	0	0	0.7	0.7	0	0	0	0	0	0	0	0	0	0	0	0	0	0	0	0	0	0	0	0	0	0	0	0	0	0	0	0	0	0	0	0	0	0	0	0	0	0	0
M7	0	0	0.7	0	0.8	0	0	0	0	0.8	1	0.9	0.8	0.8	0	0	0	0	0	0	0	0	0	0	0	0	0	0	0	0	0	0	0	0	0	0	0	0	0	0	0	0	0	0	0	0	0	0	0	0
M8	0	0.7	0	0.9	0.5	0	0.5	0.5	0.8	0	0	0	0	0	0.8	0	0	0	0	0	0	0	0	0	0	0	0	0	0	0	0	0	0	0	0	0	0	0	0	0	0	0	0	0	0	0	0	0	0	0
M9	1	0.6	0	0	0.6	0	0	0	0	0	0	0	0	0	0.5	0	0	0	0	0	0	0	0	0	0	0	0	0	0	0	0	0	0	0	0	0	0	0	0	0	0	0	0	0	0	0	0	0	0	0
M10	0	0	0	0	0.7	0.8	0	0	0	0	0	1	0.9	0.7	0	0	0	0	0	0	0	0	0	0	0	0	0	0	0	0	0	0	0	0	0	0	0	0	0	0	0	0	0	0	0	0	0	0	0	0
M11	0.6	0	0.7	0.9	0	0	0.8	0	0	0	0	0	0	0	0	0	0	0	0	0	0	0	0	0	0	0	0	0	0	0	0	0	0	0	0	0	0	0	0	0	0	0	0	0	0	0	0	0	0	0
M12	0	0	0	0	0	0	0	0	0	0	0	0	0	0	0	0	0	0	0	0	0	0	0	0	0	0	0	0	0	0	0	0	0	0	0	0	0	0	0	0	0	0	0	0	0	0	0	0	0	0
M13	0	0	0	0	0	0	0	0	0	0	0	0	0	0	0	0	0	0	0	0	0	0	0	0	0	0	0	0	0	0	0	0	0	0	0	0	0	0	0	0	0	0	0	0	0	0	0	0	0	0
M14	0	0	0	0	0	0	0	0	0	0	0	0	0	0	0	1	0	0	0.5	0	0.8	0	0	0	0.9	0	0	0	0	0	0	0	0	0	0	0	0	0	0	0	0	0	0	0	0	0	0	0	0	0
M15	0	0	0	0	0	0	0	0	0	0	0	0	0	0	0	0	0	0.7	0.6	0	0	1	0	0	0	0.9	0.8	0	0	0	0	0	0	0	0	0	0	0	0	0	0	0	0	0	0	0	0	0	0	0
M16	0	0	0	0	0	0	0	0	0	0	0	0	0	0	0	0	0	0	0	0.7	0	0	0.7	0	0	0.8	0	0	0	0	0	0	0	0	0	0	0	0	0	0	0	0	0	0	0	0	0	0	0	0
M17	0	0	0	0	0	0	0	0	0	0	0	0	0	0	0	0	0.5	1	0	0	1	0	0	0.5	0	0.8	0	0	0	0	0	0	0	0	0	0	0	0	0	0	0	0	0	0	0	0	0	0	0	0
M18	0	0	0	0	0	0	0	0	0	0	0	0	0	0	0	0.7	0	0	0	0.9	0	0.9	0	0.5	0	0	0	0	0	0	0	0	0	0	0	0	0	0	0	0	0	0	0	0	0	0	0	0	0	0
M19	0	0	0	0	0	0	0	0	0	0	0	0	0	0	0	0.6	0.6	0.6	0	0	0	0.8	0	0	0.8	0	1	0	0	0	0	0	0	0	0	0	0	0	0	0	0	0	0	0	0	0	0	0	0	0
M20	0	0	0	0	0	0	0	0	0	0	0	0	0	0	0	0.7	0.8	0	1	0	0.6	0	0.6	0	0.6	0	0	0	0	0	0	0	0	0	0	0	0	0	0	0	0	0	0	0	0	0	0	0	0	0
M21	0	0	0	0	0	0	0	0	0	0	0	0	0	0	0	0	0	0	0	0	0	0	0	0	0	0	0	0.8	0.9	0.9	0	0	0	0	0	0	0	0	0	0	0	0	0	0	0	0	0	0	0	0
M22	0	0	0	0	0	0	0	0	0	0	0	0	0	0	0	0	0	0	0	0	0	0	0	0	0	0	0	0.7	1	0.9	0	0	0	0	0	0	0	0	0	0	0	0	0	0	0	0	0	0	0	0
M23	0	0	0	0	0	0	0	0	0	0	0	0	0	0	0	0	0	0	0	0	0	0	0	0	0	0	0	0	0	0	0.8	1	0.5	0.9	0	0	0	0	0	0	0	0	0	0	0	0	0	0	0	0
M24	0	0	0	0	0	0	0	0	0	0	0	0	0	0	0	0	0	0	0	0	0	0	0	0	0	0	0	0	0	0	0.7	0.6	0.7	0.9	0	0	0	0	0	0	0	0	0	0	0	0	0	0	0	0
M25	0	0	0	0	0	0	0	0	0	0	0	0	0	0	0	0	0	0	0	0	0	0	0	0	0	0	0	0	0	0	0	0	0	0	0.5	1	0.9	0.9	0.8	0	0	0	0	0	0	0	0	0	0	0
M26	0	0	0	0	0	0	0	0	0	0	0	0	0	0	0	0	0	0	0	0	0	0	0	0	0	0	0	0	0	0	0	0	0	0	1	0	0.7	0.9	0	0	0	0	0	0	0	0	0	0	0	0
M27	0	0	0	0	0	0	0	0	0	0	0	0	0	0	0	0	0	0	0	0	0	0	0	0	0	0	0	0	0	0	0	0	0	0	0	0	0	0	0	0.9	0.8	0.5	0.7	0.9	0.9	0.9	0	0	0	0
M28	0	0	0	0	0	0	0	0	0	0	0	0	0	0	0	0	0	0	0	0	0	0	0	0	0	0	0	0	0	0	0	0	0	0	0	0	0	0	0	0	0.9	0.9	0.8	0	0	0.8	0	0	0	0
M29	0	0	0	0	0	0	0	0	0	0	0	0	0	0	0	0	0	0	0	0	0	0	0	0	0	0	0	0	0	0	0	0	0	0	0	0	0	0	0	0	0	0	0	0	0	0	0.7	0.7	0.77	0
M30	0	0	0	0	0	0	0	0	0	0	0	0	0	0	0	0	0	0	0	0	0	0	0	0	0	0	0	0	0	0	0	0	0	0	0	0	0	0	0	0	0	0	0	0	0	0	0.8	0.6	0.96	0.53

Figure 13. Data set No. 24

i/j	P1	P2	P3	P4	P5	P6	P7	P8	P9	P10	P11	P12	P13	P14	P15	P16	P17	P18	P19	P20	P21	P22	P23	P24	P25
M1	0	0	0	0	0	0	0	0.5	1	0	0	0	0	0	0	0	0	0	0	0	0.8	0	0	0	0
M2	0	0	0	0	0	0	0	0	0	0	0	0	0	0	0	0	0.8	0	0.9	0.9	0	0	0	0	0
M3	0.8	0	0	0.6	0	0	0	0	0	0	0	0	0	0	0	0.9	0	0	0	0	0	0	0	0	0
M4	0	0	0	0	0	0	0	0	0	0	0	0	0	0	0	0	0.7	0	0	0	0	0.6	0	0	0
M5	0	0.6	0.6	0	0	0	0	0	0	0	0.5	0	0	0	0	0	0	0	0	0	0	0	0	0	0
M6	0	0.7	0	0	0	0	0	0	0	0	0	0	0	0	0	0	0.6	0	0	0	0	0.7	0	0	0
M7	0	0	0	0	0	0	0	0	0	0	0	0	0	0	0	0	0	0	0.9	0	0	0	0	0	0
M8	0.8	0	0	1	0	0	0	0	0	0	0	0	0	0	0	0	0	0	0	0	0	0	0	0	0
M9	0	0	0	0	0	0	0	0.9	0	0	0	0	0	0	0	0	0	0	0	0	0.7	1	0	0	0
M10	0	0.9	0.5	0	0	0	0	0	0	0	0	0	0	0	0	0	0	0	0	0	0	0	0.5	0.5	0
M11	0	0	0	0	0	0	0	0.6	0.5	0	0	0	0	0	0	0	0	0	0	0	0	0	0.7	0	0.7
M12	0	0	0	0	0	0	0	0	0	0	0	0	0	0	0	0	0	0	0.9	0.7	0	0	0.6	0.7	0.9
M13	0	0.6	0.8	0	0	0	0	0.6	1	0	0.5	0	0	0	0	0	0	0	0	0	0	0	0	0	0
M14	0	0	0	0	0	0	0	0	0	0	0	0	0	0	0	0	0	0	0.6	1	0	0	0	0	0
M15	0	0	0	0	0	0	0	0	0	0.6	0	0.6	0	0	0	0	0	0.7	0	0	0	0	0	0	0
M16	0	0	0	0	0	0	0	0	0	0	0	0	0	0	0	0	0	0	0	0	0.5	0	0	0	0
M17	0	0	0	0	0.9	0	0.8	0	0	0	0	0.5	0	0	0	0	0	0	0	0	0	0	0	0	0
M18	0	0	0	0	0.9	0.9	0	0	0	0	0	0	0	0	0	0	0	0	0	0	0	0	0	0	0
M19	0.5	0	0	0	0	0	0	0	0	0	0	0	0	0	0	0	0	0	0	0	0.7	0	0	0	0
M20	0	0	0	0	0.9	0.8	0.6	0	0	0	0	0.6	0	0	0	0	0	0	0	0	0	0	0	0	0
M21	0	0	0	0	0.9	0	0.8	0	0	0	0	0	0	0.8	0	0	0	0	0	0	0	0	0	0	0
M22	0	0	0	0	0	0	0	0	0	0	0	0	0.9	0	0.7	0	0	0	0	0	0	0	0	0	0
M23	0	0	0	0	0	0.6	0	0	0	0	0	0	0	0	0	0	0	0	0	0	1	0	0	0	0
M24	0	0	0	0	0	0	0	0	0	0	0	0	0.8	0	0.5	0	0	0	0	0	0	0	0	0	0
M25	0.6	0	0	0.8	0	0	0	0	0	0	0	0	0	0	0	0.7	0	0	0	0	0	0	0	0	0
M26	0	0	0	0	0	0	0	0	0	0.6	0	0	0.6	0	0	0	0.8	0	0	0.8	0	0	0	0	0
M27	0	0	0	0	0	0	0	0	0	0	0	0	1	0	0.6	0	0	0	0	0	0	0	0	0	0
M28	0	0	0	0	0	0	0	0	0	0.8	0	0	0	0.7	0	0	0	0.6	0	0	0	0	0	0	0
M29	0	0	0	0	0	0	0	0	0	0	0	0	0	0	0	0	0.7	0	0	0	0	0	0.9	0.6	0.6
M30	0	0	0	0	0	0	0	0	0	0	0	0	0	0	0	0.5	0	0	0	0	0	0	0	0	0

i/j	P26	P27	P28	P29	P30	P31	P32	P33	P34	P35	P36	P37	P38	P39	P40	P41	P42	P43	P44	P45	P46	P47	P48	P49	P50
M1	0	0	0	0	0	0	0	0	0.9	0	0	0	0	0	0	0	0	0.8	0	0	0	0	0	0	0
M2	0	0	0	0	0	0	0	0	0	0	0	0	0.9	0	0	0	0	0	0	0	1	0	0	0	0
M3	0	0	0	0	0	0	0	0.5	0.5	0	0	0	0	1	0	0	0	0	0	0	0	0	0	0.83	0
M4	0	0	0	0	0	0	0.5	0	0.5	0	0	0	0	0	0	0	0	0.7	0	0	0	0	0	0	0
M5	0	0	0	0	0	0	0	0	0	0	0	0	0	0	0	0	0	0	0	0	0	0	0	0	0
M6	0	0	0	0	0	0	0	0	0	0	0	0	0	0.5	0	0	0	0	0	0	0	0	0.6	0	0
M7	0	0	0.7	0.7	0.7	0	0	0	0	0	0	0	0	0	0	0.8	0	0	0	0	0	0	0	0	0
M8	0	0	0	0	0	0	0	0.9	0	0	0	0	0	0	0	0	0.8	0	0	0.8	0	0	0	0	0
M9	0	0	0	0	0	0	0	0	0	0	0.9	0	0	0	0	0.9	0	0	0	0	0	0	0	0.7	0
M10	0	0	0	0	0	0	0	0	0	0	0	0.8	0	0.8	0	0	0	0	0	0	1	0.6	0	0	0
M11	0	0	0	0	0	0	0	0	0	0	0.9	0	0	0	0	1	0	0	0	0	0	0	0	0	0
M12	0	0	0	0	0	0	0	0	0	0	0	0	0	0	0	0	0	0	0.8	0	0	0	0	0	0
M13	0	0	0	0	0	0	0	0	0	0	0	0	0	0	0.5	0	0	0	0.9	0	0	0	0	0	0
M14	0	0	0	0	0	0.9	0.9	0	0	0	0	0	0	0	0	0	0	0	0.8	0	0	0	0.7	0	0
M15	0	0	0	0	0	0	0	0	0	0	0	0	0	0	0.9	0	0	0	0	0	0.7	0	0	0	0.98
M16	0	0	0	0	0	0.7	0.9	0	0	0.5	0	0	0	0.8	0	0	0	0	0	0	0	0	0	0	0
M17	0	0	0.9	0	0.8	0	0	0	0	0	0	0.5	0	0	0	0	0	0	0.7	0	0	0	0	0	0.92
M18	0.8	0.7	0	0	0	0	0	0	0	0	0	0	0	0	0	0	0	0	0	0	0	0	0	0	0
M19	0.6	0	0.5	0	0.7	0	0	0	0	0	0	0	0	0	0	0	0	0.9	0	0	0.7	0.7	0	0	0.77
M20	1	0.5	0	0.9	0	0	0	0	0	0	0	0	0	0	0	0	0	0	0	0	0	0	0	0	0
M21	0	0	0	0	0	0	0	0	0	1	0	0	0	0.5	0	0	0	0	0	0.8	0	0	0	0	0
M22	0	0	0	0	0	0	0	0	0	0	0	0	0	0	0	0	0	0	0	0	0	0	0	0	0
M23	0	0	0	0	0	0.9	0	0	0.8	0	0	0	0	0.7	0	0	0	0	0	0	0.7	0	0	0	0
M24	0	0	0	0	0	0	0	0	1	0	0	0	0	0	0	1	0	0	0	0	0	0	0	0	0
M25	0	0	0	0	0	0	0	0.9	0	0	0.9	0	0	0	0.5	0	0	0	0	0.6	0	0	0	0	0
M26	0	0	0	0	0	1	0	0	0.7	0.8	0	0	0	0	0	0	0	0	0	0	0	0	0	0	0
M27	0	0	0	0	0	0	0	0	0.6	0	0	0	0	0	0	0	0	0.6	0	0	0	0	0	0	0
M28	0	0	0	0	0	0	0	0	0	0	0	0	0	0	0.8	0	0	0	0	0.8	0	0	0	0	0
M29	0	0	0	0	0	0	0	0	0	0	0	0	0	0	0	0	0	0	0	0	0	0	0	0	0
M30	0	0	0	0	0	0	0	0	0	0.5	0	0	0	0.8	0	0	0	0	0	0	0	0	0	0	0

Chapter 6
Design of Cellular Manufacturing System Using Non-Traditional Optimization Algorithms

P. Venkumar
Kalasalingam University, India

K. Chandra Sekar
Sardar Raja College of Engineering, India

ABSTRACT

For the last many years a lot of study has been done on design of Cellular Manufacturing System (CMS). Cellular Manufacturing is an application of Group Technology (GT) philosophy in which similar parts are identified and grouped together to take advantage of their similarities in design and manufacturing. The design of CMS involves three stages i) grouping of parts and production equipments into cells (Cell Formation), ii) allocation of the machine cells to the areas within the shop floor and iii) layout of the machines within each cell. In recent years non-traditional optimization algorithms/techniques have fascinated scientists and engineers all over the world. Particularly in complex dynamic environments, these algorithms/techniques are needed to explore beyond the vicinity of existing knowledge. These algorithms have the ability to think and learn from own experience. They are called Meta heuristics because they perform considerable search before terminating to provide a good solution to the problem. Popular Meta heuristics are Genetic Algorithms (GA), Simulated Annealing (SA), Tabu Search (TS), Artificial Neural Networks (ANN), Artificial Immune System (AIS), and Sheep Flock Heredity Algorithm (SFHA). In this chapter the implementation of Meta heuristics for the design of Cell Formation problem in CMS is discussed.

DOI: 10.4018/978-1-61350-047-7.ch006

INTRODUCTION

Cellular manufacturing system (CMS) is synonymous with terms such as group technology (GT), cell system and cellular production systems. Cellular manufacturing system is an application of group technology, which is a solution to the problems of batch manufacturing. Here, machines are divided into cells and components are divided into the same number of families in such a way that all the components in each family can be completely processed by a particular cell. The component families are formed based on their design and production characteristics. The design of CMS involves three stages i) grouping of parts and production equipments into cells, ii) allocation of the machine cells to the areas within the shop floor and iii) layout of the machines within each cell.

Group technology ideas were first systematically presented by Burbidge (1963) following the pioneering work of Mitrofanov (1959). The literature on cell formation can be broadly classified in two ways – one based on techniques used for cell formation and other one the way the cell formation problem is modeled. Crama and Oosten (1996) made a study on various models available for Cell Formation problems (CFP). The concept of production flow analysis was introduced by Burbidge (1963). The aim of the technique as stated by Burbidge (1971) is finding the families of components and associated groups of machines for group layout by a progressive analysis of the information in route cards. The main disadvantage with implementation of PFA was the manual work involved in grouping parts and machines. Burbidge (1975) introduced a holistic approach to GT called Production Flow Analysis. It discussed the production situation and recommended a systematic solution to the problems of batch production. Burbidge (1977) introduced a two dimensional representation with a tick mark used to indicate the visit of a component to a machine. The method uses hand computations, which limits its applicability.

The array-based clustering methods are based on the Part-Machine Incidence Matrix (PMIM). In the PMIM, rows and columns indicate machines and parts, respectively. Each column of PMIM is an array of "0–1" numbers, which indicates the set of machines that produce each part. The well-known array-based clustering methods are: the Bond Energy Algorithm (BEA) by McCormick et al. (1972), the Rank Order Clustering method (ROC) by King (1980) and the direct clustering algorithm (DCA) by Chan and Milner (1982). These methods, group parts and machines are regardless of the production volume, operational sequences, production cost, inventory and other limitations in the production system, which is a main problem for them.

The hierarchical clustering based methods are defined by an input data set to determine similarity or distance function and determine a hierarchy of clusters or partitions. The Single Linkage Clustering (SLC) dendogram proposed by McAuley (1972) uses measure of similarity between machines. This model, that is based on the mathematical coefficient uses the distance matrix to determine machine groups. But the major cause for drawback of SLC is the "chaining problem". Therefore, researchers improved it by adding more inputs, such as production volumes (proposed by Seifoddini (1989) who uses Average Linkage Clustering (ALC) algorithm). Yasuda and Yin (2001) proposed a method on dissimilarity measure for solving CFP.

Graph partition approach represents the machines as vertices and the similarity between machines as the weights of the arcs. Rajagopalan and Batra (1975) suggested the use of graph theory to form machine groups. Chandrasekaran and Rajagopalan (1986a) proposed an ideal seed nonhierarchical clustering algorithm for cellular manufacturing. Ballakur and Steudel (1987) developed graph searching algorithms which select a key machine or component according to a pre-specified criterion. Vohra et al (1990) presented a non-heuristic network approach to from manufacturing cells with minimum intercellular interactions. Srinivasan (1994) presented an approach using minimum spanning tree for the machine cell formation problem. A minimum spanning tree for machines is constructed and the seeds to cluster components are generated from this tree. Veeramani and Mani (1996) described a polynomial-time algorithm based on a graph theoretic approach for optimal cluster formation called as vertex-tree graphic matrices.

Mathematical programming approaches are classified under integer programming (Kusiak 1987, Co and Araar 1988), dynamic programming (Ballakur and Steudel 1987), goal programming (Shafer and Rogers, 1993a), and linear programming (Boctor 1991). Kusiak (1987) developed clustering problem known as p-median model. Choobineh (1988) uses a sequential approach forming part families in the first stage and then a cost based mathematical programming method to allocate machines to part families to form cells. Rajamani et al. (1990) developed integer programming models to form cells sequentially as well as simultaneously. Solimanpur et al. (2004) presented the inter-cell layout problem that was discussed a mathematical formulation for material flow between the cells. The problem is modeled as a Quadratic Assignment Problem (QAP). Zahir Albadawi et al (2005) have developed a mathematical model using eigen value matrix for cell formation problems.

Above said conventional approaches for cell formation are focused on reducing exceptional elements and computational burden using zero-one PMIM. The major limitations of these approaches lie in the fact that real life production factors like operational time, sequence of operations, lot size of the parts etc. are not considered resulting in inefficient cells. Since cell formation problems are non-polynomially complete in nature (Nair and Narendran 1999), it is difficult to obtain solutions that satisfy all constraints. Therefore, it is expected to make use of simple but efficient computing techniques called heuristic algorithms.

Heuristic algorithms are used to provide quick approximate solutions to "hard" combinatorial optimization problems. They do not guarantee optimal solutions but can give the optimum sometimes. They are used only when the problem is to be solved which belongs to the NP category, which means that there is no algorithm that exists which can solve the problem optimally using polynomial time algorithms. Polynomial time algorithms can solve the given problem with polynomial times that are polynomial functions of problem parameters.

Heuristic algorithms are also called approximate algorithms that run in polynomial time and provide quick and acceptable solutions. A heuristic algorithm is called an approximate algorithm where the performance of the heuristic is assessed in terms of worst and average case behavior. Heuristics can either be generic or problem specific. Generic heuristics can be applied to any hard problem and guarantee a certain level of performance. Heuristics can also be classified by construction heuristics or improvement heuristics depending on whether they are constructed from the problem or whether they are improved by an existing heuristic solution.

These are general heuristics that have been developed in the last two decades. They are called Meta heuristics because they perform considerable search before terminating to provide a good solution to the problem. Popular Meta heuristics are Genetic Algorithms (GA), Simulated Annealing (SA), Tabu Search (TS), Artificial Neural Networks (ANN), Artificial Immune System (AIS) and Sheep Flock Heredity Algorithm (SFHA).

In this chapter, the following Meta heuristics, which have been reportedly successful in solving a wide variety of search and optimization problems in sciences, engineering, and commerce, are described:

1. Genetic Algorithm (GA)
2. Simulated Annealing Algorithm (SA)

3. Tabu Search (TS)
4. Artificial Neural Networks (ANN)
5. Artificial Immune System (AIS)
6. Sheep Flock Heredity Algorithm (SFHA)

The remainder of this chapter is organized as follows. The next section presents the representation of CMS problem. Section three describes the implementation of GA to CMS problem. Section four presents the implementation of SA to CMS problem. Section five describes the implementation of TS to CMS problem. Section six describes the implementation of ANN to CMS problem. Section seven presents the implementation of AIS to CMS problem. Section eight describes the implementation of SFHA. Experimentation of Meta heuristic algorithms on CMS problems and comparison of results were discussed in section nine. The final section includes some general conclusions and discussions for the future work prospects.

PROBLEM REPRESENTATION

The problem in cellular manufacturing consists of information from the route card, arranged in the form of one-zero matrix. The column of the matrix represents components and the row of the matrix represents machines. This one-zero matrix is also called as incidence matrix (See Figure 1). A entry of '1' in (i,j)th position indicates that the ith machine is used for processing the jth component. If the entry is 'zero' or 'blank', it indicates that the component does not require that machine for processing.

Representation plays a key role in the development of algorithms. A problem can be solved once it is represented in the form of solution string. In the problem, each gene represent whether the machine or part is in that cell or not.

Each cell or the string is coded as seen in Table 1. Decoded information is shown in Table 2.

After representation, initial solution can be generated using a random numbers and the generated solution is subjected to iterations or generations.

GENETIC ALGORITHM

Introduction

Genetic Algorithm (GA) is computerized search and optimization algorithms based on the mechanics of natural genetics and natural selection. GA

Figure 1. Incidence matrix (Boctor, 1991)

MACHINES \ Parts	1	2	3	4	5	6	7	8	9	10	11	12	13	14	15	16	17	18	19	20	21	22	23	24	25	26	27	28	29	30
1	1			1			1			1			1																	
2						1		1						1	1			1		1		1		1	1	1				
3		1	1				1			1					1						1									1
4		1	1	1			1					1						1			1		1					1	1	1
5	1					1		1		1								1		1					1				1	
6					1	1				1					1		1		1		1	1		1				1		
7		1	1		1		1	1				1		1	1	1	1					1	1	1		1				
8			1									1			1		1	1			1			1			1		1	
9	1			1																	1		1						1	
10					1			1						1				1	1	1				1		1				1
11		1			1		1		1					1	1				1		1	1				1				
12	1	1									1		1		1			1		1			1		1		1		1	
13		1			1				1				1	1		1	1			1		1	1						1	1
14	1	1							1	1	1					1			1	1									1	
15					1			1											1		1						1			1
16	1									1	1	1																		

Table 1.

CELL NO	MACHINES LIST	COMPONENT LIST
CELL 1	0010001010011100	110110001111101011111101011010
CELL 2	1101110101100011	001001110000010100000010100101

Table 2.

CELL NO	MACHINES LIST	COMPONENT LIST
CELL 1	3,7,9,12,13,14	1,2,4,5,9,10,11,12,13,15,17,18,19,20,21,22,24,26,27,29
CELL 2	1,2,4,5,6,8,10,11, 15,16	3,6,7,8,14,16,23,25,28,30

is a search technique for global optimization in a search space. As the name suggests they employ the concepts of natural selection and genetics using past information. They direct the search with much expected improved performance and achieve fairly consistent and reliable results. The traditional methods of optimization and search don't fare well over a broad spectrum of problem domain.

GA attempts to mimic the biological evolution process for discovering good solutions. They are based on a direct analogy to Darwinian natural selection and mutations in biological reproduction and belong to a category of heuristics known as randomized heuristics that employ randomized choice operators in their search strategy and do not depend on complete prior knowledge of the features of the domain. These operators have been conceived through abstractions of natural genetic mechanisms such as crossover and mutation and have been cast into algorithmic forms. Holland envisaged the concept of these algorithms in the mid-sixties since then. It has been applied in diverse areas such as music generation, genetic synthesis, strategy planning and also to address business problems such as traveling salesman problem, production planning and scheduling problem, facility location problem and cell design problems. Among these problems, cell design problem is reported that it is equivalent to two traveling salesman problems (Lenstra, 1974). GAs is different from traditional optimization and search technique in the following ways (Goldberg, 1989).

1. It works with a coding of parameters, not with parameter themselves.
2. GA searches from population of points, not a single point.
3. GA uses information of fitness function not derivatives or other auxiliary knowledge.
4. GA uses probabilistic rules rather than deterministic rules.

Underlying Principles of GA

To really appreciate the technique, the analogy to the biological systems must be understood. Moreover, the GA uses many of the same terms which biologists use. The nature of a living organism is described by the specific structure of the DNA molecules, which are present in the cell. The DNA is really information coded chemically and can be thought as of very long strings of bits. One such string is called chromosome and each bit is called a gene.

When a cell in an organism reproduces, it first duplicates its DNA. The cell reproduced may be just like the parent or it may be different. The variation is introduced by two factors.

GA Operators

The operation of GA begins with a population of random strings representing design and decision variables. Thereafter each string is evaluated to find the objective value. The population is then operated by 3 main operators- reproduction, crossover and mutation. The new population is further evaluated and tested for termination. If the termination criterion is met, GA process stops otherwise the above cycle is followed until the termination criterion met. One such cycle is called a generation.

Reproduction

Reproduction is usually the first operator applied on population. Reproduction selects good strings in a population and forms a mating pool. That is why the reproduction operator is sometimes known as the selection operator. There exist a number of reproduction operators in GA literature and Rank selection method is used for reproduction. The individuals in the population are ranked according to fitness, and the expected value of each individual depends on its rank rather than on its absolute fitness.

Ranking avoids giving for the largest share of offspring to a small group of highly fit individuals, and thus reduces the selection pressure when the fitness variance is high. It also keeps up selection pressure when the fitness variance is low: the ratio of expected values of individuals is ranked i+1 where i will be the same although their absolute fitness difference are high or low. The linear ranking method proposed by Baker is as follows: Each individual in the population is ranked in increasing order of fitness from 1 to N. The expected value of each individual 'i' in the population at time 't' is given by

$$\mathit{Expected\ Value}(i,t) = (\min) + (\max - \min) \times \frac{rank(i,t) - 1}{N - 1}$$

Where

N = Sample Size,
Min = 0.4,
Max = 1.6

After calculating the expected value of each rank, reproduction is performed using Monte Carlo simulation by employing random numbers.

Reproduction will select against those strings which are in subsequent generations. n strings, n random numbers between zero and one are created at random. Then a string that represents a chosen random in the cumulative probability range for the string is copied to the mating pool. By this way, the string with a higher fitness value will represent a larger range in the cumulative probability values and therefore has a higher probability of being copied into the mating pool. On the other hand, a string with a smaller fitness value represents a smaller range in cumulative probability values and has a smaller probability of being copied into the mating pool. This is the reproduction operator.

Crossover

In the crossover, new strings are created by exchanging information among strings of the mating pool. In crossover operator two strings are picked from the mating pool at random and some portions of the strings are exchanged between the strings. The two strings participating in the crossover operation are known as parent strings and the resulting strings are known as child strings.

In single point crossover, one crossover point is selected, binary string from beginning of chromosome to the crossover point is copied from one parent, and the rest is copied from the second parent Single point crossover operation is explained in Figure 3.

In Two point crossover, two crossover points are selected, binary string from beginning of chromosome to the first crossover point is copied from one parent, the part from the first to the

Figure 2. Block diagram of GA

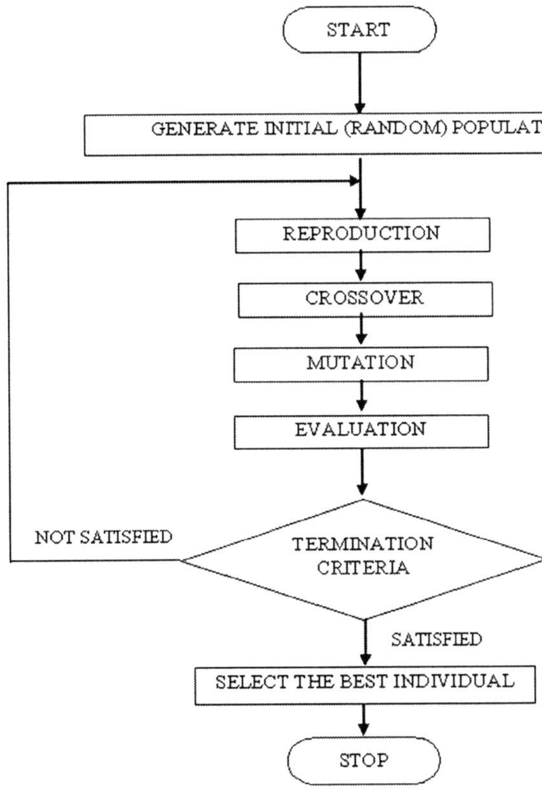

Figure 3. Single point crossover

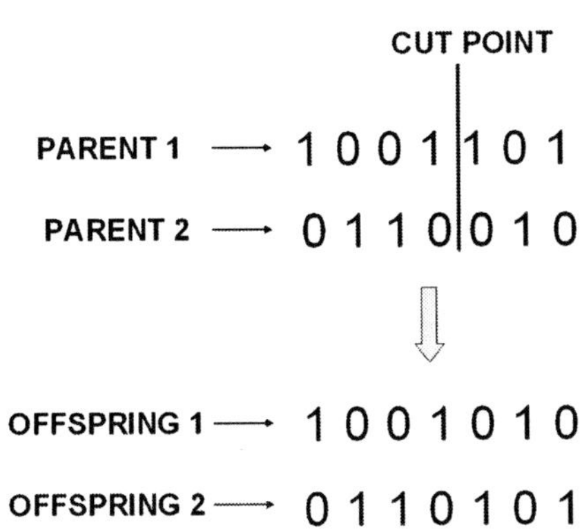

Figure 4. Two point crossover

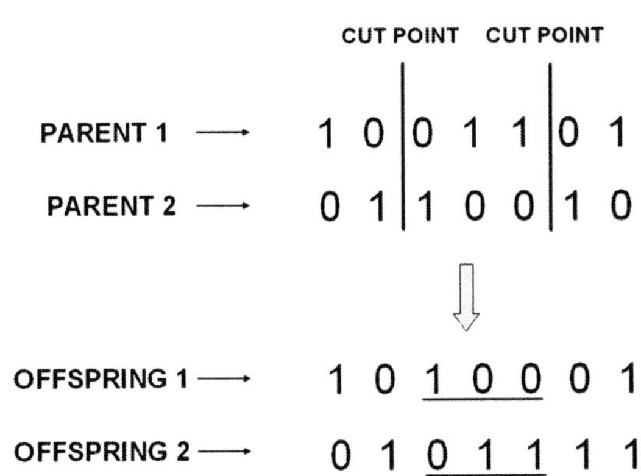

second crossover point is copied from the second parent and the rest is copied from the first parent.. Two-point crossover operation is explained in Figure 4.

As told, GA needs an initial population as the input. Then the GA operates on the population and derives a new population in one generation. Here the new population may contain bad solutions also and these are accepted. This result in

Figure 5. Inversion mutation

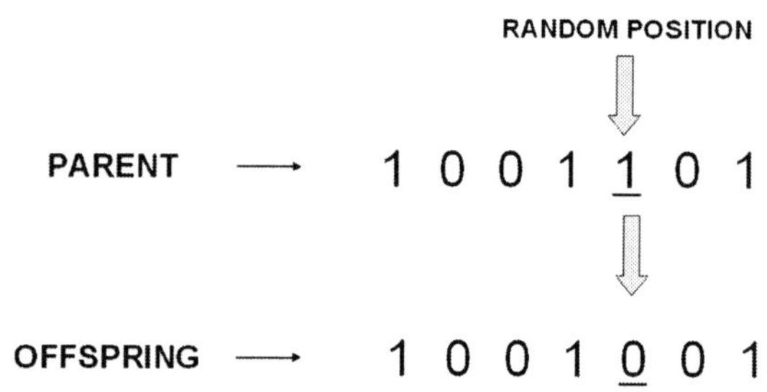

Figure 6. Exchange mutation

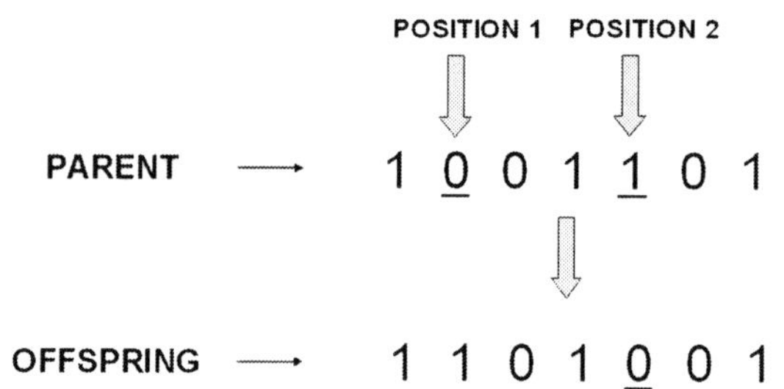

finding out the global optimum rather than getting bogged down at a local optimum. Thus GA is an effectively and efficiently search through a complex search space.

Mutation

Mutation is also done randomly for each gene and it depends upon another parameter called mutation probability. Here one gene is selected at random and the mutation operation is performed. The mutation operation may consist of any of four operators given below.

1. Shifting: It is nothing, but a gene or a machine going out from one cell and residing in another cell.
2. Inversion: In this, a machine comes out from one cell and goes to another cell, while a machine from latter cell comes to the former cell (Figure 5).
3. Creating is a process in which a machine goes from one cell and creates a new cell and resides there.
4. Exchange: In this, two sites selected randomly, swap operation carried out between these two sites (Figure 6).

Implementation

In GA a candidate solution represented by sequence of genes called chromosome. A judiciously selected set of chromosomes is called population and the population is subjected to generations.

Reproduction

In reproduction, an objective function value is computed for each string in the population and the objective is to find a string with the maximum objective function value. Since objective is minimization it is required to map it inversely and then maximize the resultant. Ranking avoids giving for the largest share of offspring to a small group of highly fit individuals, and thus reduces the selection pressure when the fitness variance is high. It also keeps up selection pressure when the fitness variance is low: the ratio of expected values of individuals ranked i+1 and i will be the same whether their absolute fitness difference are high or low. The linear ranking method proposed by Baker is as follows: Each individual in the population is ranked in increasing order of fitness from 1 to N. The expected value of each individual 'i' in the population at time 't' is given by (1).

After calculating the expected value of each rank, reproduction is performed using Monte Carlo simulation by employing random numbers. An example for Reproduction operation is given in Table 3.

Table 3. Reproduction

RANK	PROBABILITY	CUM PROBABILITY	RANDOM NUMBER	NEW RANK
1	0.02	0.02	174	7
2	0.023	0.043	177	7
3	0.027	0.07	774	17
4	0.03	0.1	662	16
5	0.033	0.133	142	6
6	0.036	0.169	684	16
7	0.039	0.208	269	9
8	0.042	0.25	851	18
9	0.046	0.3	111	5
10	0.049	0.345	165	6
11	0.052	0.397	264	9
12	0.055	0.452	952	20
13	0.058	0.51	678	16
14	0.061	0.571	973	20
15	0.064	0.635	732	16
16	0.068	0.703	752	16
17	0.071	0.774	640	16
18	0.074	0.848	258	9
19	0.077	0.925	454	13
20	0.08	1	616	15

Crossover

The crossover operator is carried out with a probability known as crossover probability (P_c). Crossover is nothing but exchange of a portion of strings at a paint called crossover site. The two strings, which take part in the crossover operation, are also selected at random. Here partial mapped crossover is performed i.e., crossover site is selected and the genes of one string between the sites are swapped with another string.

Mutation

Mutation is also done randomly for each gene and it depends upon another parameter called mutation probability (P_m). Here one gene is selected at random and the mutation operation is performed. Thus the genes are mutually interchanged.

Problem and Rectification

Sometimes the crossover and mutation results in the formation of an empty cell or violates some constraint. Sometimes crossover or mutation may result in the formation of a chromosome, which has only one machine. But the constraint is each cell should have eight machines. In such a case, the crossover operation is carried out on that chromosome until it satisfies the constraint.

Acceptance of Strings

In classical genetic algorithm, the population obtained after reproduction. Crossover, and mutation is accepted without any question. This new population forms the input for the next generation. Thus the process continues until the termination condition is reached.

The trouble with the classical GA is due to accepting inferior solutions, from one generation to another. In this process, the loss of good strings, are high.

One more important feature is that GA must accept bad solutions also, so that it can search for global solution rather than getting trapped in local solution. So care should be taken to accept bad solution also.

SIMULATED ANNEALING ALGORITHM

Introduction

Simulated Annealing is a combinatorial optimization technique based on random evaluations of the objective function. It resembles the cooling process of molten metals through annealing. The name of the method is derived from the simulation of thermal annealing of critically heated solids. A slow and controlled cooling of a heated solid ensures proper solidification with a highly ordered, crystalline state that corresponds to the lowest internal energy. Rapid cooling causes defects inside the material.

The quality of the final solution is not affected by the initial guesses; expect that the computational effort may increase with worse starting designs.

Because of the discrete nature of the function and constraint evaluations, the convergence or transition characteristics are not affected by the continuity or differentiability of the functions.

The convergence is also not influenced by the convexity status of the feasible space.

The design variables need not be positive.

The method can be used to solve mixed-integer, discrete, or continuous problems.

For problems involving behavior constraints (in addition to lower and upper bounds on the design variables), an equivalent unconstrained function is to be formulated as in the case of genetic algorithms.

Underlying Principles of SA

The simulated annealing procedure simulates this process of annealing to achieve the minimum function value in a minimization problem. Controlling a temperature – like parameter introduced with the concept of the Boltzmann probability distribution simulates the slow cooling phenomenon of annealing process. According to the Boltzmann probability distribution, a system in thermal equilibrium at a temperature T has its energy distributed probabilistically according to P(E) = exponent of (-ΔE/kT), where " k " is the Boltzmann constant. This expression suggests that a system at a high temperature has almost uniform probability of being at any energy state, but at a low temperature it has a small probability of being at a high-energy state. Therefore, controlling the temperature "T" and assuming that the search process follows the Boltzmann probability distribution can control the convergence of an algorithm. Metropolis et al (1953) suggested one way to implement the Boltzmann probability distribution in simulated thermodynamic systems. The same can be found in the function minimization context. Let us say, at any instant the current point is x^t and the function value at that point is $E(t) = f(x^t)$. Using Metropolis algorithm, we can say that the probability of the next point being at x^{t+1} depends on the difference in the function values at these two points or on $(\Delta E = E(t + 1) - E(t)$ and is calculated using the Boltzmann probability distribution:

$$P(E(t + 1)) = \min [1, \exp(-\Delta E/kT)]$$

If $\Delta E \leq 0$, this probability is one and the point x^{t+1} is always accepted. In the function minimization context, this makes sense because if the function value at x^{t+1} is better than that at x^t, the point x^{t+1} must be accepted. The interesting situation happens when $\Delta E > 0$, which implies that the function value at x^{t+1} is worse than that at x^t. According to Metropolis algorithm, there is some finite probability of selecting the point x^{t+1} even though it is a worse than the point x^t. However, this probability is not same in all situations. This probability is depends on relative magnitude of ΔE and T values. If the parameter T is large, this probability is more for points with largely disparate function values. Thus any point is almost acceptable for a larger value of T. On the other hand, if the parameter T is small, the probability of accepting an arbitrary point is small. Thus for small values of T, the points with only small deviation in function value are accepted.

The above procedure can be used in the function minimization of exceptional elements in cell formation problem. The algorithm begins with an initial point x^1 (v^1, f^1) and a high temperature T. A second point x^2 (v^2,f^2) is created at random in the vicinity of the initial point and the difference in the function values (ΔE) at these two points is calculated. If the second point has a smaller function value, the point is accepted; otherwise the point is accepted with a probability exp $(-\Delta E/T)$. This completes one iteration of the simulated annealing procedure. In the next generation, another point is created at random in the neighbourhood of the current point and the Metropolis algorithm is used to accept or reject the point. In order to simulate the thermal equilibrium at every temperature, a number of points are usually tested at a particular temperature, before reducing the temperature. The algorithm is terminated when a sufficiently small temperature is obtained or a small enough change in function values is found.

SA Operators

Using one of following procedures may create neighbourhood:

1. Inversion at one site: In this, a machine comes out from one cell and goes to another cell, while a machine from latter cell comes to the former cell.
2. Inversion at Two sites: In this, a machine and a part comes out from one cell and goes to another cell, while a machine from latter cell comes to the former cell.
3. Exchange: In this, two sites selected randomly, swap operation carried out between these two sites.
4. Single Point Crossover: one crossover point is selected, binary string from beginning of chromosome to the crossover point is copied from one parent, the rest is copied from the second parent

SA ALGORITHM

Notations

In this section, the various notations used are described.

C - counter
T - Temperature
r - repetition counter
CMAX - number of iterations to be performed a particular Temperature.
RMAX - Maximum repetition allowed.
α - Cooling rate
β - Reduction rate in Repetition
Z (current) - Objective function value for current point
Z (best) - Best Objective function value

Implementation

In SA a candidate solution represented by sequence of genes called initial string.

A simple Simulated Annealing Algorithm

```
Call initial (to obtain an initial solution)
     Store it as the current solution
     Store it as the best solution found so far
     C:=0 (initialize the stopping counter)
     Repeat until c=CMAX
               R:=RMAX
               r:=0 (initialize the repetition counter)
               T:=α*T (reduce the cooling temperature)
          Repeat until r=R
               Call neighbor (generate a neighbor solution)
                    δ:=Z (neighbor) - Z (current)
               if δ≤0 or random (0,1) ≤exp(-δ/T) then
               store the neighbor solution as the current one
                    if Z (current) < Z(best) then
                     store the current solution as the best found
                    c:=0 (reinitialize the stopping counter)
                    r:=0 (reinitialize the repetition counter)
                    R:=β*R (reduce the number of repetition)
                    End if
               End if
          End repeat
     End repeat
End
```

Neighbourhood Generation

The neighbourhood strings can be generated by inversion at one site, inversion at two sites, exchange and single point crossover methods.

Problem and Rectification

Sometimes the neighbourhood creation procedure results in the formation of an empty cell or violates some constraint. Sometimes neighbourhood creation may result in the formation of a point, which has only one machine. But the constraint is each cell should have eight machines. In such a case, the neighbourhood creation procedure is carried out on that string until it satisfies the constraint.

TABU SEARCH

Introduction

Tabu search is a Meta strategy for guiding known heuristics to overcome local optimality. Although still in its infancy, this Meta heuristic has been reported in the literature during the last few years as providing successful solution approaches for a great variety of problem areas. Tabu search (TS) has its antecedents in methods designed to cross boundaries of feasibility or local optimality standard treated as barriers, and to systematically impose and release constraints to permit exploration of otherwise forbidden regions.

The philosophy of tabu search is to derive and exploit a collection of principles of intelligent problem solving. A fundamental element underlying tabu search is the use of flexible memory. From the standpoint of tabu search, flexible memory embodies the dual processes of creating and exploiting structures for taking advantage of history (hence combining the activities of acquiring and profiting from information). TS methods operate under the assumption that a neighborhood can be constructed to identify "adjacent solutions" that can be reached from any current solution.

Underlying Principles of TS

Many solution approaches are characterized by identifying a neighbourhood of a given solution, which contains other so-called transformed solutions that can be reached in a single iteration. A transition from a feasible solution to a transformed feasible solution is referred to as a move. A starting point for tabu search is to note that such a move may be described by a set of one or more attributes (or elements), and these attributes (properly chosen) can become the foundation for creating an attribute based memory.

For example, in a zero-one integer-programming context these attributes may be the set of all possible value assignments (or changes in such assignments) for the binary variables. Then two attributes e and %, which denote that a certain binary variable is set to 1 or 0, may be called complementary to each other. Considering the number of attributes representing a move we may distinguish single-attribute moves (where every move is described by exactly one attribute) and multi-attribute moves (where every move may be described by more than one attribute).

Following a steepest descent / mildest ascent approach, a move may either result in a best possible improvement or a least possible deterioration of the objective function value. Without additional control, however, such a process can cause a locally optimal solution to be re-visited immediately after moving to a neighbour, or in a future stage of the search process, respectively. To prevent the search from endlessly cycling between the same solutions, the attribute-based memory of tabu search is structured at its first level to provide a short-term memory function, which may be visualized to operate as follows.

Imagine that the attributes of all explored moves are stored in a list, named a running list, representing the trajectory of solutions encountered. Then, related to a sub list of the running list a so-called tabu list may be introduced. Based on certain restrictions the tabu list implicitly keeps track of moves (or more precisely, salient features of these moves) by recording attributes complementary to those of the running list. These attributes will be forbidden from being embodied in moves selected in at least one subsequent iteration because their inclusion might lead back to a previously visited solution. Thus, the tabu list restricts the search to a subset of admissible moves (consisting of admissible attributes or combinations of attributes). The goal is to permit "good" moves in each iteration (Figure 7) without re-visiting solutions already encountered.

TS Operators

1. **Move Attribute:** The pair of sites of the string being swapped.
2. **History Record:** List of sites of the string classified under Tabu restrictions with record of on which iterations each of them classified as Tabu.
3. **Tabu Classification/Restriction:** The sites of string swapped in the previous iterations will not be considered for swapping
4. **Tabu Tenure:** The tabu restriction of a site is lifted after a consecutive three iterations
5. **Aspiration criterion:** Tabu restrictions are lifted for the solutions under tabu classification, with the value of the Objective, 10 percent or more, less than that of the current solution

Figure 7. Iteration in Tabu search

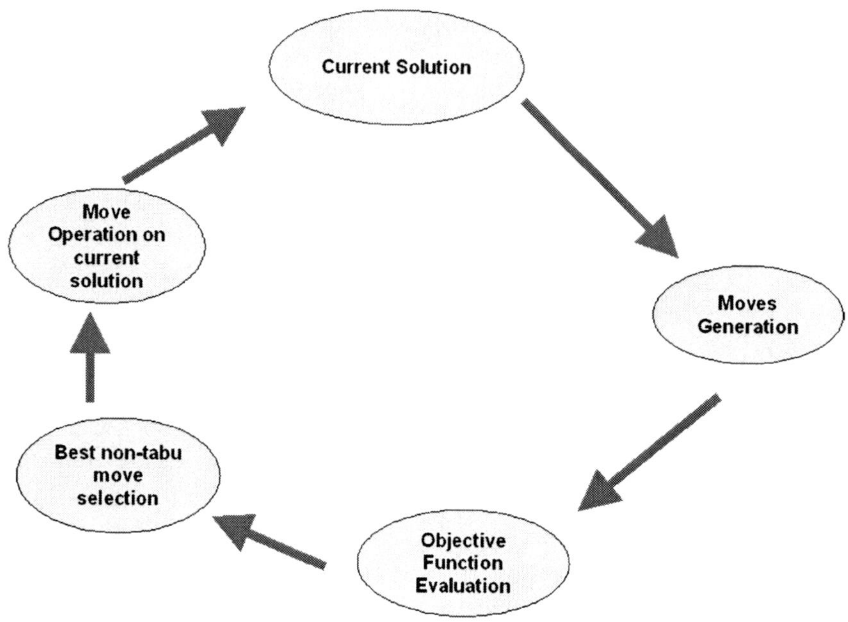

6. **Choice criterion:** The solution with the minimum Objective value among the neighboring solutions of the current solution
7. **Termination criteria:** Reaching a predefined minimum value of Objective or 50 numbers of iterations whichever occurs first

Tabu Search Algorithm

Step 1: The initial solution is stored as the present best solution and the number of inter-cell moves obtained with this solution is stored as the present best value of the objective function.

Step 2: Now each machine in all the groups is exchanged with every machine in the other groups other than the group to which it belongs. Parts are assigned to the machine groups and inter-cell moves are calculated.

Step 3: After each iteration, the best neighborhood solution with minimal inter-cell moves, is selected and taken as the initial solution for the next iteration. The machine pair exchanged is made tabu for the next t iterations, where t is the tabu tenure. If this solution is better than the present best solution, then it is updated.

Step 4: Steps 2 and 3 are repeated until the stopping criterion is reached.

Implementation

Tabu search (TS) is a Meta heuristic which helps to explore the solution space beyond the local optimum (Glover, 1989). TS strategy consists in preventing configurations of tabu list from being recognized for the next k iterations. k, called tabu tenure.

The main feature of basic TS is short-term memory.

Neighbourhood Generation

Pair wise exchanges (or swaps) used to identify moves that lead from one solution to the next. With this procedure, we can get N(N-1)/2 neighbourhood solutions where the N is size of the string (Figure 8).

Problem and Rectification

Sometimes the neighbourhood creation procedure results in the formation of an empty cell or violates some constraint. Sometimes neighbourhood creation may result in the formation of a point, which has only one machine. But the constraint is each cell should have eight machines. In such a case, the neighbourhood creation procedure is carried out on that string until it satisfies the constraint.

Artificial Neural Networks (ANN)

In its most general form a network of artificial neurons, as information processing units, is inspired by the way in which the brain performs a particular task or function of interest. Aleksander and Morton [1990] define a neural network in a broader sense such that the neural nets of the actual brain are included in the field of study and provide room for a consideration of biological findings. Their definition is as follows:

Neural computing is the study of networks of adaptable nodes, which through a process of learning from task examples, store experiential knowledge and make it available for use.

Learning algorithms are procedures used for modifying synaptic weights in an orderly fashion. Linear adaptive filter theory, which is widely applied in various fields (Haykin, [1991], uses a similar approach. However, neural networks which are inspired by the brain (where cells die and regenerate all the time) can also incorporate plasticity (ability to modify its own topology)

Figure 8. TS neighbor generation

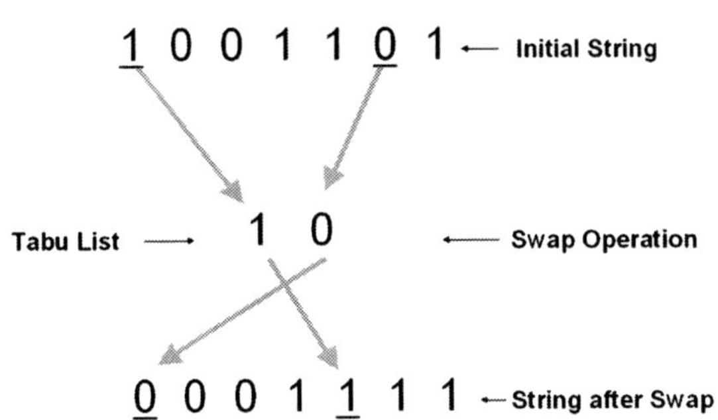

The following statement from Hecht-Nielsen [1990, p.2] defines neural networks as follows:

A neural network is a parallel, distributed information processing structure consisting of processing elements (which can possess a local memory and can carry out localized information processing operations) interconnected via unidirectional signal channels called connections. Each processing element has a single output connection that branches ("fans out") into as many collateral connections as desired; each carries the same signal – the processing element output signal. The processing element output signal can be of any mathematical type desired. The information processing that goes on within each processing element can be defined arbitrarily with the restriction that it must be completely local; that is it must depend only on the current values of the input signals arriving at the processing element via impinging connections and values stored in the processing element's local memory. Figure 9 shows a general model of a neuron with synaptic connections and the simple processing unit, which is capable of performing nonlinear transformations.

In recent years artificial neural networks (ANNs) have fascinated scientists and engineers all over the world. They have the ability to learn and recall the main functions of the human brain. A major reason for this fascination is that ANNs are "BIOLOGICALLY" inspired. They have the apparent ability to imitate the brain's activity to make decisions and draw conclusions when presented with complex and noisy information. It has been successfully applied for variety of engineering problems such as signal processing, control, pattern recognition, speech production, adaptive control manufacturing etc.

An artificial neural network is an information processing system that has certain performance characteristics in common with biological neural networks. ANN's have been developed as

Figure 9. A general model of neuron

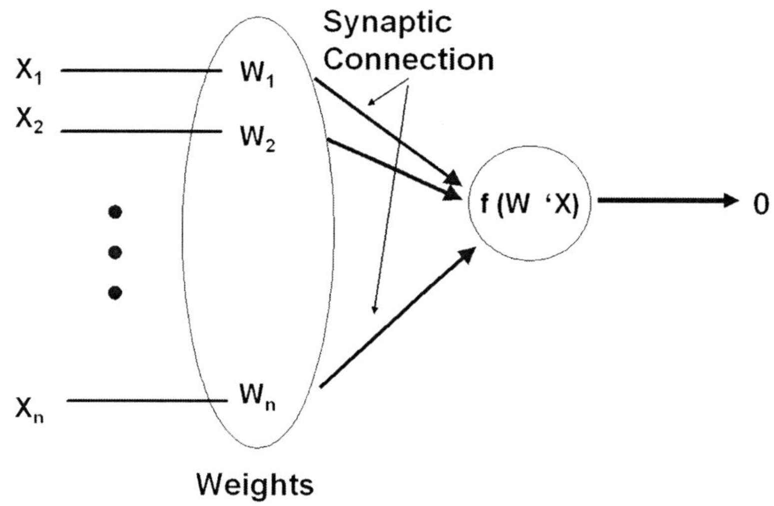

generalizations of mathematical model of human cognition or neural biology. A network is characterized by

1. Its pattern of connections between the neurons (called its architecture),
2. Its method of determining the weights on the connections (called its training, or learning algorithm) and,
3. Its activation function.

Adaptive Resonance Theory (ART1)

Introduction of ART1

The ART1 network is an unsupervised vector classifier that aspect input vectors that are classified according to the stored pattern they most resemble. It also provides for a mechanism adaptive expansion of the output layer of neurons until an adequate size is reached based on the number of classes, inherent in the observation. The ART1 network can adaptively create a new neuron corresponding to an input pattern if it is determined to be "Sufficiently" different from existing clusters. This determination called the vigilance test is incorporated into the adaptive backward network. Thus, the ART1 architecture allows the user to control the degree of similarity of patterns placed in the cluster.

Unsupervised learning algorithms try to identify several prototypes or exemplars that can serve as cluster centers. A prototype can be either one of the actual patterns or a synthesized pattern vector centrally located in the respective cluster. K-means algorithm, ISODATA algorithm, vector quantization (VQ) techniques are examples of decision –theoretical approaches for cluster formation. ART1 structure is a neural network approach for cluster formation in an unsupervised learning domain.

ART1 Philosophy

The ART1 network is an unsupervised vector classifier that accepts input vectors that are classified according to the stored pattern they most resemble. It also provides for a mechanism allowing adaptive expansion of the output layer of neurons until an adequate size reached based on the number of classes, inherent in the observation. The ART1 network can adaptively create a new neuron corresponding to an input pattern if it is determined to be "sufficiently" different from existing clusters. This determination, called the vigilance, is incorporated into the adaptive backward network. Thus, the ART1 architecture allows the user to control the degree of similarity of patterns placed in the same cluster (Figure 10).

The Stability-Plasticity Dilemma

The human brain has the ability to learn and memorize many new things in a fashion that does not necessarily cause the exiting ones to be forgotten. In order to design a truly intelligent pattern recognition machine, compatible with the human brain, it would be highly desirable to impart this ability to our models. The ability of a network to adapt and learn a new pattern well at any stage of operation is called plasticity.

Grossberg [1987] describes the stability-plasticity dilemma as follows:

How can a learning system be designed to remain plastic, or adaptive, in response to significant events and yet remain stable in response to irrelevant events? How does the system know how to switch between its stable and its plastic modes to achieve stability without chaos? In particular how can it preserve its previously learned knowledge while continuing to learn new things? What prevents the new learning from washing away the memories of prior learning?

ART1 networks attempt to address the stability-plasticity dilemma. As such ART1 provides a mechanism by which the network can learn new patterns without forgetting (or degrading) old knowledge. For example, in the context of the character recognition problem, this could be useful in contexts such as training writer specific handwriting in an on-line system or in adding new fonts to an existing off-line system without needing to retrain the network from scratch.

The incorporation of a tolerance measure (vigilance test) allows ART1 architecture to resolve the stability-plasticity dilemma. New patterns form the environment can create additional classification categories, but they cannot cause an existing memory to be changed unless the two match closely.

In a physical system, when a small vibration of proper frequency causes a large amplitude

Figure 10. Simplified ART1 architecture

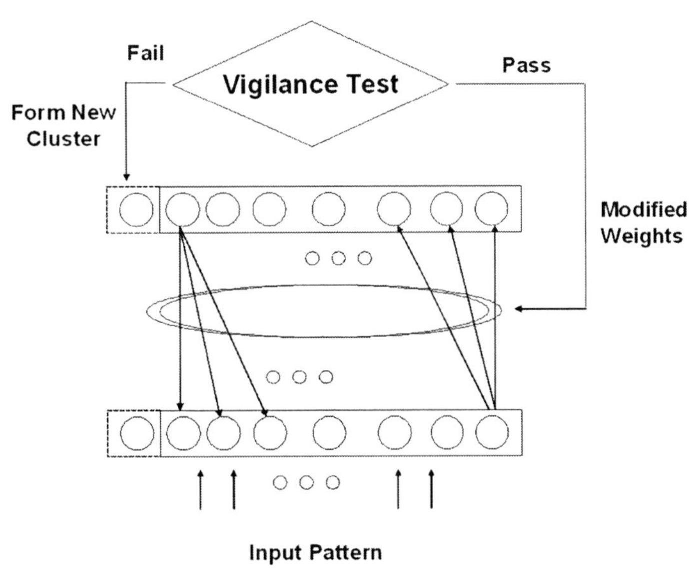

vibration, it is termed as resonance. The ART1 architecture gets its name due to the fact that the information in the form of a processing element output reverberates back and forth between layers. The neural network equivalent of resonance occurs when a proper pattern develops and a stable oscillation ensues. The pattern of activity that develops in the resonant state is called short-term memory (STM). The STM traces exist only in association with a single application of an input vector.

Learning (i.e., modification of weights) in the ART1 paradigm occurs only during the resonant period. The time required for updates in the weights between the processing elements is much longer than the time required to achieve resonance. These weights associated with the processing elements in different layers are called long-term memory (LTM) traces. The LTM traces encode information that remains a part if the network for an extended period.

ART1 Paradigm

The information that is sent to a neural network is often represented as a pattern. Every node in the network contains a representation of previously stored patterns that fit the category associated with that node. When a new pattern is presented to the ART1 network, each node competes to make a match with the new pattern. The node with the strongest match wins the competition. If the match is strong enough, the input pattern is placed into that node's grouping, whereas if the match is not very strong, then the pattern is considered unique and a new node (or category) is created for it. Different thresholds can be used to specify the classification between groupings. Since the threshold determines whether a new category is created, a different degree of clustering is obtained for each threshold. If the similarity exceeds the defined threshold, a heuristic is used to change the existing representative pattern that is used to define the category or classification of the node. The performance of the ART1 is very sensitive to the values given to the threshold and heuristic. During the optimization of the machine-part matrix, the column vectors representing the part patterns are first classified by the ART1 to obtain a series of part groups. Similar columns are grouped into adjacent areas within an intermediate matrix.

This begins the clustering of the "1" elements of the matrix next to each other. The machine row vectors are then classified and clustered in a similar manner to obtain the machine groups. Then the machine column vectors are classified and clustered in a similar manner to obtain the parts groups.

Figure 11 illustrates the sequence of events. The grouping of the rows and columns can occur simultaneously. Once the grouping is completed the resulting matrix can be inspected for bottleneck machines and exceptional machine-part cells. An additional advantage of the ART1 paradigm is that it supports on-line learning that allows new parts and machines to be immediately classified and scheduled on the shop floor and results in an intelligent manufacturing system.

Working Principle of ART1

This section discusses the ART1 network architecture and operation. ART1 inputs are binary valued as would as would be the case if the raw bit map of character image were inputs, such as gray levels for each pixel for an image.

Figure 12 shows the overall architecture of the ART1 network. The ART1 architecture consists of two layers of neurons called the comparison layer and the recognition layer. The classification decision is indicated by a single neuron in the recognition layer that fires. The neurons in the comparison layer respond to input features in the patterns, analogous to the cell groups in a sensory area of the cerebral cortex. The synaptic connections (weights) between these two layers are modifiable in both directions, according to two different learning rules. The recognition layer neurons have inhibitory connections that allow

Figure 11. Machine-part matrix formation with the ART1 Paradigm

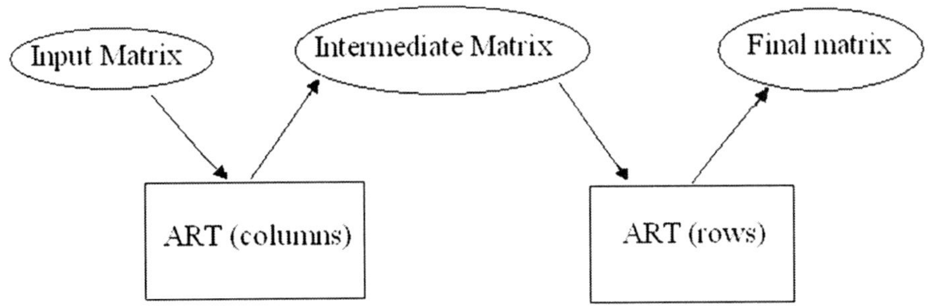

for a competition. This mechanism is common in artificial neural net architecture, inspired by the visual Neurophysiology of the biological systems. The network architecture also consists of three additional modules labeled Gain 1, Gain 2, and reset.

The attention system consists of two layers of neurons (comparison and recognition) with feed – forward and feed-backward characteristics. This system determines whether the input pattern matches one of the prototypes stored. If a match occurs, match between the bottom-up and top-down pattern on the recognition layer.

The recognition layer response to an input vector is compare to the original input vector through a mechanism termed vigilance. Vigilance provides a measure of the distance between the inputs vector and the cluster center corresponding to the firing recognition layer neuron. When vigilance falls below a preset threshold, a new category associated with the new input pattern.

The recognition layer follows the winner-take-all paradigm (this behavior is sometimes referred to as MAXNET [Kung, 1993]). If the input vector passes the vigilance, the winning neuron (the one most like the input vector) is trained such that its associated cluster center in feature space is moved toward the input vector. The recognition layer is alternately termed F2 and is top-down layer.

Figure 12. ART1 architecture

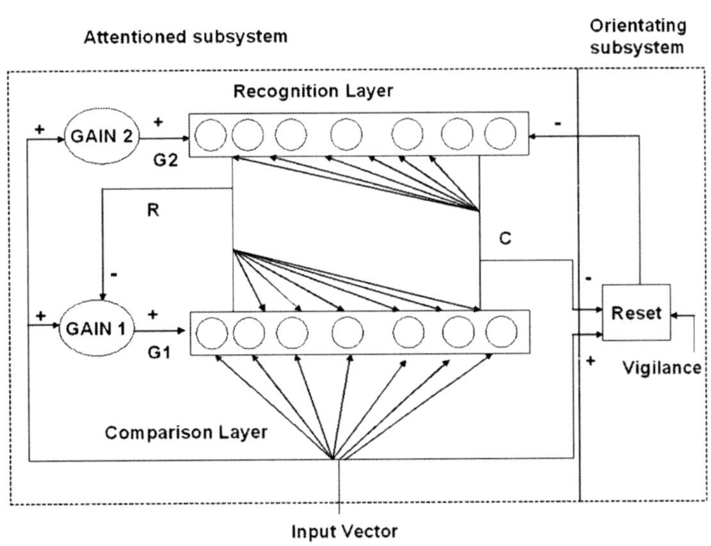

Each recognition layer neuron, j, has a real-valued weight vector B_j associated with it. This vector represents a stored exemplar pattern for a category of input patterns. Each neuron receives as input, the output of the comparison layer (vector C) through its weight vector, B_j.

The output of the recognition layer neuron, j, is given as:

$$net_j = \sum_{i=1}^{M} b_{ij} c_i$$

$$r_j = f(net_j) = \begin{cases} 1 & for\ net_j\ for\ all\ i \neq j \\ 0 & otherwise \end{cases}$$

where c_i is the output of ith comparison layer neuron; *f* is a step function and thus r_j results in a binary value. M is the number of neurons in the comparison layer.

Each neuron, i, in the comparison layer receives the following three inputs:

1. A component of the input pattern is X, i.e., xi
2. The gain G1 is a scalar (binary value); thus the same value is input to each neuron.
3. A feedback signal from the recognition layer is a weighted sum of the recognition layer outputs.

The feedback P_i through binary weights t_{ij} is given by:

$$P_i = \sum_{j=1}^{N} t_{ji} r_j \quad for\ i = 1, \ldots, M$$

where r_j is the output of the jth recognition layer neuron and N is the number of neurons in the recognition layer, T_j is the weight vector associated with the recognition layer neuron j. vector C represents the output of the comparison layer, with c_i representing the output of the ith neuron.

Gain 1 is one when the R vector is zero and the logical "OR" of the components of the input vector, X, is one, as seen in the equation.

$$G1 = (r_1 \mid r_2 \mid \ldots \mid r_N) \times X_1 \mid X_2 \mid \ldots \mid X_M)$$

Gain 2 is one when the logical OR of the components of the input vector, X, is 1, as seen in equation.

$$G2 = (X_1 \mid X_2 \mid \ldots \mid X_M)$$

The comparison layer utilizes a two-thirds rule, which states that if 2 of the 3 inputs are 1, then a 1 is output. Otherwise the result is zero. Equation shows the two-thirds rule:

$$C_j = \begin{cases} 0 & for\ G1 + X_j + P_j < 2 \\ 1 & for\ G1 + X_j + P_j < 2 \end{cases}$$

The ART process occurs in stages. Initially there is no input; thus from equation 7 we can see that G2 is zero. When an input vector, and as X, is first presented to the network, the network enter the recognition phase. The R vector feedback from the recognition layer is always set to zero at the beginning of the recognition phase. Based on the equation 6 and 7 we can see that presentation of X at this stage makes both G1 and G2 equal to 1.

As can be seen based on the initial conditions in the recognition phase, the output C of the comparison layer will be the unmodified input vector X. Thus the comparison layer passes X through to the recognition layer.

Next each neuron in the recognition layer computers a dot product between its weight vector B_j (real valued) and the C vector (which is the output of the comparison layer). The winning vector fires, inhibiting all other neurons in the recognition layer. Thus a single component r_j of the R vector will be one and all other components of R will be zero. This initiates the comparison phase.

In other words, the recognition phase results in each recognition layer neuron comparing its prototype (stored in the bottom-up weights) with the input pattern (the dot product of $B_{j \text{ and}}$ C). The mutual inhibition mechanism causes the one with the best match to fire.

During the comparison phase a determination must be made as to whether an input pattern is sufficiently similar to the winning stored prototype to be assimilated by that prototype. A test $_{\text{for}}$ this termed vigilance is performed during this phase.

In the comparison phase, the vector R is no longer zero so Gain 1 will be zero. By the two-thirds rule only neurons with simultaneous 1's in the X and P vectors will fire. Note that the weights t_{ij} are binary valued. This top-down feed back path then forces components of C to zero whenever the input vector X fails to match the stored pattern.

Let D be the number of ones in the x vector and k be the number of ones in the C vector. Then the similarity ratio, S, is simply: S=K/D

The similarity vector, S, is therefore a metric for likeness between the prototype and the input pattern. Now we must establish a criterion by which to accept or reject clusters according to this metric. The test for vigilance can be represented as follows:

$S > \rho \rightarrow$ Vigilance test passed

$S \leq \rho \rightarrow$ Vigilance test failed

If the vigilance is passed, there is no substantial difference between the input vector and the winning prototype. Thus, the required action is simply to store the input vector into the winning neuron cluster center. In this case, there is no reset signal. Therefore, when the search phase is entered, the weights for this input vector are adjusted. At this point, the operation of the network is complete.

If S is below a preset threshold, the vigilance level, then the pattern P is not sufficiently similar to the winning neuron cluster center and the firing neuron should be inhibited. The inhibition is done by the reset block, which resets the currently firing neuron throughout the duration of the current classification. This concludes the comparison phase.

The search phase is then entered and if no reset signal has been generated, the match is considered adequate and the classification is complete. Otherwise, with the firing R layer neuron disabled the R vector is once again set to zero. As a result, Gain 1 (G1) goes to one so that X once again appears on C and a different neuron in the recognition layer wins. The new winners is checked against vigilance just as before and the process repeats until either:

1. A neuron is found that matches X with a similarity above the vigilance level ($S > \rho$). The weight vectors, T_j and B_j of the firing neuron, are adjusted, or
2. All stored patterns have been tried. Then a previously unallocated neuron is associated with the pattern, and T_j and B_j are set to match the pattern.

ART1 Algorithm

Initially the weights b_{ij} are initialized to the same low value which should be

$$b_{ij} < \frac{L}{(L-1+m)}$$

where m is the number of components in the input vector and L is a constant, typically L=2.

The algorithm for the ART1 architecture is as follows:

Step 1. When an input partten, X, is presented to the network, the recognition layer selects the winner as the maximum of all the net outputs:

$$net_j = \sum_{i=1}^{M} b_{ij} c_i$$

where N is the number of neurons in the comparison layer.

Step 2. Perform the vigilance threshold test. A neuron j is declared to pass the vigilance test, if and only if,

$$\frac{net_j}{\sum_{i=1}^{N} b_{ij} c_i} > \rho$$

Where ρ is the vigilance threshold.

a. If the winner fails the test, mask the current winner and go to the step1 to select another winner.

b. Repeat the cycle (step 1 through 2a) until a winner is determined that passes the vigilance test; then go to step 4.

Step 3. If no neuron passes the vigilance test, create a new neuron to accommodate the new pattern

Step 4. Adjust the feed-forward weights for the winner neuron. Update the feed-back weights from the winner neuron to its inputs.

Step 5. The equations governing the training of the bottom-up and top-down weights are

$$b_{ij} < \frac{L_{c_i}}{(L - 1 + \sum_{c_k})}$$

$$t_{ij} = c_i$$

where c_i is the ith component of the comparison layer vector and j is the index of the winning recognition layer neuron.

ARTIFICIAL IMMUNE SYSTEM (AIS)

The natural immune system consists of a complex set of cells and molecules that protect our bodies against infections. Our bodies are under constant attack by antigens that can stimulate the adaptive immune system. The white blood cells (leukocytes) are the main operative elements of the immune system. The main characteristic of the leukocytes is the presence of surface receptor molecules capable of recognizing and binding to molecular patterns. The molecular patterns that can be recognized by the surface receptors on lymphocytes are named antigens. The portion of a surface receptor molecular on a leukocyte that binds with an antigen is generally termed as antibody. Therefore, an antibody corresponds to the portion of any leukocytes capable of recognizing a molecular pattern, and an antigen is equivalent to ant pattern that can be recognized by the antibody. The strength of binding between an antigen and an antibody is named affinity.

The general principles of AIS are colonal selection, mutation and receptor editing. Colonal selection maintains the quality of solution by triggering growth of lower affinity antibodies. Mutation is used to diversity the search process and receptor editing helps in escaping from the local optimal (Das Gupta and Gonzalez 2002). AIS has been successfully applied to different optimization problems including in pattern recognition (Carter, 2000; Carvalho and Freitas 1991) scheduling (M. Chandrasekaran et al 2006) Multiobjective optimization (K.C. Jan et al 2008), Machine loading (Nitesh Khilwani et al 2008), Machine learning (Hunt and Cooke, 1996).

The Artificial Immune System (AIS) is inspired by theoretical immunology and observed immune functions, principles and models. It is not only related to the creation of abstraction or metaphorical models of the biological immune system, it also includes theoretical immunology models being applied to tasks such as optimization, control, and autonomous robot navigation.

Applications of AIS include pattern recognition, fault and anomaly detection, data mining and classification, scheduling, machine learning, autonomous navigation, search and optimization. In this research, AIS is applied to solve the Manufacturing Cell Formation Problem. When AIS is applied to a cell formation problem, the problem can be treated as the antigen and the solution to the problem as the antibody.

AIS Algorithm

1. Initialize population (randomly)
2. Individuals (Candidate solution)
3. Evaluation (Affinity function) for all antibodies.
4. While (termination criterion not satisfied)
5. Select (Superior antibodies from parent population)
6. Cloning based on fitness value
7. Variation operators on clones (Mutation)
8. Evaluate new generated antibodies
9. Selection of superior antibodies
10. Creation of next generation population (Receptor editing)
11. End

Implementation

1. Represent the problem variable as an antibody with a string representation, which is similar to a Chromosome in genetic algorithms, and generate initial antibody population randomly. Each antibody is an M x N machine-part incidence matrix MPIM (See fig. 1) a string with a length of M + N is needed to encode. The first M bits of the string represent the sequence of machines that appear in the rows of the MPIM, while the last N bits of the string represent the sequence of parts appearing in the columns of the MPIM.
2. Calculate the individual solution (i.e.) calculate solution of each antibody by equation number 1.
3. Calculate the affinity value ρ is each antibody. Affinity value of each string is calculated from the affinity function.

$$\text{Affinity } (\rho) = \frac{1}{Exceptional\ Elements\ (Z)}$$

4. From this relation a cell that is more capable of recognizing and subsequently removing a given antigen has a higher affinity than others.
5. Antibody receives more stimulation to proliterate the immune system by a mechanism called cloning (De Castro L.N. et al 2003). Cloning is a mitotic process that produces exact copies of the parent cells. Based on the affinity value of the individual string number of clones are calculated for the population.

$$Number\ of\ clones = \frac{Individual\ affinity}{Total\ affinity} x Population\ size$$

The number of clones rounded of the integer.
6. A two phased mutation were used for the generated clones (May P. Mander K and Timmis J 2003).
   i. **Inverse Mutation:** For a string s, let i and j be randomly selected two positions in the strings. A neighbor of s is obtained by inversing the string of machines and parts between i and j positions. If the objective value of the mutated string (after inverse mutation) is smaller than that of the original string (a generated clone from an antibody), then the mutated one is stored in the place the original one. Otherwise, the

string will be mutated again with random pair wise mutation.

ii. **Pair-Wise Mutation:** Given a string s, let i and j be randomly selected two positions in the string s. A neighbor of s is obtained by interchanging the machine and parts in positions i and j. If the exceptional element value of the mutated string (after pair wise interchange mutation) is smaller than that of the original, then stored the mutated one in the place of the original. In the case where the algorithm could not find a better string after the two mutation procedure, then it stores the original one (generated clone).

7. Receptor editing after cloning and mutation processes, a percentage of the antibodies (worst %B of the whole population) in the antibody population are eliminated and randomly created antibodies are replaced with them. This mechanism allows to find new schedules that correspond to new search regions in the total search space. (De Castro LN, Von Zuben FJ 1999).

Steps 2 – 7 are repeated until either the minimize the exceptional elements or the specified maximum number of iterations has been reached.

SHEEP FLOCKS HEREDITY ALGORITHM

Sheep flock algorithm was developed by Hyunchul and Byungchul (2001). The algorithm simulates heredity of sheep flocks in a prairie. The algorithm is developed for solving a large scale scheduling problem for a period of several successive years. It is referred to as the multi-stage genetic operation can find better solutions than simple genetic algorithm.

Consider the several separated flocks of sheep in a field (Koichi Nara et al, 1999). Normally, sheep in each flock are living within their own flock under the control of shepherds. So, the genetic inheritance only occurs within the flock. In other words, some special characteristics in one flock develop only within the flock by heredity, and the sheep with high fitness characteristics to their environment breed in the flock In such a world, let us assume that two sheep flocks were occasionally mixed in a moment when shepherds looked aside as shown in Figure 4. Then, shepherd of the corresponding flocks run into the mixed flock, and separate the sheep as before. However, shepherds can not distinguish their originally owned sheep because the appearance of any sheep is the same. Therefore, several sheep of one flock are inevitably mixed with the other flocks, namely, the characteristics of the sheep in the neighboring flocks can be inherent to the sheep in other flocks in this occasion. Then, in the field, the flock of the sheep which has better fitness characteristics to the field environment breeds most. The above natural evolution phenomenon of sheep flocks can be corresponded to the genetic operations of this type of string. For this kind of string, we can define the following two kinds of genetic operations:

- Normal genetic operations between strings
- Genetic operations between sub-strings within one string

This type of genetic operation is referred to "multi-stage genetic operation". Sheep algorithm is used because of the following;

- It is a multi-stage genetic operation, can find better solutions than those of the simple genetic algorithm.
- Algorithm shows reasonable combination of local and global search.
- The method is effectively applied to planning problems for multiple years, and the method is tested by the real scale generator maintenance scheduling problem.

Algorithm

```
    Begin
Initialize the population
Stage 1:
    Select the parent (Initialize the Random Population Say 10 Nos)
    Sub Chromosome level crossover
Set sub chromosome level crossover probability
If Population probability is less than or equal to sub chromosome level prob-
ability
        Perform Sub chromosome level crossover
Else Retain the old string
Sub chromosome level mutation
Set sub chromosome probability
If population probability is less than or equal to sub chromosome mutation
probability
        Perform sub chromosome level mutation
Else retain the same string
Stage 2:
Select two strings from population
Chromosome level crossover
Set crossover probability
If Population probability is less than or equal to crossover probability
    Perform chromosome level crossover
Else retain the same string
Chromosome level mutation
Set mutation probability
If Population probability is less than or equal to mutation probability
    Perform chromosome level mutation
Else retain the same string
End if terminal condition satisfied
```

EXPERIMENTATION

The Meta heuristics were implemented on ten 16 X 30 sized benchmark problems (Boctor 1991). The data sets are used as inputs, are given in Appendix I. Objective function is used to evaluate the goodness of the cell formation. Minimization of exceptional elements (Boctor 1991) is considered as the objective function.

$$z = \sum_{i=1}^{m}\sum_{j=1}^{n} a_{ij}\left(\sum_{k=1}^{G}\left|x_{ik} - y_{jk}\right|\right) / 2$$

Where

K= cell index
G= number of manufacturing cell
m=number of machines
n= total number of parts
x_{ik} = binary value indicating the machine i is assigned to cell k

Figure 13. Sheep flocks heredity algorithm

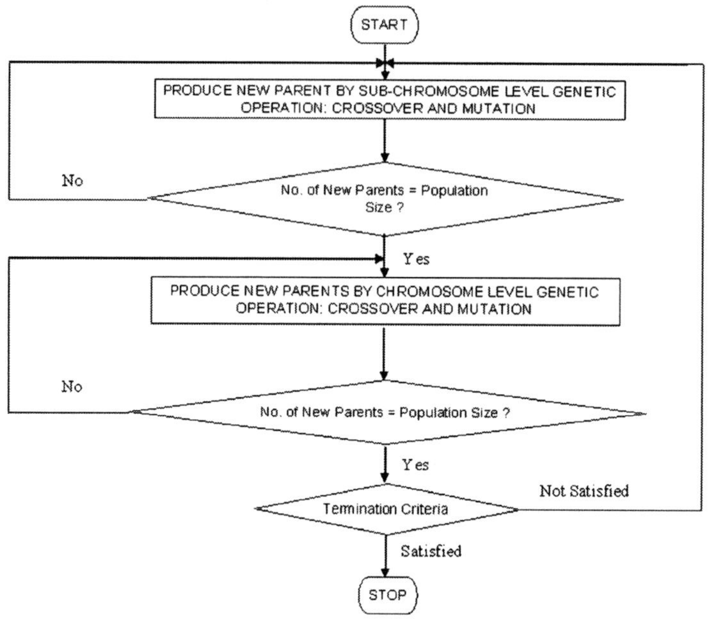

y_{jk}= binary value indicating the part j is assigned to cell k
a_{ij} = element of machine part incident matrix
X_{ik} =1 if machine type is assigned to cell k.
=0, otherwise
Y_{jk} =1, if part/component j is assigned to cell k
=0, otherwise

Parameter Selection

Based on the sensitivity analysis carried out, the parameters of the Algorithms for the considered problem structures the following values were found more effective and satisfactory.

Genetic Algorithm

Population size (N) = 20
Number of generations (GN) = 1000
Crossover probability (p_c) = 0.6
Mutation probability (p_m) = 0.2

Simulated Annealing Algorithm

CMAX = 320
RMAX = 64
$\alpha = 0.8$
T = 64
B = 0.8

Tabu Search

Number of iterations: 100
Tabu History Length: 3
tabu tenure = 5

Artificial Neural Network

Vigilance threshold (ρ) = 0.1

Artificial Immune System

Population Size = 20
Elimination Percentage = 0.5

Table 4. Result comparison

Problem No	GA		SA		TS		ANN		AIS		SFHA	
	Z	CPU Time	Z	CPU Time	Z	CPU Time	Z	CPU Time	Z	CPU Time	Z	CPU Time
1	11	1.78	11	1.34	11	1.8	11	1.81	11	0.235	11	3.61
2	3	1.8	3	1.35	6	2.91	6	1.99	3	0.2	3	3.61
3	1	1.78	1	1.36	4	2.73	4	1.28	1	1.32	1	3.72
4	13	1.77	13	1.36	13	2.56	13	1.56	13	0.188	13	3.58
5	4	1.8	4	1.34	8	2.56	8	1.8	4	1.74	4	3.63
6	2	1.84	2	1.34	5	2.35	4	1.82	2	1.688	2	3.58
7	4	1.79	4	1.36	5	2.13	5	1.53	4	1.53	4	3.61
8	5	1.81	5	1.39	10	2.9	10	1.3	5	0.25	5	3.59
9	5	1.81	5	1.36	8	2.9	8	1.34	5	0.2	5	3.66
10	5	1.92	5	1.36	8	2.23	8	1.39	5	0.266	5	3.66

Figure 14. Objective value comparison

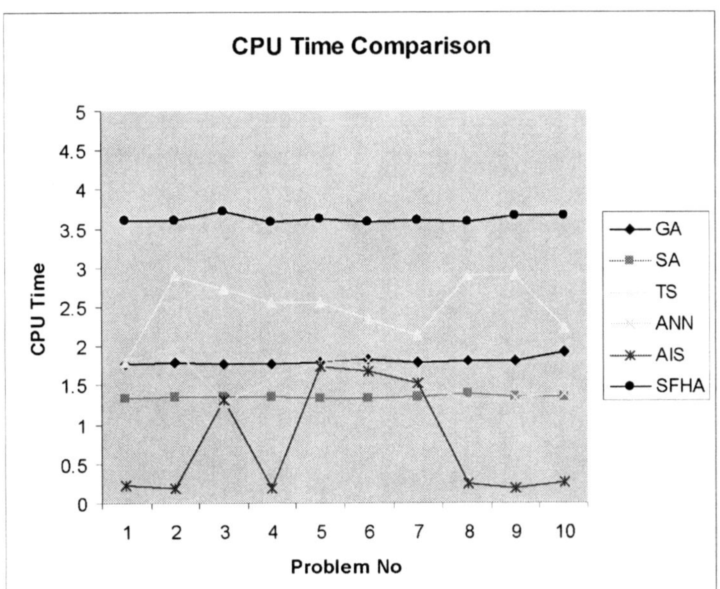

Sheep Flocks Heredity Algorithm

Population Size = 20
Number of generations =1000
Chromosome level crossover probability = 0.6
Chromosome level mutation probability = 0.2
Sub chromosome level crossover probability = 0.1
Sub chromosome level mutation probability = 0.001

The results obtained from these algorithms are shown in the Table 4. Figure 14 shows the comparison of objective values (i.e. number of exceptional elements) for these algorithms. GA, SA, AIS and SFHA results were same for all the

Figure 15. CPU time comparison

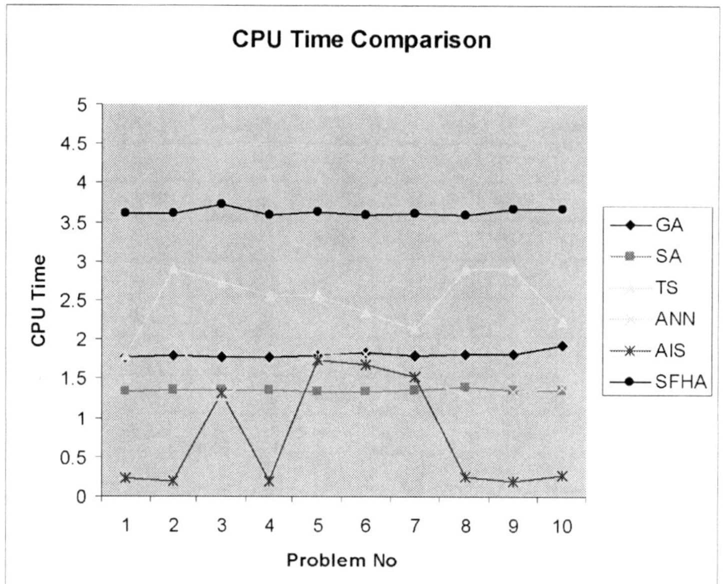

problems. But TS & ANN yields worst results when compared with others. It is observed that, TS & ANN yields better results for well structured datasets and it yields worst results for ill-structured datasets. Figure 15 shows the comparison of computational time (CPU Time in Sec) for the algorithms. Computational time for SFHA is minimum when compared with other algorithms and it yields better results in objective value. SA also takes minimum computational time when compared with GA, TS and AIS. It also yields minimum objective value for all the problems.

CONCLUSION

This chapter describes the implementation of non-traditional optimization techniques for CMS. In this chapter the Meta heuristics Genetic Algorithms, Simulated Annealing Algorithm, Tabu Search, Artificial Neural Networks, Artificial Immune System and Sheep Flock Heredity Algorithm were discussed in detail. Implementation of these algorithms to design CMS is also described. The cell formation problem in GT using Meta heuristics with an objective of minimizing exceptional elements, up to 16 X 30 sized matrices are solved. It is observed that SFHA and SA perform better than the other Meta heuristic algorithms. For most of the problems SFHA and SA results in solutions with fewer exceptional elements. The computational time is also less in SFHA and SA when compared to the other Meta heuristics. TS & ANN yields better results for well structured datasets and it yields worst results for ill-structured datasets. All algorithms are coded with C++ language. The datasets are tested with the Pentium IV 900MHz systems. The Meta heuristic algorithms can be implemented on the problems which includes operating costs, machine duplication and providing alternate path routing, fractional cell, processing time of the machines and part demand for cell formation.

REFERENCES

Albadawi, Z., Bashir, H. A., & Chen, M. (2005). A mathematical approach for the formation of manufacturing cells. *Computers & Industrial Engineering*, *48*(1), 3–21. doi:10.1016/j.cie.2004.06.008

Aleksander, I., & Morton, H. (1990). *An introduction to neural computing*. London, UK: Chapman & Hall.

Ballakur, A., & Steudel, H. J. (1987). A within cell utilization based heuristic for designing cellular manufacturing systems. *International Journal of Production Research*, *25*(5), 639–655. doi:10.1080/00207548708919868

Boctor, F. (1991). A linear formulation of the machine part cell formation problem. *International Journal of Production Research*, *29*(2), 343–356. doi:10.1080/00207549108930075

Burbidge, J. L. (1963). Production flow analysis. *Production Engineer, December*, 742-752.

Burbidge, J. L. (1971). Production flow analysis. *Production Engineering*, *50*(4/5), 139–152.

Burbidge, J. L. (1975). *The introduction of group technology*. London, UK: Heinemann Press.

Burbidge, J. L. (1977). A manual method of production flow analysis. *Production Engineering*, *56*, 34–38. doi:10.1049/tpe.1977.0129

Carpenter, G. A., & Grossberg, S. (1987). ART 2: Self-organization of stable category recognition codes for analog input patterns. *Applied Optics*, *26*, 4919–4930. doi:10.1364/AO.26.004919

Carter, J. H. (2000). The immune system as a model for pattern recognition and classification. *Journal of the American Medical Informatics Association*, *7*(1), 28–41. doi:10.1136/jamia.2000.0070028

Carvalho, D. R., & Freitas, A. A. (1991). An immunological algorithm for discovering small – Disjunct rules in data mining. In Specter, et al (Eds.), *Proceedings of the Genetic and Evolutionary Computation Conference,* (pp. 401–404).

Chan, H. M., & Milner, D. A. (1982). Direct clustering algorithm for group formation in cellular manufacturing. *Journal of Manufacturing Systems*, *1*, 65–75. doi:10.1016/S0278-6125(82)80068-X

Chandrasekaran, M., Asokan, P., Kumanan, S., Balamurugan, T., & Nickolas, S. (2006). Solving job scheduling problems using artificial immune system. *International Journal of Advanced Manufacturing Technology*, *31*, 580–593. doi:10.1007/s00170-005-0226-3

Chandrasekharan, M. P., & Rajagopalan, R. (1986a). An ideal seed nonhierarchical clustering algorithm for cellular manufacturing. *International Journal of Production Research*, *24*(2), 451–464. doi:10.1080/00207548608919741

Choobineh, F. (1988). A frame work for the design of manufacturing systems. *International Journal of Production Research*, *26*, 1161–1172. doi:10.1080/00207548808947932

Co, H. C., & Araar, A. (1988). Configuring cellular manufacturing system. *International Journal of Production Research*, *26*(9), 1511–1522. doi:10.1080/00207548808947962

Crama, Y., & Oosten, M. (1996). Models for machine part grouping in cellular manufacturing. *International Journal of Production Research*, *34*(6), 1693–1713. doi:10.1080/00207549608904991

Das Gupta, D., & Gonzadez, F. (2002). An immunity based technique to characterize intrusions in computer networks. *IEEE Transactions on Evolutionary Computation*, *6*(3).

Glover, F., & Greenberg, H. J. (1989). New approaches for heuristic search a bilateral linkage with artificial intelligence. *European Journal of Operational Research*, *39*, 119–130. doi:10.1016/0377-2217(89)90185-9

Goldberg, D. E. (1989). *Genetic algorithms in search, optimization, and machine learning*. Reading, MA: Addison–Wesley.

Hetch–Nielson, R. (1990). *Neuro computing*. Reading, MA: Addison–Wesley.

Hunt, J. E., & Cooke, D. E. (1996). Learning using an artificial immune system. *Journal of Network and Computer Applications*, *19*, 189–212. doi:10.1006/jnca.1996.0014

Khilwani, N., Prakash, A., Shankar, R., & Tiwari, M. K. (2008). Fast clonal algorithm. *Engineering Applications of Artificial Intelligence*, *21*, 106–128. doi:10.1016/j.engappai.2007.01.004

King, J. R. (1980). Machine-component grouping in production flow analysis: An approach using a rank order-clustering algorithm. *International Journal of Production Research*, *18*, 213–232. doi:10.1080/00207548008919662

Koichi, N., Tomomi, T., & Hyunchul, K. (1999). A new evolutionary algorithm based on sheep flocks heredity model and its application to thermal generator maintenance scheduling problem. *IEEE Transactions*, *4*, 503–508.

Kung, S. Y. (1993). *Digital neural networks*. Englewood Cliffs, NJ: Prentice Hall.

Kusiak, A. (1987). The generalized group technology concept. *International Journal of Production Research*, *25*(4), 561–569. doi:10.1080/00207548708919861

Lenstra, J. K. (1974). Clustering a data array and the traveling salesman problem. *Operations Research*, *22*, 413–414. doi:10.1287/opre.22.2.413

Leung, H., & Haykin, S. (1993). Rational function neural network. [MIT Press.]. *Neural Computation*, *5*(6), 928–938. doi:10.1162/neco.1993.5.6.928

McCormick, W. T., Schweitzer, J. P. J., & White, T. W. (1972). Problem decomposition and data reorganization by a clustering technique. *Operations Research*, *20*(5), 993–1009. doi:10.1287/opre.20.5.993

Metropolis, N., Rosenbluth, A., Rosenbluth, M., Teller, A., & Teller, E. (1953). Equation of State Calculations by Fast Computing Machines. *The Journal of Chemical Physics*, *21*(6), 1087–1092. doi:10.1063/1.1699114

Mitrofanov, S. P. (1959). *Scientific principles of group technology, part I*. Boston, MA: National Lending Library of Science and Technology. (in Russian)

Nair, G. J., & Narendran, T. T. (1999). ACCORD: A bicriterion algorithm for cell formation using ordinal and ratio level data. *International Journal of Production Research*, *37*(3), 539–556. doi:10.1080/002075499191661

Rajagopalan, R., & Batra, J. L. (1975). Design of cellular production system –A graph theoretic approach. *International Journal of Production Research*, *13*(6), 567–579. doi:10.1080/00207547508943029

Rajamani, D., Singh, N., & Aneja, Y. P. (1990). Integrated design of cellular manufacturing systems in the presence of alternate process plans. *International Journal of Production Research*, *28*, 1541–1554. doi:10.1080/00207549008942811

Seifoddini, H. (1989). Duplication process in machine cells formation in group technology. *IIE Transactions*, *21*(4), 382–388. doi:10.1080/07408178908966245

Solimanpur, M., Vrat, P., & Shankar, R. (2004). Ant colony optimization algorithm to the inter-cell layout problem in cellular manufacturing. *European Journal of Operational Research*, *157*(3), 592–606. doi:10.1016/S0377-2217(03)00248-0

Srinivasan, G. (1994). A clustering algorithm for machine cell formation in group technology using minimum spanning trees. *International Journal of Production Research*, *32*(9), 2149–2158. doi:10.1080/00207549408957064

Tan, K. C., Goh, C. K., Mamun, A. A., & Ei, E. Z. (2008). An evolutionary artificial immune system for multi objective optimization. *European Journal of Operational Research*, *187*, 371–392. doi:10.1016/j.ejor.2007.02.047

Veeramani, D., & Mani, K. (1996). A polynomial-time algorithm for optimal clustering in a special class of {0, 1}-matrices. *International Journal of Production Research*, *34*(9), 2587–2611. doi:10.1080/00207549608905046

Vohra, T., Cheng, D., & Chang, J. L. (1990). A network model for cell formation in CMS. *International Journal of Production Research*, *28*(11), 2075–2084. doi:10.1080/00207549008942854

Yasuda, K., & Yin, Y. (2001). A dissimilarity measure for solving the cell formation problem in cellular manufacturing. *Computers & Industrial Engineering*, *39*(1-2), 1–17. doi:10.1016/S0360-8352(00)00060-7

APPENDIX

Table 5. Problem no: 1

	Parts																													
MACHINES	1	2	3	4	5	6	7	8	9	10	11	12	13	14	15	16	17	18	19	20	21	22	23	24	25	26	27	28	29	30
1	1			1			1			1			1																	
2						1		1						1	1			1		1		1		1	1	1				
3		1	1				1			1					1						1									1
4		1	1	1			1					1						1			1		1					1	1	1
5	1					1		1		1								1		1					1				1	
6					1	1				1					1		1		1		1	1		1				1		
7		1	1		1		1	1				1		1	1	1	1					1	1	1		1				
8			1									1			1		1	1			1			1			1		1	
9	1			1																	1		1						1	
10					1			1						1				1	1	1				1		1				1
11		1			1		1		1					1	1				1		1	1				1				
12	1	1									1		1		1			1		1			1		1		1		1	
13		1			1				1				1	1		1	1			1		1	1						1	1
14	1	1							1	1	1					1			1	1									1	
15					1			1											1		1						1			1
16	1									1	1	1																		

Table 6. Problem no: 2

		Parts																													
		1	2	3	4	5	6	7	8	9	10	11	12	13	14	15	16	17	18	19	20	21	22	23	24	25	26	27	28	29	30
MACHINES	1					1	1		1		1	1	1	1													1				
	2				1	1				1	1	1	1	1		1	1		1								1				
	3	1	1		1													1													
	4			1															1			1		1	1	1		1	1		
	5		1			1	1		1	1	1																1		1		
	6			1																	1	1		1	1			1		1	1
	7							1													1			1	1	1		1	1	1	
	8				1	1				1	1		1	1						1							1				
	9											1																			
	10	1	1												1			1													
	11					1	1		1	1	1	1		1						1							1				
	12			1															1		1	1		1					1	1	1
	13			1				1											1			1	1		1			1			1
	14					1		1			1	1	1				1			1											
	15	1													1	1															
	16	1	1															1													

Table 7. Problem no: 3

	Parts																													
MACHINES	1	2	3	4	5	6	7	8	9	10	11	12	13	14	15	16	17	18	19	20	21	22	23	24	25	26	27	28	29	30
1																	1									1		1		
2				1																1				1						1
3	1					1							1		1		1		1		1	1								
4	1		1			1	1		1	1			1	1	1		1				1	1								
5		1			1				1							1		1						1			1			
6		1									1					1								1			1			
7	1		1				1		1	1			1		1	1	1				1	1					1			
8							1																1		1			1	1	
9																							1		1	1			1	
10								1				1								1										1
11				1								1								1										1
12				1				1												1										
13	1	1			1											1		1						1			1			
14																			1				1			1		1	1	
15											1					1		1						1			1			
16																							1					1	1	

Table 8. Problem no: 4

		Parts																													
		1	2	3	4	5	6	7	8	9	10	11	12	13	14	15	16	17	18	19	20	21	22	23	24	25	26	27	28	29	30
MACHINES	1						1	1									1			1								1	1	1	
	2		1							1				1	1				1			1	1				1	1			
	3	1			1		1					1					1	1								1			1	1	
	4		1							1									1												
	5	1			1		1					1					1			1				1	1					1	
	6	1			1						1	1	1							1				1					1	1	
	7		1	1			1				1				1		1	1										1		1	1
	8													1					1												
	9	1					1				1													1					1		1
	10							1		1												1	1				1			1	
	11	1	1			1			1					1					1	1						1					
	12			1	1								1				1	1						1	1				1		1
	13								1						1												1				
	14												1			1			1			1				1					1
	15					1			1				1				1		1		1		1			1	1				
	16					1	1								1				1								1				

Table 9. Problem no: 5

	Parts																													
MACHINES	1	2	3	4	5	6	7	8	9	10	11	12	13	14	15	16	17	18	19	20	21	22	23	24	25	26	27	28	29	30
1																	1	1						1	1			1		
2																1		1							1	1		1		
3			1		1		1		1		1			1					1											
4					1		1				1			1																
5	1			1				1		1					1		1								1					1
6		1								1		1	1					1	1		1	1	1							
7																	1	1						1	1	1		1		
8			1		1						1			1															1	
9			1		1		1		1	1	1			1											1					
10	1			1				1																						1
11				1				1																						1
12	1		1		1		1		1		1			1																
13		1				1				1		1	1						1	1	1		1				1			1
14		1					1			1		1			1				1	1	1	1							1	
15																1	1							1	1	1		1		
16						1				1		1	1							1	1	1	1				1		1	

Table 10. Problem no: 6

MACHINES	Parts																													
	1	2	3	4	5	6	7	8	9	10	11	12	13	14	15	16	17	18	19	20	21	22	23	24	25	26	27	28	29	30
1				1	1						1		1				1	1				1								1
2					1	1		1	1						1					1	1									
3										1													1				1	1		
4																							1	1		1	1			
5	1		1		1	1		1	1	1					1				1		1									
6		1												1		1									1			1	1	
7					1	1		1	1						1				1		1									
8	1		1		1	1		1	1										1	1	1									
9																							1	1		1	1			
10	1		1		1			1							1						1									
11				1			1				1	1	1				1	1				1								1
12							1							1		1									1			1	1	
13	1										1			1		1									1			1	1	
14																							1	1						
15		1			1											1									1			1	1	
16		1												1		1									1			1	1	

Table 11. Problem no: 7

		Parts																													
		1	2	3	4	5	6	7	8	9	10	11	12	13	14	15	16	17	18	19	20	21	22	23	24	25	26	27	28	29	30
MACHINES	1	1					1	1	1					1	1	1	1														
	2			1		1					1						1							1	1			1		1	1
	3	1							1	1				1		1		1													
	4																				1	1					1		1		
	5	1	1				1	1	1	1					1		1	1													
	6	1	1					1		1				1	1		1	1	1		1										
	7				1								1						1	1			1								
	8	1	1				1	1	1	1					1	1	1	1													
	9																				1	1				1			1		
	10	1					1	1		1				1	1	1	1	1													
	11				1	1							1						1	1			1								
	12												1						1	1											
	13	1					1	1	1	1				1	1		1	1													
	14			1											1									1	1	1		1		1	1
	15			1																		1	1			1	1				
	16					1					1													1	1		1			1	1

Table 12. Problem no: 8

		Parts																													
		1	2	3	4	5	6	7	8	9	10	11	12	13	14	15	16	17	18	19	20	21	22	23	24	25	26	27	28	29	30
MACHINES	1		1						1					1	1	1	1	1		1	1								1	1	1
	2						1															1			1	1	1	1			
	3		1						1							1		1	1	1									1	1	1
	4	1								1		1		1																	
	5						1						1									1	1	1		1		1		1	
	6	1				1		1	1			1		1						1						1					
	7										1		1									1	1		1		1	1			
	8	1			1			1				1																			
	9	1							1			1			1						1			1						1	1
	10				1	1				1		1		1																	
	11														1	1	1		1	1	1								1		
	12			1			1				1		1					1					1		1	1	1				
	13																1		1	1										1	
	14			1			1				1		1										1	1		1		1			
	15										1				1			1		1										1	
	16					1			1						1	1	1	1	1		1								1		1

Table 13. Problem no: 9

	Parts																													
MACHINES	1	2	3	4	5	6	7	8	9	10	11	12	13	14	15	16	17	18	19	20	21	22	23	24	25	26	27	28	29	30
1									1										1							1		1	1	
2		1	1	1	1		1	1				1	1		1															
3			1	1						1	1				1	1	1							1						1
4					1	1	1	1				1	1																	
5														1				1			1	1	1		1					
6																		1			1	1			1					
7	1		1	1					1	1	1				1	1	1							1						1
8				1										1				1			1				1				1	
9									1											1						1	1	1	1	
10	1									1	1			1				1			1	1	1		1					
11	1		1	1						1	1					1	1							1					1	1
12			1	1						1						1	1							1					1	
13														1				1			1		1		1					
14	1		1								1				1	1	1		1					1						1
15	1		1	1						1					1	1	1							1						1
16									1										1	1						1	1	1	1	

Table 14. Problem no: 10

		Parts																													
		1	**2**	**3**	**4**	**5**	**6**	**7**	**8**	**9**	**10**	**11**	**12**	**13**	**14**	**15**	**16**	**17**	**18**	**19**	**20**	**21**	**22**	**23**	**24**	**25**	**26**	**27**	**28**	**29**	**30**
MACHINES	**1**	1		1	1			1		1		1				1	1											1			
	2										1				1									1	1				1	1	
	3			1											1																1
	4											1		1					1	1			1			1		1			
	5								1		1													1	1					1	1
	6	1		1	1	1	1			1		1				1	1														
	7						1		1		1				1														1		1
	8	1	1	1	1	1	1	1		1		1															1				
	9	1	1	1	1		1	1				1				1	1														
	10	1	1	1		1	1			1		1				1	1														
	11								1		1													1					1	1	1
	12	1						1								1	1														
	13			1							1									1				1	1					1	1
	14												1	1				1	1	1	1		1			1		1			
	15								1	1					1									1					1		
	16												1								1						1				

Chapter 7
Similarity-Based Cluster Analysis for the Cell Formation Problem

Riccardo Manzini
University of Bologna, Italy

Riccardo Accorsi
University of Bologna, Italy

Marco Bortolini
University of Bologna, Italy

ABSTRACT

This chapter illustrates the cell formation problem (CFP) supported by similarity based methods. In particular, problem oriented indices are based on several factors which play an important role in the determination of the value of similarity between two generic machines, e.g. the number of machines visited by each part, the sequence of manufacturing operations, the production quantity for each part, et cetera. A numerical example illustrates the basic steps for the implementation of an effective hierarchical procedure of clustering machines into manufacturing cells and parts/products into families of parts. Literature presents many indices, but a few significant case studies and instances not useful to properly compare them and support the best choice given an operating context, i.e. a specific production problem. As a consequence the authors illustrate an experimental analysis conducted on a literature problem oriented instance to compare the performance of different problem settings and define best practices and guidelines for professional and practitioners.

DOI: 10.4018/978-1-61350-047-7.ch007

INTRODUCTION

Group technology (GT) is a manufacturing philosophy for the identification of similar parts and grouping them to take advantages from their similarities in design and manufacturing (Manzini et al. 2010). A special application of GT is *cellular manufacturing* (CM), defined as a hybrid system including the advantages of both flexible and mass production approaches. CM can be defined as an application of GT that involves grouping machines based on the parts manufactured by them. The design of a CM system is called the *cell formation* (CF) problem and includes also the definition of families of "parts", i.e. products and components, assigned to the groups of manufacturing resources, called "machines".

Since 1966 when the first contribution on CM and its topics was published (Yin and Yasuda 2006), the large number of advantages presented by CM compared to batch production (generally implemented in the so-called *functional layouts* or *job shop* systems) have been widely discussed in the literature, e.g. inventory level reduction, production lead time reduction, reduced set-up times, etc. The main difference between a traditional job shop environment and a CM environment is in the grouping and layout of machines: in a job shop system, machines are grouped on the basis of their functional similarities; in a CM environment each cell is dedicated to the manufacture of a specific part family, and the machines in each cell have dissimilar functions (Heragu 1997).

An effective approach to forming manufacturing cells and introducing families of similar parts, consequently increasing production volumes and machine utilization, is the use of similarity coefficients in conjunction with clustering procedures.

Recent studies and applications on cluster analysis (CA) to industrial problems and applications are illustrated by Manzini and Bindi (2009) in transportation issues, Bindi et al. (2009) in warehousing and storage systems, Manzini et al. (2006) and (2001) in GT and CM.

Object of this chapter is the introduction, illustration and application of a cluster based systematic procedure for the design of a CM system by the adoption of *general purpose* and *problem oriented similarity* indices.

A general design of a CM system consists of the following three basic activities (Papaioannou and Wilson, 2010):

1. part families formation usually formed according to their processing requirements;
2. machine groups formation. These groups are usually called "manufacturing cells" and "clusters";
3. part families assignment to cells.

Three different strategies to execute these activities can be applied:

1. *Part family identification (PFI) strategy.* Part families are formed first and then machines grouped into families in accordance to the part families formation;
2. *Machine group identification (MGI) strategy.* Manufacturing cells are first created and then parts are allocated to cells;
3. *Part family/machine grouping (PF/MG) strategy.* Part families and manufacturing cells are formed simultaneously.

This chapter adopts the second strategy. As a consequence, this chapter illustrates a systematic procedure for the cell formation problem, i.e. the allocation of machines to cells. The number of cells to be formed is not known in advance. In a second decision step the assignment of manufacturing parts to the previously defined clusters is executed in accordance with a known processing sequence.

The simultaneous parts and machines clustering processes is usually based on the minimization of intercell movement of parts (Stawowy 2004) which specifically deals with the CF problem and methods. In other words, the object is to minimize the interactions between manufacturing cells,

Figure 1. CF solution methods classification

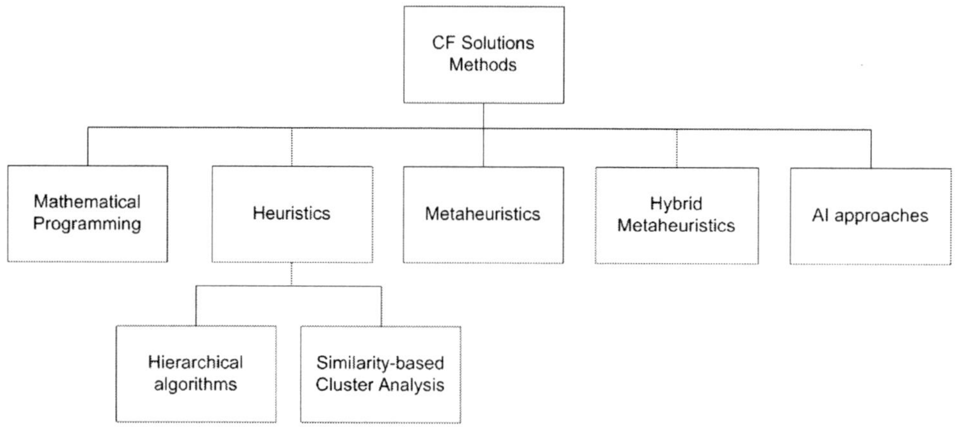

where an interaction occurs if a part requires machines belonging to two or more cells. The degree of interaction between manufacturing cells is measured by the number of the "exceptional elements" as illustrated below in the discussion of the efficiency in the formation process.

The remainder of this chapter is organized as follows: Section 2 presents a literature review on CM and CF problems, Section 3 illustrates the proposed similarity based hierarchical clustering process based on the application of a *threshold level of group similarity* as introduced by Manzini et al. (2010). In particular, Section 3 presents both a set of general purpose similarity indices and a set of problem oriented indices. Section 4 reports the most important clustering performance evaluation metrics. Section 5 illustrates a few significant numerical examples, and Section 6 discusses about the results obtained by an experimental analysis conducted on an instance of the literature adopting different settings of the decision problem. Finally Section 7 presents conclusions and further research.

LITERATURE REVIEW

A survey on CF problem methodologies is presented by Papaioannou and Wilson (2010). They distinguish three main categories: informal/visual methods, part coding analysis methods, and production based methods which can be further classified as follows (see Figure 1):

- cluster analysis (CA)
- graph partitioning approach
- mathematical programming method
- heuristic and metaheuristic algorithms
- artificial intelligent methodologies.

In particular, the CA, which is the adopted set of methods for the CF problem and CM in this chapter, groups either objects or entities into clusters such that the generic cluster is made of individual with high degree of homogeneity, i.e. "natural association", but different clusters have very little association between them.

The proposed CA is supported by the application of "agglomerative methods" where the process of grouping starts with singleton clusters and merges them into larger sets. The systematic approach proposed in this chapter is similarity based, adopts cluster analysis and heuristic agglomerative and hierarchical algorithms.

Figure 2. Systematic procedure for the similarity based hierarchical clustering

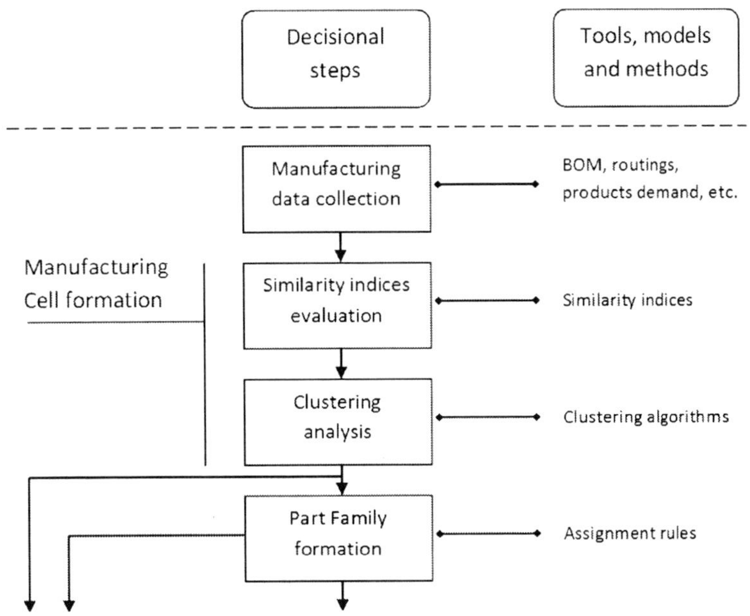

SIMILARITY BASED HIERARCHICAL CLUSTERING PROCESS

The clustering process of parts and machines for CM presented and applied in this chapter is based on the adoption of similarity indices both "general purpose" and "problem oriented". The main decisional steps are cited by Manzini et al. (2010) and are discussed in this section in detail. Figure 2 illustrates the proposed systematic procedure for the formation of manufacturing cells and homogeneous part families. This procedure takes inspiration from the *cluster analysis (CA)*, as a set of statistics based techniques and tools designed for grouping items and partitioning a set of elements in different research areas, e.g. economics, medicine, biology, etc., and is made of the following main steps.

Step 1: Manufacturing Data Collection

It deals with the analysis of product mix including all parts and components, both *product structure*, e.g. the bill of materials (BOM), and *production process*, e.g. the manufacturing work cycle, frequently called routing. The level of detail for the collection of data is significantly influenced by the adopted similarity index for the CA. In particular, in presence of a *general purpose index* the necessary data generally refer to the assignment of products and components, the so called "parts", to the manufacturing resources, the so called "machines". The adoption of a *product oriented similarity index* justifies the collection of several product data, e.g. expected production demand, manufacturing process unit time eventually including the set-up time, sequence of visited machines, existence of alternative manufacturing routines, tools assignment and availability, production costs, etc.

The basic and elementary data collection activity is usually supported by the construction of the well known *part-machine incidence matrix* (PMIM) whose generic entry a_{ik} is defined as follows:

Figure 3. Similarity coefficients by Yin and Yasuda (2006)

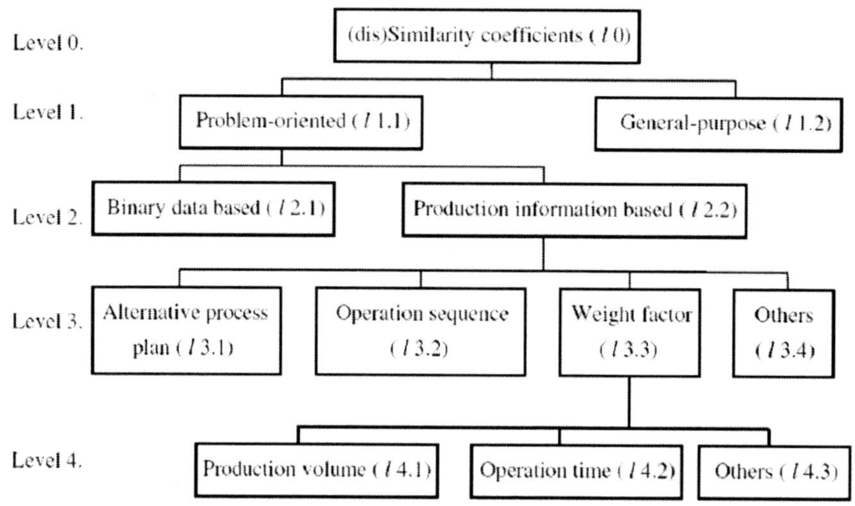

$$a_{ik} = \begin{cases} 1 \text{ if part i visits machine k} \\ 0 \text{ otherwise} \end{cases} \qquad (1)$$

where

$i=1,..,I$ part index
$k=1,..,K$ machine index.

The transpose of the part-machine matrix is known as the *machine-part incidence matrix* (MPIM).

An example of a MPIM is illustrated in Table 3. Obviously, the collection of data for the evaluation of problem oriented indices is more onerous, but the performance of the clustering process is expected to be higher than in presence of not problem oriented indices.

Step 2: Similarity Index Evaluation

The clustering activity is a process of grouping and set-partitiong items in homogeneous and disjunctive groups, called "clusters". The generic cluster of parts is called "part-family", while the cluster of machines is called "manufacturing cell". A similarity index refers to a generic pair of items of the same typology in the beginning mix of parts and machines to be partitioned. The index measures the degree to which two items, e.g. two different machines, need to belong to the same cluster, i.e. a manufacturing cell of homogeneous machines.

The literature presents a very large number of similarity coefficients. A taxonomy for these indices is presented by Yin and Yasuda (2006). They distinguish the previously cited two distinct main groups of coefficients*: problem-oriented* (*l*1.1 in Figure 3) and *general purpose* (*l*1.2). Problem-oriented measures are specifically designed for application to manufacturing problems in industry, while general purpose are used in many disciplines, e.g. medicine, sociology, biology, economics, decision science, etc.

Table 1 lists a not exhaustive set of general purpose indices S_{ij} as proposed by the literature and based on the following assumptions: given two machines i and j, a is the number of parts visiting both machines, b is the number of parts visiting machine i but not j, c is the number of parts visiting machine j but not i, d is the number

Table 1. General purpose similarity indices

Code	Coefficient	Range	S_{ij}
B	Baroni-Urbani and Buser	0–1	$[a+(ad)^{1/2}]/[a+b+c+(ad)^{1/2}]$
H	Hamann	-1 to 1	$[(a+d)-(b+c)]/[(a+d)+(b+c)]$
J	Jaccard	0–1	$a/(a+b+c)$
O	Ochiai	0–1	$a/[(a+b)(a+c)^{1/2}]$
R	Rogers and Tanimoto	0–1	$(a+d)/[a+2(b+c)+d]$
RM	Sarker and Islam (Relative matching)	-1 to 1	$[a+(ad)^{1/2}]/[a+b+c+d+(ad)^{1/2}]$
SK	Sokal and Sneath	0–1	$2(a+d)/[2(a+d)+b+c]$
SO	Sorenson	0–1	$2a/(2a+b+c)$
SI	Sokal and Michener (Simple matching)	0-1	$(a+d)/(a+b+c+d)$
RR	Russel and Rao	0-1	$a/(a+b+c+d)$

of parts visiting neither machine *i*, *j*. These coefficients are applied in the numerical example and experimental analysis illustrated in this chapter.

Jaccard - J is the most commonly used general-purpose similarity coefficient in the literature. Jaccard similarity coefficient between machine *i* and machine *j* is defined as follows (McAuley, 1972):

$$S_{ij} = \frac{a}{a+b+c}, \quad 0 \le S_{ij} \le 1 \qquad (2)$$

Problem oriented similarity coefficients can be classified into binary data based (*l*2.1 in Figure 3) and production information based (*l*2.2) similarity coefficients. The similarity coefficients at the level *l*2.1 only consider assignment information (a part need or need not a machine). Yin and Yamada (2006) present a list of binary data problem oriented indices. The level 3 (*l*3), see Figure 3, introduces different manufacturing factors for cell formation, e.g. machine requirement, machine setup, utilization, workload, alternative routings, machine capacities, operation sequences, set-up cost and cell layout. Finally level 4 (*l*4) introduces different weights for the evaluation of weighted similarity coefficients in order to adjust the strength of matches or misses between objects pairs to reflect the resemblance value more realistically (Yin and Yamada, 2006).

A few examples of problem oriented measures are reported below and are based on the following basic assumptions: n is the number of parts, $k=1,\dots,n$ the generic part, m the number of machines and $i,j = 1,\dots, m$ the generic machine (Alhourani and Seifoddini 2007). Then:

$$x_{ijk} = \begin{cases} 1 \text{ if part } k \text{ visits both machine } i \text{ and machine } j \\ 0 \end{cases}$$

$$y_{ijk} = \begin{cases} 1 \text{ if part } k \text{ visits either machine } i \text{ or machine } j \\ 0 \end{cases}$$

S - Seifoddini (1987). He modifies the Jaccard's coefficient incorporating the production volume as follows:

$$S_{ij} = \frac{\sum_{k=1}^{n} x_{ijk} \cdot m_k}{\sum_{k=1}^{n} y_{ijk} \cdot m_k + \sum_{k=1}^{n} x_{ijk} \cdot m_k} \qquad (3)$$

where m_k planned production volume during a predefined period of time for part type *k*.

GS - Gupta and Seifoddini (1990). They extend the Jaccard's similarity coefficient to incorporate the effect of the operational time, operational sequence and production volume. The proposed index is:

$$S_{ij} = \frac{\sum_{k=1}^{n}(x_{ijk} \cdot T_{ijk} + \sum_{o=1}^{n_k} z_{ko}) \cdot m_k}{\sum_{k=1}^{n}(x_{ijk} \cdot T_{ijk} + \sum_{o=1}^{n_k} z_{ko} + y_{ijk}) \cdot m_k} \quad (4)$$

where n_k number of times the part type k visits both machines in row, i.e. sequentially

$$z_{ko} = \begin{cases} 1 \text{ if part } k \text{ visits both machines } i \text{ and } j \text{ in row} \\ 0 \end{cases}$$

$$T_{ijk} = \frac{\min\left\{T_{ik}, T_{jk}\right\}}{\max\left\{T_{ik}, T_{jk}\right\}} \quad (5)$$

T_{ik} global time spent by part type k on machine i
T_{jk} global time spent by part type k on machine j.

SH - Seifoddini and Hsu (1994). They propose a weighted similarity coefficient with the aim of eliminating *improper machine assignment* by giving a higher weight to components having common operations on both the machines.

$$S_{i,j} = \frac{\sum_{k=1}^{n} f_{bk} \cdot x_{ijk}}{\sum_{k=1}^{n} f_{bk} \cdot x_{ijk} + \sum_{k=1}^{n} f_{ek} \cdot y_{ijk}} \quad (6)$$

where

f_{bk} weighting factor for parts visiting both machines i and j
f_{ek} weighting factor for parts visiting either machine i or j but not both.

In the numerical example and experimental analysis illustrated below, the adopted value of f_{bk} and f_{ek} are respectively 0.6 and 0.4.

N - Nair and Narendran (1998). They propose a similarity measure as the ratio of the sum of the movements common to the machines i and l, and the sum of the total number of movements to and from machines i and l.

$$s_{il} = \frac{c_i + c_l}{t_i + t_l} \quad (7)$$

$$t_i = \sum_{j=1}^{n}\sum_{p=1}^{n_{ji}} w_j t_{jip} \quad (8)$$

$$c_{il} = \sum_{j=1}^{n}\sum_{p=1}^{n_{jil}} w_j c_{jilp} \quad (9)$$

where

m number of machines
n number of parts
n_{ji} number of times part j visits the ith machine
n_{jil} number of times part j visits the ith and jth machine
b_{jip} operation sequence number if the part j visits its machine i for pth time ($0 \leq p \leq n_{ji}$); zero otherwise
b_{jlp} operation sequence number if the part j visits its machine 1 for pth time ($0 \leq p \leq n_{ji}$); zero otherwise
r_j maximum number of operations for part j

$$t_{jip} = \begin{cases} 0 \ if \ b_{jip} = 0 \\ 1 \ if \ b_{jip} = 1 \ or \ r_j \\ 2 \text{ otherwise} \end{cases} \qquad c_{jilp} = \begin{cases} 0 \ if \ b_{jip} = 0 \ or \ b_{jlp} = 0 \\ 1 \ if \ b_{jip} = 1 \ or \ r_j \\ 2 \text{ otherwise} \end{cases} \quad (10)$$

Equation (8) accounts for the total number of movements to and from the machine i. Equation (9) quantifies the total number of movements to and from machine i made by all parts that visit both machine i and machine l.

Step 3: Clustering Analysis and Manufacturing Cells Formation

The Clustering step deals with the grouping of machines into different homogeneous clusters so that machines in each cluster have high values of correlation. The clustering analysis is supported by the application of a grouping algorithm, which is a hierarchical heuristic for the partitioning of machines into disjunctive cells. There are usually substantial differences between the machine groups obtained by clustering, but the individuals within an individual machine group are similar because they are similarly visited by different parts/components. The clustering process is generally supported by one of the following well known hierarchical algorithms (Mosier 1989): Complete Linkage Method (CLINK) also known as farthest neighbor (fn) clustering, Single Linkage Method (SLINK) also known as nearest neighbor (*nn*) clustering, Unweighted Pair-Group Method using Arithmetic Average (UPGMA), Weighted Pair-Group Method using Arithmetic Average (Average Linkage), and Unweighted Pair-Group Method using Centroid (UPGMC).

In particular, the authors (of this chapter) choose to illustrate the *fn* clustering and the *nn* clustering algorithm (Aldenderfer and Blashfield, 1984). The generic algorithm is based on a scheme that erases rows and columns in a similarity matrix S that collects the values of similarity for each couple of items, as old clusters are merged into new ones by the degradation of the level of similarity within each under construction cluster. The similarity matrix S (dimension $N x N$) contains all the correlations among the machines $s(i,j)$, calculated as described in the previous step. To the clusterings are assigned sequence numbers $0, 1, \ldots, (n-1)$, and $L(k)$ is the level of similarity of the k-th clustering. A cluster with sequence number m is denoted (m) and the correlation between clusters (r) and (s) is denoted $s[(r),(s)]$. In particular, the *nn* algorithm is composed of the following steps:

1. Begin with the disjoint clustering having level $L(0) = 1$ and sequence number $m = 0$.

Find the least dissimilar pair of clusters in the current clustering, say pair (r) and (s), according to:

$$s\left[\left(r\right),\left(s\right)\right] = \max_{i,j}\left\{s\left[\left(i\right),\left(j\right)\right]\right\} \tag{11}$$

where the maximum is over all pairs of clusters in the current clustering.

Increment the sequence number: $m = m + 1$. Merge the groups of items (r) and (s) into a single cluster to form the next clustering m. Set the level of this clustering to:

$$L(m)=s[(r),(s)] \tag{12}$$

Update the similarity matrix S by deleting the rows and columns corresponding to clusters (r) and (s) and adding a row and column corresponding to the newly formed cluster. The level of correlation between the new cluster, denoted (r,s) and the old generic cluster (k) is defined as:

$$s[(k), (r,s)] = \max\ \{s[(k),(r)], s[(k),(s)]\} \tag{13}$$

If all the machines are in one cluster, stop. Else, go to point 2.

The *fn* algorithm is based on the same scheme and equation (13) in point 4 is modified as follows:

$$s[(k), (r,s)] = \min\ \{s[(k),(r)], s[(k),(s)]\} \tag{14}$$

The dendrogram is the graphical representation of the process of degradation of the similarity level as the result of the grouping process executed by the clustering algorithm (Sokal and Sneath 1968, McAuley 1972). Figure 4 exemplifies a dendro-

Figure 4. Dendrogram - Simple Matching and farthest neighbour, [1-11] nodes

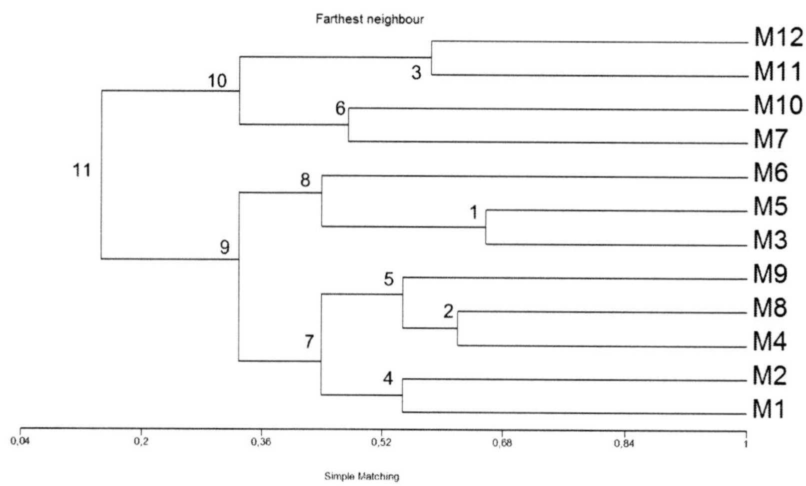

gram generated by the execution of the clustering process to an instance of example.

As a consequence the generic algorithm classifies the machines on the basis of the similarity of the manufacturing characteristics and hierarchically executes aggregations of machines into clusters reducing progressively the level of similarity, i.e. homogeneity, within the under construction heterogeneous clusters.

The result of the clustering process does not only depend on the rule adopting for grouping but also on the threshold level of similarity adopted for grouping, i.e. the minimal admissible value of similarity within the generic cluster as discussed by Manzini et al. (2010). In particular, they demonstrate that a best performing percentile based level of threshold similarity exists to optimize the cell formation process in presence of general purpose similarity metrics. The percentile based threshold value is a range of group similarity measurements which cuts the dendrogram at the percentile number of aggregations identified by the clustering rule, as follows:

$$T\_value_{\%_p} \in \left] simil\left\{\left\lceil \%_p \times N_{nodes} \right\rceil\right\}, simil\left\{\left\lfloor \%_p \times N_{nodes} \right\rfloor\right\}\right[\quad (15)$$

where

$\%_p$ percentile of aggregations, expressed as a percentage

N_{nodes} number of nodes generated by the clustering algorithm

$Simil\{N\}$ similarity value which corresponds to the node N.

The basic idea of this criterion is properly illustrated by the example discussed in Section 5. Figure 4 presents a dendrogram generated by the application of clustering analysis. *m1,..,m12* are the identifications of machine items. The number within the diagram (1,..,12) identify the aggregations (called nodes) ordered in agreement with the similarity measurements. In particular, low numbers identify aggregations between under construction clusters characterized by a high level of similarity (very similar clusters). Table 5 reports the list and configuration of nodes as generated by the application of the clustering algorithm in agreement with

Figure 4. Assuming the 80° percentile of aggregation (i.e. $\%_p=0.80$) as follows (see Equation (15)):

$$T\_value_{80^\circ} \in \left]simil\left\{\left\lceil 0.80\times 12\right\rceil\right\}, simil\left\{\left\lfloor 0.80\times 12\right\rfloor\right\}\right[= \\ = \left]simil\{10\}, simil\{9\}\right[= \left]0.329,\ 0.329\right[= 0.329$$

Similarly, for a percentile value of 20° (i.e. $\%_p=0.20$):

$$T\_value_{20^\circ} \in \left]simil\left\{\left\lceil 0.2\times 12\right\rceil\right\}, simil\left\{\left\lfloor 0.2\times 12\right\rfloor\right\}\right[= \\ = \left]simil\{3\}, simil\{2\}\right[= \left]0.585,\ 0.622\right[$$

Step 4: Part Families Formation

Given a generic solution for cells, a part may have to visit more than one group of machines before it is completed. Consequently, the generic part has to be assigned to the manufacturing cell with the minimum number of inter-group journeys. Another way of reducing the number of inter-group journeys is to duplicate machines, but it can be very expensive in terms of space and monetary costs.

Part families can be formed concurrently with the cell/group machine formation (CF steps illustrated above), or otherwise executed after the cells have been defined. In particular, the second of these hypotheses is adopted in this paper, and in particular a hierarchical heuristic rule is applied to assign parts to manufacturing cells. The main steps are:

- Given a configuration of the disjunctive groups of machines (i.e. manufacturing cells) name them as $c=1,..,C$. Then quantify the following measurements for each part i and each manufacturing cell c in accordance with the working cycle of part i:
  - number of intra-cell movements: ICM_{ic};
  - number of tasks executed in the cell: $NTask_{ic}$;
  - processing time of i in c: $Time_{ic}$.
  - Assign part i to cell c^* where:

$$ICM_{ic^*} > \max_{\substack{c=1,..,C \\ c\neq c^*}} \left\{ICM_{ic}\right\} \tag{16}$$

If c^* does not exist than GO to STEP III.

- Assign part i to cell c' shown in Equation 17 (Box 1). If c' does not exist than GO to STEP IV.
- Assign part i to cell c'' shown in Equation 18 (Box 1). If c'' does not exist than assign part i randomly to a cell in the last equation shown in Box 1.

A result of the assignment of parts to the manufacturing cells is the so-called *block-diagonal incidence matrix* shown in Figure 5 as the result of the application of the proposed procedure to the instance illustrated in Section 5.

Step 5: Plant Layout Configuration

This step deals with the determination of the location of each manufacturing resource (machines and human resources) in the production area. Layout decisions are significantly influenced by the configuration of cells and part families in CM systems, but they are omitted in this chapter. Nevertheless a few significant performance measures on layout results are quantified as clearly explained below.

CLUSTERING PERFORMANCE EVALUATION

Sarker (2001) presents, discusses, and compares the most notable measurements of grouping efficiency in CM. The measurements adopted in the following illustrated experimental analysis are based on the following definitions:

Box 1.

$NTask_{ic'} > \max\limits_{\substack{c=1,\dots,C \\ c\neq c'}} \left\{ NTask_{ic} : ICM_{ic} = \max\limits_{\substack{c^*=1,\dots,C \\ c^*\neq c}} \left\{ ICM_{ic^*} \right\} \right\}$	(17)
$Time_{ic''} > \max\limits_{\substack{c=1,\dots,C \\ c\neq c''}} \left\{ Time_{ic} : ICM_{ic} = \max\limits_{\substack{c^*=1,\dots,C \\ c^*\neq c}} \left\{ ICM_{ic^*} \right\} \wedge NTask_{ic} = \max\limits_{\substack{c'=1,\dots,C \\ c'\neq c}} \left\{ Ntask_{ic'} \right\} \right\}$	(18)
$\left\{ c = 1,\dots,C : ICM_{ic} = \max\limits_{\substack{c^*=1,\dots,C \\ c^*\neq c}} \left\{ ICM_{ic^*} \right\} \wedge NTask_{ic} = \max\limits_{\substack{c'=1,\dots,C \\ c'\neq c}} \left\{ NTask_{ic'} \right\} \wedge Time_{ic} = \max\limits_{\substack{c''=1,\dots,C \\ c''\neq c}} \left\{ Time_{ic''} \right\} \right\}$	

Figure 5. Block-diagonal matrix. Simple matching, farthest neighbour & 75° percentile.

	p11	p12	p16	p17	p5	p6	p13	p14	p15	p18	p19	p9	p10	p1	p2	p3	p4	p7	p8
Cell 1																			
m11		1	1	1			1	1	1										
m12	1	1	1	1			1												
Cell 2																			
m7		1		1	1	1	1	1		1	1				1	1	1		
m10					1	1	1	1	1	1	1								
Cell 3																			
m6	1				1	1				1		1	1		1			1	1
m3												1	1						1
m5												1							1
Cell 4																			
m1					1									1	1	1	1		
m2															1	1			1
m9					1	1						1		1		1	1	1	1
m4												1	1	1	1	1	1	1	1
m8						1						1	1	1	1	1		1	1

- *lock*. This is a submatrix of the machine-part matrix composed of rows representing a part family and columns representing the related machine cell.
- *Void*. This is a "zero" element appearing in a diagonal block (see Figure 4).
- *Exceptional element*. This is a "one" appearing in off-diagonal blocks (see Figure 4). The exceptional element causes inter-cell movements.

A set of CM measurements of performance quantified in the proposed experimental analysis is now reported and discussed. (high) and (low) labels refer to the expected values for best performing the CF and CM.

Problem Density: PD

$$PD = \frac{\text{number of "ones" in the incidence matrix}}{\text{number of elements in the incidence matrix}} \tag{19}$$

Global Inside cells density: ICD (high)

$$ICD = \frac{\text{number of "ones" in diagonal blocks}}{\text{number of elements in diagonal blocks}} \tag{20}$$

Exceptional elements: EE (low)

EE = Number of exceptional elements in the off-diagonal blocks (21)

Ratio of non-zero elements in cells: REC

$$REC = \frac{\text{total number of "ones"}}{\text{number of elements in diagonal blocks}} \tag{22}$$

Grouping Efficiency: η (Sarker 2001) (high)

It is a weighted average of two functions and it is defined as:

$$\eta = q\eta_1 + (1-q)\eta_2$$
$$\eta_1 = \frac{e_d}{\sum_{r=1}^{k} M_r N_r} \tag{23}$$
$$\eta_2 = 1 - \frac{e_o}{mn - \sum_{r=1}^{k} M_r N_r}$$

where

e_d number of 1s in the diagonal blocks
e_o number of 1s in the off-diagonal blocks
k number of diagonal blocks
M_r number of machines in the rth cell
N_r number of components in the rth part-family
q weighting factor (0≤q≤1) that fixes the relative importance between voids and inter-cell movements. If q=0.5 both get the same importance: this is the value adopted in the numerical example and in the experimental analysis illustrated in sections 5 and 6.

Quality Index: QI (Seifoddini and Djassemi 1994, 1996) (high)

It is a measure of independence of machine-component groups. High values of QI are expected in presence of high independency. QI is defined as:

$$QI = 1 - \frac{ICW}{PW} \tag{24}$$

where

ICW total intercellular workload
PW total plant workload.
ICW and PW can be defined as:

$$ICW = \sum_{c=1}^{C} \sum_{i=1}^{m} \left[Y_{ic} \left(\sum_{k=1}^{n} (1 - Z_{kc}) X_{ik} m_k T_{ik} \right) \right] \tag{25}$$

$$PW = \sum_{i=1}^{m} \sum_{k=1}^{n} X_{ik} m_k T_{ik} \tag{26}$$

where n is the number of parts, $k=1,\dots,n$ the generic part, m the number of machines and $i,j=1,\dots,m$ the generic machine. This is the notation previously introduced

$$Y_{ic} = \begin{cases} 1 \text{ if machine } i \text{ is assigned to cell } c \\ 0 \text{ otherwiswe} \end{cases}$$
$$Z_{kc} = \begin{cases} 1 \text{ if part } k \text{ is assigned to cell } c \\ 0 \text{ otherwiswe} \end{cases}$$
$$X_{ki} = \begin{cases} 1 \text{ if part } k \text{ has operation on machine } i \\ 0 \text{ otherwiswe} \end{cases}$$

m_k volume of part k
T_{ki} processing time of part k on machine i

QI measures the number of intracellular movements which ask to be maximized minimizing intercellular ones.

Now the authors introduce a new grouping efficiency based on QI as previously defined.

Grouping Efficiency based on QI: ηQI (high)

$$\eta_{QI}=q\eta_1+(1-q)QI \tag{27}$$

The adopted value of weighting factor q is:

$$q=\frac{\sum_{r=1}^{k} M_r N_r}{mn} \tag{28}$$

Grouping Efficacy: τ (Kumar and Chandrasekharan 1990) (high)

Group efficacy can be quantified by the application of the following equation:

$$\tau=\frac{e-e_0}{e+e_v} \tag{29}$$

where

e total number of "ones" in the matrix (i.e. the total number of operations)

$e_0=EE$ number of exceptional elements (number of "ones" in the off-diagonal blocks)

e_v number of voids (number of "zeros" in the diagonal blocks).

Grouping measure: ηG (Miltenburg and Zhang 1991) (high)

It gives higher values if both the number of voids and exceptional elements are fewer, and it is defined as:

$$\begin{aligned}\eta_g &= \eta_u - \eta_m\\ \eta_u &= \frac{e_1}{e_1+e_v}\\ \eta_m &= \frac{e_o}{e}\end{aligned} \tag{30}$$

where

η_u ratio of the number of 1s to the number of total elements in the diagonal block (this is the inside cell density - ICD)

η_m ratio of exceptional elements to the total number of 1s in the matrix

e_l number of 1s in the diagonal block.

Group technology efficiency: GTE (Nair and Narendran 1998) (high)

It is defined as the ratio of the difference between the maximum number of inter-cell travels possible and the number of inter-cell travels actually required to the maximum number of inter-cell possible:

$$\begin{aligned}GTE &= \frac{I-U}{I}\\ I &= \sum_{j=1}^{n}(r_j-1)\\ U &= \sum_{j=1}^{n}\sum_{s=1}^{r_j-1} xl_{js}\end{aligned} \tag{31}$$

where

- maximum number of inter-cell travels

U number of inter-cell movements required by the system

r_j maximum number of operations for component j

$$xl_{js}=\begin{cases}0 \text{ if operations } s,\ s+1 \text{ are performed in the same cell}\\ 1 \text{ otherwise}\end{cases}$$

Table 2. Manufacturing input data, De Witte (1980)

Part	Volume	Work Cycle	Processing Time
p1	2	m1, m4, m8, m9	20, 15, 10, 10
p2	3	m1, m2, m6, m4, m8, m7	20, 20, 15, 15, 10, 25
p3	1	m1, m2, m4, m7, m8, m9	20, 20, 15, 25, 10, 15
p4	3	m1, m4, m7, m9	20, 15, 25, 15
p5	2	m1, m6, m10, m7, m9	20, 15, 20, 25, 15
p6	1	m6, m10, m7, m8, m9	15, 50, 25, 10, 15
p7	2	m6, m4, m8, m9	15, 15, 10, 15
p8	1	m3, m5, m2, m6, m4, m8, m9	30, 50, 20, 15, 15, 10, 15
p9	1	m3, m5, m6, m4, m8, m9	30, 50, 15, 15, 10, 15
p10	2	m3, m6, m4, m8	30, 15, 15, 10
p11	3	m6, m12	15, 20
p12	1	m11, m7, m12	40, 25, 20
p13	1	m11, m10, m7, m12	40, 50, 25, 20
p14	3	m11, m7, m10	40, 25m 50
p15	1	m11, m10	40, 50
p16	2	m11, m12	40, 20
p17	1	m11, m7, m12	40, 25m 20
p18	3	m6, m7, m10	15, 25, 50
p19	2	m10, m7	50, 25

Bond efficiency: BE (high)

This is an important index because depends on both the within-cell compactness (by the ICD) and the minimization of inter-cell movements by the GTE. It is defined as:

$$BE=q\cdot IDC+(1-q)\cdot GTE \quad (32)$$

The adopted value of weight q in the experimental analysis is 0.5.

NUMERICAL EXAMPLE

This section presents a numerical example which relates to a problem oriented instance presented by De Witte (1980) and made of 19 parts and 12 machines. Manufacturing input data are reported in Table 2.

Table 3 reports the *12x19 machine-part* incidence matrix useful for the evaluation of a general purpose similarity index.

A General Purpose Evaluation

This section presents the results obtained by the application of a general purpose similarity index in cluster analysis for the cell formation problem.

Table 4 reports the result of the evaluation of the general purpose index known as Simple Matching (SI) and defined in Table 1. Figure 4 shows the dendrogram generated by the application of the fn combined with the SI similarity coefficient. In particular a sequence of numbers is explicitly reported in figure for each node of the diagram. The generic node corresponds to a

Table 3. Machine-part incidence matrix

	p1	p2	p3	p4	p5	p6	p7	p8	p9	p10	p11	p12	p13	p14	p15	p16	p17	p18	p19
m1	1	1	1	1	1														
m2		1	1					1											
m3								1	1	1									
m4	1	1	1	1			1	1	1	1									
m5								1	1										
m6		1			1	1	1	1	1	1	1							1	
m7		1	1	1	1	1						1	1	1			1	1	1
m8	1	1	1			1	1	1	1	1									
m9	1		1	1	1	1	1	1	1										
m10					1	1							1	1	1			1	1
m11												1	1	1	1	1	1		
m12											1	1	1			1	1		

specific aggregation ordered in agreement with the similarity metric and the adopted hierarchical rule. The list of nodes and aggregations, the related values of similarity, and the number of objects per group are also reported in Table 5. The obtained number of nodes is 11.

Now it is possible to define a partitioning of the available set of machines by the identification of a cut value, the so called "cutting threshold similarity value". The adopted level of homogeneity within the generic cluster is the percentile-based threshold measure discussed in Section 3.

Given the dendrogram in Figure 4 and assuming a threshold percentile cut value equal to 20°, the corresponding range of similarity is (0.585, 0.622) as demonstrated in Section 3. The obtained configuration of the manufacturing cells (nine different cells are obtained) is:

Cell 1 (single machine): M12
*Cell 2 (*single *machine): M11*
*Cell 3 (*single *machine): M10*
Cell 4 (single machine): M7
*Cell 5 (*two *machines): M3, M5*
Cell 6 (single machine): M9
Cell 7 (two machines*): M8, M4*
*Cell 8 (*single *machine): M2*
Cell 9 (single machine): M1

In case a cut value corresponds to one or more nodes generated by the hierarchical process of aggregation, it is possible to include (exclude) the node in the formation of cells. In particular, assuming a level of threshold similarity equal to 80°, two alternative configurations can be obtained as the result of inclusion/exclusion of one or more nodes of the dendrogram located in correspondence of the cutting level:

- Including node 10 and node 9
  Cell 1 (four machines): M12, M11, M10 and M7
  Cell 2 (eight machines): M6, M5, M3, M9, M8, M4, M2, M1.
- Not including node 10 and node 9
  Cell 1 (two machines): M12, M11
  Cell 2 (two machines): M10, M7
  Cell 3 (3 machines): M6, M5, M3
  Cell 4 (5 machines): M9, M8, M4, M2, M1.

The second column in Table 8 reports the obtained values of the performance evaluation for the case study object of this numerical example adopting the Simple Matching similarity index, the

Table 4. Simple matching similarity matrix

	m1	m2	m3	m4	m5	m6	m7	m8	m9	m10	m11	m12
m1	1.0000											
m2	0.5479	1.0000										
m3	0.4021	0.5479	1.0000									
m4	0.5118	0.5118	0.5118	1.0000								
m5	0.4389	0.5847	0.6576	0.4750	1.0000							
m6	0.3292	0.4021	0.4750	0.4389	0.4389	1.0000						
m7	0.4021	0.3292	0.1826	0.2195	0.2195	0.2556	1.0000					
m8	0.4389	0.5118	0.5118	0.6215	0.4750	0.5118	0.2194	1.0000				
m9	0.5118	0.4389	0.4389	0.5479	0.4750	0.4389	0.2924	0.5479	1.0000			
m10	0.3292	0.3292	0.3292	0.1465	0.3653	0.3292	0.4750	0.2194	0.2924	1.0000		
m11	0.2924	0.3653	0.3653	0.1826	0.4021	0.1465	0.3653	0.1826	0.1826	0.4389	1.0000	
m12	0.3292	0.4021	0.4021	0.2194	0.4389	0.2556	0.3292	0.214	0.2194	0.3292	0.5847	1.0000

Table 5. List and configuration of nodes generated by fn rule & SI similarity coefficient

Node	Group 1	Group 2	Simil.	Objects in Group
1	M3	M5	0.658	2
2	M4	M8	0.622	2
3	M11	M12	0.585	2
4	M1	M2	0.548	2
5	Node 2	M9	0.548	3
6	M7	M10	0.475	2
7	Node 4	Node 5	0.439	5
8	Node 1	M6	0.439	3
9	Node 7	Node8	0.329	8
10	Node 6	Node 3	0.329	4
11	Node 9	Node 10	0.146	12

fn heuristic, and the cutting threshold percentile value equal to 75°.

A Problem Oriented Evaluation

Table 6 reports the result of the evaluation of the problem oriented similarity coefficient as proposed by Nair and Narendran (1998). Figure 6 shows the dendrogram as the result of the application of the fn clustering rule. The generic node of the dendrogram corresponds to a specific aggregation ordered in agreement with the adopted similarity metric and the adopted hierarchical rule. The list of nodes and aggregations, the related values of similarity, and the number of objects in group are also reported in Table 7 as the result of the application of the fn rule and Nair and Narendran (1998) problem oriented similarity coefficient. The obtained number of nodes is 11.

Figure 6 reports the dendrogram obtained by the application of the fn clustering heuristic rule and the "Nair and Narendran" problem oriented similarity coefficient to the literature instance of interest.

Assuming $\%_p$=80°:

$$T\_value_{80°} \in \left]simil\left\{\lceil 0.80\times 11\rceil\right\}, simil\left\{\lfloor 0.80\times 11\rfloor\right\}\right[= \left]simil\{9\}, simil\{8\}\right[= \left]0.095,\ 0.222\right[$$

The obtained configuration of the manufacturing cells (four different cells are obtained) is:

Cell 1 (two machines): M11, M12
Cell 2 (two machines): M7, M10

Table 6. Nair & Narendran similarity matrix

	m1	m2	m3	m4	m5	m6	m7	m8	m9	m10	m11	m12
m1	1.0000											
m2	0.5000	1.0000										
m3	0.0000	0.2220	1.0000									
m4	0.6920	0.5000	0.4210	1.0000								
m5	0.0000	0.2860	0.6670	0.2350	1.0000							
m6	0.3450	0.3480	0.3640	0.5450	0.2000	1.0000						
m7	0.5620	0.3080	0.0000	0.3890	0.0000	0.4620	1.0000					
m8	0.5000	0.5560	0.4710	0.8570	0.2670	0.6450	0.2940	1.0000				
m9	0.6670	0.2220	0.2350	0.7150	0.2670	0.4520	0.4720	0.6150	1.0000			
m10	0.1670	0.0000	0.0000	0.0000	0.0000	0.3870	0.7060	0.0770	0.2310	1.0000		
m11	0.0000	0.0000	0.0000	0.0000	0.0000	0.0000	0.4000	0.0000	0.0000	0.4550	1.0000	
m12	0.0000	0.0000	0.0000	0.0000	0.0000	0.2310	0.2070	0.0000	0.0000	0.0950	0.5880	1.0000

Figure 6. Dendrogram by the application of Nair & Narendran similarity coefficient and the farthest neighbour

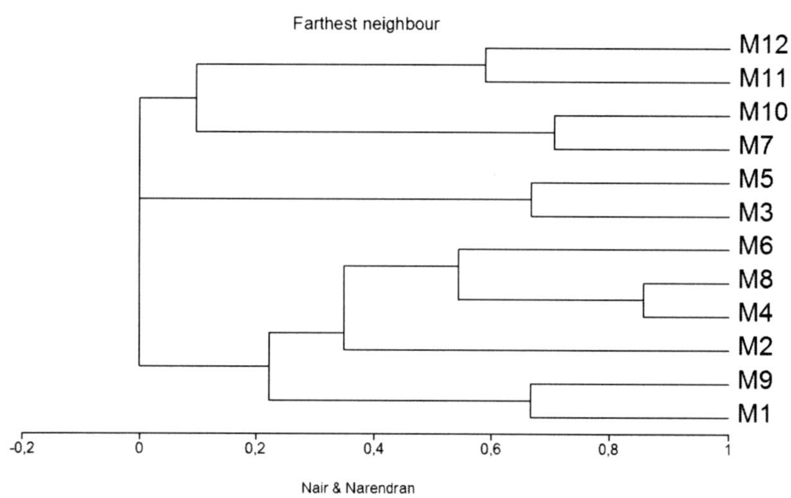

Cell 3 (two machines): M3, M5
Cell 4 (six machines): M6, M8, M4, M2, M9, M1
Assuming $\%_p$=20°:

$$T\_value_{20^\circ} \in \left]simil\left\{\left[0.20\times 11\right]\right\}, simil\left\{\left[0.20\times 11\right]\right\}\right[=$$
$$= \left]simil\left\{3\right\}, simil\left\{2\right\}\right[= \left]0.667,\ 0.706\right[$$

The obtained configuration of the manufacturing cells (eleven different cells are obtained) is:

Single machine cells: Cell 1(M12), Cell 2(M11), Cell 4(M5), Cell 5(M3), Cell 6(M3), Cell7(M6), Cell 9(M2), Cell 10(M9), Cell 11(M1)

Table 7. List and configuration of nodes generated by the fn rule & Nair and Narendran (1998) similarity coefficient

Node	Group 1	Group 2	Simil.	Objects in Group
1	M4	M8	0.857	2
2	M7	M10	0.706	2
3	M1	M9	0.667	2
4	M3	M5	0.667	2
5	M11	M12	0.588	2
6	Node 1	M6	0.545	3
7	M2	Node 6	0.348	4
8	Node 3	Node 7	0.222	6
9	Node 2	Node 5	0.095	4
10	Node 8	Node 4	0	8
11	Node 10	Node 9	0	12

Table 8. Performance evaluation of numerical example; 75° percentile

Similarity index	ID	Simple matching	Nair & Narendran
Problem Density	PD	0.329	0.329
Inside Cells Density	ICD	0.705	0.828
REC	REC	0.962	1.293
Exceptional Element	EE	20	27
Grouping Efficiency [%]	η	60.2	57.9
Grouping Efficiency QI [%]	η_{QI}	68.8	72.8
Group Technology Efficiency [%]	GTE	61.9	45.5
Bond Efficiency [%]	BE	66.2	66.1
Group Efficacy [%]	τ	82.9	84.5
Grouping measure	η_G	0.438	0.468

Double machines cells: Cell 3 (M7, M10), Cell 8 (M8, M4).

The third column in Table 8 reports the obtained values of the performance evaluation for the case study object of this numerical example adopting the "Nair and Narendran" similarity index, the fn heuristic, and the cutting threshold percentile value equal to 75°.

Which is the best similarity index? It is not correct to try to reply to this question as is, because previous sections demonstrate that there are different factors affecting the performance of the system configuration: the similarity index, the clustering rule, the threshold cutting value of similarity, and the part assignment rule. As a consequence it is useful to measure the simultaneous effects generated by different combinations of these critical factors. Next section presents an experimental analysis conducted on the instance proposed by De Witte (1980) comparing the performance obtained adopting general purpose and problem oriented similarity metrics.

Figure 7. Block-diagonal matrix. Nair and Narendran, farthest neighbour.& 75° percentile.

	p13	p14	p15	p18	p19	p11	p12	p16	p17	p1	p2	p3	p4	p5	p6	p7	p8	p9	p10
Cell 1																			
m7	1	1		1	1		1		1		1	1	1	1	1				
m10	1	1	1	1	1									1	1				
m3																	1	1	1
m5																	1	1	
Cell 2																			
m11	1	1	1				1	1	1										
m12	1					1	1	1	1										
Cell 3																			
m1										1	1	1	1	1					
m9										1		1	1	1	1	1	1	1	
m2											1	1					1		
m6				1		1					1			1	1	1	1	1	1
m4										1	1	1	1			1	1	1	1
m8										1	1	1			1	1	1	1	1

Table 9. What-if analysis, factors and levels

	general purpouse	problem oriented
Similarity Coefficient	J, SI, H, B, SO, R, SK, O, RM, RR	S, GS, SH (f_{bk}=0.6;f_{ek}=0.4), N
Rule	CLINK, ALINK, SLINK	
Percentile	10°, 25°, 40°, 50°, 75°	

Figure 8. Main effects plot for grouping measure

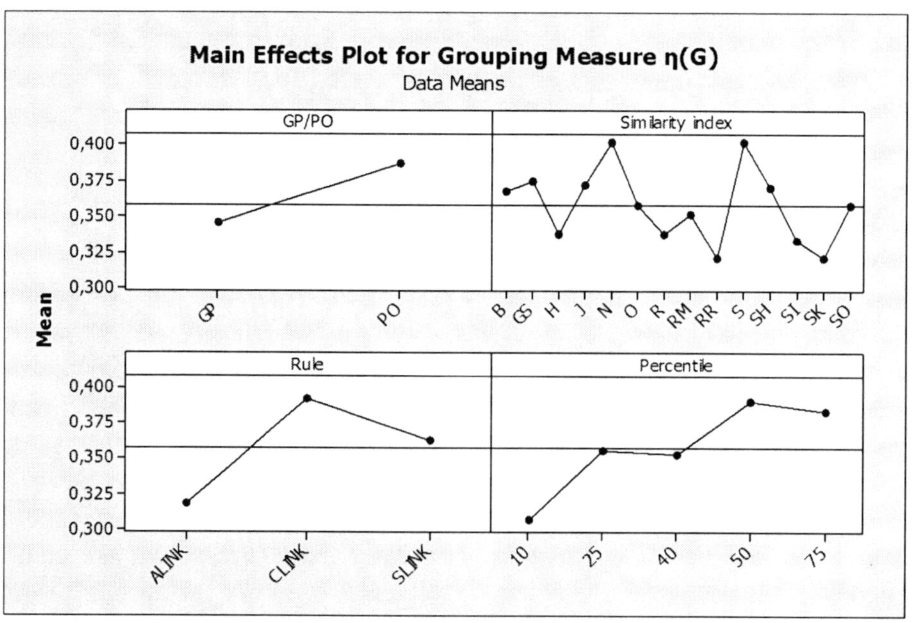

Figure 9. Main effects plot for grouping efficacy

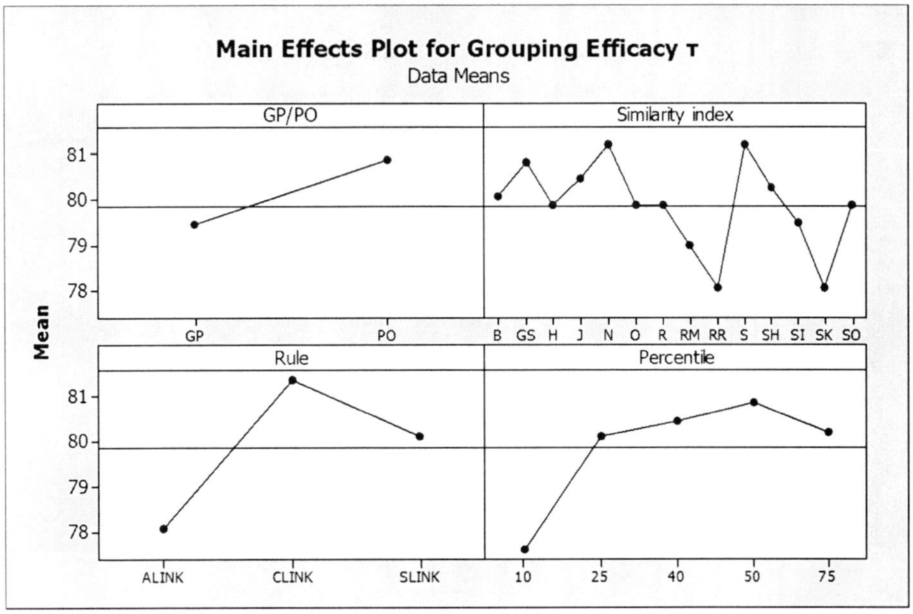

EXPERIMENTAL ANALYSIS

This section presents the results obtained by the application of the proposed systematic procedure to cell formation and parts assignment to cells (*part family formation*), as the result of different settings of the similarity and hierarchical procedure as illustrated in previous sections. This what-if analysis is applied to the problem oriented instance introduced by De Witte (1980) and reported in Table 2. This analysis represents the first step to identify the best combination of values, called *levels*, for the parameters, called *factors*, of the decision problem.

Table 9 reports the adopted levels for each factor in the experimental analysis.

Figures 8 to 10 present the *main effects plot* (Minitab ® Statistical Software Inc.) for the following performance indices: η_G, called η(G) in figures, τ, BE.

Similarity indices perform in a different way in terms of η_G, τ, BE. In particular problem oriented (PO) perform better than general purpose (GP). Clink rule and percentile threshold value equal to 50° (or 75°) seem to be the best levels to set the clustering algorithm. The best performing indices are Seiffoddini - S (1987) and Nair and Narendran - N (1998).

Figure 11 shows that the number of exceptional elements significantly depends on the adopted threshold value of group similarity, but the adopted similarity index is not important.

η_{QI}, called η(QI) in Figure 12, has an anomalous trend if compared with previous graphs.

Figure 13 shows the trend of the EE for different values of couples of factors, and the importance of the percentile threshold value of group similarity.

Similarly, Figure 14 shows the importance of threshold value of similarity and clink rule for grouping items.

CONCLUSION AND FURTHER RESEARCH

This chapter illustrates the CFP as supported by the similarity based manufacturing clustering,

Figure 10. Main effects plot for bond efficiency

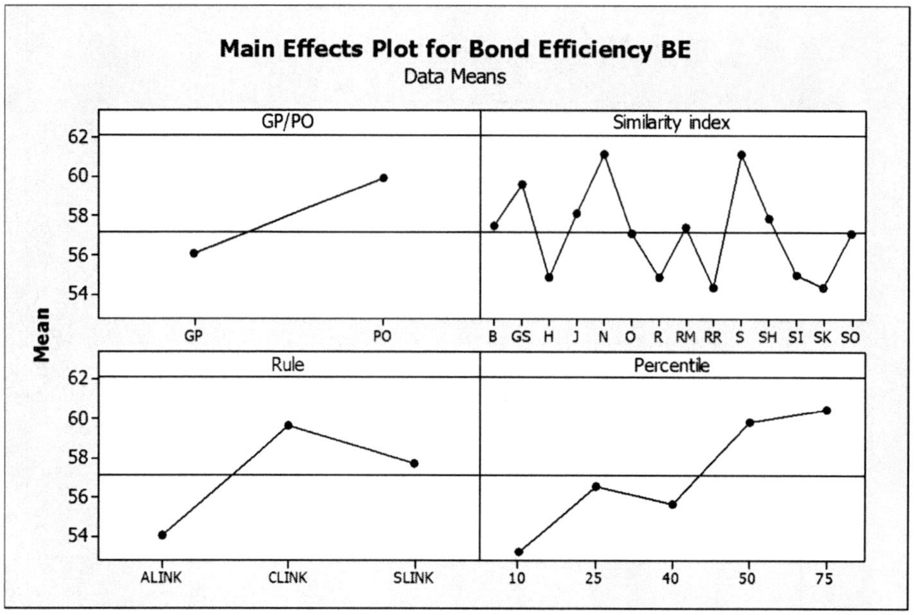

Figure 11. Main effects plot for exceptional elements

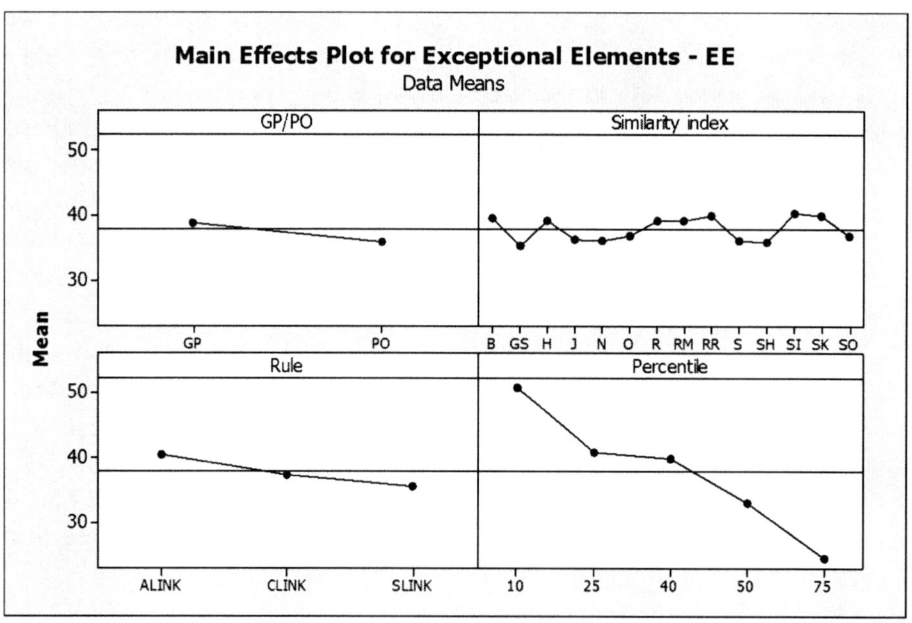

and a hierarchical and systematic procedure for supporting managers in the configuration of cellular manufacturing systems by the application of cluster analysis and similarity indices. In particular, both general purpose and problem oriented indices are illustrated and applied. The experimental analysis conducted on a literature problem oriented case study represents the first

Figure 12. Main effects plot for grouping efficacy based on QI

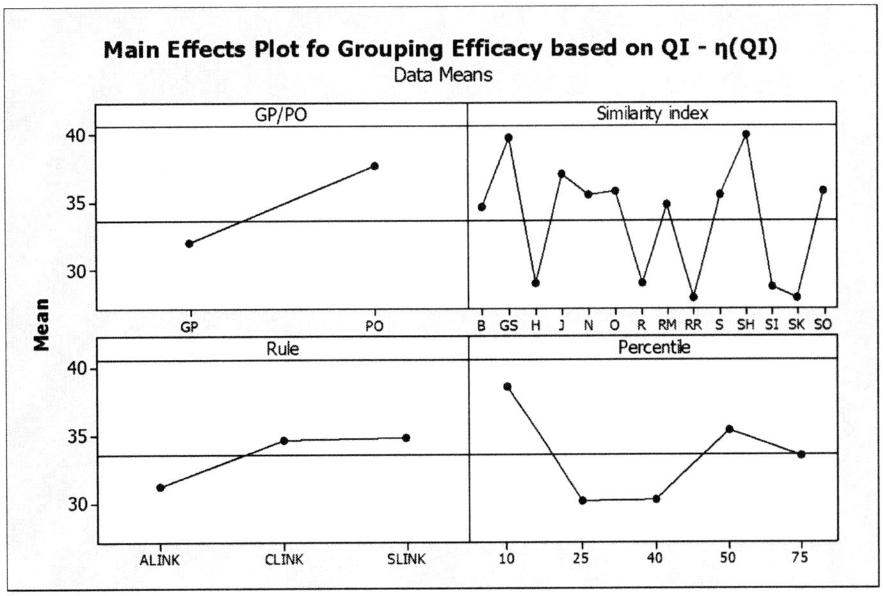

Figure 13. Exceptional elements for couples of factors

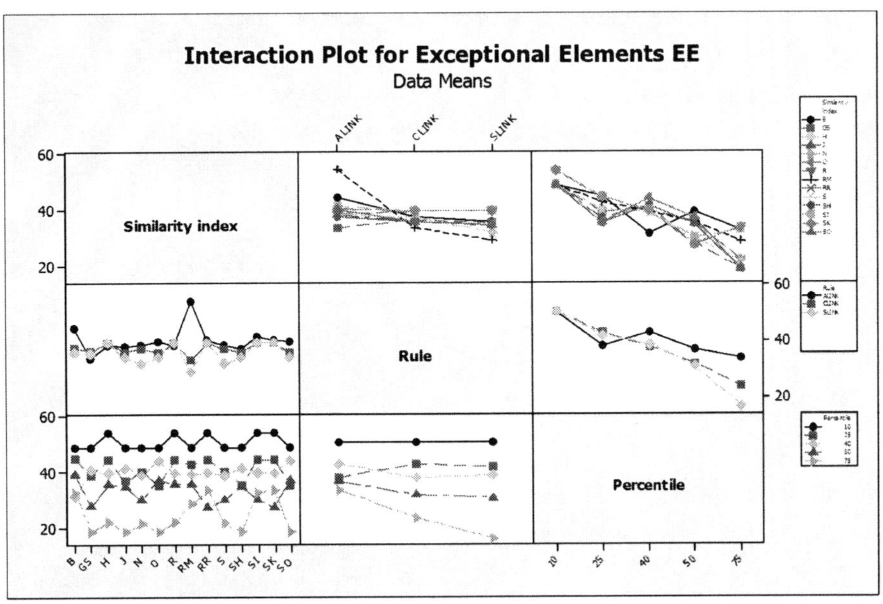

basis for the identification of the best setting of the cell formation problem and supporting decision models and tools.

For the first time, this chapter successfully applies the threshold group similarity index to problem oriented similarity environment. The threshold value was introduced by the authors in a previous study on general purpose indices evaluation (Manzini et al. 2010).

Figure 14. Interaction plot for τ

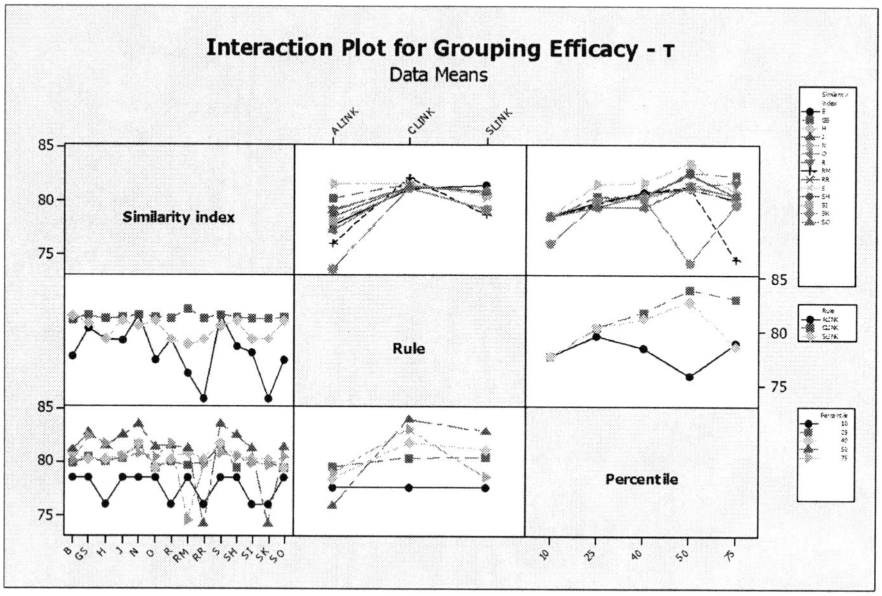

This chapter confirms the importance of this threshold cut value for the dendrogram when it is explained in percentile on the number of nodes.

Further research is expected to improve the experimental analysis including more case studies and applications. Finally it is important to improve the critical process of part family formation and the decisions regarding the duplication of machines and resources in different manufacturing cells in order to minimize intercellular flows.

REFERENCES

Aldenderfer, M. S., & Blashfield, R. K. (1984). *Cluster analysis*. (Sage University Paper series on Quantitative Applications in the Social Sciences, 1984, No. 07-044), Beverly Hills, CA: Sage.

Alhourani, F., & Seifoddini, H. (2007). Machine cell formation for production management in cellular manufacturing systems. *International Journal of Production Research*, *45*(4), 913–934. doi:10.1080/00207540600664144

Bindi, F., Manzini, R., Pareschi, A., & Regattieri, A. (2009). Similarity-based storage allocation rules in an order picking system. An application to the food service industry. *International Journal of Logistics Research and Applications*, *12*(4), 233–247. doi:10.1080/13675560903075943

De Witte, J. (1980). The use of similarity coefficients in production flow analysis. *International Journal of Production Research*, *18*, 503–514. doi:10.1080/00207548008919686

Gupta, T., & Seifoddini, H. (1990). Production data based similarity coefficient for machine-component grouping decisions in the design of a cellular manufacturing system. *International Journal of Production Research*, *28*, 1247–1269. doi:10.1080/00207549008942791

Heragu, S. (1997). *Facilities design*. Boston, MA: PWS Publishing Company.

Kumar, C. S., & Chandrasekharan, M. P. (1990). Grouping efficacy a quantitative criterion for goodness of block diagonal forms of binary matrices in group technology. *International Journal of Production Research*, *28*(2), 233–243. doi:10.1080/00207549008942706

Manzini, R., & Bindi, F. (2009). Strategic design and operational management optimization of a multi stage physical distribution system. *Transportation Research Part E, Logistics and Transportation Review*, *45*, 915–936. doi:10.1016/j.tre.2009.04.011

Manzini, R., Bindi, F., & Pareschi, A. (2010). The threshold value of group similarity in the formation of cellular manufacturing system. *International Journal of Production Research*, *48*(10), 3029–3060. doi:10.1080/00207540802644860

Manzini, R., Persona, A., & Regattieri, A. (2006). Framework for designing and controlling a multicellular flexible manufacturing system. *International Journal of Services and Operations Management*, *2*, 1–21. doi:10.1504/IJSOM.2006.009031

McAuley, J. (1972). Machine grouping for efficient production. *Production Engineering*, *51*, 53–57. doi:10.1049/tpe.1972.0006

Mosier, C. T. (1989). An experiment investigating the application of clustering procedures and similarity coefficients to the GT machine cell formation problem. *International Journal of Production Research*, *27*(10), 1811–1835. doi:10.1080/00207548908942656

Nair, G. J., & Narendran, T. T. (1998). CASE: A clustering algorithm for cell formation with sequence data. *International Journal of Production Research*, *36*, 157–179. doi:10.1080/002075498193985

Papaioannou, G., & Wilson, J. M. (2010). The evolution of cell formation problem methodologies based on recent studies (1987-2008): Review and directions for future research, (vol. 206, pp. 509-521).

Sarker, B. R. (2001). Measures of grouping efficiency in cellular manufacturing systems. *European Journal of Operational Research*, *130*, 588–611. doi:10.1016/S0377-2217(99)00419-1

Seifoddini, H. (1987). Incorporation of the production volume in machine cell formation in group technology applications. *Proceedings of the 9th International Conference on Production Research ICPR*, (pp. 2348-2356).

Seifoddini, H., & Djassemi, M. (1994). Analysis of efficiency measures for block diagonal machine-component charts. *Proceedings of the 16th International Conference on Computers and Industrial Engineering*, Ashikaga, Japan.

Seifoddini, H., & Djassemi, M. (1996). The thresold value of a quality index for formation of cellular manufacturing systems. *International Journal of Production Research*, *34*(12), 3401–3416. doi:10.1080/00207549608905097

Sokal, R. R., & Sneath, P. H. A. (1968). *Principles of numerical taxonomy*. San Francisco, CA: W. H. Freeman.

Stawowy, A. (2004). Evolutionary strategy for manufacturing cell design. *Omega: The International Journal of Management Science*, *34*, 1–18. doi:10.1016/j.omega.2004.07.016

Yin, Y., & Yasuda, K. (2006). Similarity coefficient methods applied to cell formation problem: A taxonomy and review. *International Journal of Production Economics*, *101*, 329–352. doi:10.1016/j.ijpe.2005.01.014

Chapter 8
An Estimation of Distribution Algorithm for Part Cell Formation Problem

Saber Ibrahim
University of Sfax, Tunisia

Bassem Jarboui
University of Sfax, Tunisia

Abdelwaheb Rebaï
University of Sfax, Tunisia

ABSTRACT

The aim of this chapter is to propose a new heuristic for Machine Part Cell Formation problem. The Machine Part Cell Formation problem is the important step in the design of a Cellular Manufacturing system. The objective is to identify part families and machine groups and consequently to form manufacturing cells with respect to minimizing the number of exceptional elements and maximizing the grouping efficacy. The proposed algorithm is based on a hybrid algorithm that combines a Variable Neighborhood Search heuristic with the Estimation of Distribution Algorithm. Computational results are presented and show that this approach is competitive and even outperforms existing solution procedures proposed in the literature.

INTRODUCTION

The principle objective of Group Technology is to reduce the intercellular flow of parts and to provide an efficient grouping of machines into cells. The main contribution in this chapter is to develop an efficient clustering heuristic based on evolutionary algorithms and to apply the proposed heuristic for Machine Part Cell Formation Problem which includes the configuration and capacity management of manufacturing cells. We propose to apply a novel population based evolutionary algorithm called Estimation of Distribution Algorithm in order to form part families and machine cells simultaneously.

DOI: 10.4018/978-1-61350-047-7.ch008

The objective of the proposed heuristic is to minimize exceptional elements and to maximize the goodness of clustering and thus the minimization of intercellular movements. In order to guarantee the diversification of solutions, we added an efficient technique of local search called Variable Neighborhood Search at the improvement phase of the algorithm. Many researchers have combined local search with evolutionary algorithms to solve this problem. However, they did not apply yet the Estimation of Distribution Algorithm for the general Group Technology problem. Furthermore, we have used a modified structure of the probabilistic model within the proposed algorithm.

In order to quantify the goodness of the obtained solutions, we present two evaluation criteria namely the percentage of exceptional elements and the grouping efficacy. A comparative study was elaborated with the most known evolutionary algorithms as well as the well known clustering methods.

LITERATURE REVIEW

A wide body of publications has appeared on the subject of Group Technology (GT) and Cellular Manufacturing Systems (CMS). The history of approaches that tried to solve this problem began with the classification and coding schemes. Several authors have proposed various ways trying to classify the methods of Cell Formation Problem. It includes descriptive methods, cluster analysis procedures, graph partitioning approaches, mathematical programming approaches, artificial intelligence approaches and other analytical methods.

Burbidge (1963) was the first who developed a descriptive method for identifying part families and machine groups simultaneously. In his work "Production Flow Analysis" (PFA). Burbidge has proposed an evaluative technique inspired from an analysis of the information given in route cards to find a total division into groups, without any need to buy additional machine tools.

Then, researchers applied array based clustering techniques which used a binary matrix A called "Part Machine Incidence Matrix" (PMIM) as input data. Given i and j the indexes of parts and machines respectively, an entry of 1 (a_{ij}) means that the part i is executed by the machine j whereas an entry of 0 indicates that it does not. The objective of the array based techniques is to find a block diagonal structure of the initial PMIM by rearranging the order of both rows and columns. Thus, the allocation of machines to cells and the parts to the corresponding families is trivial. McCornick et al. (1972) were the first who applied this type of procedure to the CFP. They developed the Bond Energy Analysis (BEA) which seeks to identify and display natural variable groups and clusters that occur in complex data arrays. Besides, their algorithm seeks to uncover and display the associations and interrelations of these groups with one another. King (1980) developed the Rank Order Clustering (ROC). In ROC algorithm, binary weights are assigned to each row and column of the PMIM. Then, the process tries to gather machines and parts by organizing columns and rows according to a decreasing order of their weights. Chan and Milner (1981) developed the Direct Clustering Algorithm (DCA) in order to form component families and machine groups by restructuring the machine component matrix progressively. A systematic procedure is used instead of relying on intuition in determining what row and column rearrangements are required to achieve the desired result. King & Nakornchai (1982) improved the ROC algorithm by applying a quicker sorting procedure which locates rows or columns having an entry of 1 to the head of the matrix. Chandrasekharan & Rajagopalan (1986a) proposed a modified ROC called MODROC, which takes the formed cells by the ROC algorithm and applies a hierarchical clustering procedure to them. Later, other array based clustering techniques are proposed namely the Occupancy Value method of Khator & Irani (1987), the Cluster Identification Algorithm (CIA) of Kusiak & Chow (1987) and the Hamiltonian Path Heuristic of Askin et al. (1991).

McAuley (1972) was the first who suggested similarity coefficient to clustering problems. He applied the Single Linkage procedure to the CF problem and used the coefficient of Jaccard which is defined for any pair of machines as the ratio of the number of parts that visit both machines to the number of parts that visit at least one of these machines. Then, some other clustering techniques are developed namely Single Linkage Clustering (SLC), Complete Linkage Clustering (CLC), Average linkage Clustering (ALC) and Linear Cell Clustering (LCC). Kusiak (1987) proposed a linear integer programming model maximizing the sum of similarity coefficients defined between two parts

The category that is the most used in literature in recent years is heuristics and metaheuristics. Such heuristics are based essentially on Artificial Intelligence approaches including Genetic Algorithms (GA), Simulated Annealing (SA), Tabu Search (TS), Evolutionnary Algorithms (EA), neural network and fuzzy mathematics. In what follows we present some research papers that used this type of heuristics for designing CM systems.

Boctor (1991) developed the SA approach to deal with large-scale problems. Sofianopoulos (1997) proposed a linear integer formulation for CF problem and employed the SA procedure to improve the solution quality taking as objective the minimization of inter-cellular flow between cells. Caux et al. (2000) proposed an approach combining the SA method for the CF problem and a branch-and-bound method for the routing selection. Lozano et al. (1999) presented a Tabu Search algorithm that systematically explores feasible machine cells configurations determining the corresponding part families using a linear network flow model. They used a weighted sum of intra-cell voids and inter-cellular moves to evaluate the quality of the solutions. Solimanpur et al. (2003) developed an Ant colony optimization algorithm to solve the inter cell layout problem by modelling it as a quadratic assignment problem. Kaparthi et al. (1993) proposed an algorithm based on neural network for the part machine grouping problem. Xu & Wang (1989) developed two approaches of fuzzy cluster analysis namely fuzzy classification and fuzzy equivalence in order to incorporate the uncertainty in the measurement of similarities between parts. They presented also a dynamic part-family assignment procedure using the methodology of fuzzy pattern recognition to assign new parts to existing part families.

Recently many researchers have focused on the approaches based on AI for solving the part-machine grouping problem. Venugopal & Narendran (1992a) proposed a bi-criteria mathematical model with a solution procedure based on a genetic algorithm. Joines et al. (1996) presented an integer programming solved using a Genetic Algorithm to solve the CF problem. Zhao & Wu (2000) presented a genetic algorithm to solve the machine-component grouping problem with multiple objectives: minimizing costs due to inter-cell and intra-cell part movements; minimizing the total within cell load variation; and minimizing exceptional elements. Gonçalves & Resende (2002) developed a GA based method which incorporates a local search to obtain machine cells and part families. The GA is responsible for generating sets of machines cells and the mission of the local search heuristic is to construct sets of machine part families and to enhance their quality. Then, Gonçalves & Resende (2004) employed a similar algorithm to find first the initial machine cells and then to obtain final clusters by applying the local search. Mahdavi et al. (2009) presented a GA based procedure to deal with the CF problem with nonlinear terms and integer variables. Stawowy (2006) developed a non-specialized Evolutionary Strategy (ES) for CF problem. His algorithm uses a modified permutation with separators encoding scheme and unique concept of separators movements during mutation. Andrés

& Lozano (2006) applied for the first time the Particle Swarm Optimization (PSO) algorithm to solve the CF problem respecting the objective the minimization of inter-cell movements and imposing a maximum cell size.

ESTIMATION OF DISTRIBUTION ALGORITHM

It was first introduced by Mühlenbein & Paaß (1996). The Estimation of Distribution Algorithm belongs to Evolutionary Algorithms family. It adopts probabilistic models to reproduce individuals in the next generation, instead of crossover and mutation operations. This type of algorithms uses different techniques to estimate and sample the probability distribution.

The probabilistic model is represented by conditional probability distributions for each variable. This probabilistic model is estimated from the information of the selected individuals in the current generation and selects good individuals with respect to their fitness. This process is repeated until the stop criterion is met. Such a reproduction procedure allows the algorithm to search for optimal solutions efficiently. However, it considerably decreases the diversity of the genetic information in the generated population when the population size is not large enough. For this reason, the incorporation of a local search technique is encouraged in order to enhance the performance of the algorithm.

As a result, the Estimation of Distribution Algorithm can reach best solutions by predicting population movements in the search space without needing many parameters. The main steps in this procedure are shown in the following pseudo code:

Estimation of Distribution Algorithm

1. Initialize the population according to some initial distribution model.
2. Form P' individuals from the current population using a selection method.
3. Build a probability model p(x) from P' individuals using both the information extracted from the selected individuals in the current population and the previously built model.
4. Sample $p(x)$ by generating new individuals from the probability model and replace some or all individuals in the current population.
5. End the search if stop criteria are met, otherwise return to Step 2.

This method can be divided into two different classes. The first class assumes that there are no dependencies between variables of the current solution during the search. These are known as non-dependency Estimation of Distribution Algorithms: Population Based Incremental Learning (Baluja, 1994) and Univariate Marginal Distribution Algorithm (Mühlenbein & Paaß, 1996). The second class takes into account these variable dependencies: Mutual Information Maximization for Input Clustering (De Bonet et al., 1997), Bivariate Marginal Distributional Algorithm (Pelikan & Mühlenbein, 1999), Factorized Distribution Algorithm (Mühlenbein et al., 1999) and the Bayesian Optimization Algorithm (Pelikan et al., 1999a).

Generally, non-dependency algorithms are expected to have a worse modelling ability than the ones with variable dependencies (Zhang et al., 2004). But combining heuristic information or local search with non-dependency algorithms can compensate for this disadvantage.

Univariate EDAs

This category assume that each variable is independent; it means that the algorithm do not consider any interactions among variables in the solution. As a result, the probability model distribution,$p(x)$, becomes simply the product of Univariate marginal probabilities of all variables in the solution and expressed as follows:

$$p(x) = \prod_{i=1}^{1} p(x_i)$$

Due to the simplicity of the model of distribution used, the algorithms in this category are computationally inexpensive, and perform well on problems with no significant interaction among variables.

In what follows, we present the well-known works related to this category.

Population Based Incremental Learning

It was proposed by Baluja (1994). The algorithm starts with initialisation of a probability vector. In each iteration, it updates and samples the probability vector to generate new solutions. The main steps in this procedure are shown in the following pseudo code:

Population Based Incremental Learning

1. Initialise a probability vector $p=\{p_1,p_2,...,p_n\}$with 0.5 at each position. Here, each pi represents the probability of 1 for the ith position in the solution.
2. Generate a population P of M solutions by sampling probabilities in p.
3. Select set D from P consisting of N promising solutions.
4. Estimate univariate marginal probabilities p(xi) for each xi.
5. For each i, update pi using $p_i=p_i+\lambda(p(x_i)-p_i)$
6. For each i, if mutation condition passed, mutate pi using

$p_i=p_i(1-\mu)$+randon (0 or 1)μ.

7. End the search if stop criteria are met, otherwise return to Step 2.

Univariate Marginal Distribution Algorithm

Univariate Marginal Distribution Algorithm was proposed by Muhlenbein & Paaß (1996). We note that this category can be seen as a variant of Population Based Incremental Learning when λ=1 and μ=0 Different variants of Univariate Marginal Distribution Algorithm have been proposed, and the mathematical analysis of their workflows has been carried out (Muhlenbein, 1998; Muhlenbein et al., 1999; Gonzalez et al., 2002). The main steps in this procedure are shown in the following pseudo code:

Univariate Marginal Distribution Algorithm

1. Generate a population *P* composed of *M* solutions.
2. Select a set *P'* from *P* consisting of *N* promising solutions.
3. Estimate univariate marginal probabilities $p(x)$ from *P'* for each x_i.
4. Sample$p(x)$to generate *M* new individual and replace *P*.
5. End the search if stop criteria are met, otherwise return to Step 2.

Bivariate EDAs

In contrast with Univariate case, the probability model contains factors involving the conditional probability of pairs of interacting variables. This class of algorithms performs better in problems, where pair-wise interaction among variable exists.

In what follows, we present the well-known works related to this category.

Mutual Information Maximization for Input Clustering

The Mutual Information Maximization for input clustering uses a chain model of probability distribution (de Bonet et al., 1997) and it can be written as:

$$p(x) = \prod_{i=1}^{1} p(x_{\pi_1} | x_{\pi_2}) p(x_{\pi_2} | x_{\pi_3}) \ldots p(x_{\pi_{n-2}} | x_{\pi_{n-1}}) p(x_{\pi_n})$$

where $\Pi=\{\pi_1,\pi_2,\ldots,\pi_n\}$ is a permutation of the numbers $\{1,2,\ldots,n\}$ used as an ordering for the pair wise conditional probabilities. At each iteration, the algorithm first tries to learn the linkage. Then, the algorithm uses a greedy algorithm to find a permutation Π that does not always give accurate model. Once the permutation Π is learnt, the algorithm estimates the pair wise conditional probabilities and samples them to get next set of solutions.

Combining Optimizers with Mutual Information Trees

The Combining Optimizers with Mutual Information Trees proposed by Baluja & Davies (1997, 1998) also uses pair-wise interaction among variables. The model of distribution used by this algorithm can be written as follows:

$$p(x) = \prod_{i=1}^{1} p(x_i | x_j)$$

where, x_j is known as parent of x_i and x_i is known as a child of x_j. This model is more general than the chain model used by Mutual Information Maximization for input clustering as two or more variables can have a common parent.

Bivariate Marginal Distribution Algorithm

It was proposed by (Pelikan & Muhlenbein, 1999) as an extension to Univariate Marginal Distribution Algorithm. The model of distribution used by Bivariate Marginal Distribution Algorithm can be seen as an extension to the Combining Optimizers with Mutual Information Trees model and can be written as follows:

$$p(x) = \prod_{x_k \in Y} p(x_k) \prod_{x_i \in \{X|Y\}} p(x_i | x_j)$$

where, $Y \subseteq X$ represents the set of root variables.

As a result, Bivariate Marginal Distribution Algorithm is a more generalised algorithm in this class and can cover both univariate interaction as well as bivariate interaction among variables.

Multivariate EDAs

The model of probability distribution becomes more complex than the one used by univariate and bivariate Estimation of Distribution Algorithms. Any algorithm considering interaction between variables of order more than two can be placed in this class. As a result, the complexity of constructing such model increases exponentially to the order of interaction making it infeasible to search through all possible models.

In what follows, we present the well-known works related to this category.

Extended Compact Genetic Algorithm

The Extended Compact Genetic Algorithm has been proposed by Harik (1999) as an extension to the Compact Genetic Algorithm. The model of distribution used in the Extended Compact Genetic Algorithm, is distinct from other previously described models as they only consider the marginal probabilities and do not include conditional probabilities. Also, it assumes that a variable appearing in a set of interacting variables cannot appear in another set. The model of distribution used by the Extended Compact Genetic Algorithm can be written as follows:

$$p(x) = \prod_{k \in m} p(x_k)$$

where, m is the set of disjoint subsets in n and $p(x_k)$ is the marginal probability of set of variables x_k in the subset k.

Factorised Distribution Algorithm

The Factorised Distribution Algorithm was proposed by Muhlenbein et al. (1999) as an extension to the Univariate Marginal Distribution Algorithm. The probability $p(x)$, for such linkage, can be expressed in terms of conditional probabilities between sets of interacting variables. In general, the Factorised Distribution Algorithm requires the linkage information in advance, which may not be available in a real world problem.

Bayesian Optimization algorithm

The Bayesian Optimization algorithm was proposed by Pelikan et al. (1999a). The probabilistic model $p(x)$ is expressed in terms of a set of conditional probabilities as follow:

$$p(x) = \prod_{i=1}^{n} p(x_i | \pi_i)$$

where, π_i is a set of variables having conditional interaction with x_i. Also no variable in π_i can have x_i or any children of x_i as their parent.

An extension to the Bayesian Optimization algorithm called hierarchical Bayesian Optimization algorithm has also been proposed by Pelikan & Goldberg (2000). The idea is to improve the efficiency of algorithm by using a Bayesian network with a local structure (Chickering et al., 1997) to model the distribution and a restricted tournament replacement strategy based on work of Harik (1994) to form the new population.

Estimation of Bayesian Network Algorithm

The Estimation of Bayesian Network Algorithm was proposed by Etxeberria & Larranaga (1999) and Larranaga et al., (2000) and also uses Bayesian networks as its model of probability distribution.

The algorithm has been applied for various optimisation problems, such as graph matching (Bengoetxea et al., 2000, 2001b,), partial abductive inference in Bayesian networks (de Campos et al., 2001), job scheduling problem (Lozano et al., 2001b), rule induction task (Sierra et al., 2001), travelling salesman problem (Robles et al., 2001), partitional clustering (Roure et al., 2001), Knapsack problems (Sagarna & Larranaga, 2001).

Learning Factorised Distribution Algorithm

The Learning Factorised Distribution Algorithm was proposed by Muhlenbein & Mahnig (1999b) as an extension to the Factorised Distribution Algorithm. The algorithm does not require linkage in advance. In each iteration, it computes a bayesian network and samples it to generate new solutions. The main steps in the Bayesian Optimization algorithm (BOA), the Estimation of Bayesian Network Algorithm (EBNA) and the Learning Factorised Distribution Algorithm (LFDA) procedures are shown in the following pseudo code:

BOA, EBNA and LFDA

1. Generate population P of M solutions
2. Select N promising solution from P.
3. Estimate a Bayesian network from selected solutions.
4. Sample Bayesian network to generate M new individual and replace P.
5. End the search if stop criteria are met, otherwise return to Step 2.

Markov Network Factorised Distribution Algorithm and Markov Network Estimation of Distribution Algorithm

The Markov Network Factorised Distribution Algorithm and the Markov Network Estimation of Distribution Algorithm were proposed by Santana (2003a, 2005). They used Markov network (Pearl, 1988; Li, 1995) as the model of distribution for $p(x)$. The first algorithm uses a technique called junction graph approach, while the second one uses a technique called Kikuchi approximation to estimate a Markov network.

PROBLEM STATEMENT

Manufacturing Cell Formation consists of grouping, or clustering, machines into cells and parts into families according to their similar processing requirements. The most known and efficient idea to achieve the objective of cell formation is to convert the initial Part Machine Incidence Matrix to a matrix that has a diagonal block structure. Among this process, entries with a '1' value are grouped to form mutually independent clusters, and those with a '0' value are arranged outside these clusters. Once a block diagonal matrix is obtained, machine cells and part families are clearly visible. However, the process engenders intercellular movements that require extra cost or time due to the presence of some parts that are processed by machines not belonging to its corresponding cluster. These parts are called Exceptional Elements. As a result, the objective of the block diagonalization is to change the original matrix into a matrix form minimizing Exceptional Elements and maximizing the goodness of clustering.

For cell formation problem, this matrix can be regarded as a binary matrix A which shows the relationship between any given m machines and p parts. Rows and columns represent respectively machines and parts. Each element in the matrix is usually represented by the binary entries a_{ij} where an entry of 1 indicates that a part i is processed by the corresponding machine j while an entry of 0 means a contrary situation. In Figure 1, we illustrate an (5×7) incidence matrix of King & Nakornchai (1982).

Figure 2 provides a block diagonal form for the initial matrix illustrated above. The obtained matrix has not any intercellular movement which means that it represents the optimal solution for the given matrix with 2 cells and 3 machines per cell.

In this chapter, we will deal with two efficient evaluation criteria namely the Grouping Efficacy (GE) and the Percentage of Exceptional Elements (PE). The Grouping Efficacy, proposed by Kumar & Chandrasekharan (1990), is considered one of the best criteria which distinguish ill-structured matrices from well-structured ones when the matrix size increases and it is expressed as follows:

Figure 1. King & Nakornchai (1982) initial matrix

		Parts						
		1	2	3	4	5	6	7
Machines	1	0	1	0	1	1	1	0
	2	1	0	1	0	0	0	0
	3	1	0	1	0	0	0	1
	4	0	1	0	1	0	1	0
	5	1	0	0	0	0	0	1

Figure 2. A block diagonal matrix with no exceptional elements

		Parts						
		1	3	7	2	4	5	6
Machines	3	1	1	1	0	0	0	0
	2	1	1	0	0	0	0	0
	5	1	0	1	0	0	0	0
	4	0	0	0	1	1	0	1
	1	0	0	0	1	1	1	1

$$GE = \frac{e(X) - e_0(X)}{e(X) + e_v(X)}$$

Where:

$e_0(X)$: Number of Exceptional Elements in the solution X,
e: Number of 1's in the Part Machine Incidence Matrix,
$e_v(X)$: Number of voids in the solution X.

The second evaluation criterion is called the "Percentage of Exceptional Elements (PE)" is developed by Chan & Milner (1982) and expressed as follows:

$$PE = \frac{e_0(X)}{e} \times 100.$$

Some other performance measurements can be used to evaluate manufacturing cell design results. In what follows, we presents some of them.

The Grouping Efficiency which is developed by Chandrasekaran & Rajagopalan (1989). It expresses the goodness of the obtained solutions and depends on the utilization of machines within cells and inter-cell movements. This indicates that there are no voids and no exceptional elements in the diagonal blocks which imply a perfect clustering of parts and machines. Although grouping efficiency was widely used in the literature, it has an important limit which is the inability of discrimination of good quality grouping from bad one. Indeed, when the matrix size increases, the effect of 1's in the off-diagonal blocks becomes smaller, and in some cases, the effect of inter-cell moves is not reflected in grouping efficiency.

The Machine Utilization Index (MUI) which is defined as the percentage of the time that the machines within cells are being utilized most effectively and it is expressed as follows:

$$MUI = \frac{e}{\sum_i (m_i \times p_i)}$$

where m_i indicates the number of machines in cell i and p_i indicates the number of parts in cell i.

The Group technology efficiency which is defined as the ratio of difference between maximum number of inter-cell travels possible and number of inter-cell travels actually required by the system to the maximum number of inter-cell travels possible.

The Group efficiency which is defined as the ratio of difference between total number of maximum external cells that could be visited and total number of external cells actually visited by all parts to total number of maximum external cells that could be visited.

The Global efficiency is defined as the ratio of the total number of operations that are performed within the suggested cells to total number of operations in the systems.

PROPOSED EDA FOR MPCF PROBLEM (EDA-CF)

Solution Representation and Initial Population

Generally, for a Cell Formation Problem, a solution is represented by an m-dimensional vector $X=[x_1,x_2,\ldots,x_m]$ where x_i represents the corresponding assignment of the machine i to the specified cell. The problem consists in creating partitions of the set of the m machines assignments into a given number of cells. The created solutions must respect all the constraints defined in Section 3.3. We choose to generate the initial population randomly following a uniform distribution.

Selection

The goal is to allow individuals to be selected more often to reproduce. We adopt the truncated selection procedure to create new individuals: in each iteration, we select randomly P_I individuals from the 50% of the best individuals in the current population. These P_I individuals will be reproduced in the next generation using the probabilistic model to form new individuals.

Probabilistic Model and Creation of New Individuals

After the selection phase, a probabilistic model is applied to the P_I selected individuals in order to generate new individuals.

The probabilistic model provides the assignment probability of the machine *i* to cell *j* and expressed as follows:

$$P_{ij} = \frac{number\ of\ times\ where\ machine\ i\ appears\ in\ cell\ j\ +\ \varepsilon}{number\ of\ selected\ individuals\ +\ C \times \varepsilon}$$

where, ε>0 is a factor which guarantees that the model provides a probability $P_{ij} \neq 0$.

Replacement

The replacement represents the final step in our search procedure. It is based on the following idea: when a new individual is created, we compare it to the worst individual in the current population and we retain the best one.

Fitness Function

A fitness function is used for evaluating the aptitude of an individual to be kept or to be used for reproducing new individuals in the next generation. In the proposed algorithm, we used two fitness functions F_1 and F_2 to perform the objectives of minimizing the percentage of Exceptional Elements and maximizing the Grouping Efficacy respectively.

Let m_i be the number of machines assigned to the cell *i*. we define F_1 and F_2 as follows:

$$F_1(X)=e_0(X)+Pen(X)$$

and

$$F_2(X)=GE(X)-Pen(X).$$

where:

$$Pen(X) = \alpha_1 \sum_{i=1}^{C} \max\left\{0, m_i - k_{\max}\right\} + \alpha_2 \sum_{i=1}^{C} \max\left\{0, 1 - m_i\right\}$$

expressed the distance between the solution *X* and the feasible space.

This penalty under-evaluate the fitness of solution *X* when *X* violate the constraint of the problem. i.e a penalty value is encountered either when the number of assigned machines exceeds the capacity of a cell or when machines are assigned to a number of cells that exceeds the fixed number of cells *C*.

Variable Neighborhood Search Algorithm

Variable Neighborhood Search is a recent metaheuristic for combinatorial optimization developed by Mladenović & Hansen (1997). The basic idea is to explore different neighborhood structures and to change them within a local search algorithm to identify better local optima with shaking strategies. The main steps in this procedure are shown in the following pseudo code:

Variable Neighborhood Search
Select the set of neighborhood structures $N_k, k=\{1,2,\ldots,n_{\max}\}$ that will be used in the search, find an initial solution *X*, choose a stopping condition.

Repeat the following steps until the stopping condition is met:

Set k=1

Repeat the following steps until all neighborhood structures are used:

1. Shaking: generate a point X' at random from k^{th} neighborhood of X ($X' \in N_k(X)$)
2. Local Search: apply some local search method with X' as initial solution; denote with X'' the obtained local optimum.
3. Move or not: if this local optimum X'' is better than the incumbent, or if some acceptance criterion is met, move there ($X \leftarrow X''$), and set k=1 ; otherwise, set $k \leftarrow k+1$.

Local Search Procedure

Generally, obtaining a local minimum following a neighborhood structure does not imply that we obtain a local optimum following another one. For this reason, we choose to use two local search procedures which are based on two different neighborhood structures. The first neighborhood structure consists to select one machine and to insert it in a new cell. The second consists to select two machines from two different cells and to swap them.

Then, we apply these two local search procedures iteratively until there is no possible improvement to the current solution.

Shaking Phase

The main idea consists to define a set of neighbourhood structures that allow to obtain a distance equal to k between the solution X and the new neighbour solution X'. This distance can be defined by the number of differences between the two vectors X and X'. Then, we define N_k as the neighbourhood structure given by applying randomly k insertion moves.

COMPARATIVE STUDY

In order to show the competitiveness of the proposed EDA-CF algorithm, we provide in this section a comparative study with the well known approaches that treated Cell Formation problem. During all experiments, the proposed algorithm is coded using C++ and run on a computer Pentium IV with 3.2 GHz processor and 1GB memory.

Test Data Set

In order to evaluate the goodness of clusters obtained from the clustering heuristic for MPCF problem, 30 problems taken from the literature were tested. These data sets include a variety of sizes, a range from 5 machines and 7 parts to 40 machines and 100 parts, difficulties, and well structured and ill structured matrices.

For all instances, the initial matrix is solved by Estimation of Distribution Algorithm method and then improved by the Variable Neighborhood Search procedure. Then, the cells are formed and the machine layout in each cell is obtained optimally.

Table 1 shows the different problems and their characteristics. The columns illustrate respectively the sources of data sets, the problem size, the number of cells C, the maximum number per cell, k_{max} and the matrix density. All problems can be easily accessed from the references and they are transcribed directly from the original article they appeared. The appendix gives the block diagonal matrices for the improved solutions by the proposed algorithm. The maximum number of permissible cells C has been set equal to the best known number of cells as found in literature. The following equation expressed the density of the initial binary matrix and which informs about haw the one's elements are distributed inside the matrix.

$$\frac{\sum_{i}^{m}\sum_{j}^{n} a_{ij}}{m \times n}$$

Table 1. Test problems from cellular manufacturing literature

No.	References	Size	C	k_{max}	Density
1	King & Nakornchai, 1982	5×7	2	4	0.400
2	Waghodekar & Sahu, 1984	5×7	2	5	0.5714
3	Seifoddini, 1989	5×18	2	12	0.5111
4	Kusiak & Cho, 1992	6×8	2	6	0.2987
5	Kusiak & Chow, 1987	7×11	3	4	0.2250
6	Boctor, 1991	7×11	3	4	0.2044
7	Seifoddini & Wolfe, 1986	8×12	3	5	0.6100
8	Chandrasekharan & Rajagopalan, 1986	8×20	3	9	0.2400
9	Chandrasekharan & Rajagopalan, 1986	8×20	2	11	0.3067
10	Mosier & Taube, 1985	10×10	3	4	0.3223
11	Chan & Milner, 1982	10×15	3	5	0.3646
12	Stanfel, 1985	14×14	5	6	0.1726
13	McCormick et al., 1972	16×24	6	7	0.2240
14	King, 1980	16×43	5	13	0.1831
15	Mosier & Taube, 1985	20×20	5	5	0.2775
16	Carrie, 1973	20×35	4	10	0.1957
17	Boe & Cheng, 1991	20×35	5	8	0.2186
18	Chandrasekharan & Rajagopalan, 1989 - 1	24×40	7	8	0.1365
19	Chandrasekharan & Rajagopalan, 1989 - 2	24×40	7	8	0.1354
20	Chandrasekharan & Rajagopalan, 1989 - 3	24×40	7	8	0.1437
21	Chandrasekharan & Rajagopalan, 1989 - 4	24×40	9	8	0.1365
22	Chandrasekharan & Rajagopalan, 1989 - 5	24×40	9	7	0.1375
23	Chandrasekharan & Rajagopalan, 1989 - 6	24×40	9	7	0.1365
24	McCormick et al., 1972	27×27	4	12	0.2977
25	Kumar & Vanelli, 1987	30×41	11	6	0.1041
26	Stanfel, 1985	30×50	12	7	0.1033
27	Stanfel, 1985	30×50	11	7	0.1113
28	King & Nakornchai,1982	36×90	9	27	0.0935
29	McCormick et al., 1972	37×53	2	35	0.4895
30	Chandrasekharan & Rajagopalan, 1987	40×100	10	6	0.1041

Comparative Study

In this section, we evaluate the proposed algorithm by comparing it with the best results obtained by several well known algorithms respecting to the Grouping Efficacy and the Percentage of Exceptional Elements measures. In all tests, the proposed EDA-CF algorithm has proved its competitiveness against the best available solutions respecting to the same required number of cells.

As a stop condition to our algorithm, we fixed the maximal computational time to 5 seconds and the maximal number of iteration of Variable Neighborhood Search algorithm to 3.

The values of the following parameters are fixed as: ε=0,1; α_1=50; α_2=500; P=200 and P_1=3.

Comparison Respecting the Grouping Efficacy Measure

In this subsection we perform a comparative study with the best algorithm presented in the literature. These algorithms can be classified into two categories. The first category corresponds to the based population algorithm including Genetic Algorithm (GA) of Onwubolu & Mutingi (2001), Grouping Genetic Algorithm (GGA) of Brown & Sumichrast (2001), Evolutionary Algorithm (EA) of Gonçalves & Resende (2004) and Hybrid Grouping Genetic Algorithm (HGGA) of James et al. (2007). The second category represents the clustering based methods including ZODIAC of Chandrasekharan & Rajagopalan (1987), GRAFICS of Srinivasan & Narendran (1991), MST-Clustering Algorithm of Srinivasan (1994). Table

Table 2. Summary of GE performance evaluation results

No	Size	C	GA	GGA	EA	HGGA	ZODIAC	GRAFICS	MST	EDA-CF	CPU
1	5×7	2	-	82.35	73.68	82.35	73.68	73.68	-	73.68	0.000
2	5×7	2	62.50	69.57	52.50	69.57	56.52	60.87	-	69.57	0.000
3	5×18	2	77.36	79.59	79.59	79.59	-	-	-	79.59	0.000
4	6×8	2	76.92	76.92	76.92	76.92	-	-	-	76.92	0.000
5	7×11	2	50.00	60.87	53.13	60.87	39.13	53.12	-	58.62	0.000
6	7×11	3	70.37	70.83	70.37	70.83	-	-	-	70.37	0.000
7	8×12	5	-	69.44	68.30	69.44	68.30	68.30	-	68.30	0.000
8	8×20	3	85.25	85.25	85.25	85.25	85.24	85.24	85.24	85.25	0.000
9	8×20	2	55.91	55.32	58.72	58.72	58.33	58.33	58.72	58.72	0.000
10	10×10	3	72.79	75.00	69.86	75.00	70.59	70.59	70.59	70.59	0.000
11	10×15	3	92.00	92.00	92.00	92.00	92.00	92.00	-	92.00	0.015
12	14×24	4	63.48	72.06	69.33	72.06	64.36	64.36	64.36	70.51	0.015
13	16×24	6	-	51.58	52.58	52.75	32.09	45.52	48.70	51.96	0.046
14	16×43	4	86.25	55.48	54.86	57.53	53.76	54.39	54.44	54.86	0.031
15	20×20	5	34.16	40.74	42.96	43.18	21.63	38.26	-	43.18	1.232
16	20×35	4	66.30	77.02	76.22	77.91	75.14	75.14	75.14	76.27	0.078
17	20×35	5	44.44	57.14	58.07	57.98	-	-	-	57.98	0.093
18	24×40	7	100.00	100.00	100.00	100.00	100.00	100.00	100.00	100.00	0.031
19	24×40	7	85.11	85.11	85.11	85.11	85.10	85.10	85.10	85.11	0.092
20	24×40	7	73.51	73.51	73.51	73.51	37.85	73.51	73.51	**76.97**	5.171
21	24×40	7	37.62	52.41	51.97	53.29	20.42	43.27	51.81	**72.92**	0.732
22	24×40	7	34.76	46.67	47.06	48.95	18.23	44.51	44.72	**53.74**	0.670
23	24×40	7	34.06	45.27	44.87	47.26	17.61	41.67	44.17	**48.95**	0.233
24	27×27	6	-	52.53	54.27	54.02	52.14	47.37	51.00	**54.98**	7.260
25	30×41	14	40.96	61.39	58.48	63.31	33.46	55.43	55.29	45.22	0.562
26	30×50	13	48.28	57.95	59.66	59.77	46.06	56.32	58.70	59.43	0.447
27	30×50	14	37.55	50.00	50.51	50.83	21.11	47.96	46.30	50.78	1.406
28	36×90	17	-	43.78	42.64	46.35	32.73	39.41	40.05	45.94	5.094
29	37×53	2	-	52.47	56.42	60.64	52.21	52.21	-	55.43	4.318
30	40×100	10	83.90	82.25	84.03	84.03	83.92	83.92	83.66	83.81	7.421

2 reports the results obtained by the proposed algorithm and these algorithms such that their results were taken from the original citations.

As seen in Table 2, in all the benchmark problems, the grouping efficacy of the solution obtained by the proposed method is either better than that of other methods or it is equal to the best one. We note that the solutions obtained by the GA method for problems 1, 7, 13, 24, 28 and 29 were not available. In five problems, namely 20, 21, 22, 23 and 24, the grouping efficacy of the solution obtained by the proposed method is better than that of all other methods. In other words, the proposed method outperforms all the other methods and the best solutions for these problems are reported in this paper for the first time. In eleven problems, namely 2, 3, 9, 15 and 17, the solution obtained by the proposed method is as good as the best solution available in the literature. In five problems, namely 4, 8, 10, 18 and 19, all the methods have obtained the same grouping efficacy.

Comparing with clustering methods, it is clear that the results obtained by the proposed algorithm are either equal or better than ZODIAC, GRAFICS and MST methods in all cases except for the problems 25 and 30. More specifically, the EDA-CF obtains for 6 (23%) problems values of the grouping efficacy that are equal to the best ones found in the literature by the three compared clustering methods and improves the values of the grouping efficacy for 19 (73%) problems.

Comparison Respecting the Percentage of Exceptional Elements Measure

Table 3 provides a comparison of the proposed algorithm against the best reached results available in literature. The comparison was done respecting to the Percentage of Exceptional Elements criteria. PE^a represents the best-known Percentage of Exceptional Elements found in the literature.

We note that the compared solutions for problems 15, 16, 17, 24, 26, 27, 28 and 29 were not available. The results shows that in all the benchmark problems, the number of exceptional elements of the solution obtained by the proposed method is either better than the best reached values or it is equal to the best ones. In 11 problems, namely 3, 6, 7, 12, 13, 14, 19, 20, 21, 22 and 23 the *PE* of the solution obtained by the EDA-CF is better than that of all other methods. In other words, the proposed method outperforms all the other methods. In nine problems, namely 1, 4, 5, 8, 9, 10, 11, 18 and 25, all the methods have obtained the same Percentage of Exceptional elements.

CONCLUSION

Cellular manufacturing is a production technique that leads to increase productivity and efficiency in the production floor. In this chapter, we have presented the first Estimation of Distribution Algorithm (EDA) method to solve the Machine Part Cell Formation Problem. Detailed numerical experiments have been carried out to investigate the EDAs' performance. Although the EDA approach

Table 3. Comparison between the obtained results and the best-known results respecting to the PE criterion

No.	size	C	Problem Source	PE	CPU	PE [a]
1	5×7	2	King & Nakornchai, 1982	0.000	0.000	0.000
2	5×7	2	Waghodekar & Sahu, 1984	0.150	0.000	0.125
3	5×18	2	Seifoddini, 1989	**0.000**	0.000	0.1957
4	6×8	2	Kusiak & Cho, 1992	0.0909	0.000	0.0909
5	7×11	2	Kusiak & Chow, 1987	0.1304	0.000	0.1304
6	711	3	Boctor, 1991	**0.0952**	0.000	0.1905
7	8×12	5	Seifoddini & Wolfe, 1986	**0.1714**	0.000	0.2857
8	8×20	3	Chandrasekharan & Rajagopalan, 1986	0.1475	0.000	0.1475
9	8×20	3	Chandrasekharan & Rajagopalan, 1986	0.2967	0.000	0.2967
10	10×10	3	Mosier & Taube, 1985	0.000	0.000	0.000
11	10×15	3	Chan & Milner, 1982	0.000	0.015	0.000
12	14×24	4	Stanfel, 1985	**0.0328**	0.015	0.1639
13	16×24	8	McCormick et al., 1972	**0.3721**	0.031	0.4302
14	16×43	4	King, 1980	**0.2063**	0.031	0.2222
15	20×20	6	Mosier & Taube, 1985	0.3693	0.078	-
16	20×35	5	Carrie, 1973	0.1985	0.031	-
17	20×35	5	Boe & Cheng, 1991	0.1764	0.062	-
18	24×40	7	Chandrasekharan & Rajagopalan, 1989 - 1	0.000	0.031	0.000
19	24×40	7	Chandrasekharan & Rajagopalan, 1989 - 2	**0.0308**	0.451	0.0769
20	24×40	7	Chandrasekharan & Rajagopalan, 1989 - 3	**0.1087**	5.171	0.1527
21	24×40	7	Chandrasekharan & Rajagopalan, 1989 - 4	**0.0992**	0.732	0.1527
22	24×40	7	Chandrasekharan & Rajagopalan, 1989 - 5	**0.2652**	0.670	0.3740
23	24×40	7	Chandrasekharan & Rajagopalan, 1989 - 6	**0.2824**	0.233	0.4214
24	27×27	6	McCormick et al., 1972	0.2350	0.203	-
25	30×41	14	Kumar & Vanelli, 1987	0.1094	0.219	0.1094
26	30×50	13	Stanfel, 1985	0.2754	3.109	-
27	30×50	14	Stanfel, 1985	0.1225	0.406	-
28	36×90	17	King & Nakornchai,1982	0.1254	0.969	-
29	37×53	3	McCormick et al., 1972	0.000	0.109	-
30	40×100	10	Chandrasekharan & Rajagopalan, 1987	0.0907	7.421	0.0857

does not require any problem-specific information, the use of sensible heuristics can improve the optimisation and speed up convergence. For this reason, we used the Variable Neighborhood Search (VNS) procedure in the improvement phase of the algorithm. The results from test cases presented here have shown that the proposed EDA-CF algorithm is very a competitive algorithm comparing with the previously published metaheuristics applied to the same problem. It has been shown that the EDAs provide efficient and accurate solutions for the test cases. The results are promising and encourage further studies on other versions of the Group Technology problems where we can introduce sequence data, machine utilization and routings.

REFERENCES

Andrés, C., & Lozano, S. (2006). A particle swarm optimization algorithm for part–machine grouping. *Robotics and Computer-integrated Manufacturing*, *22*, 468–474. doi:10.1016/j.rcim.2005.11.013

Askin, R. G., Creswell, J. B., Goldberg, J. B., & Vakharia, A. J. (1991). A Hamiltonian path approach to reordering the part-machine matrix for cellular manufacturing. *International Journal of Production Research*, *29*, 1081–1100. doi:10.1080/00207549108930121

Baluja, S. (1994). *Population-based incremental learning: A method for integrating genetic search based function optimization and competitive learning*. (Technical Report CMU-CS, 94-163). Computer Science Department, Carnegie Mellon University.

Baluja, S., & Davies, S. (1997). Using optimal dependency-trees for combinatorial optimization: Learning the structure of the search space. In *Proceedings of the 1997 International Conference on Machine Learning*.

Baluja, S., & Davies, S. (1998). *Fast probabilistic modeling for combinatorial optimization*. In AAAI-98.

Bengoetxea, E., Larranaga, P., Bloch, I., & Perchant, A. (2001b). Solving graph matching with EDAs using a permutation–based representation. In Larranaga, P., & Lozano, J. A. (Eds.), *Estimation of distribution algorithms. A new tool for evolutionary computation*. Kluwer Academic Publishers. doi:10.1007/978-1-4615-1539-5_12

Bengoetxea, E., Larranaga, P., Bloch, I., Perchant, A., & Boeres, C. (2000). *Inexact graph matching using learning and simulation of Bayesian networks. An empirical comparison between different approaches with synthetic data*. In Workshop Notes of CaNew2000: Workshop on Bayesian and Causal Networks: From Inference to Data Mining, fourteenth European Conference on Artificial Intelligence, ECAI2000. Berlin.

Boctor, F. (1991). A linear formulation of the machine-part cell formation problem. *International Journal of Production Research*, *29*(2), 343–356. doi:10.1080/00207549108930075

Brown, E., & Sumichrast, R. (2001). CF-GGA: A grouping genetic algorithm for the cell formation problem. *International Journal of Production Research*, *36*, 3651–3669. doi:10.1080/00207540110068781

Burbidge, J. L. (1963). Production flow analysis. *Production Engineering*, *42*, 742–752. doi:10.1049/tpe.1963.0114

Caux, C., Bruniaux, R., & Pierreval, H. (2000). Cell formation with alternative process plans and machine capacity constraints: A new combined approach. *International Journal of Production Economics*, *64*(1-3), 279–284. doi:10.1016/S0925-5273(99)00065-1

Chan, H. M., & Milner, D. A. (1982). Direct clustering algorithm for group formation in cellular manufacture. *Journal of Manufacturing Systems*, *1*, 65–75. doi:10.1016/S0278-6125(82)80068-X

Chandrasekharan, M. P., & Rajagopalan, R. (1986a). MODROC: An extension of rank order clustering for group technology. *International Journal of Production Research*, *24*(5), 1221–1264. doi:10.1080/00207548608919798

Chandrasekharan, M. P., & Rajagopalan, R. (1987). ZODIAC: An algorithm for concurrent formation of part-families and machine-cells. *International Journal of Production Research*, *25*(6), 835–850. doi:10.1080/00207548708919880

Chandrasekharan, M. P., & Rajagopalan, R. (1989). Groupability: Analysis of the properties of binary data matrices for group technology. *International Journal of Production Research*, *27*(6), 1035–1052. doi:10.1080/00207548908942606

Chickering, D., Heckerman, D., & Meek, C. (1997). A Bayesian approach to learning Bayesian networks with local structure. In *Proceedings of Thirteenth Conference on Uncertainty in Artificial Intelligence,* (pp. 80–89). (Technical Report MSRTR- 97-07), Microsoft Research, August, 1997.

De Bonet, J., Isbell, C. L., & Viola, P. (1997). MIMIC: Finding optima by estimating probability densities. *Advances in Neural Information Processing Systems*, *9*, 424–430.

De Campos, L. M., Gamez, J. A., Larranaga, P., Moral, S., & Romero, T. (2001). Partial abductive inference in Bayesian networks: An empirical comparison between GAs and EDAs. In Larranaga, P., & Lozano, J. A. (Eds.), *Estimation of distribution algorithms. A new tool for evolutionary computation*. Kluwer Academic Publishers. doi:10.1007/978-1-4615-1539-5_16

Etxeberria, R., & Larranaga, P. (1999). Optimization with Bayesian networks. In *Proceedings of the Second Symposium on Artificial Intelligence*. Adaptive Systems. CIMAF 99, (pp. 332-339). Cuba.

Gonçalves, J., & Resende, M. (2004). An evolutionary algorithm for manufacturing cell formation. *Computers & Industrial Engineering*, *47*, 247–273. doi:10.1016/j.cie.2004.07.003

Goncalves, J. F., & Resende, M. (2002). *A hybrid genetic algorithm for manufacturing cell formation. Technical report*. Rapport.

Gonzalez, C., Lozano, J. A., & Larranaga, P. (2002). Mathematical modelling of UMDAc algorithm with tournament selection: Behaviour on linear and quadratic functions. *International Journal of Approximate Reasoning*, *31*(3), 313–340. doi:10.1016/S0888-613X(02)00092-0

Harik, G. (1994). *Finding multiple solutions in problems of bounded difficulty*. Tech. Rep. IlliGAL Report No. 94002, University of Illinois at Urbana-Champaign, Urbana, IL.

Harik, G. (1999). *Linkage learning via probabilistic modeling in the ECGA*. Tech. Rep. IlliGAL Report No. 99010, University of Illinois at Urbana-Champaign.

Harik, G., Lobo, F., & Goldberg, D. E. (1998). *The compact genetic algorithm*, (pp. 523-528). (IlliGAL Report No. 97006).

James, T. L., Brown, E. C., & Keeling, K. B. (2007). A hybrid grouping genetic algorithm for the cell formation problem. *Computers & Operations Research*, *34*, 2059–2079. doi:10.1016/j.cor.2005.08.010

Joines, J. A., Culbreth, C. T., & King, R. E. (1996). Manufacturing cell design: An integer programming model employing genetic algorithms. *IIE Transactions*, *28*(1), 69–85. doi:10.1080/07408179608966253

Kaparthi, S., Suresh, N. C., & Cerveny, R. P. (1993). An improved neural network leader algorithm for part-machine grouping in group technology. *European Journal of Operational Research*, *69*, 342–355. doi:10.1016/0377-2217(93)90020-N

Khator, S. K., & Irani, S. A. (1987). Cell formation in group technology: A new approach. *Computers & Industrial Engineering*, *12*, 131–142. doi:10.1016/0360-8352(87)90006-4

King, J. R. (1980). Machine-component grouping formation in group technology. *International Journal of Management Science*, *8*(2), 193–199.

King, J. R., & Nakornchai, V. (1982). Machine-component group formation in group technology: Review and extension. *International Journal of Production Research*, *20*(2), 117–133. doi:10.1080/00207548208947754

Kumar, K. R., & Chandrasekharan, M. P. (1990). Grouping efficacy: A quantitative criterion for block diagonal forms of binary matrices in group technology. *International Journal of Production Research*, *28*(2), 233–243. doi:10.1080/00207549008942706

Kusiak, A. (1987). The generalized group technology concept. *International Journal of Production Research*, *25*, 561–569. doi:10.1080/00207548708919861

Kusiak, A., & Chow, W. S. (1987). Efficient solving of the group technology problem. *Journal of Manufacturing Systems*, *6*(2), 117–124. doi:10.1016/0278-6125(87)90035-5

Larranaga, P., Etxeberria, R., Lozano, J. A., & Pena, J. M. (2000). Combinatorial optimization by learning and simulation of Bayesian networks. In *Proceedings of the Sixteenth Conference on Uncertainty in Artificial Intelligence*, (pp. 343–352). Stanford.

Li, S. Z. (1995). *Markov random field modeling in computer vision*. Springer-Verlag.

Lozano, J. A., Sagarna, R., & Larranaga, P. (2001b). Solving job scheduling with estimation of distribution algorithms. In Larranaga, P., & Lozano, J. A. (Eds.), *Estimation of distribution algorithms. A new tool for evolutionary computation* (pp. 231–242). Kluwer Academis Publishers. doi:10.1007/978-1-4615-1539-5_11

Lozano, S., Adenso-Diaz, B., Eguia, I., & Onieva, L. (1999). A one step tabu search algorithm for manufacturing cell design. *The Journal of the Operational Research Society*, *50*, 509–516.

Mahdavi, I., Paydar, M. M., Solimanpur, M., & Heidarzade, A. (2009). Genetic algorithm approach for solving a cell formation problem in cellular manufacturing. *Expert Systems with Applications*, *36*, 6598–6604. doi:10.1016/j.eswa.2008.07.054

McAuley, J. (1972). Machine grouping for efficient production. *Production Engineering*, *51*(2), 53–57. doi:10.1049/tpe.1972.0006

McCormick, W. T. Jr, Schweitzer, P. J., & White, T. W. (1972). Problem decomposition and data reorganization by a cluster technique. *Operations Research*, *20*(5), 993–1009. doi:10.1287/opre.20.5.993

Mladenoviç, N., & Hansen, P. (1997). Variable neighborhood search. *Computers & Operations Research*, *24*, 1097–1100. doi:10.1016/S0305-0548(97)00031-2

Muhlenbein, H. (1998). The equation for response to selection and its use for prediction. *Evolutionary Computation*, *5*(3), 303–346. doi:10.1162/evco.1997.5.3.303

Muhlenbein, H., & Mahnig, T. (1999b). FDA - A scalable evolutionary algorithm for the optimization of additively decomposed functions. *Evolutionary Computation, 7*, 353–376. doi:10.1162/evco.1999.7.4.353

Muhlenbein, H., Mahning, T., & Ochoa, A. (1999). Schemata, distributions and graphical models in evolutionary optimization. *Journal of Heuristics, 5*, 215–247. doi:10.1023/A:1009689913453

Muhlenbein, H., & Paaß, G. (1996). From recombination of genes to the estimation of distribution. Binary parameters. *Lecture Notes in Computer Science, 1411. Parallel Problem Solving from Nature, PPSN, IV*, 178–187.

Onwubolu, G. C., & Mutingi, M. (2001). A genetic algorithm approach to cellular manufacturing systems. *Computers & Industrial Engineering, 39*(1–2), 125–144. doi:10.1016/S0360-8352(00)00074-7

Pearl, J. (1988). *Probabilistic reasoning in intelligent systems*. Palo Alto, CA: Morgan Kaufman Publishers.

Pelikan, M., Goldberg, D. E., & Cantu-Paz, E. (1999a). BOA: The Bayesian optimization algorithm. In Banzhaf, W., Daida, J., Eiben, A. E., Garzon, M. H., Pelikan, V., & Goldberg, D. E. (Eds.), *Hierarchical problem solving by the Bayesian optimization algorithm. IlliGAL Report No. 2000002*. Urbana, IL: Illinois Genetic Algorithms Laboratory, University of Illinois at Urbana-Champaign.

Pelikan, P., & Muhlenbein, H. (1999). The bivariate marginal distribution algorithm. In Roy, R., Furuhashi, T., & Chandhory, P. K. (Eds.), *Advances in soft computing-engineering design and manufacturing* (pp. 521–535). London, UK: Springer.

Robles, V., de Miguel, P., & Larranaga, P. (2001). Solving the travelling salesman problem with estimation of distribution algorithms. In Larranaga, P., & Lozano, J. A. (Eds.), *Estimation of distribution algorithms. A new tool for evolutionary computation*. Kluwer Academic Publishers.

Roure, J., Sanguesa, R., & Larranaga, P. (2001). Partitional clustering by means of estimation of distribution algorithms. In Larranaga, P., & Lozano, J. A. (Eds.), *Estimation of distribution algorithms. A new tool for evolutionary computation*. Kluwer Academic Publishers.

Sagarna, R., & Larranaga, P. (2001). Solving the knapsack problem with estimation of distribution algorithms. In Larranaga, P., & Lozano, J. A. (Eds.), *Estimation of distribution algorithms. A new tool for evolutionary computation*. Kluwer Academis Publishers.

Santana, R. (2003a). A Markov network based factorized distribution algorithm for optimization. *Proceedings of the 14th European Conference on Machine Learning (ECMLPKDD 2003); Lecture Notes in Artificial Intelligence, 2837*, (pp. 337–348). Berlin, Germany: Springer-Verlag.

Santana, R. (2005). Estimation of distribution algorithms with Kikuchi approximation. *Evolutionary Computation, 13*, 67–98. doi:10.1162/1063656053583496

Sierra, B., Jimenez, E., Inza, I., Larranaga, P., & Muruzabal, J. (2001). Rule induction using estimation of distribution algorithms. In Larranaga, P., & Lozano, J. A. (Eds.), *Estimation of distribution algorithms. A new tool for evolutionary computation*. Kluwer Academic Publishers. doi:10.1007/978-1-4615-1539-5_15

Sofianopoulou, S. (1997). Application of simulated annealing to a linear model for the formation of machine cells in group technology. *International Journal of Production Research, 35*, 501–511. doi:10.1080/002075497195876

Solimanpur, M., Vrat, P., & Shankar, R. (2003). Ant colony optimization algorithm to the inter-cell layout problem in cellular manufacturing. *European Journal of Operational Research*, *157*(3), 592–606. doi:10.1016/S0377-2217(03)00248-0

Srinivasan, G. (1994). A clustering algorithm for machine cell formation in group technology using minimum spanning trees. *International Journal of Production Research*, *32*, 2149–2158. doi:10.1080/00207549408957064

Srinivasan, G., & Narendran, T. T. (1991). GRAFICS - A non hierarchical clustering-algorithm for group technology. *International Journal of Production Research*, *29*(3), 463–478. doi:10.1080/00207549108930083

Stawowy, A. (2006). Evolutionary strategy for manufacturing cell design. *OMEGA: The International Journal of Management Science*, *34*(1), 1–18. doi:10.1016/j.omega.2004.07.016

Venugopal, V., & Narendran, T. T. (1992a). A genetic algorithm approach to the machine component grouping problem with multiple objectives. *Computers & Industrial Engineering*, *22*(4), 469–480. doi:10.1016/0360-8352(92)90022-C

Xu, H., & Wang, H. P. (1989). Part family formation for GT applications based on fuzzy mathematics. *International Journal of Production Research*, *27*(9), 1637–1651. doi:10.1080/00207548908942644

Zhang, Q., Sun, J., Tsang, E., & Ford, J. (2004). Hybrid estimation of distribution algorithm for global optimisation. *Engineering Computations*, *2*(1), 91–107. doi:10.1108/02644400410511864

Zhao, C., & Wu, Z. (2000). A genetic algorithm for manufacturing cell formation with multiple routes and multiple objectives. *International Journal of Production Research*, *38*(2), 385–395. doi:10.1080/002075400189473

APPENDIX

Table 4. Problem 20

	4	16	7	14	23	24	9	10	17	2	5	11	19	6	8	12	15	18	3	20	1	13	21	22
8	1	1																						
19	1	1																						
21	1	1																						
28	1																							
37		1															1							
38	1	1																						
39	1	1																						
3			1	1	1	1																		
25			1	1	1																			
32		1	1	1	1	1																		
6							1	1	1															
7							1	1	1															
20								1	1							1								
29							1		1															
40							1	1	1															
10										1	1	1	1										1	
13										1		1	1											
14										1	1	1	1											
22										1	1	1	1											
35											1	1	1											
36											1	1	1											
4														1	1	1	1	1						
5											1			1	1	1	1	1						
18														1	1	1	1	1						
26								1						1	1		1	1						
27														1	1	1	1	1						
30														1	1	1		1						
2						1													1	1				
11																			1	1				
12																			1	1				
15																			1	1				
23																			1	1				
24																1			1	1				
31									1										1	1				
34																			1	1				
1																					1	1	1	1

continues on following page

Table 4. Continued

	4	16	7	14	23	24	9	10	17	2	5	11	19	6	8	12	15	18	3	20	1	13	21	22
9																					1	1	1	1
16																					1	1	1	1
17																					1	1		1
33										1											1	1	1	1

Table 5. Problem 21

	3	20	6	8	12	15	18	1	13	21	22	2	5	11	19	4	16	9	10	17	7	14	23	24
2	1	1																						1
11	1	1																						
12	1	1												1										
15	1	1																				1		
23	1	1																						
24	1	1			1																			
31		1																		1				
34	1	1																						
4			1	1	1	1	1																	
5			1	1	1	1	1						1											
18			1		1	1	1																	
26			1	1		1	1												1					
27			1	1	1	1	1																	
30			1	1	1		1				1													
1						1			1	1	1													
9								1	1	1	1													
16								1	1	1	1											1		
17								1	1		1													
33								1	1	1	1	1						1						
10										1		1	1	1	1									
13												1		1	1									
14												1	1		1									
22												1	1	1	1									
35													1	1	1									
36												1	1	1	1									
8																1	1							
19		1						1								1	1							
21																1	1							
28																1								
37						1											1							

continues on following page

Table 5. Continued

	3	20	6	8	12	15	18	1	13	21	22	2	5	11	19	4	16	9	10	17	7	14	23	24
38			1													1	1							
39																1	1							
6																		1	1	1				
7																		1	1	1				
20					1														1	1				
29																		1		1				
40															1			1	1	1	1			
3																					1	1	1	1
25																					1	1	1	
32																	1				1	1	1	1

Table 6. Problem 22

	3	20	2	5	11	19	6	8	12	15	18	1	13	21	22	10	4	16	7	24	9	17	14	23
2	1	1																		1				
11	1	1																						
12	1	1																						
15	1	1																						
23	1	1																						
24	1	1							1															
31		1																				1		
34	1	1																						
10			1	1	1	1								1										
13			1		1	1																		
14			1	1		1																	1	
22			1	1	1	1																		
35				1	1	1																		
36			1	1	1	1																		
4								1	1	1	1													
5				1			1	1	1	1	1													
18							1		1	1	1													
26							1	1		1	1					1								
27							1	1	1	1	1													
30							1	1	1		1				1									
1										1			1	1	1									
9												1	1	1										
16												1	1	1	1								1	

continues on following page

Table 6. Continued

	3	20	2	5	11	19	6	8	12	15	18	1	13	21	22	10	4	16	7	24	9	17	14	23
17												1	1		1									
33			1									1	1	1	1						1			
6																1					1			
20									1							1						1		
8																	1	1						
19		1										1					1	1						
21																	1	1						
28																	1							
37										1								1						
38							1										1	1						
39																	1	1						
32																		1	1	1				1
7																1					1	1		
29																					1	1		
40						1													1		1	1		
3																				1			1	1
25																			1				1	1

Table 7. Problem 23

	1	21	3	20	7	14	23	24	9	4	16	10	2	5	11	19	6	8	17	12	15	18	13	22
9	1	1																						
33	1	1							1				1											
2			1	1				1																
11				1						1												1		
12			1	1										1									1	
15			1	1											1									
23			1	1											1									
34			1	1														1						
3		1				1	1	1																
25				1	1	1	1																	
32	1				1		1	1			1													
6									1			1												
29									1										1					
39									1									1						
40					1				1							1								
8										1	1													
19	1			1						1	1													

continues on following page

Table 7. Continued

	1	21	3	20	7	14	23	24	9	4	16	10	2	5	11	19	6	8	17	12	15	18	13	22
21										1	1													
28										1														
37											1										1			
38										1							1							
20												1							1	1				
24			1									1								1				
10		1												1	1		1							
13							1						1		1	1								
14						1							1	1		1					1			
22													1	1	1					1				
35											1			1	1	1							1	
36											1		1	1		1								
26											1	1					1	1			1			
30																	1	1				1		1
7																			1					
31				1															1					
4																		1		1	1	1		
5														1				1		1	1	1		
18																	1			1		1		
27																1				1	1	1		
1		1																			1		1	1
16		1				1																	1	1
17	1																		1				1	1

Table 8. Problem 24

	6	8	18	2	15	19	9	4	7	14	13	22	1	21	23	24	17	3	20	10	12	16	5	11
4		1	1																		1			
5		1	1		1																		1	
18	1		1																		1			
26	1	1			1															1		1		
30	1	1	1									1												
38	1	1						1																
13				1		1			1						1									
14				1	1	1				1														
27			1		1	1															1			
36				1		1																	1	
6							1													1				

continues on following page

Table 8. Continued

	6	8	18	2	15	19	9	4	7	14	13	22	1	21	23	24	17	3	20	10	12	16	5	11
29							1										1							
39		1					1															1		
4						1	1		1															
11			1					1		1									1					
25									1	1									1					
28								1	1															
1					1						1	1		1										
16		1								1	1	1												
35						1					1												1	
9						1							1	1						1				
33				1			1						1	1										
3														1	1	1								
32									1						1	1						1		
7																	1							
17											1		1				1							
31													1				1		1					
2																1		1	1					
12							1				1				1			1	1				1	
15																		1	1					1
23																		1	1					1
34		1														1		1	1					
20																	1			1	1			
24																		1		1	1			
8								1														1		
19													1						1			1		
21								1														1		
37					1																	1		
10	1													1									1	1
22				1																	1		1	1

Chapter 9
Cellular or Functional Layout?

Abdessalem Jerbi
University of Sfax, Tunisia

Hédi Chtourou
University of Sfax, Tunisia

ABSTRACT

The cellular layout has been compared to the traditional functional layout using multiple comparison methodologies that either lack objectivity or are highly time-consuming. The main purpose of this chapter is to propose a novel and objective methodology. Hence, a critical analysis of ten comparison studies is followed by the presentation of the layouts simulation models. Subsequently, the proposed comparison methodology is described. Following this methodology, simulations are conducted according to a plan of experiments developed from Taguchi standard orthogonal arrays. Consequently, results, expressed in Signal to Noise ratios, are analyzed using ANOVA. Next, a mathematical model is derived by interpolation between the factors and interactions effects. This model must be validated by the confirmation test, otherwise the comparison methodology should be reiterated while considering new interactions. This cycle should be reiterated as much as necessary to obtain a valid mathematical model. The proposed comparison methodology has been applied with success on an academic manufacturing system.

INTRODUCTION

The increased competition within industry has resulted in manufacturing companies spending considerable effort to improve flexibility and responsiveness to meet customer needs. Cellular manufacturing, a facet of group technology, has emerged as one of the major techniques being used for the improvement of manufacturing competitiveness. A large number of empirical, analytical and simulation studies have been devoted to compare the cellular layout (CL) to the classical functional layout (FL). Simulation-based comparative studies constitute the mainstream of this research field. Varied results were reported by these comparative simulation studies. Indeed, different researches found the FL always superior to the CL with regard to all used performance measures (Jensen,

DOI: 10.4018/978-1-61350-047-7.ch009

Malhotra, & Philipoom, 1996; Morris & Tersine, 1990, 1994). Further researches reported that the CL is superior to the FL in all operating conditions (Pitchuka, Adil, & Ananthakumar, 2006; Shafer & Charnes, 1992). Finally, other simulation studies showed that every layout could outperform the other in different particular experimental conditions (Faizul huq, Douglas, & Zubair, 2001; Farrington & Nazametz, 1998; Li, 2003; Shafer & Charnes, 1995; Suresh & Meredith, 1994). The divergence in the studies conclusions is referred to as the "cellular manufacturing paradox" (Shambu, Suresh, & Pegels, 1996). In fact, Agarwal and Sarkis (1998) and Shambu et al. (1996) reviewed a number of FL-CL comparative studies. However they did not identify any objectivity flaws responsible for the conflicting conclusions. Indeed, they simply reported the major findings of some published studies without any critical objectivity assessment.

Actually, methodologies used by comparison studies vary widely but can be classified into three groups. In the first group, authors used the one-factor-at-a-time method. So the two layouts are first compared for one manufacturing context considered as the "base model". Then, other experiments are carried out in order to test the robustness of the layout choice obtained in the base model. Every experiment corresponds to the modification of a single operating factor (Morris & Tersine, 1990, 1994). In the second group authors considered only some specific combinations of the studied factors settings without any justification (Faizul huq et al., 2001; Li, 2003; Suresh & Meredith, 1994). In the third group authors used the full factorial design technique in order to study the effect of all factors (Farrington & Nazemetz, 1998; Jensen et al., 1996; Pitchuka et al., 2006; Shafer & Charnes 1992, 1995). Methodologies belonging to the two first groups undoubtedly lack objectivity in the choice of the experimentation conditions. Therefore, they do not permit to attach any statistical confidence level to their conclusions. In addition, they do not provide any information about factor interaction. The third group methodology is highly time-consuming. In addition, it is impractical when the number of factors to study is large.

This chapter essentially focuses on the development of an objective FL-CL comparison. It first highlights the lacks of objectivity of the main published FL-CL simulation-based comparison studies in order to explain the origin of their conflicting conclusions. Then it deals with the development of comprehensive FL and CL simulation models using the widely used commercial simulation software Arena 7.0. Finally, it presents the framework of a methodology, based on the coupling of the Taguchi method of experiment design (TM) and simulation. This methodology can be easily applied to any manufacturing context and provides trustworthy results with a minimum experimentation effort.

The remainder of this chapter is organized as follows. The next section presents a taxonomy of the key factors used in the main published FL-CL comparison simulation studies. The foremost used performance measures are also presented in this section. Finally it presents and analyses the findings of a number of relevant studies. The third section presents some general simulation features, needed for modeling both layouts. Then, it respectively gives details of the developed FL and the CL simulation models. Section four gives a general presentation of the objective comparison methodology and then presents a comprehensive academic case study depicting its application. The final section includes some general conclusions and discusses future work prospects.

COMPARATIVE STUDIES FRAMEWORK

Main Experimental Factors

General Manufacturing System (MS) Characteristics

Every MS is characterized by a number of machines arranged either into departments in the functional layout, or else, into manufacturing cells in the cellular layout. Following the FL structure, the shop is composed of *d* departments *Di* (*i*=1, …, *d*). Each of them includes *Mn* functionally equivalent machines. In contrast, the CL is composed of *c* independent manufacturing cells *Cj* (*j*=1, …, *c*). Each cell is a cluster of *Mf* different machines dedicated to a number of similar part types. Furthermore, every MS is designed for a demand pattern comprising different products. Products are identified by two indicators, which are the type (*t*) and the family (*f*). Products are grouped into families according to the similarity of their manufacturing process. Each product type requires a number of manufacturing operations (*mopt*).

Degree of Decomposability of the Part Machine Matrix (DD)

This degree translates the feasibility of the decomposition of the MS into independent cells. In fact, the more the product/machine matrix is diagonal, the more the decomposability is feasible. This degree is negatively correlated to the density of off-diagonal elements.

Batch Size (BS)

Products are generally manufactured and transferred in batches in order to reduce machine setup and transport between machines. Numerous authors included *BS* in there comparison studies as a variable factor and demonstrated that the combination of small batch sizes with an efficient scheduling rule results in the improvement of the cellular layout performances. Most authors used the same batch size for both cellular and functional layouts.

Demand Rate (DEMAND)

The demand rate is mainly expressed by the batch inter-arrival times (*IAT*) in the MS. A large part of authors generated this time by common probabilistic distributions. Others used constant *IAT*. Besides, some authors focus only on the stability of this factor without changing its average value.

Transfer Time (TT)

This parameter corresponds to the interdepartmental travel times in the FL. They are often modeled using appropriate probabilistic laws. In the CL these times correspond to the durations of intra-cell moves. Generally, they are very small compared to those in the FL.

Transfer Mode (TM)

Because of the considerable interdepartmental distances in the FL products are generally transferred by batches in order to reduce transfer costs. Some studies also used this transfer mode between same-cell machines whereas others make use of operations overlapping. This mode exploits the proximity of same- cell machines to allow simultaneous execution of different operations on parts of the same batch.

Flow Direction (FLOW)

A number of authors included the flow direction within a cell as an experimental factor. This factor has two possible levels: "unidirectional" or "backtracking allowed".

Scheduling Rules (RULE)

Part batches arriving at a department or a cell are put in a waiting queue until the required machine becomes idle. These batches are then sequenced in order to establish the order in which they will be processed. This order is specified by the use of standard scheduling rule such as "First Come First Served" (FCFS), "Shortest Process Time" (SPT), "Earliest Due Date" (EDD) or else, "Repetitive Lots" (RL). The limited versions of the first three rules, FCFS-L, SPT-L and EDD-L are used in order to avoid the duplication of machines setups for the same product type. Finally, the RL rule selects batches of the same type that the one just processed in order to minimize setups.

Processing Time (PT) and Set up Time (ST)

As for the *IAT*, most studies generally modeled both times by independent probabilistic laws. On the other hand, other studies formulated *ST* as a fraction of *PT*.

Set up Time Reduction Factor (δ)

This factor materializes one of the most key advantages of the CL. Indeed, part types of a same family need generally similar setups. Hence, if a machine is set up for a part type and then should be set for a same-family part type, the nominal setup time for the second part is reduced by the δ factor.

Performance Measures

Work in Process (WIP)

WIP is one of the most popular performance measures used in the FL-CL comparative studies (Farrington & Nazametz, 1998; Jensen et al., 1996; Li, 2003; Morris & Tersine, 1990, 1994; Shafer & Charnes, 1992, 1995; Suresh & Meredith, 1994). It essentially characterizes the fluidity of the material flow in the system.

Mean Flow Time (MFT)

MFT constitutes the other most popular measure used in FL-CL comparative studies (Faizul huq et al., 2001; Farrington & Nazametz, 1998; Jensen et al., 1996; Li, 2003; Morris & Tersine, 1990; Shafer & Charnes, 1992, 1995; Suresh & Meredith, 1994). It also characterizes the fluidity of the material flow in the system. The *MFT* is the average time that every batch remains in the system in order to be manufactured.

Due Date Related Measures

Researchers used essentially Mean Tardiness (*MT*) and Mean Earliness (*ME*) as due date related performance measures (Farrington & Nazametz, 1998; Jensen et al., 1996). *MT* is the average over all tardy jobs of the difference between delivery date and the promised due date. *ME* is similarly obtained for all early jobs. Other researchers used the percentage of tardy jobs (*TARDY*) and the percentage of early jobs (*EARLY*).

Other Measures

FL-CL comparative studies consider several other performance measures. The system *Throughput*, considered as productivity measure, is the average number of parts exiting the system by time unit (Faizul huq et al., 2001; Morris & Tersine, 1994). It is also used for detecting the attainment of steady state indicator in a simulation run. Besides, some studies used the operator utilization rate (*OPUR*) (Morris & Tersine, 1994), the average machine utilization rate (*MUR*) (Farrington & Nazametz, 1998; Morris & Tersine, 1994; Shafer, & Charnes, 1995), the mean "queue" waiting time (Pitchuka et al., 2006) or the average *ST/PT* ratio (Li, 2003) as performance indicators. The first two measures must be maximized to ensure a high degree of resource exploitation but the third and fourth measures should be minimized to improve the efficiency of the MS.

Comparative Studies Findings

By means of four simulation experiments, Morris and Tersine (1990) examined the influence of the ratio *ST/PT, TT, DEMAND* stability and parts *FLOW* within cells on the performance of CLs compared to FLs. In this comparative study, the performances were measured using *MFT* and *WIP*. Results demonstrate that in the quasi totality of the tested contexts, the FL always outperforms the CL and generates smaller *MFT* and *WIP*. Besides, comparison results reveal that the ideal context for CL must be characterized by a high *ST/PT* ratio, a stable *DEMAND*, a unidirectional *FLOW* and a substantial *TT* between process departments.

To find out the operating conditions under which the CL outperform the FL, Shafer and Charnes (1992) investigated 24 combinations of *DD*, *mopt*, *PT* and *BS*. As for the previous study, authors used the same performance measures. They found the CL superior to the FL in all operating conditions according to both performance measures.

Morris and Tersine (1994) extended the results of their first comparison study (Morris & Tersine, 1990) by investigating the impact of a dual resource constrained shop on the performances of CL and FL using three operator scheduling rules in the CL. Simulation observations were collected for four performance measures including the mean *Throughput*, the *WIP*, the *MUR* and the *OPUR*. Results reveal that the FL outperformed the CL on all of the used performance measures regardless of the operator scheduling rule.

Authors investigate the sensitivity of their results relatively to changes in shop congestion level and changed respectively the *IAT* and *OPUR* in two other experiments. It appeared that the FL still outperforms the CL.

Besides, Suresh and Meredith (1994) aimed to overcome the loss of pooling synergy in the CL. Hence, they used simulation in order to compare the CL to an efficiently operated FL (EFL) using average *MUR*, *WIP* and *MFT* to assess the two layout's performance measures. The EFL is characterized by an optimal *BS*, a reduced *TT* and part-family-oriented scheduling rules. The main experimental factors involved in this study were *PT*, *ST*, *BS*, δ and *IAT*. First, every experimental factor was tested separately. Then all the experimental factors were tested together. The FL was found to be superior to the CL for large batch sizes (*BS*>32). However, for relatively small *BS*, the CL could outperform the FL if δ is smaller than 0.2. Comparison results do not change when the variability of *PT*, *ST* or *IAT* were separately reduced. On the other hand, if all factor effects were combined, the CL outperformed the FL even for small *BS*.

Shafer and Charnes (1995) used simulation to study a manufacturing context inspired from Morris and Tersine (1990). In fact, they used the same levels of the following factors: *t*, *f*, *c*, *d*, *Mn*, *Mf* and *mopt*. Authors aimed to compare alternative loading procedures for CL and FL in a variety of operating environments defined by combinations of 4 factors: *FLOW*, *TT*, labor constraints and MS congestion level. The third factor was modeled using two levels of the operator number while the last factor was modeled through the variation of the *PT*. Besides, each layout was investigated using two loading policies. For the FL the first policy permitted machine dedication while the second did not. On the other hand, for CL the first policy restricted the processing to only one batch at a time in a cell and the second allowed the processing of different batches at the same time. Both policies authorized CL operations overlapping. The authors used *MFT* and *WIP* in a two stage comparison methodology. In the first stage, labor constraints were not considered. In the second stage, a constraint was imposed on labor allowing only 8 operators to the whole shop in both configurations. It is worth noting here that the presence of one operator is required during setups and processing operations. The first stage simulation results demonstrate that the two layouts were equivalent regarding *WIP* while the CL generated lower *MFT* than the FL. In contrast, in the second comparison stage the FL showed lower *MFT* than the CL. The authors justified this result by the labor constraint effect on the CL. Indeed, according to the authors the labor constraints handicaps more seriously the CL since it reduces the operations overlapping possibilities while its effect on the

FL is not significant since the departments have only 3 machines in average.

Another study by Jensen et al. (1996) assessed the FL and CL performances through *MFT*, *WIP* *MT*, *ME* and *TARDY*. They based their study on a full- factorial experimental plan involving layout type, *RULE*, *DEMAND* variability and δ as experimental factors. To determine the influence of each factor on the studied performance measures, the authors analyzed their simulation results by ANOVA. Aside from the layout type, the most influent factor was found to be *DEMAND* variability followed by δ and *RULE*. Then, the authors performed a pairwise comparison of *RULE*. Results demonstrate that SPT-L and EDD were the best performing rules regarding *MFT* and *MT* respectively. Finally, they compared layouts using the best found *RULE*. The results of this final step revealed that the FL was always superior to the CL with regard to all performance measures.

As for Farrington and Nazemetz (1998), their comparative study is based on a three-factor-full-factorial experimental plan. The three experimental factors were the layout type, the *PT* variability and the *IAT* variability. It's worth noting here that the high variability level was associated to a small *BS* and vice versa. They assessed the two layouts using different performance measures, namely *MFT*, *WIP*, *TARDY*, *MUR* and a number of others less common measures. Comparison results prove that the FL is superior to the CL in a context defined by a high variability of *PT* and a low variability of *IAT*. But, when both factors show high variability, the performances of the two layouts are close. Besides, The CL outperforms the FL in all remaining conditions.

Faizul huq et al. (2001) presented in their comparison study a straightforward two-factor-full-factorial simulation plan using the *MFT* and the *Throughput*. The two studied factors were *BS* and δ. For the sake of objectivity, the authors used the EFL concept. ANOVA investigation showed that the two layout *Throughput* performances were not significantly different. In deed, the two layouts presented significant differences in some of the studied combinations only in terms of *MFT*. In fact, The CL outperformed the FL only for small *BS* and very large δ. In all other conditions, the FL was clearly superior.

Regarding Li (2003), the author used *MFT* and *WIP* to explore the superiority domains of both layouts in a diversity of contexts. These contexts are defined by the *FLOW*, the *TM*, the variability of *PT*, the variability of *ST* and finally δ. The performance measures results analysis showed that the major factor in establishing the superiority of one of the two layouts is δ. Hence, the CL outperformed the FL at high level of δ and the FL was the best layout in the low δ region. Both layouts showed equivalent performance measures for intermediate value of δ.

The last reviewed study, done by Pitchuka et al. (2006), compared FL to CL using a four-factor-full-factorial experimental plan featuring *PT*, *ST*, *BS* and *IAT*. The authors considered only the "queue" waiting time as performance measure. It was shown that the CL can outperform the FL in the majority of the studied contexts. Indeed, in the CL numerous work centers generated inferior "queue" times to those of the corresponding work centers in the FL.

Objectivity Assessment

Conditions Favoring FL

Jensen et al. (1996), Pitchuka et al. (2006) and Shafer and Charnes (1992) considered very low *TT* which implicitly advantage the FL, since one of the main advantages of the CL is time saving by locating machines required to manufacture a part close to each other. On the other hand, Jensen et al. (1996), Morris and Tersine, (1990, 1994) and Pitchuka et al. (2006) used a CL with no operations overlapping allowed in part processing. This does not permit to take advantage of CL benefits. Moreover, Farrington and Nazametz (1998) stated that they chose not to reduce the *ST* in the CL context. Their motivation was to avoid any biases in favor of the CL. But, by doing so, they favored the FL since they eliminated one of the main advantages of the CL.

Conditions Favoring CL

The study of Shafer and Charnes (1992) is obviously biased in favor of the CL. In deed, the authors consider single-machine departments. So, they eliminate the main and probably the only benefit of this type of layout: the pooling synergy effect between same department machines. Consequently, the results were clearly in favor of the CL even with the assumption of null transfer times advantaging the FL. Regarding Li (2003), the study featured unidirectional cell *FLOW* by duplicating the necessary machines to avoid backtracking. This indirectly led to the reduction of the cell number. The machine duplication within cells biased the comparison results in favor of the CL. In fact, this attaches to the CL the main advantages of the FL which is the synergy between functionally equivalent machines. As for Suresh and Meredith (1994), they used FL *TT* relatively very high compared to the *PT*. This probably advantage the CL and make clear why its performance are superior to the performance of the FL in almost all the testing contexts even though no operations overlapping has been used in the CL.

Other Conditions

This category essentially includes the lack of vital information about the used experimental factor settings as well as key elements defining the manufacturing contexts. Indeed, even if Morris and Tersine, (1990, 1994) provided in their studies the material handling equipment speed, they did not mention any distances between departments or machines. These distances are required in order to evaluate the *TT* in the two layouts. On the other hand, Farrington and Nazametz (1998) and Shafer and Charnes (1992) did not mention the *RULE* they used. In addition, Farrington and Nazametz (1998) failed to report numerous key experimental factors such as *mopt*, *IAT* and *PT*. Despite its established importance, Jensen et al. (1996) did not use the *BS* as an experimental factor neither did they mention its constant value used throughout the investigation.

Other lacks of important data are included in this category, particularly technical simulation-related information such as the replication length (Farrington & Nazametz, 1998; Li, 2003) and the warm-up period length (Farrington & Nazametz, 1998). On the other hand, numerous incongruities appear in different comparative studies. For example, the difference between the two shop configurations of the CL studied by Li (2003) is not clear. Indeed, in the figures illustrated by the authors, the arrows indicating the products *FLOW* show that there is no backtracking flow even in the CL with backtracking flow allowed. These incongruities are more serious in Faizul huq et al. (2001) study. Indeed, despite stating that no inter-cell moves were allowed, the authors defined the inter-cell travel time by a uniform law.

The use of inappropriate MS data appears especially in the study of Faizul huq and al. (2001). Indeed, the major flaw of this study is the definition of the manufacturing context. In fact, they used the same routings for the same product types. This generated three identical manufacturing cells. More gravely, the use of single-product families annuls any setup operation in the cell except for the initial setups. Hence, the factor δ becomes irrelevant and any results showing the importance of this factor are seriously questionable.

SIMULATION MODELS

Basic Simulation Features

FL and CL layouts simulation models are developed using the commercial simulation software

Arena 7.0 (Kelton, Sadowski, & Sadowski, 2002). This simulation tool integrates all the needed simulation functions including animation, analysis of input and output data. Every MS model consists of the four main components: manufacturing orders launching and attribute assignation, part transfer, part manufacturing and statistics collection.

Manufacturing Orders Launching and Attribute Assignation

Manufacturing orders (MOs), being batches of parts of the same type, are launched by "Create" modules. Every "Create" module defines batches *IAT* following the used probabilistic rule in addition to their *BS*. A specific "Create" module is dedicated for each part type. As soon as parts are lunched, they pass through an "Assign" module where characteristics are attributed to them. These characteristics are either time-related such as *PT* and *ST*, or also identification indicators such as part's type as well as part's family and factors necessary to the MS piloting like part's routing or "Sequence".

Part Transfer

Parts are transferred, either individually or in batches, between physical locations modeled by the "Station" modules, in which they should undergo the required manufacturing steps. These locations are either machines in CL or departments in FL. Transfer are carried out by "Route" modules permitting to prescribe destinations as well as transfer times. These modules use "Sequence" attribute of the transferred parts in order to prescribe the next destination. The "Sequence" corresponds to the part routing expressed as stations list.

Two manufacturing strategies could be followed for the parts transfer in the shops: "with operations overlapping" or "without operations overlapping". In the first strategy, parts of the same batch could be processed simultaneously on different machines of a department or a cell. In the second strategy, all parts of the same batch are processed on the same machine of the cell or department before being transferred collectively to the next machine or department. In all cases, batches must be split by "Separate" modules before accessing any machine. Batch reconstitution for transfer is performed using "Batch" modules.

Part Manufacturing

Every machine is modeled by a "Process" module, associated to a "Station" module and a "Resource" module. The "station" module determines the physical location of the machine and the "Resource" module represents the capacity and the availability of the machine itself. In fact, the "Process" module seizes the associated resource for the required period of time and then releases it. So, the machine becomes idle and available again for manufacturing another part. The machine resource is seized during a period of time that corresponds to the *PT* of the part being processed and eventually the required *ST* if the machine was set for a different part type. The *ST*, when relevant, is weighed by the setup reduction factor δ whenever the part type belongs to the family of the last processed one.

Statistics Collection

Before leaving the MS, every batch must go through an "Assign" module in which the parameters defined as performance measures are computed and updated. The acquired data is then stored in an Excel file using a "Readwrite" module for eventual treatment and analysis.

Functional Layout Model

The functional layout model is composed of three sections: "MOs launching", "Departments" and "System exit" (see Figure 1).

MOs are launched by "Create" modules dedicated each part type. Each "Create" module

Figure 1. FL model

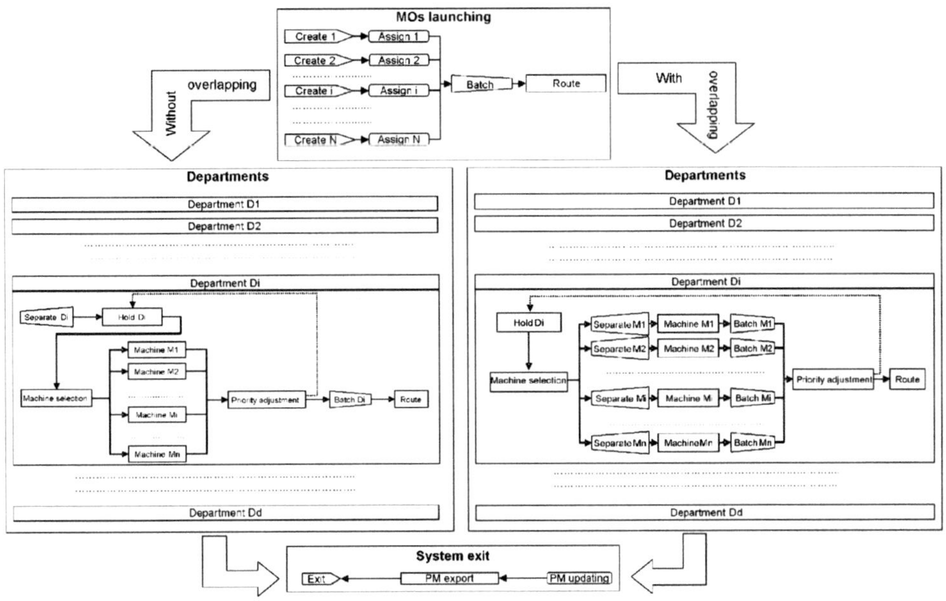

is coupled to an "Assign" module. The generated parts are then grouped into batches and routed to their first manufacturing step's department. A batch arriving at a department is made waiting in a queue modeled by a "Hold" module. This module is governed by a priority rule that could be FCFS, SPT or any other priority rule. When at least one of the department machines becomes available, the "Hold" module releases the prioritized batch from the waiting queue. The released batch is then transferred to the "Machine selection" sub-model that selects one among the available machines. The logic of this sub-model is coherent with the waiting queue priority rule.

When operations overlapping are not allowed, every batch is split once it reaches the assigned machine. Hence, each batch can be treated only by a single machine. On the other hand, if operations overlapping are permitted, parts batches are split before accessing the department queue. So, parts become independent and could be dispatched to several machines of the same department to be processed simultaneously. In both cases, batches are gathered by a "Batch" module right after processing and before the transfer to next manufacturing step. The combination of the operations overlapping strategy, the machine selection process and the waiting queue priority rule define the shop scheduling policy.

Cellular Layout Model

The CL model is composed of "*c*" sub-models corresponding to the "*c*" MS cells. Each sub-model is composed of three sections: "MOs launching", "Machine cells" and "Cell exit".

As for the FL, MOs are launched by "Create" and "Assign" modules dedicated to each part type. The generated parts are then grouped in batches before being routed to the general cell queue. Such a queue holds part batches until their first routing step machine becomes available. In addition, each machine has its own waiting queue. Both queues are governed by the same priority rule.

If operations overlapping are allowed, batches are split just before leaving the cell general queue. Hence, every part can follow its routing without waiting for the other batch parts. Batches are finally regrouped just before the cell exit. In contrast, if operations overlapping are not implemented, every batch is split when it reaches the machine next machine on its routing. Batches are regrouped once their processing is accomplished. Then, they are transferred towards the following machine or to the system exit.

THE OBJECTIVE COMPARISON METHODOLOGY

Overview

The objective comparison methodology (OCM) aims essentially at the development of a mathematical model permitting to predict the superiority of one layout or the other. It is the product of the application of Taguchi method of experiment design. Hence, the OCM is mainly composed of 3 main phases:

- Phase 1: Choosing MS parameters and setting their levels
- Phase 2: Construction of the experiments plan, results analysis and development of the mathematical model
- Phase 3: Refinement of the simulation plan and improvement of the mathematical model

Each of these phases is composed of one or several stages. Some stages should be reiterated several times.

Phase 1: Choosing Levels of the Manufacturing System Parameters

In the first phase of the OCM the manager must choose the MS parameters as well as their levels. Generally, every MS can be characterized by three types of parameters: signal factors, control factors and noise factors.

Figure 2. CL model

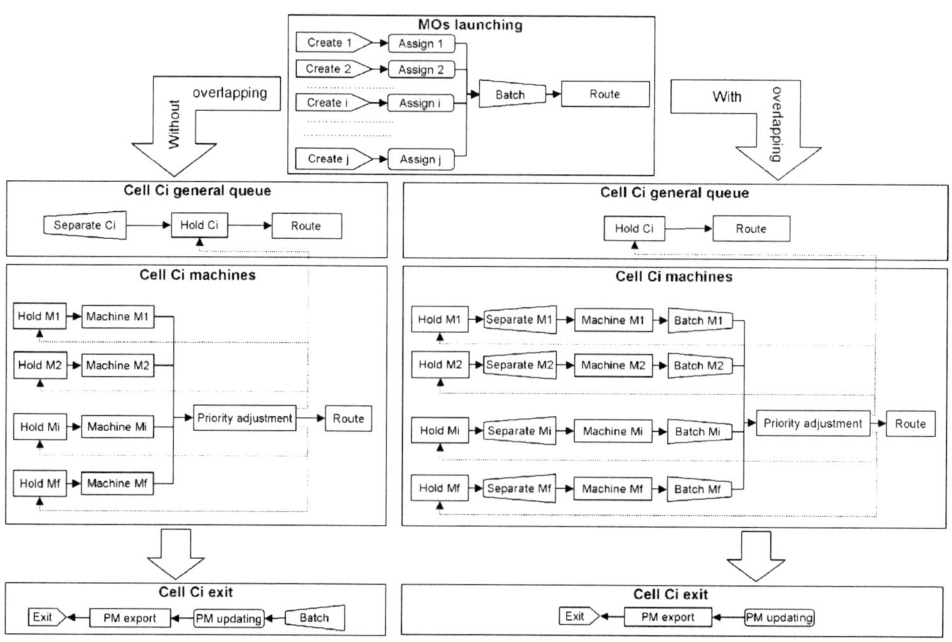

Figure 3. Overview of the OCM

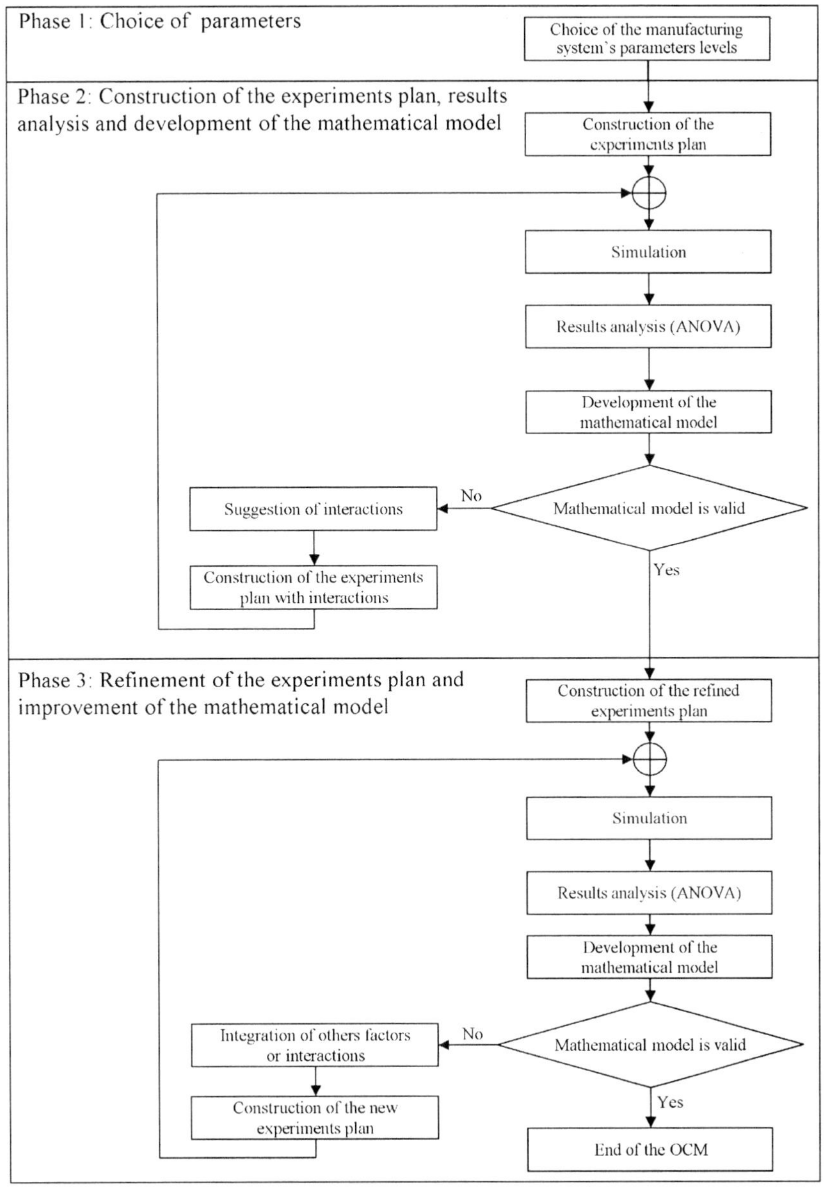

Signal Factors (SF)

Signal factors are factors that are expected to affect the average response. In addition, these factors identify the manufacturing context and are kept constant in every application of the OCM. This category includes the department's number *d*, the cell's number *c*, the number of equivalent machines in every department *Mn* and the number of different machines in every cell *Mf*. The four other signal factors are the number of part families *f*, the number of part types by family *t*, the number of manufacturing operations *mopt* and the existence or no of inter-cell moves.

Control Factors (CF)

As for the signal factors, control factors can affect the average response but, more importantly, can affect the extent of the variability about the average response. These factors are to be varied throughout the simulation plan. This category includes the *ST,* the *PT*, the *TT,* the *IAT* and theδ. The three other Control factors are the *BS*, the *RULE* and the *TM.* For more objectivity of comparison results, *ST*, *TT* and *PT* are put into the following ratio forms *ST/PT* and *TT/PT*. Indeed, *ST* and *TT* being nonproductive activities, these ratios are used to compare them to *PT* which is a productive activity. In addition to the studied CFs, several factor interactions (CFI) could also be investigated in every application of the OCM. CFI between CF_x and CF_y is here noted $CF_x xCF_y$.

Noise Factors (NF)

Noise factors are difficult or even impossible to control. Some of these factors could have a direct influence on the MS performances. Hence, instead of controlling them, the methodology aims at determining a solution in terms of CF that is robust relatively to unpredictable variations of *NF*.

Phase 2: Construction of the Experiments Plan, Results Analysis and Development of the Mathematical Model

The main purpose of the second phase of the OCM is to develop the mathematical model. This model gives an interpretation of the SM parameters effect's on the performances of the two layouts. It is developed through the following stages:

- Stage 1: An initial plan of experiments is constructed using standard OAs developed by Taguchi (Taguchi, Elsayed, & Hsiang, 1989). This plan is a set of experiments (simulations) where several CFs levels are varied from an experiment to another. It permits to considerably minimize the experimental effort.
- Stage 2: Simulations are conducted and performance measures of the two layouts are collected. The performance measures are expressed using the signal to noise ratio (*S/N*). This ratio is an essential indicator of the ability of the system to perform robustly in the presence of some noise effect (Park, 1998). There are three type of *S/N* ratios: lower-the-better (LB), nominal-the-best (NB), and higher-the-better (HB). In the OCM, the HB type *S/N* is used.
- L is better than FL, it is proposed to maximize the HB type S/N characterizing the MFT ratio MFTFL/MFTCL
- Stage 3: Simulations results are then analyzed by the analysis of variance method (ANOVA). The ANOVA establishes the relative significance of CFs in terms of their percentage contribution to the response (Phadke, 1989; Ross, 1996). The relative significance of CFs is translated by the Fischer factor "*F*" (Montgomery, 2001). The ANOVA also estimates the variance of error.
- Stage 4: The mathematical model is developed by interpolating the CFs effects. The validity of the developed mathematical model is then verified through the confirmation experiment. This experiment consists of adopting in an extra simulation experiment the best levels of CFs. If the average of the results of the confirmation experiment is within the limits of the confidence interval (CI) of the predicted result, then the mathematical model is considered confirmed (Kiefer, 1977). Hence the OCM can move to the following phase. Otherwise, interactions between CFs are taken in account in a new model. The second phase of the OCM is then reiterated from the third stage. This cycle should be reiterated

as much as necessary to get a valid mathematical model. In each iteration, the insignificant interactions must be eliminated and replaced by other interactions.

Phase 3: Refinement of the Simulation Plan and Improvement of the Mathematical Model

The purpose of this phase is to refine the simulation plan and to improve the mathematical model developed in the second phase. So, in this plan, only the most significant CFs and CFIs are considered. Besides, for each CF, additional levels are investigated to study the non-linearity effect of the process factors. This phase is very similar to the second phase. Indeed, it essentially includes the same main stages. Only the choice of factors and interactions to integrate in the mathematical model is different. Once the improved mathematical model developed, its validity is tested.

Academic Case Study

The studied MS is inspired from the comparison study of Morris and Tersine (1990). This MS is composed by 30 machines grouped in 8 departments in the FL and 5 cells in the CL. It is also characterized by 30 part types grouped in 5 families. Every part family is composed of 6 part types. Each part type requires from 2 to 6 production operations. In addition, no inter-cell moves are required.

The FL and CL simulation models were developed using the ARENA commercial software. Observations were then collected for two performance measures: *MFT* and *Throughput*. The second measure is used solely for warm up period detection. The results show that a warm up period of 200000 minutes is needed. The models can then be run for 800000 minutes.

Choice of MS Parameters

The CFs are here studied using two levels each as depicted in Table I. It is worth noting that the original level corresponds to the level initially used in the MS.

Initial plan of Experiments

Each of the two level CFs has 1 degree of freedom (DOF). Hence, the total degree of freedom (TDOF) required for the studied seven CFs is 8 [=7×1+1]. As per Taguchi's method the total experiments number of the selected OA must be greater than or equal to the TDOF, an $L_8(2^7)$ OA was selected for the initial experiments plan (Taguchi et al., 1989). This OA has seven columns and eight experiment-runs (rows). The seven CFs are assigned to the OA columns as depicted in Figure 4 (stages 1&2). Every suggested experiment by the OA is then run for 2 replications in order to compute the *S/N* ratios. Results are shown in Figure 4 (stages 1&2). The results of ANOVA indicate that only the CFs *ST/PT*, *TT/PT*, *BS* and *IAT* are statistically significant (Figure 4-stage 3). Figure 4 (stage4), that depicts the main effects of the CFs, confirms these remarks. In this figure, the importance of the CF is expressed by its slope.

Based on the computed *S/N* ratios, the mathematical model is developed by linear interpolation. In this model, every CF can take one of two values: 1 or 2, depending on the chosen parameter level:

$$S / N = -1.53 + 0.70 \times (ST / PT) - 0.16 \times BS + 0.38 \times (TT / PT) - 0.16 \times IAT + 0.07 \times TM + 0.09 \times \delta - 0.08 \times RULE \quad (1)$$

Then, the confirmation experiment considers the maximum value of *S/N* ratio to choose optimum levels of the CFs. Hence the chosen levels are ST/PT_2, TT/PT_2, IAT_1, BS_1, δ_2, TM_2 and $RULE_1$ where X_i is the ith level of the control factor *X*.

Figure 4. OCM application: Initial experiments plan

STAGES 1 & 2 : ORTHOGONAL ARRAY CONSTRUCTION AND SIMULATION RESULTS

Orthogonal array								Results		
								$\frac{MFT_{FI}}{MFT_{CL}}$		S/N
Exp	$ST/PT^{(1)}$	$TT/PT^{(2)}$	$IAT^{(3)}$	$\delta^{(4)}$	$TM^{(5)}$	$RULE^{(6)}$	$BS^{(7)}$	Rep1	Rep2	
1	1	1	1	1	1	1	1	0.93	0.92	-0.69
2	1	1	1	2	2	2	2	0.93	0.90	-0.77
3	1	2	2	1	1	2	2	0.93	0.91	-0.71
4	1	2	2	2	2	1	1	0.97	0.96	-0.31
5	2	1	2	1	2	1	2	0.97	0.98	-0.24
6	2	1	2	2	1	2	1	0.98	0.98	-0.14
7	2	2	1	1	2	2	1	1.04	1.05	0.37
8	2	2	1	2	1	1	2	1.06	1.02	0.33

STAGE 3: RESULTS OF THE ANALYSIS OF VARIANCE

Parameter	*SS*	*DOF*	*MS*	*Pooling*	*DOFp*	*MSp*	*F*
ST/PT	0.99	1	0.99		1	0.99	73.77
BS	0.05	1	0.05		1	0.05	3.63
TT/PT	0.29	1	0.29		1	0.29	21.77
IAT	0.05	1	0.05		1	0.05	4.04
TM	0.01	1	0.01	x	–	–	–
δ	0.02	1	0.02	x	–	–	–
RULE	0.01	1	0.01	x	–	–	–
ERROR	0.00	0		0.04	3	0.01	
TOTAL	1.42	7			7		

Notes:
SS: Sum of Squared deviations *DOF*: Degree Of Freedom *MS*: Mean Square *F*: F-value.

STAGE 4: MATHEMATICAL MODEL DEVELOPMENT

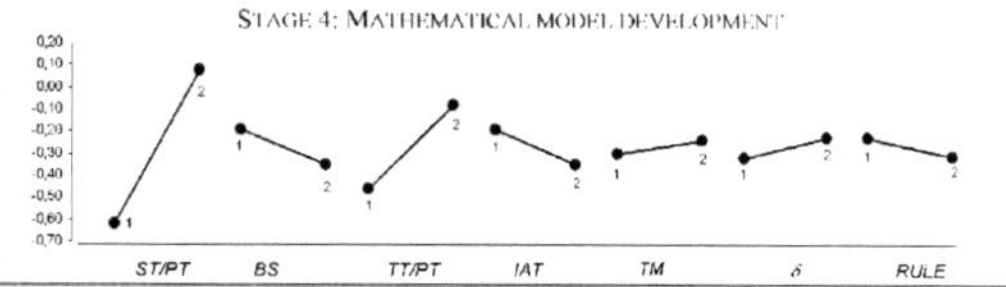

STAGE 5: MATHEMATICAL MODEL CONFIRMATION RESULTS

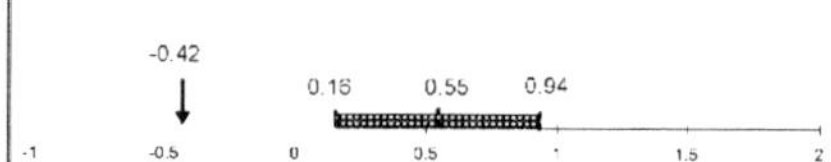

In this case, the expected result in terms of *S/N* ratio is 0.55 Db. The computed 95% confidence interval is equal to CI = ±0.39 Db. Therefore, the expected result should lie between 0.16 Db and 0.94 Db. As depicted in Figure 4 (stage5) the best expected response of -0.42 Db obtained by the confirmation experiment is outside the limits of the CI. The mathematical model is hence considered invalid. Additional analysis and experimentation are needed.

Simulation Plan with Interactions

Two additional iterations were needed to obtain a valid mathematical model. Only the results of the second iteration are depicted here. Based on the first iteration simulation plan ANOVA results, the simulation plan in the second iteration considers *TT/PT×RULE*, *ST/PT×RULE*, *ST/PT×BS*, *BS×RULE*, *TT/PT×BS*, *IAT×RULE* and *IAT×BS* as CFIs. Each of these CFIs has 1 DOF. The required TDOF is then equal to14 [=7×1+6×1+1]. Hence, the $L_{16}(2^{15})$ is the OA to use. This OA has fifteen columns and sixteen experiment-runs. The factors were assigned to the $L_{16}(2^{15})$ OA using the linear graphs displayed in the Figure 5 (stages1&2). This figure also shows the associated simulation results.

ANOVA results indicate that only the CFs *BS*, *ST/PT*, *IAT* and δ are statistically significant (Figure 5- stage 3). It also demonstrates that only the CFIs *TT/PT×RULE*, *TT/PT×BS* and *ST/PT×BS* are statistically significant. Figure 5 (stage4) illustrates the main effects of the CFs and CFIs. In this figure the importance of a CFI is expressed by the slope difference between the interaction two curves. The mathematical model is then developed:

$$\begin{aligned} S/N = {} & -4.44 + 1.70 \times BS - 3.12 \times (TT/PT) + 0.54 \times RULE \\ & + 3.26 \times (ST/PT) - 0.14 \times TM + 3.45 \times IAT - 0.51 \\ & \times \delta + 1.26 \times (TT/PT) \times RULE - 0.73 \times BS \times RULE \\ & + 1.08 \times (TT/PT) \times BS - 0.89 \times IAT \times BS - 0.36 \\ & \times (ST/PT) \times RULE - 1.20 \times (ST/PT) \times BS - 0.75 \\ & \times IAT \times RULE \end{aligned} \quad (2)$$

The levels of the CFs in the confirmation experiment are as follows: BS_1, TT/PT_2, $RULE_1$, ST/PT_2, TM_1, IAT_2, δ_1. Two confirmation trials were conducted and results show that the developed mathematical model is valid (Figure 5-stage5).

Refinement of the Simulation Plan and Improvement of the Mathematical Model

The refined simulation plan considers the control factors *BS*, *ST/PT*, *IAT* and δ in addition to the CFI

Figure 5. OCM application: Simulation plan with interaction (Second iteration)

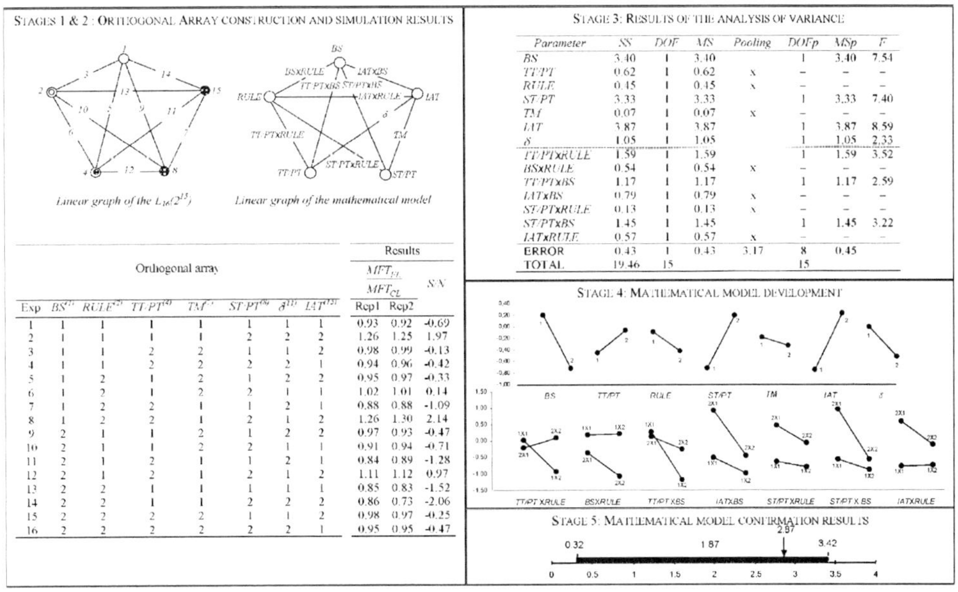

ST/PT×BS. In addition to the two studied levels, each of the three CFs was analyzed by way of a third level. This additional level corresponds to the original level as depicted in Table 1. Hence, the required TDOF is 13 [=4×2+1×4+1]. So, the $L_{27}(3^{13})$ OA was selected for the refined simulation plan and the CFs were assigned to this array using the linear graphs displayed in the Figure 6 (stages1&2). This figure depicts also the OA and the simulation results. It's worth noting that the original levels of the unused CFs in the refined plan (*RULE*, *TM* and *TT/PT)* are chosen (Table 1).

The analysis of simulation results shows that only *ST/PT* and *BS* are significant (Figure 6-stage3). This observation is confirmed by the Figure 6 (stage4) that illustrates the main effects of the CFs and CFIs. The developed mathematical model is written as follows:

$$S/N = -3.83 - 0.59\times(ST/PT)^2 - 0.15\times BS^2 - 0.12 \times IAT^2 - 0.01\times\delta^2 - 0.11\times(ST/PT)^2\times BS^2 + 0.47 \times(ST/PT)^2\times BS + 0.43\times(ST/PT)\times BS^2 - 2.17 \times(ST/PT)\times BS + 3.94\times(ST/PT) + 0.65\times BS + 0.68\times IAT - 0.06\times\delta \quad (3)$$

Table 1. Control factors

CF	Original Level	Level 1	Level 2
ST/PT	3	1	5
IAT	Exp (525) mn	Exp (420) mn	Exp (630) mn
δ	0.35	0.2	0.5
BS	38	25	50
RULE	*RL*	*RL*	*FCFS*
TM	With operations overlapping	With operations overlapping	Without operations overlapping
TT/PT	0.8 for FL ; 0.3 for CL	0.4 for FL ; 0.15 for CL	1.2 for FL ; 0.45 for CL

Table 2. Level combinations giving layout superiority

		IAT_1			IAT_2			IAT_3		
		δ_1	δ_2	δ_3	δ_1	δ_2	δ_3	δ_1	δ_2	δ_3
ST/PT_1	BS_1	FL	FL	FL	FL	FL	FL	FL	FL	FL
	BS_2	FL	FL	FL	FL	FL	FL	FL	FL	FL
	BS_3	FL	FL	FL	FL	FL	FL	FL	FL	FL
ST/PT_2	BS_1	CL	CL	CL	CL	CL	CL	CL	CL	CL
	BS_2	FL	FL	FL	CL	FL&CL	FL	CL	CL	FL&CL
	BS_3	FL	FL	FL	FL	FL	FL	FL	FL	FL
ST/PT_3	BS_1	CL	CL	CL	CL	CL	CL	CL	CL	CL
	BS_2	CL	CL	CL	CL	CL	CL	CL	CL	CL
	BS_3	FL	FL	FL	CL	CL	CL	CL	CL	CL

The confirmation experiment shows that the developed mathematical model is valid (Figure 6-stage5).

The MS manager can use this mathematical model to determine the best layout of its MS machines. He can also investigate the effect of the change of one or several CFs levels on performances of the two layouts. In fact, if the computed *S/N* ratio value is negative then the FL is the outperforming layout. In contrary, if the predicted *S/N* ratio value is positive then the CL outperforms the FL. Finally, the two layouts performances are considered equivalents if the *S/N* ratio value predicted by the mathematical model is close to zero. Table 2 depicts the CL and FL superiority contexts expressed as combinations of the CFs.

Figure 6. OCM application: Refined simulation plan

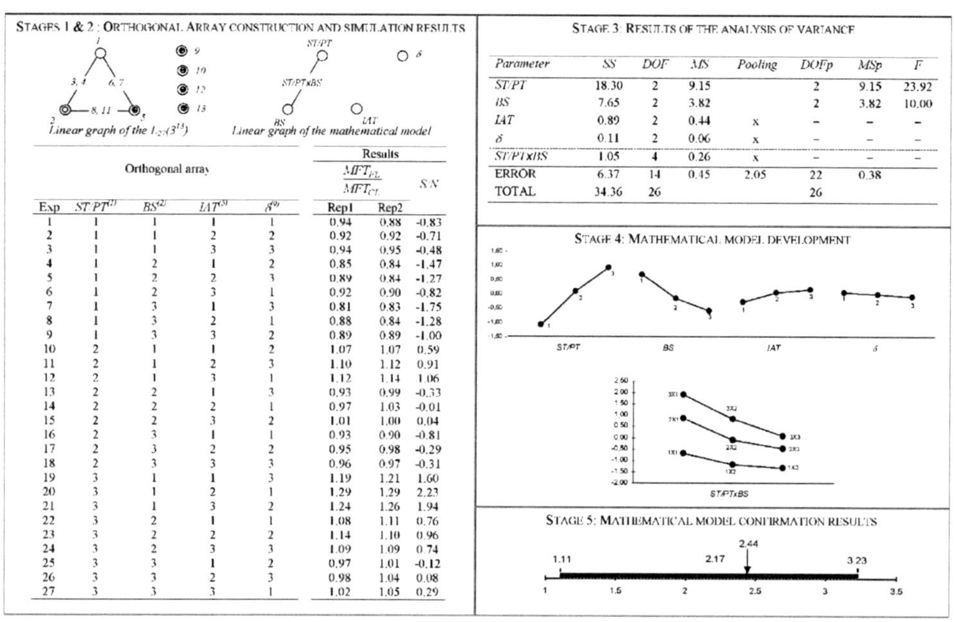

This table can be used by the manager to determine the more effective layout for every one of the 81 possible level combinations of the four considered CFs. In deed, the intersection between the line that represents the combination of the *ST/PT* and *BS* levels and the column corresponding to the *IAT* and δ levels gives the best performing layout. For example, the CL is the best layout for the following CFs levels combination: ST/PT_2, BS_1, IAT_2 and δ_2.

The mathematical model can also be used to predict the best layout for "intermediate levels" of the CFs *ST/PT*, *BS*, *IAT* andδ. Indeed unlike the *TM* and *RULE* CFs which are discrete and can be investigated only for specified levels, *ST/PT*, *BS*, *IAT* andδ are continuous factors. For example for the following setting combination: $ST/PT_{1.5}$, $BS_{2.2}$, $IAT_{1.4}$ and $\delta_{2.4}$ the FL outperforms the CL. In this case, le level X_i of the CF X is obtained by linear interpolation between the different levels of this CF.

CONCLUSION

This chapter presents an objective methodology for comparing functional and cellular layouts. This methodology aims to help MS managers choosing the appropriate layout for their manufacturing system. The developed methodology is based on the Taguchi method for the design of experiments and results analysis combined to discrete event simulation. This method permits, through a minimal experimental effort, to reliably evaluate the effect of each MS parameters on the system performances. It also reveals the possible interactions between MS parameters. The goal of this methodology is the development of a mathematical model predicting the superiority of one of the two layouts. In fact, once developed and validated, the mathematical model can be used by the MS manager to predict the *S/N* ratios for any combination of the MS parameters. The sign of the predicted *S/N* ratio indicates the best layout. The model can also be exploited to interpolate the results between the studied levels of continuous parameters such as batch inter arrival time or batch size. An academic case study showed the capacity of this methodology for choosing the best layout for a MS.

The developed methodology can find direct applications in the industry. However, many aspects of the comparison methodology should undergo further developments. The first task is the enlargement of the application scope to other control factors such as various levels of the number of operators or different degrees of the operator's qualification. In addition, in order to minimize the effort provided by the MS manager, the automation of coupling between the simulator and the analyze software is also projected. This should increase the chance of the proposed methodology to be successfully applied and validated on real cases.

REFERENCES

Agarwal, A., & Sarkis, J. (1998). A review and analysis of comparative performance studies on functional and cellular manufacturing layouts. *Computers & Industrial Engineering*, *34*(1), 77–89. doi:10.1016/S0360-8352(97)00152-6

Faizul huq, Douglas, A. H., & Zubair, M.M. (2001). A simulation analysis of factors influencing the flow time and through-put performance of functional and cellular layouts. *Integrated Manufacturing Systems, 12* (4), 285-295.

Farrington, P. A., & Nazametz, J. W. (1998). Evaluation of the performance domain of cellular and functional layouts. *Computers & Industrial Engineering*, *34*(1), 91–101. doi:10.1016/S0360-8352(97)00153-8

Jensen, J. B., Malhotra, M. K., & Philipoom, P. R. (1996). Machine dedication and process flexibility in a group technology environment. *Journal of Operations Management*, *14*(1), 19–39. doi:10.1016/0272-6963(95)00030-5

Kelton, W. D., Sadowski, R. P., & Sadowski, D. A. (2002). *Simulation with Arena* (2nd ed.). New York, NY: McGraw-Hill.

Kiefer, J. (1977). Conditional confidence statements and confidence estimators (with discussion). *Journal of the American Statistical Association*, *72*, 789–827. doi:10.2307/2286460

Li, J. (2003). Improving the performance of job shop manufacturing with demand-pull production control by reducing set-up/processing time variability. *International Journal of Production Economics*, *84*(3), 255–270. doi:10.1016/S0925-5273(02)00447-4

Montgomery, D. C. (2001). *Design and analysis of experiments* (5th ed.). New York, NY: Wiley.

Morris, J. S., & Tersine, R. J. (1990). A simulation analyses of factors influencing the attractiveness of group technology cellular layouts. *Management Science*, *36*(12), 1567–1578. doi:10.1287/mnsc.36.12.1567

Morris, J. S., & Tersine, R. J. (1994). A simulation comparison of process and cellular layouts in a dual resource constrained environment. *Computers & Industrial Engineering*, *26*(4), 733–741. doi:10.1016/0360-8352(94)90008-6

Park, S. H. (1998). *Robust design and analysis for quality engineering*. London, UK: Chapman & Hall Publication.

Phadke, M. S. (1989). *Quality engineering using robust design*. Englewood Cliffs, NJ: P.T.R. Prentice Hall.

Pitchuka, L. N., Adil, G. K., & Ananthakumar, U. (2006). Effect of the conversion of the functional layout to a cellular layout on the queue time performance: Some new insights. *International Journal of Advanced Manufacturing Technology*, *31*(5-6), 594–601. doi:10.1007/s00170-005-0219-2

Ross, P. J. (1996). *Taguchi techniques for quality engineering*. New York, NY: McGraw-Hill.

Shafer, S. M., & Charnes, J. M. (1992). Cellular versus functional layouts under a variety of shop operating conditions. *Decision Sciences*, *24*(3), 665–681. doi:10.1111/j.1540-5915.1993.tb01297.x

Shafer, S. M., & Charnes, J. M. (1995). A simulation analyses of factors influencing loading practices in cellular manufacturing. *International Journal of Production Research*, *33*(1), 279–290. doi:10.1080/00207549508930148

Shambu, G., Suresh, N. C., & Pegels, C. C. (1996). Performance evaluation of cellular manufacturing systems: A taxonomy and review of research. *International Journal of Operations & Production Management*, *16*(8), 81–103. doi:10.1108/01443579610125796

Suresh, N. C., & Meredith, J. R. (1994). Coping with the loss of pooling synergy in cellular manufacturing systems. *Management Science*, *40*(4), 466–483. doi:10.1287/mnsc.40.4.466

Taguchi, G., Elsayed, E., & Hsiang, T. (1989). *Quality engineering in production systems*. New York, NY: McGraw-Hill.

Section 2
Production Planning and Scheduling in Cellular Manufacturing Environment

Chapter 10
Cell Loading and Family Scheduling for Jobs with Individual Due Dates

Gürsel A. Süer
Ohio University, USA

Emre M. Mese
D.E. Foxx & Associates, Inc., USA

ABSTRACT

In this chapter, cell loading and family scheduling in a cellular manufacturing environment is studied. What separates this study from others is the presence of individual due dates for every job in a family. The performance measure is to minimize the number of tardy jobs. Family splitting among cells is allowed but job splitting is not. Even though family splitting increases number of setups, it increases the possibility of meeting individual job due dates. Two methods are employed in order to solve this problem, namely Mathematical Modeling and Genetic Algorithms. The results showed that Genetic Algorithm found the optimal solution for all problems tested. Furthermore, GA is efficient compared to the Mathematical Modeling especially for larger problems in terms of execution times. The results of experimentation showed that family splitting was observed in all multi-cell solutions, and therefore, it can be concluded that family splitting is a good strategy.

INTRODUCTION

Cell Loading is a decision making activity for planning the production in a Cellular Manufacturing System (CMS) including more than one manufacturing cell. The products are assigned to the manufacturing cells where they can be processed. This assignment is done based on the demand, processing times and due dates of the products and the production capacity and capability of the manufacturing cells (Süer, Saiz, Dagli & Gonzalez, 1995 and Süer, Saiz, &

DOI: 10.4018/978-1-61350-047-7.ch010

Gonzalez, 1999). Family Sequencing is a task of deciding the order by which product families are processed in a particular cell as determined by the Cell Loading process. In this chapter, a product family can be split and they can be sequenced in the same cell or different cells. Obviously, each time a new family starts in a cell, a new setup is required. Finally, Family Scheduling consists of determining start times and completion times of the product families and the individual products based on the family sequence established. Typically in a complex cellular system, we need to address Cell Loading, Family Sequencing and Family Scheduling tasks all in a satisfactory manner to obtain the desired results in terms of the selected performance measure.

In this study, we are considering minimizing the number of tardy job as the performance measure. Even though the problem has been observed in a shoe manufacturing company, it is applicable to many cellular systems. The products are grouped into families based on their processing similarity. On the other hand, products in a family might have different due dates. The overall objective of this chapter is to solve cell loading and product sequencing problem in such a multi-cell environment. To accomplish this, we propose two different approaches to tackle this complex problem namely, mathematical modeling and genetic algorithms. An experiment is carried out using both approaches and later the results are compared and a sensitivity analysis is also performed with respect to due dates and setup times.

BACKGROUND

Group Technology (GT) is a general philosophy where similar things are grouped together and handled all together. GT is established upon a common principle that most of the problems can be grouped based on their similarities and then a single solution can be found to the entire group of problems to save time and effort. This general concept has been also applied to the manufacturing world. This approach increases productivity by reducing work-in-progress inventory and improves delivery performance by reducing leadtimes, thus helping manufacturing companies to be more competitive. Thus, a Cellular Manufacturing System can be specified as an application of GT to the manufacturing system design (Askin & Standridge, 1993). Cellular Manufacturing System aims to obtain the flexibility to produce a high or moderate variety of low or moderate demand products with high productivity. CMS is a type of manufacturing system that consists of manufacturing cell(s) with dissimilar machines needed to produce part family/families. Generally, the products grouped together form a product family. The benefits of CMS are lower setup, smaller lot sizes, lower work-in-process inventory and less space, material handling, and shorter throughput time, simpler work flow (Suresh & Kay, 1998).

In this chapter, the performance measure used is minimizing the number of tardy products (n_T). If a product is completed after its due date, then it is considered as tardy product. If product is completed before its due date, then the tardiness for this product will be zero (early or on-time product). Therefore, tardiness for a product takes a value of zero or positive, $T_i = \max\{0, c_i - d_i\}$; where T_i is the tardiness for product (i), c_i is the completion time of product (i), and d_i is the due date for product (i). The number of tardy jobs is computed as $n_T = \sum_{i}^{n} g(T_i)$ where g(x) = 1 if x > 0, and zero otherwise.

The problem has been observed in a shoe manufacturing company where twelve product families have been already defined. There are multiple cells and the most critical component of each cell is the rotary injection molding machine. Even though Rotary Molding Machine is a single machine, scheduling shoes on that machine resembles to a parallel machine scheduling problem as it can hold multiple pairs of molds/shoes at any time. The Rotary Molding Machine is defined in detail in Section 3.

Several researchers focused on cell loading problem [Süer, Saiz, Dagli & Gonzalez, (1995) and Süer, Saiz, & Gonzalez (1999a)] developed simple cell loading rules to minimize number of tardy jobs. Süer, Vazquez, & Cortes (2005) developed a hybrid approach of Genetic Algorithms and local optimizer to minimize nT in a multi-cell environment. Süer, Arikan, & Babayigit (2008) and (2009) focused on cell loading subject to manpower restrictions and developed fuzzy math models to minimize nT and total manpower levels. A few works have been also reported where both cell loading and product sequencing tasks are carried out. Süer & Dagli (2005) and Süer, Cosner & Patten (2009) discussed models to minimize makespan, machine requirements and manpower transfers. Yarimoglu (2009) developed math model and genetic algorithms to minimize manpower shortages in cells with synchronized material flow. However, these work ignored setup times between products and families.

Some other researchers focused on group scheduling problem with a single machine or a single manufacturing cell. Nakamura, Yoshida & Hitomi (1978) focused on minimizing total tardiness and considered sequence-independent group setup. Hitomi & Ham (1978) also considered sequence-independent setup times for a single machine. Ham, Hitomi, Nakamura & Yoshida (1979) developed a branch-and-bound algorithm for the optimal group and job sequence to minimize total flow time with the minimum number of tardy jobs. Pan and Wu (1998) considered a single machine scheduling problem to minimize mean flow time of all jobs subject to due date satisfaction. They categorized the jobs into groups without family splitting. Gupta and Chantaravarapan (2008) studied the single machine scheduling problem to minimize total tardiness considering group technology. Individual due dates and independent family setup times have been used in their problem with no family splitting.

This paragraph summarizes the work done in the past which focused on scheduling jobs on the Rotary Injection Molding Machines. Süer. Santos, & Vazquez (1999b) have developed a three-phase Heuristic Procedure to minimize makespan in the Rotary Molding Machine scheduling problem. Subramanian (2004) has attempted to solve this problem as a part of the cell loading and scheduling process. The objective was to minimize makespan and unlimited availability of the molds was assumed. Later, Urs (2005) introduced limited mold availability into the problem for the same objective. The most recent research was done by Süer, Subramanian, and Huang (2009) includes some heuristic procedures and mathematical models for cell loading and scheduling problem.

The most important feature of the scheduling problem studied in this chapter is the presence of individual due dates for every job even in the same family (no common due dates), and family splitting is allowed to minimize the number of tardy jobs. To the best knowledge of the authors, this real problem observed in a cellular environment has not been addressed in the literature before. As a result, we decided to tackle this complex problem here and propose multiple solution approaches to deal with it.

THE PROBLEM STUDIED

This section discusses the problem studied in detail.

Family Splitting vs. Setup Times

Typically, in cellular manufacturing similar products are grouped together and processed together as a family to reduce the number of setups and thus total setup time. However, literature does not address the possibility of having different due dates for the jobs in the same family. Even though it is important to reduce the setup times, it is also important to meet the customer due dates.

There is a natural conflict between meeting due dates of jobs versus reducing total setup times between families. This point can be illustrated in Figure 1 where there are two families and two jobs in each family.

If we do not split (preempt) the families, we get two jobs in the second family. On the other hand, if family splitting is allowed then number of tardy jobs is reduced to one even if the number of setups increases from two to four. We can observe that when all of the jobs for a family are scheduled all together, setup times are reduced. However, this may also force several other jobs in the following families to be delayed and increase the possibility of becoming tardy. On the other hand, when a family is split several times, the number of setups increases thus reducing the productive time and hence may adversely affect the number of tardy products. This study attempts to find a balance between family splitting and meeting due date such that the total number of tardy products is minimized. As mentioned before, among the published papers in the literature, there is no work reported about Cell Loading and Family Scheduling subject to Individual Due Dates with group splitting allowed considering more than one manufacturing cell.

Case Study

This section describes the problem in depth. The following subsections explain the important features of the problem.

Products

This problem was observed in a shoe manufacturing plant. Products have five attributes; Gender, Size, Sole Type, Color, and Material. For shoes manufactured for men (Male (M)), there are 18 different sizes, 2 sole types (Full Shot (FS), Mid Sole (MS)), 4 colors (Black (B), Dark Green (G), Honey (H), Nicotine (N)), and 3 materials (Polyurethane (PU), Polyvinyl chloride (PVC), Thermo Plastic Rubber (TPR)). For shoes manufactured for women (Female (F)), there are 13 different sizes and the remaining attributes (Sole Types, Colors, and Materials) are similar to those of males. Besides these product types, there are also different upper designs that will be referred as models from now on. Each model will have its own identification designation (Model ID).

Figure 1. Family splitting not allowed vs. allowed

	Setup Fam1	Job1 Fam1	Job2 Fam1	Setup Fam2	Job3 Fam2	Job4 Fam2
ci	2	5	10	12	16	22
di		6	19		12	20
Ti		0	0		4	2

a) Family splitting not allowed

	Setup Fam1	Job1 Fam1	Setup Fam2	Job3 Fam 2	Setup Fam1	Job2 Fam1	Setup Fam2	Job4 Fam2
ci	2	5	7	11	13	18	20	26
di		6		12		19		22
Ti		0		0		0		4

b) Family splitting allowed

Cells/Minicells

There are six manufacturing cells in the plant and they are independent from each other (machine sharing, and thus inter-cell transfers are not allowed). In the plant, every manufacturing cell includes Lasting Minicell, Rotary Molding Machine Minicell (RMMM), and Finishing/Packing Minicell as shown in Figure 2. Lasting Minicells prepare the shoes for injection molding process. Rotary Molding Machine Minicells inject the materials into the molds. Finishing/Packing Minicells remove extra materials from the injected shoes, finish the shoes, and also pack the shoes.

Rotary Molding Machine

This study focuses only on scheduling Rotary Molding Machine Minicells (the bottleneck of the manufacturing cell). The Rotary Molding Machine has a capacity of six pair molds as shown in Figure 3. In Figure 2, P1 is the injection station where the material is injected inside the mold. P2, P3, P4, and P5 are the cooling off stations, so that worker can handle the shoes. P6 is the loading and unloading station that the worker removes the pair injected and cooled off, and then loads the new pair that will be injected in the injection station. The Rotary Molding Machine is rotated anti-clockwise, so it is rotated exactly one position at the end of every cycle time.

Injection time is defined as the time required for injecting the material inside the mold. The injection time is affected by the size of the shoe, i.e. larger shoe sizes need longer injection times, because the material injected by the Rotary Molding Machine per minute is constant. Because of the schedule of products in the specific cell, dif-

Figure 2. Manufacturing cells in the shoe manufacturing plant

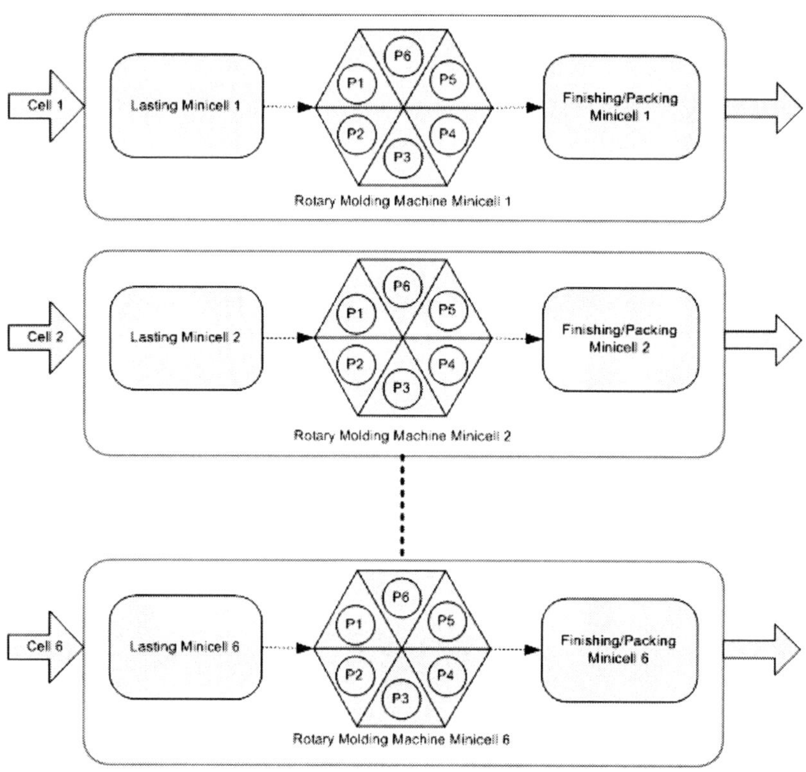

Figure 3. The rotary molding machine minicell

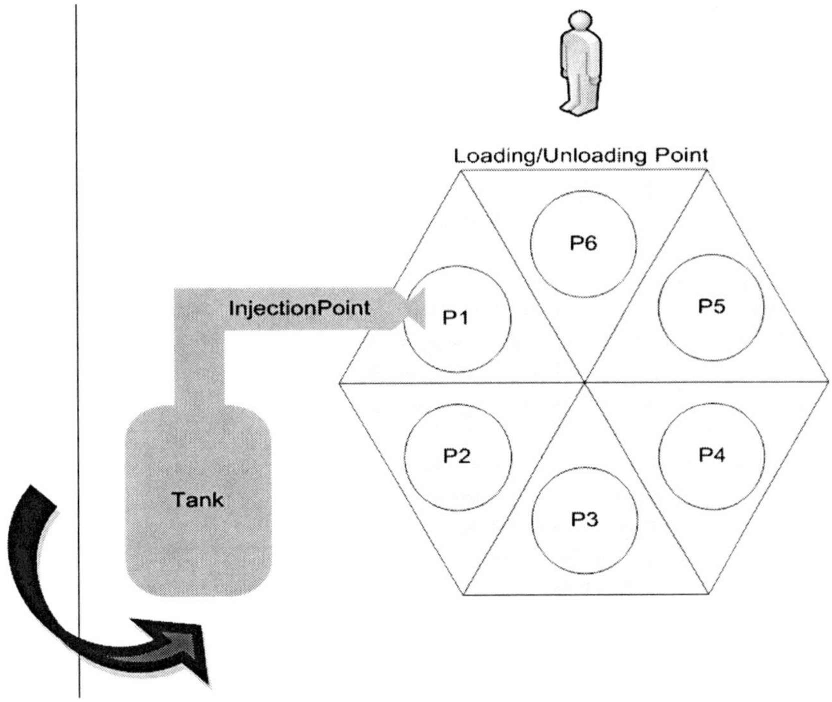

ferent sizes can be run in the Rotary Molding Machine at the same time. When this happens, the cycle time is set to the injection time of the biggest size (maximum injection time).

Product Families

A representation code is formed as "MC" to form and identify the product families. In the MC code form: M denotes the Material (PU: U, PVC: P, TPR: T), and C denotes the Color (Black: B, Dark Green: G, Honey: H, Nicotine: N). There are 12 product families (= 4 colors * 3 material types) as shown in Figure 4.

In this study, all of the sizes of a specific order (with the same Model ID, Gender, Sole Type, Material, Color, and Due Date) is called a job. Different sizes of a job can have different demand.

Figure 4. Structure of families

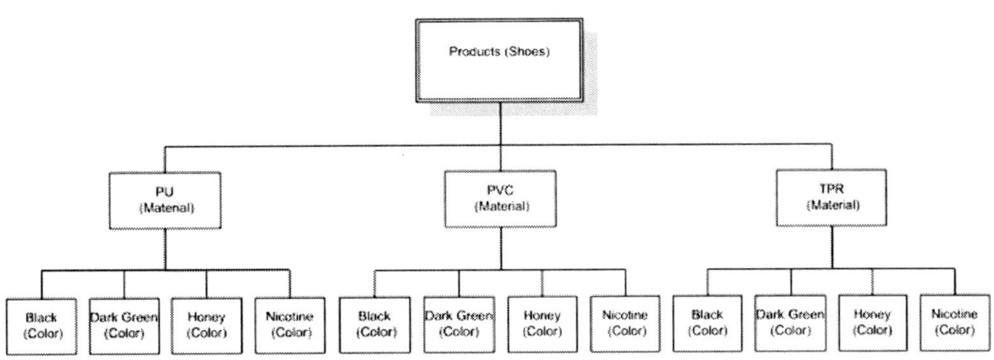

All of the sizes included in a job are assumed to have the same due date. The reason for this is that the entire job will have to be shipped to the customer all together.

Example: Family Formation

An example of customer orders that consists of 32 jobs is presented in Table 1. From the customer orders in Table 1, the families can be obtained as shown in Table 2.

Table 1. An example of customer orders

Job No.	Model ID	Gender	Sole Type	Material	Color	Size	Code	Total Demand	Due Date
1	K	F	FS	TPR	Black	5, 6, 7, …, 12	TB	269	10
2	U	M	FS	PVC	Dark Green	5, 6, 7, …, 15	PG	688	12
3	C	M	FS	TPR	Black	5, 6, 7, …, 15	TB	1045	22
4	C	M	FS	TPR	Black	5, 6, 7, …, 15	TB	208	11
5	L	F	MS	PU	Black	5, 6, 7, …, 12	UB	881	20
6	T	M	MS	PU	Black	5, 6, 7, …, 15	UB	831	17
7	T	M	MS	PU	Black	5, 6, 7, …, 15	UB	277	13
8	O	F	FS	PVC	Dark Green	5, 6, 7, …, 12	PG	250	15
9	E	M	FS	PVC	Dark Green	5, 6, 7, …, 15	PG	636	11
10	W	M	FS	PVC	Black	5, 6, 7, …, 15	PB	384	14
11	W	M	FS	PVC	Black	5, 6, 7, …, 15	PB	329	16
12	F	F	MS	TPR	Dark Green	5, 6, 7, …, 12	TG	440	17
13	F	F	MS	TPR	Dark Green	5, 6, 7, …, 12	TG	321	11
14	N	F	MS	PU	Black	5, 6, 7, …, 12	UB	355	14
15	N	F	MS	PU	Black	5, 6, 7, …, 12	UB	255	10
16	X	M	MS	PVC	Black	5, 6, 7, …, 15	PB	788	20
17	E	F	FS	PVC	Nicotine	5, 6, 7, …, 12	PN	574	16
18	E	F	FS	PVC	Nicotine	5, 6, 7, …, 12	PN	245	12
19	Y	F	FS	PVC	Honey	5, 6, 7, …, 12	PH	456	14
20	G	M	FS	PVC	Honey	5, 6, 7, …, 15	PH	345	13
21	G	M	FS	PVC	Honey	5, 6, 7, …, 15	PH	657	16
22	O	M	FS	TPR	Honey	5, 6, 7, …, 15	TH	234	11
23	M	M	FS	PU	Nicotine	5, 6, 7, …, 15	UN	621	16
24	W	M	FS	TPR	Nicotine	5, 6, 7, …, 15	TN	206	12
25	P	F	MS	TPR	Nicotine	5, 6, 7, …, 12	TN	657	17
26	P	F	MS	TPR	Nicotine	5, 6, 7, …, 12	TN	234	13
27	H	F	MS	PU	Dark Green	5, 6, 7, …, 12	UG	329	13
28	Z	F	MS	PU	Dark Green	5, 6, 7, …, 12	UG	574	15
29	L	F	MS	PU	Honey	5, 6, 7, …, 12	UH	116	10
30	J	F	MS	PU	Nicotine	5, 6, 7, …, 12	UH	432	14
31	V	F	MS	TPR	Honey	5, 6, 7, …, 12	TH	354	13
32	R	F	MS	PU	Nicotine	5, 6, 7, …, 12	UN	230	11

Table 2. Families for orders given in Table 1

Family No.	Jobs (due dates)
1	1 (10), 3 (22), 4 (11)
2	2 (12), 8 (15), 9 (11)
3	5 (20), 6 (17), 7 (13), 14 (14), 15 (10)
4	10 (14), 11(16), 16 (20)
5	12 (17), 13 (11)
6	17 (16), 18 (12)
7	19 (14), 20 (13), 21 (16)
8	22 (11), 31 (13)
9	23 (16), 32 (11)
10	24 (12), 25 (17), 26 (13)
11	27 (13), 28 (15)
12	29 (10), 30 (14)

Example: Possible Cases

This study focuses on assigning products to Rotary Molding Machine Minicells considering their families, i.e. cell loading. Family splitting among Minicells is allowed but job splitting is not. As an illustration of this, examples of the possible minicell loading cases are shown in Figure 4. The processing times of jobs in families F1, F5, and F12 are given in Table 3. While loading minicells, the families may not be divided as shown in Figure 5a, or families may be divided in the same minicell (like preemption) as shown in Figure 5b, or a family may be assigned to multiple minicells as shown in Figure 5c.

Other Issues

The molds used in the Rotary Molding Machine for injection molding vary by size, gender, and sole type. It is assumed that there is not any restriction on the availability of molds. Therefore, the same size pairs of a job can be run on all locations of the Rotary Molding Machine simultaneously. In this study, setup times between jobs in the same family are assumed negligible. However, setup times (for material or color or both changes) between families is assumed to take 20 minutes. The jobs can be back-scheduled in the Lasting Minicells and forward-scheduled in the Finishing/Packing Minicells based on the schedule of Rotary Molding Machine Minicells. However, scheduling in Lasting Minicells and Finishing/Packing Minicells are not within the scope of this work.

PROPOSED SOLUTION TECHNIQUES

The Cell Loading and Family Scheduling problems introduced in this chapter involve constraints on the number of product families, individual due dates, machine capacity, and sequence-independent setup times. This version of the problem is more difficult than the Classical Cell Loading and Group Scheduling problem. Both mathematical model and genetic algorithms approaches are proposed. The Mathematical Model guarantees the optimal solution, but it takes too much time to find the optimal solution. The Genetic Algorithm is a much faster procedure, but it cannot guarantee the optimal solution. The performance of these two procedures is compared with respect to execution time and the frequency of the optimal solutions. Genetic Algorithms Software Application (GASA) has been coded by using C# object-oriented programming language. The

Table 3. The processing times for examples of possible minicell loading cases

Family No.	Job No.	Processing Times (hrs)
F1	J1	3
	J3	9
	J4	2
F5	J12	5
	J13	3
F12	J29	2
	J30	4

Figure 5. Examples of possible minicell loading cases

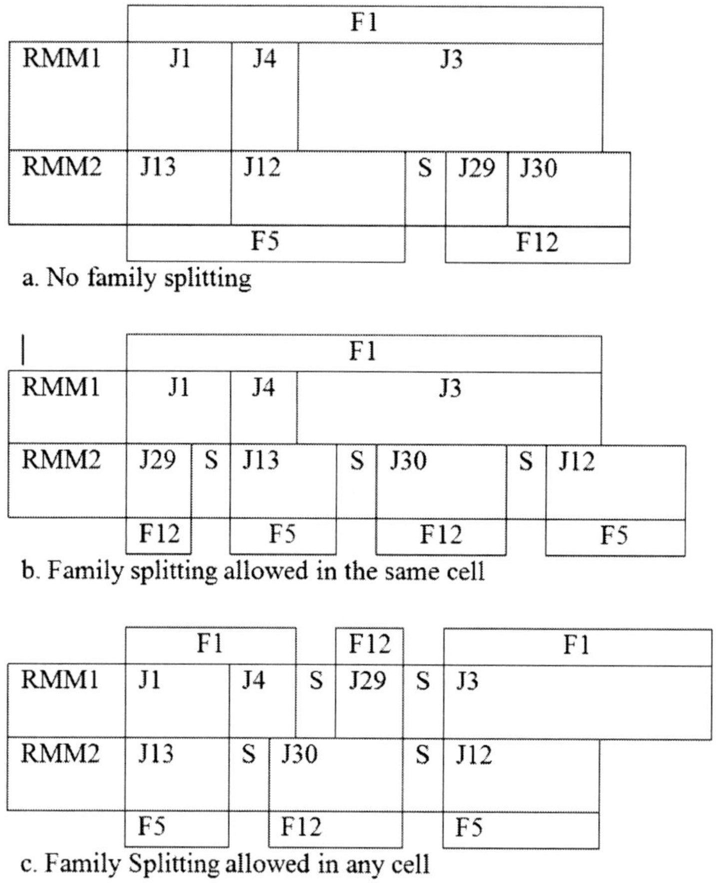

Mathematical Model is solved by using ILOG OPL 5.5. The methodology introduced in chapter is not restricted to the case study discussed here and can directly be used in similar cellular environments.

Mathematical Model

The objective function is to minimize n_T and it is given in Equation (1). Each job can be processed only once as shown in Equation (2). Equation (3) shows that each position in each cell can be assigned to at most one job. Equation (4) enforces jobs to be assigned consecutively in each cell. Equation (5) controls setup requirements. If the consecutive jobs are from different families, then this constraint adds setup between those consecutive jobs. In Equations (6), (7.a) and (7.b), If-Then constraints are used to eliminate the nonlinearity in the model. Equation (6) checks if a position is occupied by a job. If so, Equations (7.a) and (7.b) calculate the completion time of the job in that position. Equation (8) determines the tardiness value of a job. Equation (9) identifies if a job is tardy.

Notation

Indices:

i	Family index
j	Job index
k	Position index
m	Cell index

Parameters:

n	Number of jobs

n_i Number of jobs in family i
f Number of families
M Number of cells
P_{ij} Process time of job j from family i
D_{ij} Due date of job j from family i
S Setup Time
R Very big number (larger than maximum possible tardiness value)

Decision Variables:

Y_{mk} 0 if kth position in cell m is occupied, 1 otherwise.
X_{ijmk} 1 if job j from family i is assigned to the kth position in cell m, 0 otherwise.
C_{mk} Completion time of the job in kth position in cell m
T_{mk} Tardiness value of the job in kth position in cell m
W_{mk} 1 if setup is needed before the job in kth position in cell *m*, 0 otherwise.
nT_{mk} Coefficient for determining the tardiness of the job in k^{th} position in cell *m*. 1 if the job which is assigned to the kth position in cell *m* is tardy, 0 otherwise.

Objective Function:

$$\min Z = \sum_{m=1}^{M}\sum_{k=1}^{n} nT_{mk} \quad (1)$$

Subject to:

$$\sum_{m=1}^{M}\sum_{k=1}^{n} x_{ijmk} = 1 \text{ for } i=1,\dots,f,\ j=1,\dots n_i \quad (2)$$

$$\sum_{i=1}^{f}\sum_{j=1}^{ni} x_{ijmk} \leq 1 \text{ for } m=1,\dots,M,\ k=1,\dots,n \quad (3)$$

$$\sum_{i=1}^{f}\sum_{j=1}^{ni} x_{ijmk} \geq \sum_{i=1}^{f}\sum_{j=1}^{ni} x_{ijm(k+1)} \text{ for } m=1,\dots,M,\ k=1,\dots,n-1 \quad (4)$$

$$1 + W_{mk} \geq \sum_{j=1}^{n_i} x_{ijmk} + \sum_{j=1}^{n_q}\sum_{q\in(f/i)} x_{ijm(k-1)} \text{ for } m=1,\dots,M,\ k=2,\dots,n \quad (5)$$

$$\sum_{i=1}^{f}\sum_{j=1}^{ni} x_{ijmk} \leq R * (1 - Y_{mk}) \text{ for } m=1,\dots,M,\ k=1,\dots,n \quad (6)$$

$$-C_{m1} + \sum_{i=1}^{f}\sum_{j=1}^{ni} x_{ijm1} * P_{ij} \leq R * Y_{m1} \quad m=1,\dots,M \quad (7a)$$

$$C_{m(k-1)} - C_{mk} + S * W_{mk} + \sum_{i=1}^{f}\sum_{j=1}^{ni} x_{ijmk} * P_{ij} \leq R * Y_{mk} \text{ for } m=1,\dots,M,\ k=2,\dots,n \quad (7b)$$

$$C_{mk} - \sum_{i=1}^{f}\sum_{j=1}^{ni} x_{ijmk} * D_{ij} \leq T_{mk} \quad (8)$$

for $m=1,\dots,M,\ k=1,\dots,n$

$$T_{mk} \leq R^{*} nT_{nk};\ \text{for } m=1,\dots,M,\ k=1,\dots,n \quad (9)$$

Definition of Variables:

$x_{ijmk} \in \{0,1\}$; for $i=1,\dots,f,\ j=1,\dots,n_i,\ m=1,\dots,M,\ k=1,\dots,n$

$W_{mk} \in \{0,1\}$; for $m=1,\dots,M,\ k=1,\dots,n$

$nT_{mk} \in \{0,1\}$; for $m=1,\dots,M,\ k=1,\dots,n$

$Y_{mk} \in \{0,1\}$; for $m=1,\dots,M,\ k=1,\dots,n$

$C_{mk} \geq 0$; for $m=1,\dots,M,\ k=1,\dots,n$

$T_{mk} \geq 0$; for $m=1,\dots,M,\ k=1,\dots,n$

Genetic Algorithm

First, the initial population of n chromosomes is formed randomly. Then, mating partners are determined using mating strategies to perform crossover. The crossover and mutation operators are performed to generate offspring. For selecting the next generation, parents are added to the selection pool along with offspring. The next generation is selected from this pool based on their fitness function value. These steps are repeated until the number of the generations specified by the user is reached. Finally, the best chromosome obtained during the entire generations is determined as the final solution.

Chromosome Representation

The chromosome representation is used as an individual solution including genes corresponding to jobs. For each gene, the following representation code is used: (X, Y) where X denotes the Job Number and Y denotes the cell to which Job (X) is assigned. The sequence of genes in a chromosome also determines the sequence of jobs in the cells. As an illustration, an example is shown in Figure 6 where Jobs 1 & 3 are assigned to the first cell and Jobs 4 & 2 are assigned to the second cell in the order stated.

Mating

Three different mating strategies are used to determine mating pairs; Random (R), Best-Best (B-B), and Best-Worst (B-W). The reproduction probabilities of the chromosomes are calculated according to their fitness function. The next step depends on the selected mating strategy.

If the Random Mating Strategy is selected, the mating pairs are determined randomly with respect to their reproduction probabilities by using Roulette Wheel approach. In this mating strategy, each chromosome and its randomly determined partner give one offspring. If the Best-Best Mating Strategy is selected, all chromosomes are ranked with respect to their reproduction probabilities in descending order. Then, the best chromosome is paired with the second best chromosome, the third chromosome paired with the fourth chromosome and so on. In addition, the first X% of the pairs produce 3 offspring, the next Y% of the pairs give 2 offspring, and the last Z% of the pairs produce 1 offspring. If the Best-Worst Mating Strategy is selected, all chromosomes are ranked with respect to their reproduction probabilities in descending order. Then, the best chromosome is paired with the worst chromosome; the second best chromosome is paired with the second worst chromosome and so on.

Crossover

Two different strategies are used to perform the crossover operation. Those crossover strategies are Position-Based Crossover (P-B) (Syswerda, 1999) and Order Crossover (OX) Strategies (Davis, 1985). The crossover operation is applied to the identified pairs with a probability of P_C. The first parent is copied as the offspring if crossover is not performed. The crossover operator affects the sequence of jobs but not their cell assignment.

Figure 6. A Chromosome representation for a 4-job and 2-cell problem

1, 1	4, 2	3, 1	2, 2

In other words, the crossover is applied to the genes' X element and not to Y element.

Mutation

Two steps are used in the mutation operator. The first one is used for job sequence and only Reciprocal Exchange (R-E) Mutation Strategy is used [(see Gen and Cheng (1997)]. The mutation for job sequence is performed with a probability of P_{MJ}. The second step involves mutating cell assignments. In the mutation of cell assignments, two different mutation strategies are used, Random (R), and Reciprocal Exchange Mutation. The mutation of the cell assignment is performed with a probability of P_{MC}.

Selection

In this study, selection pool consists of all offspring and some of the parents. The next generation is selected from this pool. The selection from parents is a two-step process. First, the parents are ranked with respect to their reproduction probability in descending order. Then, the best P_E% parents are directly selected to advance to the selection pool. Then, the remaining (100-PE)% chromosomes are selected from the parents randomly based on their reproduction probability using Roulette Wheel Selection.

Once the selection pool is identified, the chromosomes are ranked with respect to their reproduction probability and a final selection is made from this pool to generate the next generation. In some experiments, we also allowed a certain percentage of lowest performers (Pw%) to advance automatically to the next generation to avoid immature convergence of the population.

ANALYSIS OF RESULTS

The results are grouped in four sections. 1) Genetic Algorithm Application, 2) Comparison of Mathematical Models with Genetic Algorithm, 3) Due Date Sensitivity Analysis, and 4) Setup Time Sensitivity Analysis. The experimental conditions are mentioned first and then the results obtained are discussed.

Data Sets Used

Nine data sets are used in the experimentation. The details of data sets are listed in Table 4. The data sets 1, 2, and 3 are realistic data sets obtained directly from the shoe manufacturing company. The data sets 4, 5, and 6 are relatively smaller data sets and they are generated from data sets 1, 2, and 3, respectively for only one cell. Similarly, for multiple cells; smaller data sets 7, 8, and 9, are generated from data sets 1, 2, and 3, respectively, by reducing batch sizes and due dates.

Genetic Algorithm Application

In this experiment, data sets 1, 2, and 3 are used. Initially, default parameters that are given in Table 5 are used. Then, the values of the GA parameters are changed one at a time in order to obtain better combinations. These parameters are listed in Table 6. Ten replications are performed.

The best solution is 3 tardy jobs for all three data sets. The frequencies of the best solution as well as the average values for better combinations are given in Tables 7, 8 and 9. For data set 1, four combinations stood out as the best combinations. Combination 1 has the highest frequency of the best solution. However, there is no significant difference among them according to ANOVA test results ($P=0.292> \alpha$-value=0.05). For data set 2, top six combinations are determined as given in Table 8. The combination 3 has the best frequency among all combinations. Since there is significant difference between six combinations ($P=0.008<$alpha=0.05), Fisher Test is applied. The results show that combination 5 is significantly different (worst) than other five combinations. For data set 3, top six combinations are listed in

Table 4. Details of data sets

Data Set	Number of Families	Number of Jobs	Total Processing Time (minutes)	Number of Cells	Available Production Capacity (minute/cell)
Data 1	12	41	6465	3	2400
Data 2	12	33	6603	3	2400
Data 3	12	44	7995	4	2400
Data 4	5	11	2015	1	2400
Data 5	9	12	2146	1	2400
Data 6	9	13	2077	1	2400
Data 7	6	20	3222	3	1200
Data 8	5	13	2283	3	800
Data 9	6	22	4060	4	1200

Table 5. The default parameters for GA

Setup Time:	20 minutes
Population Size:	1000
Number of Generations:	1000
Elite Ratio:	0.2
Worst Ratio:	0.1
Crossover Probability:	1
Jobs Mutation Probability:	0.1
Cells Mutation Probability:	0.7
Mating Strategy:	Random
Crossover Strategy:	Position-Based
Mutation Strategy:	Random

Table 9. Combination 2 has the largest frequency for the best solution. Since there is significant difference (P=0.034<α-value=0.05) between six top combinations, Fisher Test is applied. The results show that Combinations 1 and 2 are significantly different than Combinations 5 and 6. Combination 3 and Combination 4 are also eliminated as they are not significantly different than Combinations 5 and 6.

Comparison of Mathematical Models and Genetic Algorithms

In this section, the results of the Mathematical Models are compared with the GA results for the modified data sets. The experimental conditions

Table 6. Values of the parameters for GA

PARAMETER	VALUES			
Elite Ratio:	0.2	0.1	0.3	
Worst Ratio:	0.1	0	0.2	0.3
Crossover Probability:	1	0.7	0.5	
Jobs Mutation Probability:	0.10	0.05	0.2	
Cells Mutation Probability:	0.7	0.5	0.3	
Mating Strategy:	Random	Best-Best	Best-Worst	
Crossover Strategy:	Position-Based	Order		
Mutation Strategy:	Random	Exchange		

Table 7. The best combinations for data set 1

	Elite Ratio	Worst Ratio	Crossover Prob.	Jobs Mut. Prob.	Cells Mut. Prob.	Mating Strategy	Crossover Strategy	Mutation Strategy	Freq of Best Known Sol	Average nT
Comb. 1	0.3	0	1	0.05	0.7	B-W	P-B	R	4	3.7
Comb. 2	0.3	0	1	0.05	0.7	B-B	P-B	R	4	3.6
Comb. 3	0.3	0	1	0.05	0.7	B-W	OX	R	1	4.1
Comb. 4	0.3	0	1	0.05	0.7	B-B	OX	R	3	3.8

Table 8. The best combinations for data set 2

	Elite Ratio	Worst Ratio	Crossover Prob.	Jobs Mut. Prob.	Cells Mut. Prob.	Mating Strategy	Crossover Strategy	Mutation Strategy	Freq of Best Known Sol	Average nT
Comb. 1	0.1	0	0.7	0.05	0.5	R	P-B	R	7	3.3
Comb. 2	0.1	0	0.7	0.05	0.5	B-B	P-B	R	7	3.3
Comb. 3	0.1	0	0.7	0.05	0.5	B-W	P-B	R	**8**	**3.2**
Comb. 4	0.1	0	0.7	0.05	0.5	R	OX	R	6	3.4
Comb. 5	0.1	0	0.7	0.05	0.5	B-B	OX	R	1	4
Comb. 6	0.1	0	0.7	0.05	0.5	B-W	OX	R	5	3.5

Table 9. The best combinations for data set 3

	Elite Ratio	Worst Ratio	Crossover Prob.	Jobs Mut. Prob.	Cells Mut. Prob.	Mating Strategy	Crossover Strategy	Mutation Strategy	Frequency of Best Known Solution	Average n_T
Comb. 1	0.3	0	0.7	0.05	0.5	R	P-B	R	4	3.7
Comb. 2	0.3	0	0.7	0.05	0.5	B-B	P-B	R	**5**	**3.7**
Comb. 3	0.3	0	0.7	0.05	0.5	B-W	P-B	R	1	4.2
Comb. 4	0.3	0	0.7	0.05	0.5	R	OX	R	0	4.2
Comb. 5	0.3	0	0.7	0.05	0.5	B-B	OX	R	0	4.4
Comb. 6	0.3	0	0.7	0.05	0.5	B-W	OX	R	0	4.5

that are detailed in 1 are utilized in GA application. The Mathematical Model solutions (optimal solutions) and GA solutions for data sets 4, 5, and 6 are given in Table 10; for data sets 7, 8, and 9 are given in Table 11.

GA found the optimal solution ten times out of ten replications for both one cell and multiple cells (except data set 7). For data sets 4, 5, and 6, Mathematical Model has a better execution time compared to GA (except data set 5) due to small problem size. When data sets 7, 8, and 9 are considered, the GA has significantly better execution time compared to the Mathematical Model. This is expected since as the problem size increases; the execution time of the Mathematical Model increases dramatically. The details of data set 8

Table 10. The optimal solutions and GA results for minimizing nT for one cell

Data Set	Math Model Result (Opt. Sol.)	Decision Variable	Constraints	Math Model Execution Time (hr:min:sec)	Optimal Frequency for GA (x/10)	GA Execution Time (hr:min:sec)
Data 4	2	166	106	00:00:02	10	00:00:53
Data 5	3	193	160	00:04:56	10	00:00:57
Data 6	1	222	174	00:00:16	10	00:01:04

Table 11. The optimal solutions and GA results for minimizing nT for multiple cells

Data Set	Math Model Result (Opt. Sol.)	Decision Variable	Constraints	Math Model Execution Time (hr:min:sec)	Optimal Frequency for GA (x/10)	GA Execution Time (hr:min:sec)	Average of GA Results (n_T)
Data 7	2	1499	720	56:01:00	2	00:01:52	2.8
Data 8	4	701	425	16:21:17	10	00:01:03	4
Data 9	2	2374	1051	05:54:40	10	00:01:49	2

and corresponding Gantt chart for the mathematical model results are given in Appendix for illustration purposes.

Due Date Sensitivity Analysis

For due date sensitivity analysis, data set 1 is used. Combination 4 is selected as the parameters of GA to perform this experiment. In this experiment, due date sets of loose (Loose 1), looser (Loose 2), tight (Tight 1), and tighter (Tight 2) are generated in addition to the original due date set (Medium). The cumulative probabilities of those due date sets are given in Table 12. The results of ten replications for the due date sets are given in Table 13. As expected, the number of tardy jobs increased as due dates got reduced.

Table 12. The cumulative probabilities of different due date sets

Due Date (min.)	480	960	1440	1920	2400
Loose 2	0	0	0.33	0.67	1
Loose 1	0	0.25	0.5	0.75	1
Medium	0.2	0.4	0.6	0.8	1
Tight 1	0.25	0.5	0.75	1	0
Tight 2	0.33	0.67	1	0	0

Setup Time Sensitivity Analysis

The set up times varied from 0 to 100 for data set 1. Similarly, Combination 4 parameters were used in

Table 13. The nT results of GA for different due date sets

	DUE DATE				
Replications	Loose 2	Loose 1	Medium	Tight 1	Tight 2
1	0	0	3	6	13
2	0	0	4	7	14
3	0	0	4	8	13
4	0	0	3	6	13
5	0	0	3	6	14
6	0	0	4	7	13
7	0	1	4	7	13
8	0	0	5	7	13
9	0	0	5	7	13
10	0	0	3	6	13

this experiment. The results of GA for the various setup times are given in Table 14. The number of tardy jobs increased as the setup time increased. This was expected, however, in some cases setup times doubled (from 5 minutes to 10 minutes; from 10 min to 20 min; from 20 min to 40 min) but number of tardy jobs increased only by one. On the other hand, when setup time increased from 40 minutes to 80 minutes, the number of tardy jobs increased by three. This shows that system can tolerate increase in setup times to a certain extent, beyond which its impact will be bigger.

CONCLUSION AND FUTURE WORK

In the problem studied in this chapter, every job has individual due dates even the ones in the same family. This property of the problem completely separates this study from other cellular manufacturing scheduling problems. The reason of this complexity is because of the natural conflict between meeting due dates of jobs and reducing total setup times. If the entire family is scheduled together, then the total setup time is at minimum. But, the jobs in the consecutive families may be forced to be postponed and probably become tardy. In contrast, splitting a family several times may increase the number of setups which reduce the productive time, and finally have an adverse effect on the number of tardy jobs.

Table 14. The nT results of GA for different setup times

	Setup Time (min.)							
Replications	**0**	**5**	**10**	**20**	**40**	**60**	**80**	**100**
1	1	3	3	3	6	8	8	10
2	2	3	3	4	6	7	8	11
3	2	2	3	4	7	9	9	11
4	2	3	2	3	5	8	10	10
5	1	2	3	3	5	8	9	9
6	2	3	3	4	5	8	9	11
7	2	3	3	4	5	7	9	10
8	2	3	4	5	6	7	10	10
9	1	2	3	5	7	7	9	10
10	3	2	2	3	6	8	10	10

The Mathematical Model is one of the solution techniques which guarantee to find the optimal solution. It is not practical to solve larger problems using the mathematical models because of the complexity. As a result, there is a need to use other approaches to solve such problems. We proposed and used Genetic Algorithm approach in this Chapter.

Genetic Algorithms found the optimal solution in all problems with varying frequency. The execution time of Mathematical Model was reasonable only for small problem sizes. GA clearly outperformed Mathematical Model with respect to execution times.

The results showed that family splitting occurred in all multi-cell problems. Due to limited space, we presented only one in this chapter. The occurrence of family splitting in these problems show us that the system used the feature of family splitting since it was beneficial in terms of reducing the number of tardy jobs. Another conclusion that can be drawn is that the impact of setup times and due dates on the system performance was as expected.

We are planning to extend this work to sequence-dependent setup times in the future and also use other metata-heuristic techniques. This work can also be extended to include other performance measures and job-splitting option.

REFERENCES

Askin, R. M., & Standridge, C. R. (1993). *Modeling and analysis of manufacturing systems* (1st ed.). New York, NY: Wiley.

Davis, L. (1985). Applying adaptive algorithms to domains. In *Proceedings of the International Joint Conference on Artificial Intelligence*, (pp. 162-164).

Gen, M., & Cheng, R. (1997). *Genetic algorithms and engineering design*. New York, NY: Wiley.

Gupta, J. N. D., & Chantaravarapan, S. (2008). Single machine group scheduling with family setups to minimize total tardiness. *International Journal of Production Research*, *46*(6), 1707–1722. doi:10.1080/00207540601009976

Ham, I., Hitomi, K., Nakamura, N., & Yoshida, T. (1979). Optimal group scheduling and machining-speed decision under due-date constraints. *Transactions of the ASME: Journal of Engineering for Industry*, *101*, 128–134. doi:10.1115/1.3439485

Hitomi, K., & Ham, I. (1978). Machine loading and product-mix analysis for group technology. *Transactions of the ASME: Journal of Engineering for Industry*, *100*, 370–374. doi:10.1115/1.3439443

Nakamura, N., Yoshida, T., & Hitomi, K. (1978). Group production scheduling for minimum total tardiness. *AIIE Transactions*, *10*(2), 157–162.

Pan, J. C., & Wu, C. (1998). Single machine group scheduling to minimize mean flow time subject to due date constraints. *Production Planning and Control*, *9*(4), 366–370. doi:10.1080/095372898234091

Subramanian, A. K. (2004). *Cell loading and scheduling in a shoe manufacturing company*. Unpublished Master's Thesis, Ohio University, Athens.

Süer, G. A., Arikan, F., & Babayigit, C. (2008). Bi-objective cell loading problem with non-zero setup times with fuzzy aspiration levels in labour-intensive manufacturing cells. *International Journal of Production Research*, *46*(2), 371–404. doi:10.1080/00207540601138460

Süer, G. A., Arikan, F., & Babayigit, C. (2009). Effects of different fuzzy operators on fuzzy bi-objective cell loading problem in labor-intensive manufacturing cells. *Computers & Industrial Engineering*, *56*, 476–488. doi:10.1016/j.cie.2008.02.001

Süer, G. A., Cosner, J., & Patten, A. (2009). Models for cell loading and product sequencing in labor-intensive cells. *Computers & Industrial Engineering*, *56*, 97–105. doi:10.1016/j.cie.2008.04.002

Suer, G. A., & Dagli, C. (2005). Intra-cell transfers and cell loading in labor-intensive manufacturing cells. *Computers & Industrial Engineering*, *48*, 643–655. doi:10.1016/j.cie.2003.03.006

Süer, G. A., Saiz, M., Dagli, C., & Gonzalez, W. (1995). Manufacturing cell loading rules and algorithms for connected cells. In Kamrani, A., Parsaei, H., & Liles, D. (Eds.), *Planning, design, and analysis of cellular manufacturing systems* (pp. 97–127). New York, NY: Elsevier. doi:10.1016/S1572-4417(06)80038-0

Süer, G. A., Saiz, M., & Gonzalez, W. (1999a). Evaluation of manufacturing cell loading rules for independent cells. *International Journal of Production Research*, *37*(15), 3445–3468. doi:10.1080/002075499190130

Süer, G. A., Santos, J., & Vazquez, R. (1999b). Scheduling rotary injection molding machines. *Proceedings of the Second Asia-Pacific Conference on Industrial Engineering and Management Systems*, Kanazawa, Japan, (pp. 319-322).

Süer, G. A., Subramanian, A., & Huang, J. (2009). Heuristic procedures and mathematical models for cell loading and scheduling in a shoe manufacturing company. *Computers & Industrial Engineering*, *56*, 462–475. doi:10.1016/j.cie.2008.10.008

Süer, G. A., Vazquez, R., & Cortes, M. (2005). A hybrid approach of genetic algorithms and local optimizers in cell loading. *Computers & Industrial Engineering*, *48*, 625–641. doi:10.1016/j.cie.2003.03.005

Suresh, N. C., & Kay, J. M. (1998). Group technology & cellular manufacturing: Updated perspectives. In Suresh, N., & Kay, J. (Eds.), *Group technology and cellular manufacturing* (pp. 1–14). Kluwer Academic Publishers. doi:10.1007/978-1-4615-5467-7_1

Syswerda, G. (1991). Scheduling optimization using genetic algorithms. In Davis, L. D. (Ed.), *Handbook of genetic algorithms* (pp. 332–349). New York, NY: Van Nostrand Reinhold.

Urs, B. S. (2005). *Comparison of heuristic procedures for scheduling rotary injection molding machines*. Unpublished Master's Thesis, Ohio University, Athens.

Yarimoglu, F. (2009). *Cell loading and product sequencing subject to manpower restrictions in synchronized manufacturing cells*. Unpublished Master's Thesis, Ohio University, Athens.

APPENDIX

Optimal Solution for Data Set 8

Table 15. Data Set 8

Job Number	Family No	Processing Time	Due Date
0	PB	151.324	800
1	PB	171.471	800
2	PB	295.684	480
3	PG	116.815	160
4	PG	133.081	160
5	PH	122.016	640
6	PH	232.978	160
7	PH	87.392	800
8	PN	221.099	160
9	PN	226.336	160
10	PN	248.412	320
11	TB	127.47	800
12	TB	148.891	160

Figure 7. Gantt chart for the optimal solution for data set 8

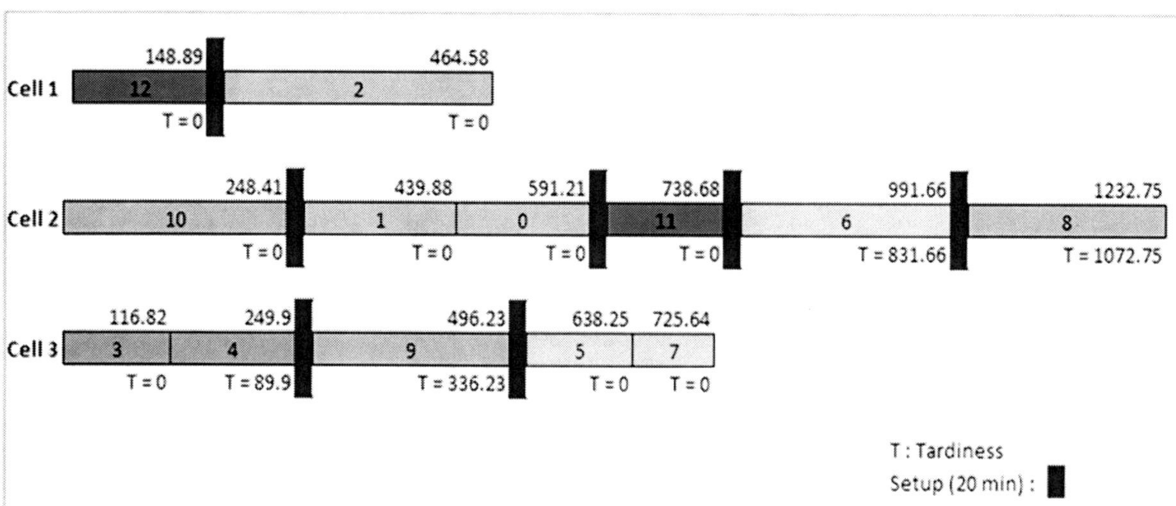

Chapter 11
Production Planning Models using Max-Plus Algebra

Arun N. Nambiar
California State University, USA

Aleksey Imaev
Ohio University, USA

Robert P. Judd
Ohio University, USA

Hector J. Carlo
University of Puerto Rico - Mayaguez, Puerto Rico

ABSTRACT

The chapter presents a novel building block approach to developing models of manufacturing systems. The approach is based on max-plus algebra. Within this algebra, manufacturing schedules are modeled as a set of coupled linear equations. These equations are solved to find performance metrics such as the make span. The chapter develops a generic modeling block with three inputs and three outputs. It is shown that this structure can model any manufacturing system. It is also shown that the structure is hierarchical, that is, a set of blocks can be reduced to a single block with the same three inputs and three output structure. Basic building blocks, like machining operations, assembly, and buffering are derived. Job shop, flow shop, and cellular system applications are given. Extensions of the theory to buffer allocation and stochastic systems are also outlined. Finally, several numerical examples are given throughout the development of the theory.

INTRODUCTION

Companies around the world are continuously striving to reduce wastes (Womack & Jones, 1996) and improve their operations in an effort to reduce their operating costs. With advances in logistics and distribution, companies are no longer restricted to a geographic region for their market resulting in increased competition. Shrinking product life-cycles, and increasing global competition make it imperative to be able to strike the proverbial iron while it is hot. Companies have to introduce

DOI: 10.4018/978-1-61350-047-7.ch011

new products that cater to the changing needs of consumers as quickly as possible (Anderson, 1997). Lead times assume increased importance and as a result production planning and scheduling becomes critical.

Often times, a company has a plethora of products that it offers to consumers. However, this compounds the scheduling problem as the company needs to determine how much of each product to make and how best to utilize the limited available resources to achieve these production targets. Researchers have developed exact and approximate solutions the various types of scheduling problems commonly encountered in the real world. Exact solutions range from exhaustive enumeration (Morton & Pentico, 1993; Temiz & Erol, 2003) and branch-and-bound techniques (Carlier & Rebai, 1996; Ladhari & Haouari, 2005) to linear programming models (Pinedo, 1995). Approximation methods (Hall, 1998; Sviridenko, 2004) include heuristic approaches such as genetic algorithm (GA) (Goldberg, 1989; Rajendran & Chaudhuri, 1992; Chen et al, 1995; Tang & Liu, 2002; Ravindran et al, 2005), simulated annealing (SA) (Osman & Potts, 1989; Ogbu & Smith, 1990; Ogbu & Smith, 1991; Ishibuchi et al 1995;Chakravarthy & Rajendran, 1999) and more recently ant colony optimization (ACO (Maniezzo & Carbonaro, 2001; Shyu et al, 2004; Ying & Liao, 2004; Rajendran & Ziegler, 2004; Rajendran & Ziegler, 2005) to name a few. Exact solutions tend to be time-consuming and computationally exhaustive.

This chapter will delve into the basics of max-plus algebra and its properties. It will then explore literature available in the areas of flow-shop scheduling, stochastic scheduling, assembly line balancing, and max-plus algebra. Specific models for flow-shops with and without buffers, batch processing, and assembly lines will be discussed. Finally, the chapter will identify other potential areas where max-plus algebra can be used to develop efficient schedules.

MAX-PLUS ALGEBRA

This algebra (The notations used and the concepts given here have been adapted from (Heidergott, 2006)) has two main operators viz. the *max* operator (maximization) which is denoted by the symbol ⊕ and the *plus* operator (addition) which is denoted by the symbol ⊗. The operators are defined as shown in Equations (1) and (2).

$$x \oplus y = \max(x, y)\ \forall x, y \in \mathbb{R}_{max} \tag{1}$$

$$x \otimes y = x + y\ \forall x, y \in \mathbb{R}_{max} \tag{2}$$

where $\mathbb{R}_{max}$ is the union of the set of real numbers and the zero element of max-plus algebra, ε=-∞, i.e. $\mathbb{R}_{max} = \mathbb{R} \cup \varepsilon$. For example

1⊕2=max(1,2)=2

1⊗2=1+2=3

The zero element and the unit element in max-plus algebra are -∞ and 0, and are denoted by εand *e*, respectively. This is shown in Equations (3) and (4).

$$x \oplus \varepsilon = x\ \forall x \in \mathbb{R}_{max} \tag{3}$$

$$x \otimes e = x\ \forall x \in \mathbb{R}_{max} \tag{4}$$

For example,

1⊕ε=max(1,-∞)=1

1⊗*e*=1+0=1

Further, we also have Equations (5) and (6)

$$x \otimes \varepsilon = \varepsilon\ \forall x \in \mathbb{R}_{max} \tag{5}$$

$$x \oplus e = x \ \forall x \in \mathbb{R}_{max} \quad (6)$$

It is clear that the commutative, associative, and distributive properties hold in max-plus.

LITERATURE REVIEW

Manufacturing systems can be grouped into 3 main categories viz. *flow-shop*, *job-shop* and *open-shop* (Baker, 1974). Uni-directional flow of parts is an important characteristic of flow-shops where all parts move in the same direction through a sequence of machines. In a job-shop, the parts do not flow in the same direction, i.e. the sequence of machines visited by each part can be different. In an open-shop, the operations on a part may be performed in any particular order as long as they are all completed. There are numerous variations of these three basic types of manufacturing systems. In a *cyclic* system, a set of parts are repeatedly processed on the machines in a sequence. The figure on the left in Figure 1 shows the flow of parts when 250 units of each of the 3 part types are required. This is also known as *batch processing*. The figure on the right in Figure 1 shows a different flow of parts in the same setting. This is *cyclic* flow where the basic sequence A-B-C is repeated 250 times to achieve the production target. This can be extended to an uneven distribution of parts as well as shown in Figure 2.

With the increasing popularity of lean principles (Womack and Jones, 1996) mass customization (Pine, 1993) and agile manufacturing (Anderson, 1997) concepts, reducing all forms of wastes is becoming one of the top priorities for manufacturing companies worldwide. Taiichi Ohno (1988) classifies wastes into seven broad categories viz. overproduction, waiting, transport, processing, inventory, motion, and defects. Using a smooth and continuous flow of parts as shown in Figure 1 and 2 helps reduce overproduction, idle times (waiting) and inventory.

Flow-Shop Scheduling

In a permutation flow-shop, the parts are processed in a cyclic sequence and the sequence of parts on all the machines is the same. Scheduling for such a system is known to be a *NP-hard* problem (Pinedo,

Figure 1. Flow of parts for uniform distribution

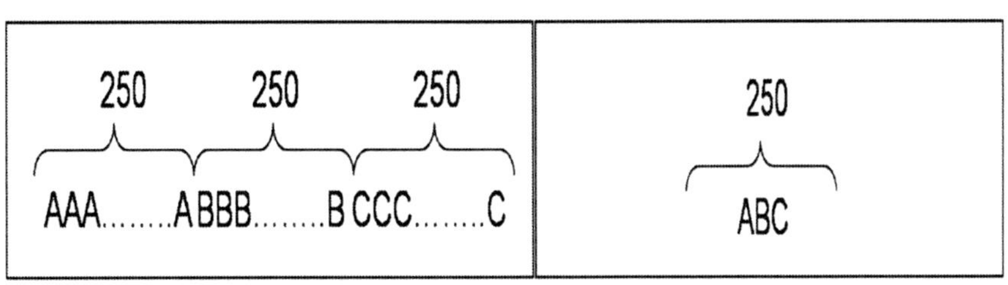

Figure 2. Flow of parts for non-uniform distribution

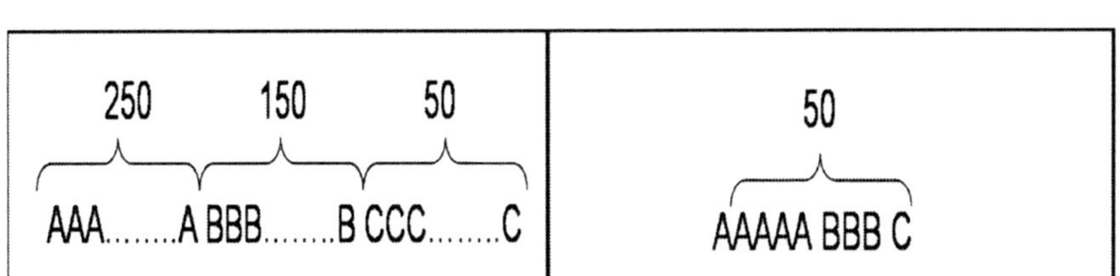

1995). Numerous enumerative techniques (Morton & Pentico, 1993; Carlier & Rebai, 1996; Temiz & Erol, 2004; Ladhari & Haouari, 2005), approximation algorithms (Hall, 1998; Sviridenko, 2004), heuristics such as genetic algorithms (Goldberg, 1989; Rajendran & Chaudhuri, 1992; Chen et al, 1995; Tang & Liu, 2002; Ravindran et al, 2005), simulated annealing (Osman & Potts, 1989; Ogbu & Smith, 1990; Ogbu & Smith, 1991; Ishibuchi et al 1995;Chakravarthy & Rajendran, 1999) and neural networks (Sabuncuoglu & Gurgun, 1996; Sabuncuoglu, 1998; Akyol, 2004; El-Bouri et al, 2005; Tang et al, 2005) have been developed. The main thrust among research (McCormick, 1989a; Matsuo, 1990; Roundy, 1992; Chauvet, 2003) in cyclic scheduling systems has been for systems with buffers. Later in this chapter, we develop a mathematical model for permutation flow-shops without buffers.

Stochastic Scheduling

Manufacturing systems are seldom static in nature. There tends to be significant variability such as machine availability, processing times, transportation times, etc. In this work, our focus is primarily on the processing time variability. In general, it has been found that stochastic scheduling for permutation flow-shop scheduling is NP-hard (Wang et al 2005). There are many different approaches developed in literature which focus on various subsets of the generic problem. Pinedo (2007) considers a single-machine scheduling problem with probabilistic processing times. However, Pinedo focuses primarily on batch processing and minimizes makespan. Gang et al. (2007) minimize the total completion times for an online scheduling problem where the release times and the processing times are unknown. The authors conclude that a non-delay algorithm provides good performance. Wang et al. (2006) minimize makespan for permutation flow shops with stochastic processing times. The authors use expected makespan as the performance measure.

Lee et al. (2005) analyze cyclic flow lines with intermediate buffers and stochastic processing times using a Markovian model. Karabati et al. (1998) investigate stochastic cyclic flow lines with synchronous transfers from one stage to another. Sung et al. (2003) develop an algebraic model to minimize the number of late jobs for a single-machine scheduling problem with uncertain processing times. Kouvelis et al. (2000) look into the two-machine flow-shop scheduling problem with uncertain processing times. Rao et al. (1999) present an algebraic approach to estimate the performance measures for cyclic systems with stochastic processing times. Other approaches to model the stochastic behavior include genetic algorithms (Wang et al 2005), mathematical programming (Balasubramanian & Grossmann, 2002), simulation-based algorithms (Honkomp, 1999). We use the mathematical model developed in this chapter to compare cycle times for stochastic systems using average processing times with the average cycle time obtained by exhaustively iterating through every possible combination of processing time values for a given range later in this chapter.

Assembly Lines

The concept of assembly lines was introduced by Henry Ford (1922). Single Model Assembly Line Balancing (SALBP) has been addressed in (Baybars, 1986; Ghosh & Gagnon,1989; Scholl & Klein, 1999; Becker & Scholl, 2006; Scholl & Becker, 2006; Sabuncuoglu et al, 2009). However, customers are no longer happy with a single model. Thus, mixed-model assembly line balancing, where multiple models are produced on the same assembly line, has been gaining more and more importance. Boysen et al. (2006, 2007, 2009a, 2009b) provide a classification schema for the mixed-model assembly line balancing problem.

Exact approaches such as branch-and-bound (Bukchin & Rabinowitch, 2006), goal programming (Choi, 2009) and heuristic approaches such as GA (Khoo & Loi, 2002; Ponnambalam et al 2000; Sabuncuoglu, 2000; Ying et al, 2009), and ACO (McMullen & Tarasewich, 2003) have been developed for assembly line balancing. Throughput assessment (Dhouib et al, 2009) and performance measures (Venkatesh & Dabade, 2008) for mixed model assembly lines have also been investigated.

Task assignment is one of the most significant aspects in assembly line balancing problem. Scholl et al. (2010) apply various restrictions on task assignment to the simple assembly line balancing problem where a single product is produced in large quantities. The authors explore restrictions such as *zoning constraints*, *resource restrictions* and *distance restrictions*. In an assembly line, since the workers perform the same task over and over again, this tends to impact the processing times for the various tasks. Toksari et al. (2010) propose an integer programming model for a simple assembly line balancing problem with deteriorating processing times. There could be inherent variability (Mirzapour et al, 2009; Weida & Xiao, 2008) in the task durations too. The sequence of parts processed in an assembly line also has a significant impact on the throughput of the line. Optimization procedures based on beam-search methods (Sabuncuoglu et al 2008) and heuristics based on simulated annealing (Fattehi & Salehi, 2009) and genetic algorithm (Cao & Ma, 2008) have been proposed to determine the sequence that minimizes the total idle costs. Battini et al. (2009) consider the same problem with finite buffers. The authors propose a branch-and-bound method to improve the performance of the assembly while minimizing the buffer capacity.

Assembly lines can also be classified into *single-sided* where the assembly operation takes place only on one side of the line and *two-sided* (Ozcan & Toklu, 2009a) where assembly operation might take place on either side of the assembly line. Integer programming models (Ozcan & Toklu, 2009a) and fuzzy goal programming models (Ozcan & Toklu, 2009b) have been proposed to solve the two-sided assembly line balancing problem. These have also been combined with heuristic approaches such as simulated annealing (Ozcan, 2010). Another variation of assembly lines is parallel workspaces (Becker & Scholl, 2009) where multiple workers perform multiple tasks simultaneously on a single part such as cars and trucks. Team-oriented (Cevikan et al, 2009) assembly lines is another possible variation. We modify the mathematical model developed in this chapter for the mixed-model assembly line balancing problem. This allows us to obtain optimal or near-optimal solutions to sufficiently large problems.

Max-Plus Algebra

Max-plus algebra has been used for discrete-event systems (Imaev & Judd, 2008, van den Boom & De Schutter, 2001a; van den Boom & De Schutter, 2001b; De Schutter & van den Boom, 2000; Takahashi et al, 2009) and event-graph modeling (Bacelli et al, 1999). It has also been applied to model railway network systems (Heidergott et al, 2006; Heidergott, 2006) for capacity assessment and delay estimation, shipbuilding lines (Hiroyuki et al, 2009), response time over ethernet (Addad & Amari, 2008), distributed computing for telecommunications (Nejad et al, 2009), robotic motion control (Lopez et al, 2009), performance analysis of public transport systems (Nait-Sidi-Moh et al, 2002). Max-plus algebra has been used to model manufacturing systems such as flow-shops (Nambiar & Judd, 2007; Nambiar & Judd, 2010; Gorji et al, 2007), reconfigurable cells (Zhu et al, 2004), lot-delivery in supply chains (Elmahi et al, 2004) and assembly lines (Carlo & Nambiar, 2008).

THE MATHEMATICAL MODEL

Block diagrams, such as state space block diagrams and transfer function block diagrams, are widely used in control theory to model the behavior of continuous-time systems. A transfer function block diagram has four basic elements: block, line segment, pick-off point and summing node. A block may model a controller, a sensor, or a whole plant. In a block diagram external input and output variables are connected to blocks by connection lines. An output of a block may also be connected to an input of another block. The interconnection of components (blocks) can be found by following the paths of signal flow along the connecting lines. One of the advantages of the transfer function representation is the simplicity of the algebraic relations between the subsystem transfer functions.

This section presents the block diagram type of model for deterministic manufacturing systems. A block can be a machine queue, a part, a manufacturing cell or a factory. Each block has the same input-output structure with three inputs and three outputs. The blocks in the block diagram are interconnected through a) part-flow interconnections, which specify flow of parts through the diagram, and b) resource-flow interconnections, which specify flow of resources through the diagram. The model is hierarchical -- it is shown how a network of blocks can be combined into one block that has the same input-output structure. Mathematically, the model is described by a set of simultaneous linear equations in max-plus algebra. Theoretical knowledge in max-plus algebra attained over the past several decades provides basis for analysis, design and control of manufacturing systems.

Manufacturing Block

Consider a manufacturing system. In order to operate, the system requires a set of parts and a set of resources. After the system is done with the parts and the resources, they are released by the system. Let $\mathbf{m}$ denote an ordered set of system's resources, such as machines, buffers, etc. Let $\mathbf{n}^{in}$ be the ordered set of parts that enter the system and let $\mathbf{n}^{out}$ be the ordered set of parts that leave the system. The order of elements in either $\mathbf{m}$, $\mathbf{n}^{in}$ or $\mathbf{n}^{out}$ can be chosen arbitrary. Let $|\mathbf{x}|$ denote the size, or number of elements in the vector or set $\mathbf{x}$. Then for $k \in \{1,2,\ldots,|\mathbf{m}|\}$, let $[\mathbf{m}]_k$ denote the k-th resource in the set $\mathbf{m}$. Similarly, for $i \in \{1,2,..,|\mathbf{n}^{in}|\}$ and $j \in \{1,2,\ldots,|\mathbf{n}^{out}|\}$, let $[\mathbf{n}^{in}]_i$ and $[\mathbf{n}^{out}]_j$ denote the i-th part in $\mathbf{n}^{in}$ and j-th part in $\mathbf{n}^{out}$, respectively. If the manufacturing process involves part assembly or disassembly then $\mathbf{n}^{in} \neq \mathbf{n}^{out}$, since during assembly several parts are needed to create a new part and during disassembly a single part is disassembled into several new parts. If there are no assembly and disassembly machines in the system then we can set $\mathbf{n}^{in} = \mathbf{n}^{out} = \mathrm{n}$.

The system can be modeled by a block with three inputs and three outputs. The inputs $\mathbf{u}$, $\mathbf{v}$ and $\mathbf{w}$ are defined as

- $[\mathbf{u}]_i$ is the time when part $[\mathbf{n}^{in}]_i$ becomes available for the system;
- $[\mathbf{v}]_j$ is the time when part $[\mathbf{n}^{out}]_j$ is removed from the system;
- $[\mathbf{w}]_k$ is the time when resource $[\mathbf{m}]_k$ becomes available for the system.

The outputs $\mathbf{x}$, $\mathbf{y}$ and $\mathbf{z}$ are defined as

- $[\mathbf{x}]_j$ is the time when part $[\mathbf{n}^{out}]_j$ is ready to leave the system;
- $[\mathbf{y}]_i$ is the time when part $[\mathbf{n}^{in}]_i$ actually enters the system;
- $[\mathbf{z}]_k$ is the time when resource $[\mathbf{m}]_k$ is "set free" by the system.

It can be seen that input and output variables are defined with respect to $\mathbf{m}$, $\mathbf{n}^{in}$ and $\mathbf{n}^{out}$. In particular, $\mathbf{m}$ is associated with $\mathbf{w}$, $\mathbf{z}$; $\mathbf{n}^{in}$ is associated with $\mathbf{u}$, $\mathbf{y}$; and $\mathbf{n}^{out}$ is associated with $\mathbf{x}$ and $\mathbf{v}$.

It is assumed that the system is deterministic, i.e. the routing of parts through the resources, the processing order of parts on the resources and the processing times of parts on the resources are known and fixed. Then its output can be described in terms of its input by the following equation in the max-plus algebra

Figure 3. Block representation of a manufacturing system: (a) describes block representation of a manufacturing system; and, (b) illustrates flow of parts and resources through the block

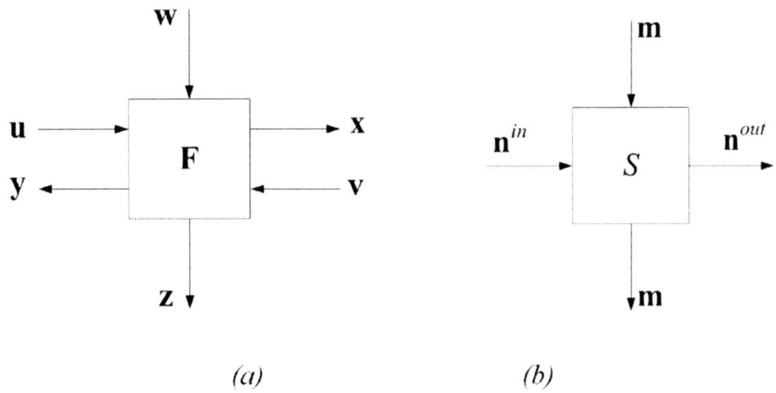

$$\begin{bmatrix} \mathbf{x} \\ \mathbf{y} \\ \mathbf{z} \end{bmatrix} = \begin{bmatrix} \mathbf{F}_{xu} & \mathbf{F}_{xv} & \mathbf{F}_{xw} \\ \mathbf{F}_{yu} & \mathbf{F}_{yv} & \mathbf{F}_{yw} \\ \mathbf{F}_{zu} & \mathbf{F}_{zv} & \mathbf{F}_{zw} \end{bmatrix} \begin{bmatrix} \mathbf{u} \\ \mathbf{v} \\ \mathbf{w} \end{bmatrix} = \mathbf{F} \begin{bmatrix} \mathbf{u} \\ \mathbf{v} \\ \mathbf{w} \end{bmatrix}, \qquad (7)$$

where **F** is a matrix that describes input-output relation -- it is called the system matrix.

The model provides an abstraction of any deterministic manufacturing system by means of block diagram having three inputs and three outputs and system matrix **F**as shown in Figure 3.

The matrix **F** and **m**, $\mathbf{n}^{in}$, $\mathbf{n}^{out}$ completely describe the model, since variables **u, v, w, x, y, z** are defined with respect to **m**, $\mathbf{n}^{in}$ and $\mathbf{n}^{out}$. Hence, the model *S* is denoted by a 4-tuple

$$S=(\mathbf{F}, \mathbf{m}, \mathbf{n}^{in}, \mathbf{n}^{out}). \qquad (8)$$

A block shown in Figure 3b illustrates flow of parts and resources through the model. The block has two inputs and two outputs. It does not contain information about timing behavior of the system. This block can be used instead of the block shown in Figure 3a in the case when we are only interested in flow of parts and resources through a network of manufacturing blocks.

Composition of Blocks

Let S_c be a system composed from a set of M manufacturing subsystems $\{S_1, S_2, \ldots, S_M\}$. Let $\mathbf{m}_c$, $\mathbf{n}_c^{in}$, $\mathbf{n}_c^{out}$ be ordered sets of resources and parts associated with S_c. Let the inputs and the outputs of S_c, namely $\mathbf{u}_c$, $\mathbf{v}_c$, $\mathbf{w}_c$ and $\mathbf{x}_c$, $\mathbf{y}_c$, $\mathbf{z}_c$, be defined with respect to $\mathbf{m}_c$, $\mathbf{n}_c^{in}$, $\mathbf{n}_c^{out}$.

Each subsystem S_i is represented by an equation of the form (9) or, specifically,

$$\begin{bmatrix} \mathbf{x}_i \\ \mathbf{y}_i \\ \mathbf{z}_i \end{bmatrix} = \begin{bmatrix} \mathbf{F}_{xu,i} & \mathbf{F}_{xv,i} & \mathbf{F}_{xw,i} \\ \mathbf{F}_{yu,i} & \mathbf{F}_{yv,i} & \mathbf{F}_{yw,i} \\ \mathbf{F}_{zu,i} & \mathbf{F}_{zv,i} & \mathbf{F}_{zw,i} \end{bmatrix} \begin{bmatrix} \mathbf{u}_i \\ \mathbf{v}_i \\ \mathbf{w}_i \end{bmatrix} = \mathbf{F}_i \begin{bmatrix} \mathbf{u}_i \\ \mathbf{v}_i \\ \mathbf{w}_i \end{bmatrix}, \qquad (9)$$

for $i \in 1,2,\ldots,M$.

The subsystems $S_1, S_2, \ldots, S_M$-- all share the system's parts and resources. It is assumed that there are no delays associated with transportation of parts or resources from S_i to S_j-- rather these delays can always be modeled by an appropriate manufacturing block or as part of S_i or S_j.

An example illustrating routing of parts and resources through subsystems is given in Figure 4. The blocks in the diagram have the form shown in Figure 3b. There are 4 parts $\{n_1, n_2, n_3, n_4\}$ and 3 resources $\{m_1, m_2, m_3\}$. The labeled arrows that

Figure 4. Interconnection of manufacturing blocks. An illustrative example.

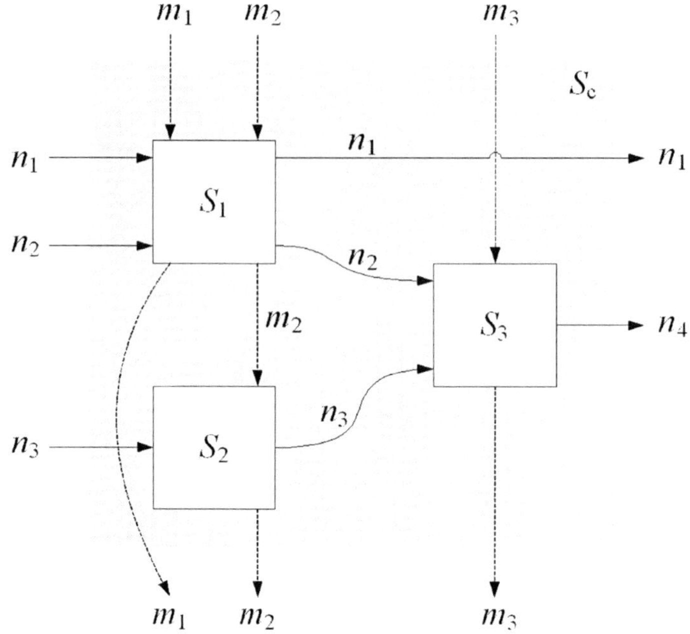

Figure 5. Interconnection of blocks: (a) part-flow interconnection and (b) machine-flow interconnection

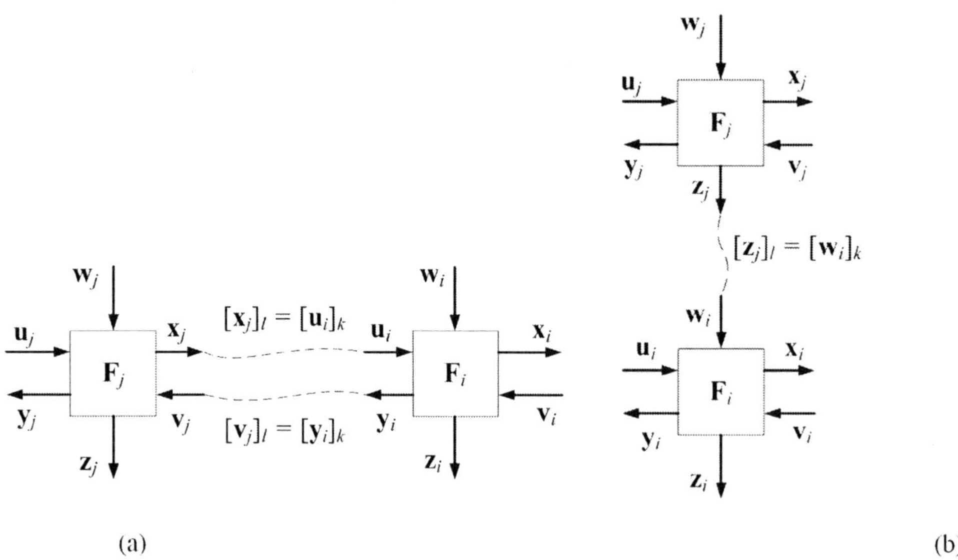

interconnect the blocks in the diagram indicate flow of parts and resources through the subsystems. Note that $\mathbf{m}=[m_1 m_2 m_3]^T$, $\mathbf{n}_{in}=[n_1 n_2 n_3]^T$ and $\mathbf{n}_{out}=[n_1 n_4]^T$. It can be seen that S_3 is an assembly operation, in which parts n_2 and n_3 are assembled by m_3 to create a new part n_4. Since there is an assembly operation in the system, we have $\mathbf{n}_{in} \neq \mathbf{n}_{out}$.

Consider part n which enters system S_i from an upstream system S_j. Suppose that $n = [\mathbf{n}_j^{out}]_l = [\mathbf{n}_i^{in}]_k$, where indexes l and k point to the location of n in $\mathbf{n}_j^{out}$ and $\mathbf{n}_i^{in}$, respectively. It is said that $[\mathbf{n}_j^{out}]_l$ is routed to $[\mathbf{n}_i^{in}]_k$, which is denoted by $[\mathbf{n}_j^{out}]_l \to [\mathbf{n}_i^{in}]_k$. Since n becomes available to S_i at the time instance when it is ready to leave S_j, we have $[\mathbf{u}_i]_k=[\mathbf{x}_j]_l$. In addition, the part n is removed from S_j when n enters $S_i S_i$, therefore $[\mathbf{v}_j]_l=[\mathbf{y}_i]_k$, as shown in Figure 5a.

Likewise, consider a resource m, which is first used by S_j and then it is used by S_i. Suppose that $m=[\mathbf{m}_j]_l=[\mathbf{m}_i]_k$, where l and k point to the location of m in $\mathbf{m}_j$ and $\mathbf{m}_i$, respectively. It is said that $[\mathbf{m}_j]_l$ is routed to $[\mathbf{m}_i]_k$, which is denoted by $[\mathbf{m}_j]_l \to [\mathbf{m}_i]_k$. Resource m is available to S_i after S_j is done using m, therefore $[\mathbf{w}_i]_k=[\mathbf{z}_j]_l$, as shown in Figure 5b.

It follows that flow of *parts* through manufacturing sub-systems is represented by horizontal interconnections (e.g., Figure 5a). We will refer to this type of interconnections as *part-flow* interconnections. Likewise, flow of *resources* through manufacturing sub-systems is represented by vertical interconnections (e.g., Figure 5b). We will refer to this type of interconnections as *resource-flow* interconnections.

Routing of parts and resources through the diagram is mathematically represented by means of part-flow and resource-flow interconnection matrices. Define

$$\tilde{\mathbf{n}}^{in} = \begin{bmatrix} \mathbf{n}_1^{in} \\ \mathbf{n}_2^{in} \\ \vdots \\ \mathbf{n}_M^{in} \end{bmatrix}, \quad \tilde{\mathbf{n}}^{out} = \begin{bmatrix} \mathbf{n}_1^{out} \\ \mathbf{n}_2^{out} \\ \vdots \\ \mathbf{n}_M^{out} \end{bmatrix}, \quad \tilde{\mathbf{m}} = \begin{bmatrix} \mathbf{m}_1 \\ \mathbf{m}_2 \\ \vdots \\ \mathbf{m}_M \end{bmatrix}.$$

Resource-flow interconnection matrices are defined as

$$[\mathbf{Q}_{in}]_{i,j} = \begin{cases} e & \text{if}[\mathbf{m}_c]_j \to [\tilde{\mathbf{m}}]_i, \\ \varepsilon & \text{otherwise;} \end{cases}$$

$$[\mathbf{Q}]_{i,j} = \begin{cases} e & \text{if}[\tilde{\mathbf{m}}]_j \to [\tilde{\mathbf{m}}]_i, \\ \varepsilon & \text{otherwise;} \end{cases}$$

$$[\mathbf{Q}_{out}]_{i,j} = \begin{cases} e & \text{if}[\tilde{\mathbf{m}}]_j \to [\mathbf{m}_c]_i, \\ \varepsilon & \text{otherwise.} \end{cases}$$

Part-flow interconnection matrices are defined as

$$[\mathbf{R}_{in}]_{i,j} = \begin{cases} e & \text{if}[\mathbf{n}_c^{in}]_j \to [\tilde{\mathbf{n}}^{in}]_i, \\ \varepsilon & \text{otherwise;} \end{cases}$$

$$[\mathbf{R}]_{i,j} = \begin{cases} e & \text{if}[\tilde{\mathbf{n}}^{out}]_j \to [\tilde{\mathbf{n}}^{in}]_i, \\ \varepsilon & \text{otherwise;} \end{cases}$$

$$[\mathbf{R}_{out}]_{i,j} = \begin{cases} e & \text{if}[\tilde{\mathbf{n}}^{out}]_j \to [\mathbf{n}_c^{out}]_i, \\ \varepsilon & \text{otherwise.} \end{cases}$$

Define

$$\tilde{\mathbf{u}} = \begin{bmatrix} \mathbf{u}_1 \\ \mathbf{u}_2 \\ \vdots \\ \mathbf{u}_M \end{bmatrix}, \quad \tilde{\mathbf{v}} = \begin{bmatrix} \mathbf{v}_1 \\ \mathbf{v}_2 \\ \vdots \\ \mathbf{v}_M \end{bmatrix}, \quad \tilde{\mathbf{w}} = \begin{bmatrix} \mathbf{w}_1 \\ \mathbf{w}_2 \\ \vdots \\ \mathbf{w}_M \end{bmatrix}.$$

Similarly define $\tilde{\mathbf{x}}$, $\tilde{\mathbf{y}}$ and $\tilde{\mathbf{z}}$. Let

$$\tilde{\mathbf{F}}_{xu} = \begin{bmatrix} \mathbf{F}_{xu,1} & \varepsilon & & \varepsilon \\ \varepsilon & \mathbf{F}_{xu,2} & & \varepsilon \\ & & \ddots & \\ \varepsilon & \varepsilon & & \mathbf{F}_{xu,M} \end{bmatrix}.$$

and similarly define $\tilde{\mathbf{F}}_{xv}$, $\tilde{\mathbf{F}}_{xw}$, $\tilde{\mathbf{F}}_{yu}$, etc. Then

Figure 6. Composition of blocks

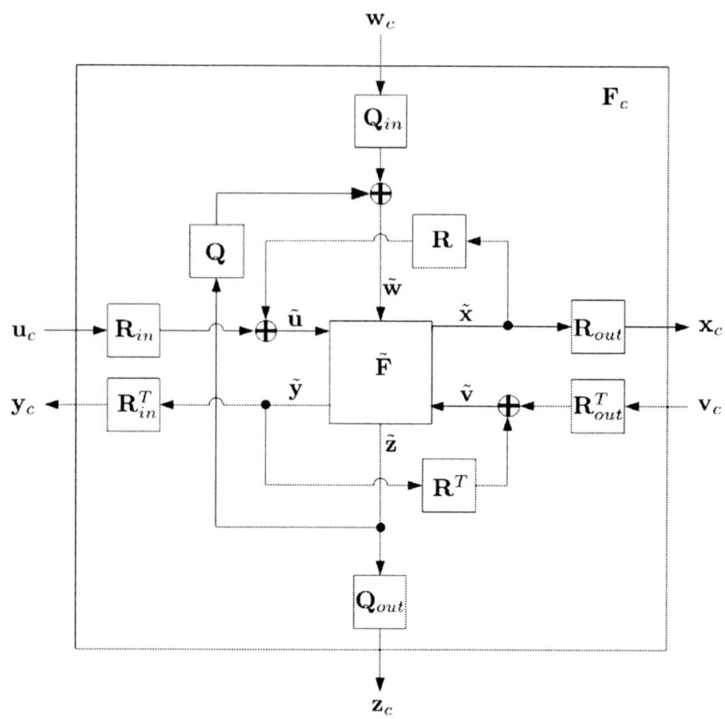

$$\begin{bmatrix}\tilde{\mathbf{x}}\\ \tilde{\mathbf{y}}\\ \tilde{\mathbf{z}}\end{bmatrix} = \begin{bmatrix}\tilde{\mathbf{F}}_{xu} & \tilde{\mathbf{F}}_{xv} & \tilde{\mathbf{F}}_{xw}\\ \tilde{\mathbf{F}}_{yu} & \tilde{\mathbf{F}}_{yv} & \tilde{\mathbf{F}}_{yw}\\ \tilde{\mathbf{F}}_{zu} & \tilde{\mathbf{F}}_{zv} & \tilde{\mathbf{F}}_{zw}\end{bmatrix}\begin{bmatrix}\tilde{\mathbf{u}}\\ \tilde{\mathbf{v}}\\ \tilde{\mathbf{w}}\end{bmatrix} = \tilde{\mathbf{F}}\begin{bmatrix}\tilde{\mathbf{u}}\\ \tilde{\mathbf{v}}\\ \tilde{\mathbf{w}}\end{bmatrix}. \qquad (10)$$

From the definition of the interconnection matrices it follows that

$$\begin{aligned}\tilde{\mathbf{u}} &= \mathbf{R}\tilde{\mathbf{x}} \oplus \mathbf{R}_{in}\mathbf{u}_c,\\ \tilde{\mathbf{v}} &= \mathbf{R}^T\tilde{\mathbf{y}} \oplus \mathbf{R}_{out}^T\mathbf{v}_c,\\ \tilde{\mathbf{w}} &= \mathbf{Q}\tilde{\mathbf{z}} \oplus \mathbf{Q}_{in}\mathbf{w}_c;\end{aligned} \qquad (11)$$

and the outputs of S_c can be expressed as

$$\begin{aligned}\mathbf{x}_c &= \mathbf{R}_{out}\tilde{\mathbf{x}},\\ \mathbf{y}_c &= \mathbf{R}_{in}^T\tilde{\mathbf{y}},\\ \mathbf{z}_c &= \mathbf{Q}_{out}\tilde{\mathbf{z}}.\end{aligned} \qquad (12)$$

Equations (10), (11) and (12) describe block diagram shown in Figure 5 and they can be used to find the system matrix $\mathbf{F}_c$.

Equation (11) can be rewritten as

$$\begin{bmatrix}\tilde{\mathbf{u}}\\ \tilde{\mathbf{v}}\\ \tilde{\mathbf{w}}\end{bmatrix} = \begin{bmatrix}\mathbf{R} & \varepsilon & \varepsilon\\ \varepsilon & \mathbf{R}^T & \varepsilon\\ \varepsilon & \varepsilon & \mathbf{Q}\end{bmatrix}\begin{bmatrix}\tilde{\mathbf{x}}\\ \tilde{\mathbf{y}}\\ \tilde{\mathbf{z}}\end{bmatrix} \oplus \begin{bmatrix}\mathbf{R}_{in} & \varepsilon & \varepsilon\\ \varepsilon & \mathbf{R}_{out}^T & \varepsilon\\ \varepsilon & \varepsilon & \mathbf{Q}_{in}\end{bmatrix}\begin{bmatrix}\mathbf{u}_c\\ \mathbf{v}_c\\ \mathbf{w}_c\end{bmatrix}; \qquad (13)$$

likewise, equation (12) can be written as

$$\begin{bmatrix}\mathbf{x}_c\\ \mathbf{y}_c\\ \mathbf{z}_c\end{bmatrix} = \begin{bmatrix}\mathbf{R}_{out} & \varepsilon & \varepsilon\\ \varepsilon & \mathbf{R}_{in}^T & \varepsilon\\ \varepsilon & \varepsilon & \mathbf{Q}_{out}\end{bmatrix}\begin{bmatrix}\tilde{\mathbf{x}}\\ \tilde{\mathbf{y}}\\ \tilde{\mathbf{z}}\end{bmatrix}. \qquad (14)$$

Substituting (10) into (13) we obtain

$$\begin{bmatrix}\tilde{\mathbf{x}}\\ \tilde{\mathbf{y}}\\ \tilde{\mathbf{z}}\end{bmatrix} = \tilde{\mathbf{F}}\begin{bmatrix}\mathbf{R} & \varepsilon & \varepsilon\\ \varepsilon & \mathbf{R}^T & \varepsilon\\ \varepsilon & \varepsilon & \mathbf{Q}\end{bmatrix}\begin{bmatrix}\tilde{\mathbf{x}}\\ \tilde{\mathbf{y}}\\ \tilde{\mathbf{z}}\end{bmatrix} \oplus \tilde{\mathbf{F}}\begin{bmatrix}\mathbf{R}_{in} & \varepsilon & \varepsilon\\ \varepsilon & \mathbf{R}^T_{out} & \varepsilon\\ \varepsilon & \varepsilon & \mathbf{Q}_{in}\end{bmatrix}\begin{bmatrix}\mathbf{u}_c\\ \mathbf{v}_c\\ \mathbf{w}_c\end{bmatrix}. \tag{15}$$

It follows that

$$\begin{bmatrix}\tilde{\mathbf{x}}\\ \tilde{\mathbf{y}}\\ \tilde{\mathbf{z}}\end{bmatrix} = \left(\tilde{\mathbf{F}}\begin{bmatrix}\mathbf{R} & \varepsilon & \varepsilon\\ \varepsilon & \mathbf{R}^T & \varepsilon\\ \varepsilon & \varepsilon & \mathbf{Q}\end{bmatrix}\right)^* \tilde{\mathbf{F}}\begin{bmatrix}\mathbf{R}_{in} & \varepsilon & \varepsilon\\ \varepsilon & \mathbf{R}^T_{out} & \varepsilon\\ \varepsilon & \varepsilon & \mathbf{Q}_{in}\end{bmatrix}\begin{bmatrix}\mathbf{u}_c\\ \mathbf{v}_c\\ \mathbf{w}_c\end{bmatrix}, \tag{16}$$

and finally, substituting (16) into (14) we get

$$\begin{bmatrix}\mathbf{x}_c\\ \mathbf{y}_c\\ \mathbf{z}_c\end{bmatrix} = \begin{bmatrix}\mathbf{R}_{out} & \varepsilon & \varepsilon\\ \varepsilon & \mathbf{R}^T_{in} & \varepsilon\\ \varepsilon & \varepsilon & \mathbf{Q}_{out}\end{bmatrix}\left(\tilde{\mathbf{F}}\begin{bmatrix}\mathbf{R} & \varepsilon & \varepsilon\\ \varepsilon & \mathbf{R}^T & \varepsilon\\ \varepsilon & \varepsilon & \mathbf{Q}\end{bmatrix}\right)^* \tilde{\mathbf{F}}\begin{bmatrix}\mathbf{R}_{in} & \varepsilon & \varepsilon\\ \varepsilon & \mathbf{R}^T_{out} & \varepsilon\\ \varepsilon & \varepsilon & \mathbf{Q}_{in}\end{bmatrix}\begin{bmatrix}\mathbf{u}_c\\ \mathbf{v}_c\\ \mathbf{w}_c\end{bmatrix} = \mathbf{F}_c\begin{bmatrix}\mathbf{u}_c\\ \mathbf{v}_c\\ \mathbf{w}_c\end{bmatrix} \tag{17}$$

Equation (17) gives general expression for the system matrix for S_c. This proves that any composition of systems represented by (7) results in a system that is also represented by (7).

Sometimes instead of explicitly specifying $\mathbf{v}_c$ it is assumed that jobs are removed from the system as soon as they are ready to leave the system. In other words machines are never blocked from outside of the system. Then $\mathbf{v}_c = \mathbf{x}_c = \mathbf{R}_{out}\tilde{\mathbf{x}}$ and we have

$$\begin{aligned}\tilde{\mathbf{u}} &= \mathbf{R}\tilde{\mathbf{x}} \oplus \mathbf{R}_{in}\mathbf{u}_c,\\ \tilde{\mathbf{v}} &= \mathbf{R}^T\tilde{\mathbf{y}} \oplus \mathbf{R}^T_{out}\mathbf{v}_c = \mathbf{R}^T\tilde{\mathbf{y}} \oplus \mathbf{R}^T_{out}\mathbf{R}_{out}\tilde{\mathbf{x}},\\ \tilde{\mathbf{w}} &= \mathbf{Q}\tilde{\mathbf{z}} \oplus \mathbf{Q}_{in}\mathbf{w}_c,\end{aligned} \tag{18}$$

which can be written as

$$\begin{bmatrix}\tilde{\mathbf{u}}\\ \tilde{\mathbf{v}}\\ \tilde{\mathbf{w}}\end{bmatrix} = \begin{bmatrix}\mathbf{R} & \varepsilon & \varepsilon\\ \mathbf{R}^T_{out}\mathbf{R}_{out} & \mathbf{R}^T & \varepsilon\\ \varepsilon & \varepsilon & \mathbf{Q}\end{bmatrix}\begin{bmatrix}\tilde{\mathbf{x}}\\ \tilde{\mathbf{y}}\\ \tilde{\mathbf{z}}\end{bmatrix} \oplus \begin{bmatrix}\mathbf{R}_{in} & \varepsilon\\ \varepsilon & \varepsilon\\ \varepsilon & \mathbf{Q}_{in}\end{bmatrix}\begin{bmatrix}\mathbf{u}_c\\ \mathbf{w}_c\end{bmatrix}. \tag{19}$$

Combining (10), (14) and (19) and after some algebraic manipulation it follows that

$$\begin{bmatrix}\mathbf{x}_c\\ \mathbf{y}_c\\ \mathbf{z}_c\end{bmatrix} = \begin{bmatrix}\mathbf{R}_{out} & \varepsilon & \varepsilon\\ \varepsilon & \mathbf{R}^T_{in} & \varepsilon\\ \varepsilon & \varepsilon & \mathbf{Q}_{out}\end{bmatrix}\left(\tilde{\mathbf{F}}\begin{bmatrix}\mathbf{R} & \varepsilon & \varepsilon\\ \mathbf{R}^T_{out}\mathbf{R}_{out} & \mathbf{R}^T & \varepsilon\\ \varepsilon & \varepsilon & \mathbf{Q}\end{bmatrix}\right)^* \otimes \tilde{\mathbf{F}}\begin{bmatrix}\mathbf{R}_{in} & \varepsilon\\ \varepsilon & \varepsilon\\ \varepsilon & \mathbf{Q}_{in}\end{bmatrix}\begin{bmatrix}\mathbf{u}_c\\ \mathbf{w}_c\end{bmatrix}. \tag{20}$$

Example

Consider a network of three manufacturing blocks, namely

$$S_1 = (\mathbf{F}_1, \mathbf{m}_1, \mathbf{n}_1^{in}, \mathbf{n}_1^{out}) = \left(\mathbf{F}_1, \begin{bmatrix}m_1\\ m_2\end{bmatrix}, \begin{bmatrix}n_1\\ n_2\end{bmatrix}, \begin{bmatrix}n_1\\ n_2\end{bmatrix}\right),$$

$$S_2 = (\mathbf{F}_2, \mathbf{m}_2, \mathbf{n}_2^{in}, \mathbf{n}_2^{out}) = \left(\mathbf{F}_2, \begin{bmatrix}m_2\end{bmatrix}, \begin{bmatrix}n_3\end{bmatrix}, \begin{bmatrix}n_3\end{bmatrix}\right),$$

$$S_3 = (\mathbf{F}_3, \mathbf{m}_3, \mathbf{n}_3^{in}, \mathbf{n}_3^{out}) = \left(\mathbf{F}_3, \begin{bmatrix}m_3\end{bmatrix}, \begin{bmatrix}n_2\\ n_3\end{bmatrix}, \begin{bmatrix}n_4\end{bmatrix}\right).$$

Consider a network of blocks comprising of a composite manufacturing block defined by

$$S_c = (\mathbf{F}_c, \mathbf{m}_c, \mathbf{n}_c^{in}, \mathbf{n}_c^{out}) = \left(\mathbf{F}_c, \begin{bmatrix} m_1 \\ m_2 \\ m_3 \end{bmatrix}, \begin{bmatrix} n_1 \\ n_2 \\ n_3 \end{bmatrix}, \begin{bmatrix} n_1 \\ n_4 \end{bmatrix}\right).$$

Flow of parts and resources through the network is given by interconnection diagram in Figure 4. The goal is to find part-flow and resource-flow interconnection matrices.

We have

$$\tilde{\mathbf{m}} = \begin{bmatrix} \begin{bmatrix} m_1 \\ m_2 \end{bmatrix} \\ \begin{bmatrix} m_2 \end{bmatrix} \\ \begin{bmatrix} m_3 \end{bmatrix} \end{bmatrix}, \quad \tilde{\mathbf{n}}^{in} = \begin{bmatrix} \begin{bmatrix} n_1 \\ n_2 \end{bmatrix} \\ \begin{bmatrix} n_3 \end{bmatrix} \\ \begin{bmatrix} n_2 \\ n_3 \end{bmatrix} \end{bmatrix}, \quad \tilde{\mathbf{n}}^{out} = \begin{bmatrix} \begin{bmatrix} n_1 \\ n_2 \end{bmatrix} \\ \begin{bmatrix} n_3 \end{bmatrix} \\ \begin{bmatrix} n_4 \end{bmatrix} \end{bmatrix},$$

$$\mathbf{m}_c = \begin{bmatrix} m_1 \\ m_2 \\ m_3 \end{bmatrix}, \quad \mathbf{n}_c^{in} = \begin{bmatrix} n_1 \\ n_2 \\ n_3 \end{bmatrix}, \quad \mathbf{n}_c^{out} = \begin{bmatrix} n_1 \\ n_4 \end{bmatrix}.$$

From the interconnection diagram in Figure 4 and using Figure 7 as a guide it follows that

Figure 7. Obtaining Q and R from Figure 4

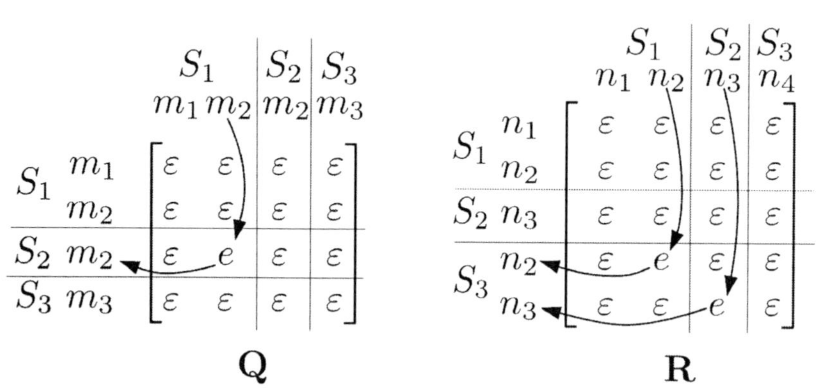

$$\mathbf{R}_{in} = \begin{bmatrix} e & \varepsilon & \varepsilon \\ \varepsilon & e & \varepsilon \\ \varepsilon & \varepsilon & e \\ \varepsilon & \varepsilon & \varepsilon \\ \varepsilon & \varepsilon & \varepsilon \end{bmatrix}, \quad \mathbf{R} = \begin{bmatrix} \varepsilon & \varepsilon & \varepsilon & \varepsilon \\ \varepsilon & \varepsilon & \varepsilon & \varepsilon \\ \varepsilon & \varepsilon & \varepsilon & \varepsilon \\ \varepsilon & e & \varepsilon & \varepsilon \\ \varepsilon & \varepsilon & e & \varepsilon \end{bmatrix},$$

$$\mathbf{R}_{out} = \begin{bmatrix} e & \varepsilon & \varepsilon & \varepsilon \\ \varepsilon & \varepsilon & \varepsilon & e \end{bmatrix},$$

$$\mathbf{Q}_{in} = \begin{bmatrix} e & \varepsilon & \varepsilon \\ \varepsilon & e & \varepsilon \\ \varepsilon & \varepsilon & \varepsilon \\ \varepsilon & \varepsilon & e \end{bmatrix}, \quad \mathbf{Q} = \begin{bmatrix} \varepsilon & \varepsilon & \varepsilon & \varepsilon \\ \varepsilon & \varepsilon & \varepsilon & \varepsilon \\ \varepsilon & e & \varepsilon & \varepsilon \\ \varepsilon & \varepsilon & \varepsilon & \varepsilon \end{bmatrix},$$

$$\mathbf{Q}_{out} = \begin{bmatrix} e & \varepsilon & \varepsilon & \varepsilon \\ \varepsilon & \varepsilon & e & \varepsilon \\ \varepsilon & \varepsilon & \varepsilon & e \end{bmatrix}.$$

BASIC BLOCKS

In this chapter timing models of basic manufacturing blocks are presented, namely the models of

1. single resource manufacturing a part;

2. assembly block;
3. disassembly block;
4. buffer models; and
5. machines with a buffer attached.

All models share the generic structure described by (7) with three inputs and three outputs. These blocks can then be used to build larger cells and manufacturing systems.

Single Machine Processing Single Part

Consider machine *m* processing part *n*. Let *t* be processing time of *n* on *m*. Suppose that the system is modeled by using equation of the form (7) having inputs *u*, *v*, *w* and outputs *x*, *y*, *z*, which are all scalars because there is only one resource and one part.

The part *n* enters the system as soon as both *m* and *n* are available, therefore

$y=u\oplus w.$

The part is ready to leave the system as soon as its processing is done on the machine, therefore

$x=t(u\oplus w).$

The machine is "set free" by the system as soon as *n* is removed from the system, therefore

$z=v.$

Thus, we have

$$\begin{bmatrix} x \\ y \\ z \end{bmatrix} = \begin{bmatrix} t & \varepsilon & t \\ e & \varepsilon & e \\ \varepsilon & e & \varepsilon \end{bmatrix} \begin{bmatrix} u \\ v \\ w \end{bmatrix}. \quad (21)$$

Block diagram model of the system is provided in Figure 8.

Assembly Machine

An assembly machine takes several input parts $\mathbf{n}_{in}$ and produces a single new part $\mathbf{n}_{out}=n$ as its output. Suppose that the system is modeled by using equation of the form (7) having inputs **u**, *v*, *w* and outputs *x*, **y**, *z*. Let *t* be the assembly time. The block diagram of the assembly machine is provided in Figure 9.

Since the assembly machine needs to wait for all the required parts before it can start processing parts, we have

Figure 8. Model of a single machine manufacturing a single part

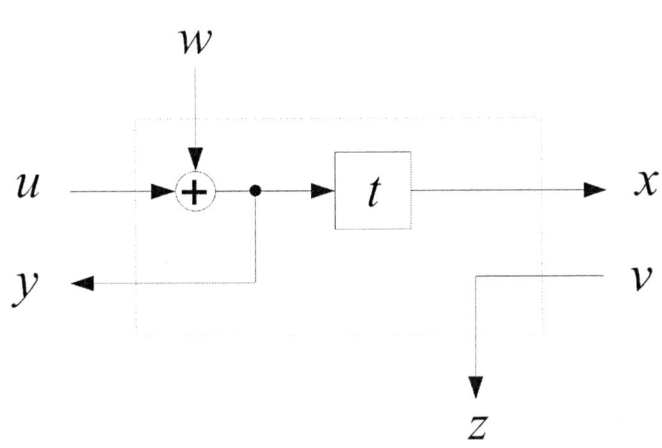

Figure 9. Model of an assembly machine

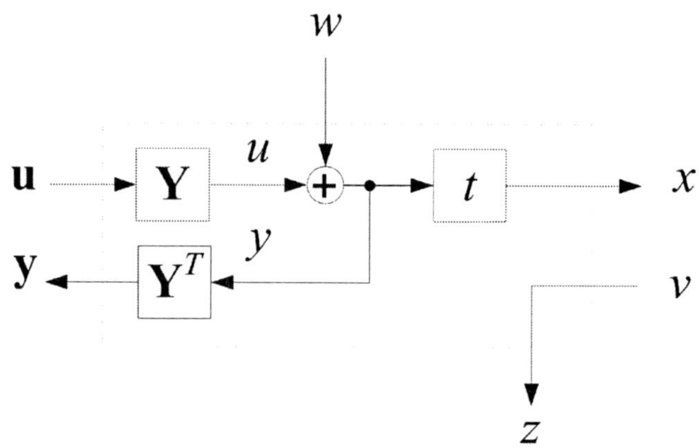

$u=[\mathbf{u}]_1\oplus[\mathbf{u}]_2\oplus\ldots\oplus[\mathbf{u}]_N=\mathbf{Y}\mathbf{u}$,

$\mathbf{y}=\mathbf{Y}^T y$,

where $\mathbf{Y}=[e\ e\ \ldots\ e]$.

Then the model is described by the following equation

$$\begin{bmatrix} x \\ \mathbf{y} \\ z \end{bmatrix} = \begin{bmatrix} \mathbf{Y}t & \varepsilon & \mathbf{Y}t \\ \mathbf{Y}^T & \varepsilon & \mathbf{Y}^T \\ \varepsilon & \varepsilon & e \end{bmatrix} \begin{bmatrix} \mathbf{u} \\ v \\ w \end{bmatrix}. \qquad (22)$$

Figure 10. Model of a disassembly machine

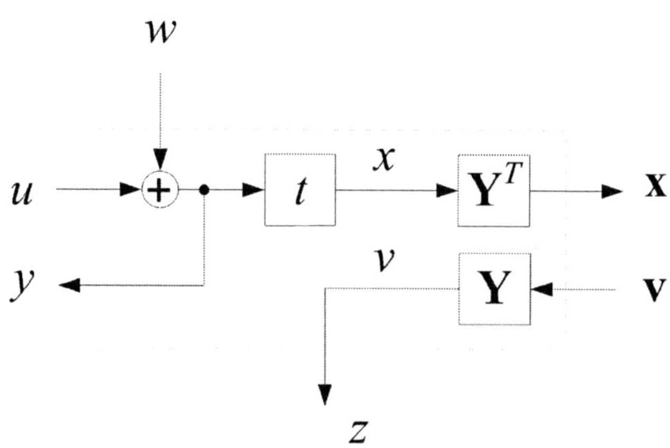

Disassembly Machine

Disassembly machine takes one part $\mathbf{n}_{in}=n$ and outputs several parts $\mathbf{n}_{out}$. Suppose that the system is modeled by using equation of the form (7) having inputs u, $\mathbf{v}$, w and outputs $\mathbf{x}$, y, z. Let t be disassembly time. The model of disassembly machine is shown in Figure 10.

After the disassembly process all the output parts are ready to leave the system at the same time, therefore

$\mathbf{x}=\mathbf{Y}^T x$

$v=\mathbf{Y}\mathbf{v}$

The model is described by

$$\begin{bmatrix} \mathbf{x} \\ y \\ z \end{bmatrix} = \begin{bmatrix} t\mathbf{Y}^T & \varepsilon & t\mathbf{Y}^T \\ \mathbf{Y} & \varepsilon & \mathbf{Y} \\ \varepsilon & e & \varepsilon \end{bmatrix} \begin{bmatrix} u \\ \mathbf{v} \\ w \end{bmatrix}. \quad (23)$$

Buffer Models

McCormick et al. (1989b) show that a buffer of unit capacity can be represented by a resource having zero processing time for jobs that enter the buffer. Therefore for buffer of unit capacity (21) becomes

$$\begin{bmatrix} x \\ y \\ z \end{bmatrix} = \begin{bmatrix} e & \varepsilon & e \\ e & \varepsilon & e \\ \varepsilon & e & \varepsilon \end{bmatrix} \begin{bmatrix} u \\ v \\ w \end{bmatrix}, \quad (24)$$

because $t=e$. Block diagram representation of (24) is provided in Figure 11a.

Consider random access buffer with unlimited capacity for storing parts. The buffer is always available to accept parts because of its unlimited capacity, therefore $w=\varepsilon$ and $z=\varepsilon$. The part enters the buffer as soon as it becomes available to the buffer, hence $y=u$. Also, the part is ready to leave the buffer as soon as it entered the buffer, and therefore $x=y=u$. Hence,

$$\begin{bmatrix} x \\ y \\ z \end{bmatrix} = \begin{bmatrix} e & \varepsilon & \varepsilon \\ e & \varepsilon & \varepsilon \\ \varepsilon & \varepsilon & \varepsilon \end{bmatrix} \begin{bmatrix} u \\ v \\ w \end{bmatrix}. \quad (25)$$

Block diagram representation of the model is shown in Figure 11b.

Machine With a Buffer

A machine preceded by an unlimited buffer can be represented by the block diagram shown in Figure 12, which can be reduced to the block diagram shown in Figure 13. This model is useful to model machines in cellular manufacture and job shops where parts are buffered before entering a machine. Then, the system is described by the following equation

$$\begin{bmatrix} x \\ y \\ z \end{bmatrix} = \begin{bmatrix} t & \varepsilon & t \\ e & \varepsilon & \varepsilon \\ \varepsilon & e & \varepsilon \end{bmatrix} \begin{bmatrix} u \\ v \\ w \end{bmatrix}. \quad (26)$$

Figure 11. Buffer models

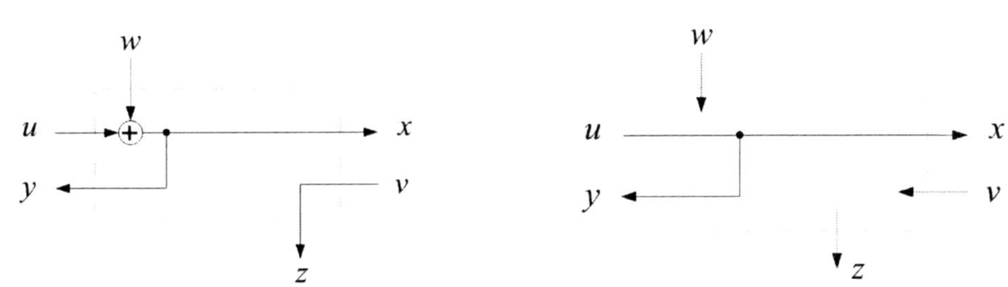

(a) Unit capacity buffer. *(b) Unlimited capacity buffer with random access.*

Figure 12. Machine with an infinite buffer in front of it processing a part (detailed block diagram)

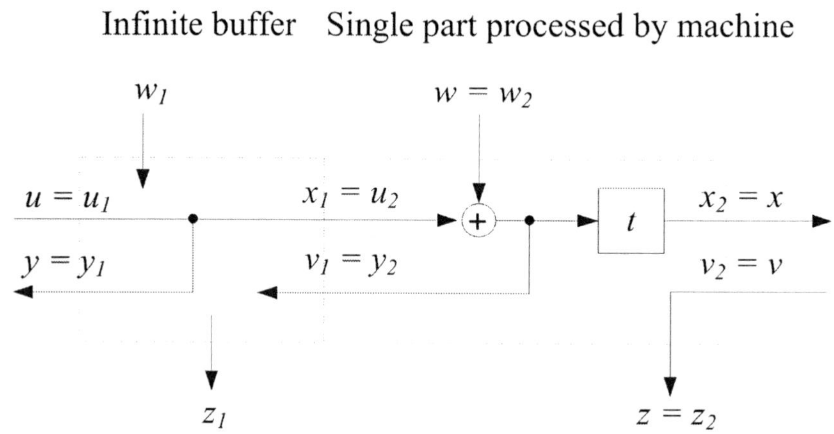

Figure 13. Machine with an infinite buffer in front of it processing a part (Reduced block diagram)

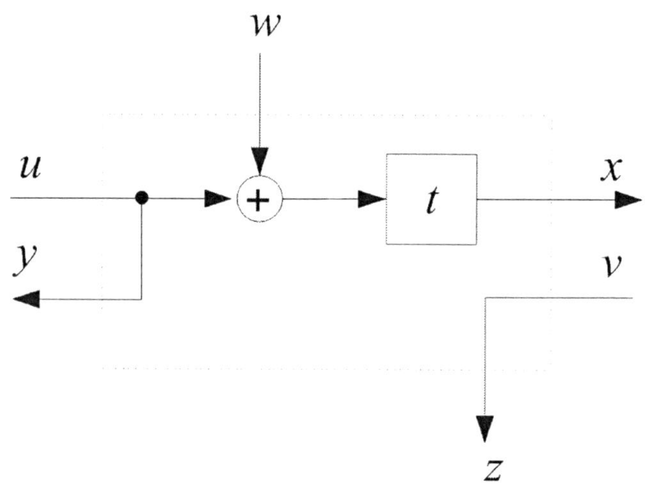

JOB SHOP EXAMPLE

Consider the job shop system shown in Figure 14 with 3 machines, m_1, m_2 and m_3, and 3 parts, n_1, n_2 and n_3. The order in which part n_i, where $i \in \{1,2,3\}$, is processed on machines is given by $\mathbf{m}_{n_i}$. The order of parts processed on machine m_i, where $i \in \{1,2,3\}$, is given by $\mathbf{n}_{m_i}$. In this example we have

$$\mathbf{n}_{m_1} = \begin{bmatrix} n_1 & n_2 \end{bmatrix},\ \mathbf{n}_{m_2} = \begin{bmatrix} n_1 & n_3 \end{bmatrix},\ \mathbf{n}_{m_3} = \begin{bmatrix} n_3 & n_2 \end{bmatrix}$$
$$\mathbf{m}_{n_1} = \begin{bmatrix} m_1 & m_2 \end{bmatrix},\ \mathbf{m}_{n_2} = \begin{bmatrix} m_1 & m_3 \end{bmatrix},\ \mathbf{m}_{n_3} = \begin{bmatrix} m_2 & m_3 \end{bmatrix}.$$

The configuration of the job shop is graphically illustrated in a routing diagram in Figure 14. In the diagram, horizontal connections describe routes of parts through the machines. For example, it can be seen that part n_1 is first processed by m_1 and then by m_2 and therefore $\mathbf{m}_{n_1} = \begin{bmatrix} m_1 & m_2 \end{bmatrix}$.

Figure 14. Interconnection diagram for job shop example

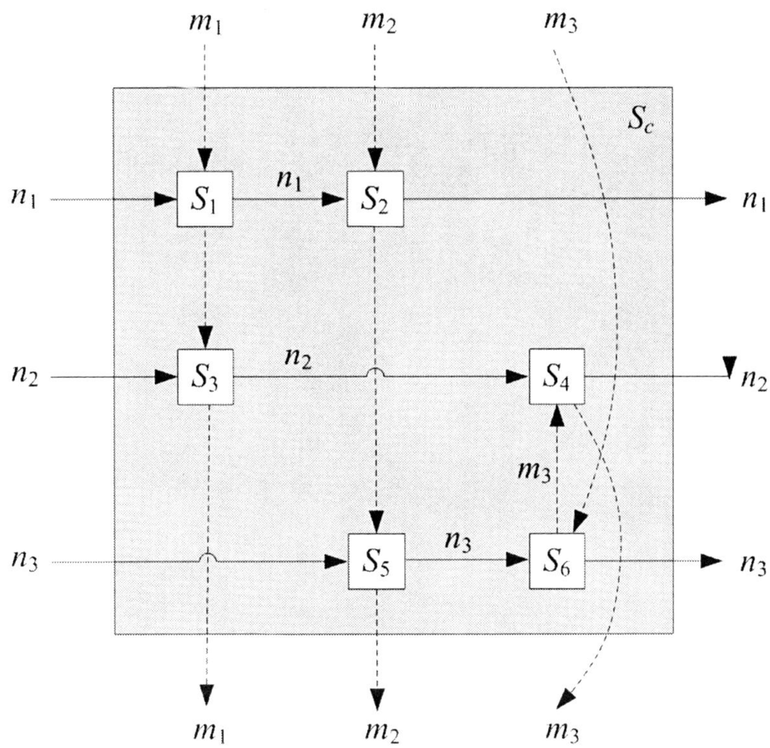

Vertical connections describe order in which parts are processed on machines. For example, it can be seen that order of parts on machine m_2 is n_1 followed by n_3, hence $\mathbf{n}_{m_2} = \begin{bmatrix} n_1 & n_3 \end{bmatrix}$. Each node S_i for $i \in \{1,2,\ldots,6\}$ represents an operation. For example, S_3 models an operation corresponding to m_1 processing n_2. It is assumed that the machines are never blocked, i.e. there is sufficient buffer storage in front of every machine. Therefore S_i is modeled by Equation (26).

Let t_i is processing time for operation S_i, where $t_1=3$, $t_2=2$, $t_3=2$, $t_4=3$, $t_5=2$, and $t_6=1$. A detailed block diagram of the system is shown in Figure 15.

We have

$$\begin{aligned} S_1 &= (\mathbf{F}_1, m_1, n_1, n_1), & S_2 &= (\mathbf{F}_2, m_2, n_1, n_1), \\ S_3 &= (\mathbf{F}_3, m_1, n_2, n_2), & S_4 &= (\mathbf{F}_4, m_3, n_2, n_2), \\ S_5 &= (\mathbf{F}_5, m_2, n_3, n_3), & S_6 &= (\mathbf{F}_6, m_3, n_3, n_3). \end{aligned}$$

It is assumed that the system is not blocked from the outside, i.e. $\mathbf{v}_c=\mathbf{x}_c$. The whole system, i.e., the job shop, is modeled by

$$S_c = (\mathbf{F}_c^{nb}, \mathbf{m}_c, \mathbf{n}_c^{in}, \mathbf{n}_c^{out}) = \left(\mathbf{F}_c^{nb}, \begin{bmatrix} m_1 \\ m_2 \\ m_3 \end{bmatrix}, \begin{bmatrix} n_1 \\ n_2 \\ n_3 \end{bmatrix}, \begin{bmatrix} n_1 \\ n_2 \\ n_3 \end{bmatrix}\right),$$

where $\mathbf{F}^{nb}$ (*nb* in $\mathbf{F}^{nb}$ stands for "non-blocking system"') describes the following relation

$$\begin{bmatrix} \mathbf{x}_c \\ \mathbf{y}_c \\ \mathbf{z}_c \end{bmatrix} = \mathbf{F}^{nb} \begin{bmatrix} \mathbf{u}_c \\ \mathbf{w}_c \end{bmatrix}.$$

The goal is to find $\mathbf{F}^{nb}$.

Figure 15. Detailed block diagram of the job shop

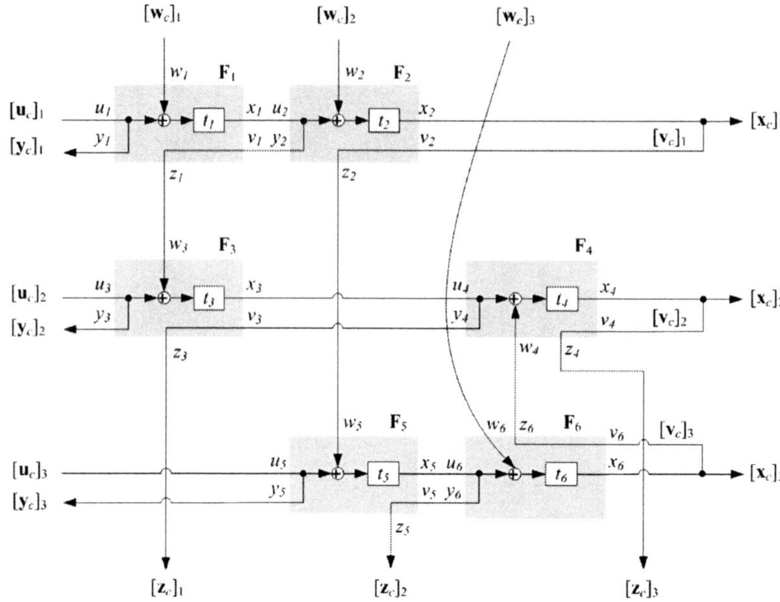

Most of the matrices derived in this section are sparse, i.e. they contain many elements that equal ε. Therefore, for simplicity, ε's in all the matrices presented in this section are replaced by dots. We have

$$\tilde{\mathbf{m}} = \begin{bmatrix} m_1 \\ m_2 \\ m_1 \\ m_3 \\ m_2 \\ m_3 \end{bmatrix}, \quad \tilde{\mathbf{n}}^{in} = \tilde{\mathbf{n}}^{out} = \begin{bmatrix} n_1 \\ n_1 \\ n_2 \\ n_2 \\ n_3 \\ n_3 \end{bmatrix},$$

$$\mathbf{m}_c = \begin{bmatrix} m_1 \\ m_2 \\ m_3 \end{bmatrix}, \quad \mathbf{n}_c^{out} = \mathbf{n}_c^{in} = \begin{bmatrix} n_1 \\ n_2 \\ n_3 \end{bmatrix}.$$

Part-flow interconnection matrices are given by

$$\mathbf{R}_{in} = \begin{bmatrix} e & . & . \\ . & . & . \\ . & e & . \\ . & . & . \\ . & . & e \\ . & . & . \end{bmatrix} \quad \mathbf{R} = \begin{bmatrix} . & . & . & . & . & . \\ e & . & . & . & . & . \\ . & . & . & . & . & . \\ . & . & e & . & . & . \\ . & . & . & . & . & . \\ . & . & . & . & e & . \end{bmatrix}$$

$$\mathbf{R}_{out} = \begin{bmatrix} . & e & . & . & . & . \\ . & . & . & e & . & . \\ . & . & . & . & . & e \end{bmatrix}$$

Resource-flow interconnection matrices are

$$\mathbf{Q}_{in} = \begin{bmatrix} e & . & . \\ . & e & . \\ . & . & . \\ . & . & . \\ . & . & . \\ . & . & e \end{bmatrix} \qquad \mathbf{Q} = \begin{bmatrix} . & . & . & . & . & . \\ . & . & . & . & . & . \\ e & . & . & . & . & . \\ . & . & . & . & . & e \\ . & e & . & . & . & . \\ . & . & . & . & . & . \end{bmatrix}$$

$$\mathbf{Q}_{out} = \begin{bmatrix} . & . & e & . & . & . \\ . & . & . & . & e & . \\ . & . & . & e & . & . \end{bmatrix}.$$

We have

$$\tilde{\mathbf{F}} = \begin{bmatrix} \tilde{\mathbf{F}}_{xu} & \tilde{\mathbf{F}}_{xv} & \tilde{\mathbf{F}}_{xw} \\ \tilde{\mathbf{F}}_{yu} & \tilde{\mathbf{F}}_{yv} & \tilde{\mathbf{F}}_{yw} \\ \tilde{\mathbf{F}}_{zu} & \tilde{\mathbf{F}}_{zv} & \tilde{\mathbf{F}}_{zw} \end{bmatrix} = \begin{bmatrix} \mathbf{P} & . & \mathbf{P} \\ \mathbf{E} & . & . \\ . & \mathbf{E} & . \end{bmatrix},$$

where

$$\mathbf{P} = \begin{bmatrix} t_1 & . & . & . & . & . \\ . & t_2 & . & . & . & . \\ . & . & t_3 & . & . & . \\ . & . & . & t_4 & . & . \\ . & . & . & . & t_5 & . \\ . & . & . & . & . & t_6 \end{bmatrix} = \begin{bmatrix} 3 & . & . & . & . & . \\ . & 2 & . & . & . & . \\ . & . & 2 & . & . & . \\ . & . & . & 3 & . & . \\ . & . & . & . & 2 & . \\ . & . & . & . & . & 1 \end{bmatrix} \tag{28}$$

Assume that the system is not blocked from the outside, i.e. $\mathbf{v}_c = \mathbf{x}_c$. Plugging in the above values into (20) we obtain

$$\begin{bmatrix} \mathbf{x}_c \\ \mathbf{y}_c \\ \mathbf{z}_c \end{bmatrix} = \begin{bmatrix} \begin{bmatrix} 5 & . & . \\ 11 & 5 & 6 \\ 8 & . & 3 \end{bmatrix} & \begin{bmatrix} 5 & 2 & . \\ 11 & 8 & 4 \\ 8 & 5 & 1 \end{bmatrix} \\ \begin{bmatrix} e & . & . \\ . & e & . \\ . & . & e \end{bmatrix} & \begin{bmatrix} . & . & . \\ . & . & . \\ . & . & . \end{bmatrix} \\ \begin{bmatrix} 5 & 2 & . \\ 7 & . & 2 \\ 11 & 5 & 6 \end{bmatrix} & \begin{bmatrix} 5 & . & . \\ 7 & 4 & . \\ 11 & 8 & 4 \end{bmatrix} \end{bmatrix} \begin{bmatrix} \mathbf{u}_c \\ \mathbf{w}_c \end{bmatrix}.$$

Assume that the parts and machines are available at time =0, then $\mathbf{u}_c = [e\ e\ e]^T$ and $\mathbf{w}_c = [e\ e\ e]^T$. Then it follows that

$$\begin{bmatrix} \mathbf{x}_c \\ \mathbf{y}_c \\ \mathbf{z}_c \end{bmatrix} = \begin{bmatrix} \begin{bmatrix} 5 \\ 11 \\ 8 \end{bmatrix} \\ \begin{bmatrix} e \\ e \\ e \end{bmatrix} \\ \begin{bmatrix} 5 \\ 7 \\ 11 \end{bmatrix} \end{bmatrix}.$$

The make span of the system is the latest time when a part leaves the system and it is equal to 11 time units.

MODELING MANUFACTURING LAYOUTS

Consider again the example presented in Figure 14. In the example S_c was obtained by composing basic blocks S_1 through S_6 in one step. There are, however, alternative methods to obtain S_c, such as *machine-based* and *part-based* modeling methods. These methods are based on dividing the system into components and then composing these components to obtain S_c. The part-based modeling approach is illustrated in Figure 16 and 17. The approach consists of two steps. In the first step the model for processing each part is derived, i.e. the models of S_{n1}, S_{n2} and S_{n3} as shown in Figure 16. Then, in the second step, S_c is obtained by "vertically" (i.e., using machine-flow interconnections) composing S_{n1}, $S_{n2}$and S_{n3}as illustrated in Figure 17.

The machine-based modeling approach is illustrated in Figure 18 and 19. In this approach, we first obtain models of each machine, S_{m1}, S_{m2} and S_{m3}, as shown in Figure 18. Then S_{m1}, S_{m2} and S_{m3} are composed "horizontally" (i.e., using part-flow interconnections) to produce S_c as shown in Figure 19. Numerical examples for part-based and machine-based modeling approaches can be found in (Imaev 2009).

A similar approach applies to cellular manufacturing systems. For example, consider a manufacturing configuration shown in Figure 20, where there are two manufacturing cells. In Cell A machines m_1 and m_2 are grouped together to perform elementary operations S_1, S_2, S_3 and S_4. In Cell B machines m_3, m_4 and m_5 are grouped together to perform elementary operations S_5, S_6, S_7 and S_8. The modeling approach for cellular manufacturing systems consists of two steps. In the first step the models of each manufacturing

Figure 16. Part base approach: obtain models for processing each part, S_{n1}, S_{n2} and S_{n3}

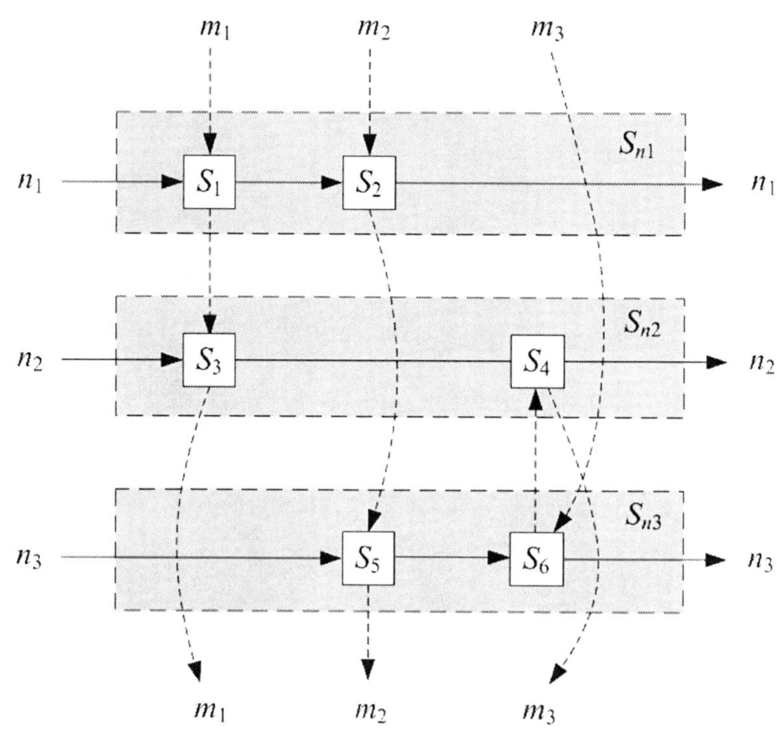

Figure 17. Part base approach: combining the part models S_{n1}, S_{n2} and S_{n3} into S_c

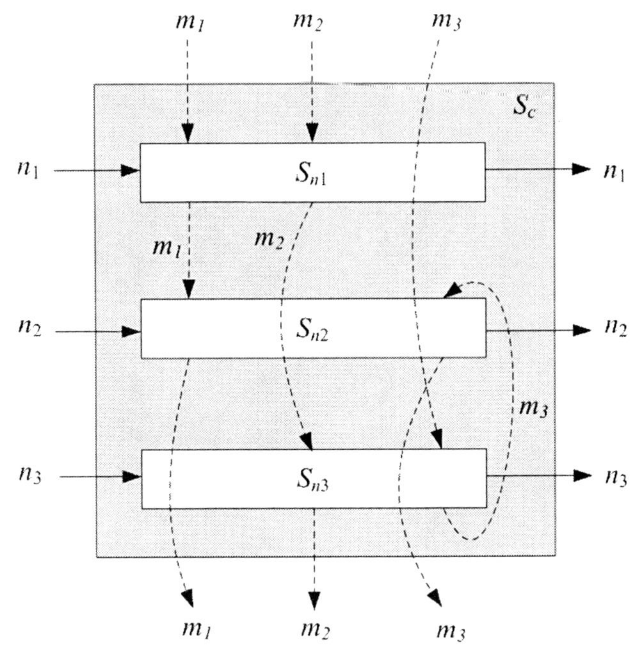

Figure 18. Machine base approach: obtain models for each machine, S_{m1}, S_{m2} and S_{m3}

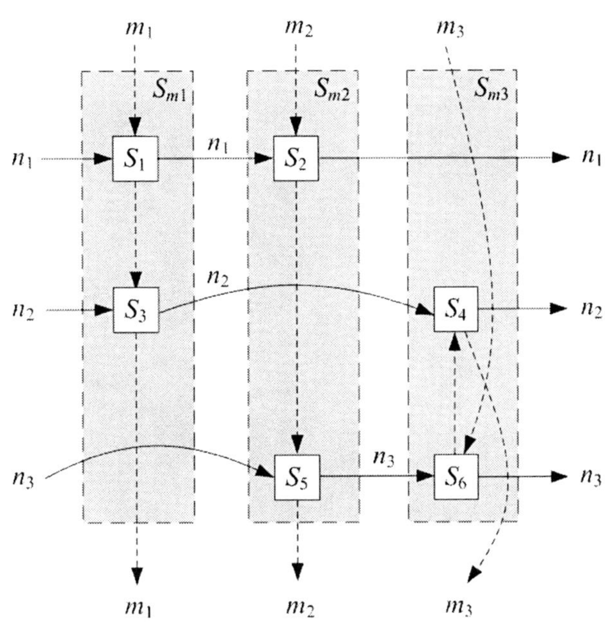

Figure 19. Machine base approach: combine the machine models S_{m1}, S_{m2} and S_{m3} into S_c

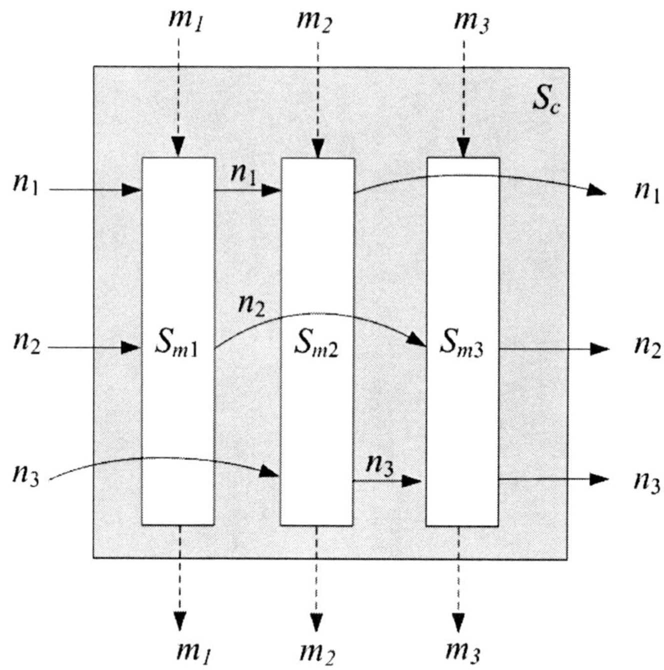

Figure 20. Modeling cellular manufacturing system

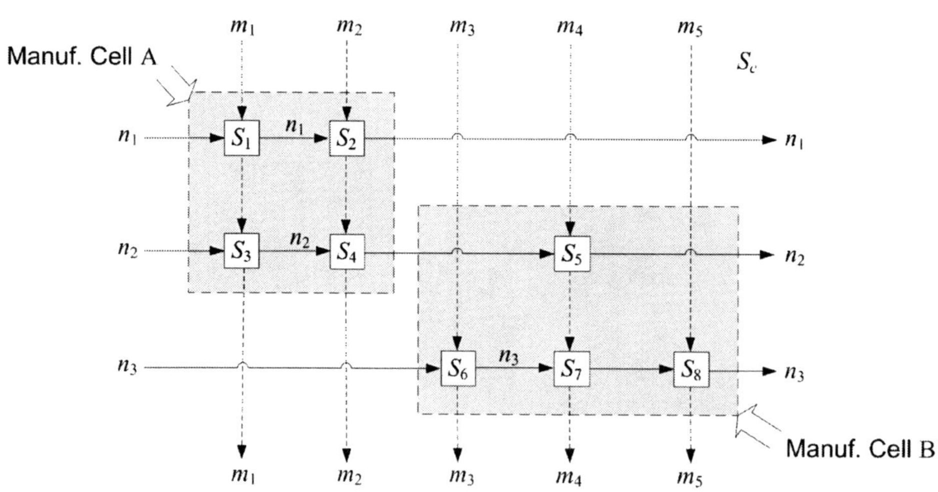

cell are derived. Scheduling heuristics may be used to optimize the performance of an individual cell. In the second step, models for Cell A and Cell B are composed to obtain the model of S_c.

Flow-Shop Scheduling

The scheduling problem for a flow shop system of *m* machines and *n* parts can best be modeled using the part based system with unit capacity machines, Equation (21). Figure 21 illustrates the block diagram of the processing of the *i*-th part. Examination of the Figure 21, shows that the part process can be modeled by the following equation

$$\mathbf{z}_i = \mathbf{A}_i \mathbf{z}_i \oplus \mathbf{B}_i \mathbf{w}_i,\ i=1,\ldots,n \tag{29}$$

where

$$\mathbf{A}_i = \begin{bmatrix} \varepsilon & \varepsilon & \cdots & \varepsilon & \varepsilon \\ [\mathbf{t}_i]_2 & \varepsilon & \cdots & \varepsilon & \varepsilon \\ \varepsilon & [\mathbf{t}_i]_3 & \cdots & \varepsilon & \varepsilon \\ \vdots & \vdots & \ddots & \vdots & \vdots \\ \varepsilon & \varepsilon & \cdots & [\mathbf{t}_i]_m & \varepsilon \end{bmatrix}$$

and

$$\mathbf{B}_i = \begin{bmatrix} [\mathbf{t}_i]_1 & e & \varepsilon & \cdots & \varepsilon \\ \varepsilon & \varepsilon & e & \cdots & \varepsilon \\ \vdots & \vdots & \vdots & \ddots & \vdots \\ \varepsilon & \varepsilon & \cdots & \varepsilon & e \\ \varepsilon & \varepsilon & \cdots & \varepsilon & \varepsilon \end{bmatrix}.$$

Equation (29) can be reduced (Bacelli et al, 1989) to

$$\mathbf{Z}_i = \mathbf{F}_i \mathbf{w}_i,\ i=1,\ldots,n, j=1,\ldots,n,\ i \neq j \tag{30}$$

where

$$\mathbf{F}_i = \mathbf{A}_i^* \mathbf{B}_i.$$

Since this is a flow-shop, all parts are processed by the machines in the same order, hence

$$\mathbf{w}_{i+1} = \mathbf{z}_i,\ i=1,2,\ldots,n\text{-}1. \tag{31}$$

Combining (30) and (31) for all n parts, yields

$$\mathbf{z}_n = \bigotimes_{i=1}^{n} \mathbf{F}_i \mathbf{w}_1 = \mathbf{F}\mathbf{w}_1 \tag{32}$$

The make span is then the maximum value of **F**. Notice that the $\mathbf{F}_i$ matrices only hold processing information and do not depend on the schedule. Flow shop scheduling is then reduced simply rearranging the order of the part models in Figure 17, which translates to rearranging the order that the $\mathbf{F}_i$ are multiplied in (32).

Buffer Allocation

In a system without buffers, a machine is considered to be blocked if it has completed the current part and is unable to off-load the part since the subsequent buffer is full. The machine is considered to be starved if it has completed the current part and is unable to begin the next part since the

Figure 21. Model of the i-th part process for a flow-shop consisting of single capacity machines

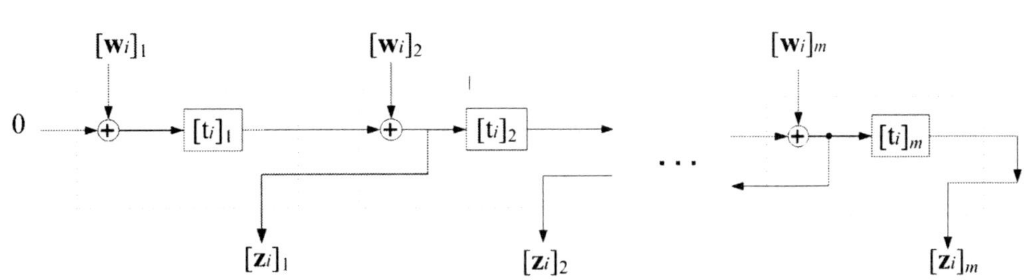

Figure 22. Operation inter-dependencies

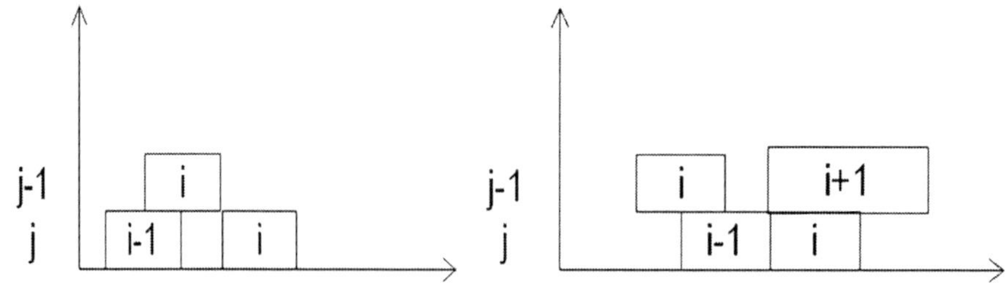

preceding buffer is empty. This is illustrated in Figure 22, left and right respectively.

In the figure on the left, machine *j* is idle for want of parts from machine *j-1*. In the figure on the right, machine *j-1* is idle because it cannot off-load its part to the next machine. Ideally, all the idle time can be eliminated by having a buffer of infinite capacity after each machine. This would keep the machines running at all times. However, it is important to bear in mind that buffers increase the work in progress and this results in increased inventory-carrying costs. Hence, it becomes imperative to allocate buffers to the best possible locations such that the overall buffer size is kept to a minimum and the throughput is increased. A heuristic algorithm has been developed (Nambiar & Judd, 2007) based on the aforementioned mathematical model to analyze the impact of having buffers on the system performance and to determine the location and size of buffers to maximize system throughput. This heuristic used the inter-dependencies between operations which result in machines getting either blocked or starved. This heuristic identified near-optimal locations for the limited number of buffer spaces available.

Stochastic Scheduling

Schedules can go awry due to unforeseen circumstances or due to the inherent variation in the processes involved. These variations have been modeled as probabilistic distributions in literature. The aforementioned model can also be used to analyze the behavior of the system with uncertain processing times. The ability to evaluate numerous instances of the problem with minimal computation time bodes well for this kind of analysis. Comparing the average cycle time obtained by evaluating every possible combination of processing times with the cycle time obtained by using the average processing time for each operation, it was observed that the cycle time obtained using average processing times was within a range of 5-10% of the average cycle times. Thus, if computation time is of essence and the accuracies of the cycle time estimates are not that significant, it might be best to use the average processing times even in cases of stochastic variability. However, if an accurate estimate of the cycle time is desired, the aforementioned model can be used to obtain the average cycle time computed through exhaustive enumeration of all possible combinations.

Assembly Line Balancing

We now consider the mixed-model assembly line balancing problem. Here, we consider completely independent models being assembled on the same assembly line. The aforementioned mathematical model can be modified (Carlo & Nambiar, 2008) to address this problem. We consider a system of *n* models and *m* workstations. We assume that the processing times or task times for each operation are given to us. These tasks can be broken down into individual units which may be then re-assigned to neighboring machines in order to balance the line. We do not consider parallel workstations or intermediate buffers in our system. We iteratively examine every possible task assignment for a given problem to determine the best solution. In order to make the computations easier and smarter, we also incorporate some efficiency such as storing intermediate matrices. We were able to exhaustively search through all possible task combinations for a system of 10 models and 10 workstations with a 15% time available for re-allocation within 89 seconds. This allows us to determine optimal task assignments for mixed-model assembly lines within reasonable amount of time.

CONCLUSION AND FUTURE RESEARCH DIRECTIONS

In today's global and interconnected world, companies are facing stiff competition from not only local competitors but also from international companies who have advantages such as cheaper labor that work in their favor. Under these circumstances, it becomes imperative for companies to continuously strive to improve their operations by eliminating all forms of wastes and introducing newer products that meet customer demands. Efficient scheduling of limited resources helps improve efficiency and reduce lead times. This becomes especially critical when companies have factories located far removed from the customer market. Efficient schedules also help effectively utilize limited resources and help with sustainability efforts. In this chapter, we looked at various mathematical models for manufacturing systems such as flow-shops, job-shops and assembly lines. These models are based on the concept of max-plus algebra which was also discussed in this chapter. The most significant advantage of these max-plus algebra-based models is that this allows us to identify optimal or near-optimal schedules with only a fraction of the computation time it would take to exhaustively explore all possible schedules. These models can also be applied to other discrete-event systems exhibiting similar behavior.

REFERENCES

Addad, B., & Amari, S. (2008). Modeling and response time evaluation of ethernet-based control architectures using timed event graphs and max-plus algebra. In *Proceedings of the 2008 IEEE International Conference on Automation Science and Engineering* (pp. 418–423). Arlington, VA.

Akyol, D. E. (2004). Application of neural networks to heuristic scheduling algorithms. *Computers & Industrial Engineering*, *46*, 679–696. doi:10.1016/j.cie.2004.05.005

Anderson, D. M. (1997). *Agile product development for mass customization*. Chicago, IL: Irwin Professional Pub.

Bacelli, F., Cohen, G., Olsder, G., & Quadrat, J.-P. (1992). *Synchronization and linearity: An algebra for discrete event systems*. Chichester, UK: John Wiley and Sons Inc.

Baker, K. R. (1974). *Introduction to sequencing and scheduling*. John Wiley.

Balasubramanian, J., & Grossmann, I. (2002). A novel branch and bound algorithm for scheduling flowshop plants with uncertain processing times. *Computers & Chemical Engineering*, *26*(1), 41–57. doi:10.1016/S0098-1354(01)00735-9

Battini, D., Faccio, M., Persona, A., & Sgarbossa, F. (2009). Balancing-sequencing procedure for a mixed model assembly system in case of finite buffer capacity. *International Journal of Advanced Manufacturing Technology*, *44*(3-4), 345–359. doi:10.1007/s00170-008-1823-8

Baybars, I. (1986). A survey of exact algorithms for the simple assembly line balancing problem. *Management Science, 32*(8), 909–932. doi:10.1287/mnsc.32.8.909

Becker, C., & Scholl, A. (2006). A survey on problems and methods in generalized assembly line balancing. *European Journal of Operational Research, 168*, 694–715. doi:10.1016/j.ejor.2004.07.023

Becker, C., & Scholl, A. (2009). Balancing assembly lines with variable parallel workplaces: Problem definition and effective solution procedure. *European Journal of Operational Research, 199*(2), 359–374. doi:10.1016/j.ejor.2008.11.051

Boysen, N., Fliedner, M., & Scholl, A. (2006). *Assembly line balancing: Which model to use when?* Jena Research Papers in Business and Economics. ISSN 1611-1311

Boysen, N., Fliedner, M., & Scholl, A. (2007). *Sequencing mixed-model assembly lines: Survey, classification and model critique*. Jena Research Papers in Business and Economics. ISSN 1864-3108

Boysen, N., Fliedner, M., & Scholl, A. (2009a). The product rate variation problem and its relevance in real world mixed-model assembly lines. *European Journal of Operational Research, 197*(2), 818–824. doi:10.1016/j.ejor.2008.06.038

Boysen, N., Fliedner, M., & Scholl, A. (2009b). Sequencing mixed-model assembly lines: Survey, classification and model critique. *European Journal of Operational Research, 192*, 349–373. doi:10.1016/j.ejor.2007.09.013

Bukchin, Y., & Rabinowitch, I. (2006). A branch-and-bound based solution approach for the mixed-model assembly line-balancing problem for minimizing stations and task duplication costs. *European Journal of Operational Research, 174*, 492–508. doi:10.1016/j.ejor.2005.01.055

Cao, Z., & Ma, S. (2008). Balancing and sequencing optimization of the mixed model assembly lines. In *Proceedings -International Symposium on computer science and computational technology,* ISCSCT 2008 (Vol. 1, p. 732 - 736).

Carlier, J., & Rebai, I. (1996). Two branch and bound algorithms for the permutation flow shop problem. *European Journal of Operational Research, 90*, 238–251. doi:10.1016/0377-2217(95)00352-5

Carlo, H. J., & Nambiar, A. N. (2008). A max-plus based algorithm for assembly line balancing problem. In J. Fowler & S. Mason (Eds.), Proceedings of the 17th Industrial Engineering Research Conference IERC (pp. 35–39). Vancouver, Canada, May 18-21, 2008.

Cevikcan, E., Durmusoglu, M. B., & Unal, M. E. (2009). A team-oriented design methodology for mixed model assembly systems. *Computers & Industrial Engineering, 56*(2), 576–599. doi:10.1016/j.cie.2007.11.002

Chakravarthy, K., & Rajendran, C. (1999). A heuristic for scheduling in a flow-shop with the bi-criteria of makespan and maximum tardiness minimization. *Production Planning and Control, 10*, 707–714. doi:10.1080/095372899232777

Chauvet, F., Herrmann, J. W., & Proth, J.-M. (2003). Optimization of cyclic production systems: A heuristic approach. *IEEE Transactions on Robotics and Automation, 19*(1), 150–154. doi:10.1109/TRA.2002.807529

Chen, C., Vempati, V. S., & Aljaber, N. (1995). An application of genetic algorithms for flow-shop problems. *European Journal of Operational Research, 80*, 389–396. doi:10.1016/0377-2217(93)E0228-P

Choi, G. (2009). A goal programming mixed-model line balancing for processing time and physical workload. *Computers & Industrial Engineering, 57*(1), 395–400. doi:10.1016/j.cie.2009.01.001

De Schutter, B., & van den Boom, T. (2000, August). Model predictive control for max-plus-linear systems. In *Proceedings of the 2000 American Control Conference* (pp. 4046–4050). Chicago, IL.

Dhouib, K., Gharbi, A., & Landolsi, N. (2009). Throughput assessment of mixed-model flexible transfer lines with unreliable machines. *International Journal of Production Economics*, *122*(2), 619–627. doi:10.1016/j.ijpe.2009.06.016

El-Bouri, A., Balakrishnan, S., & Popplewell, N. (2005). A neural network to enhance local search in the permutation flow shop. *Computers & Industrial Engineering*, *49*, 182–196. doi:10.1016/j.cie.2005.04.001

Elmahi, I., Grunder, O., & Elmoudni, A. (2004, December). A max plus algebra approach for modeling and control of lots delivery: application to a supply chain case study. In *Proceedings of the 2004 IEEE International Conference on Industrial Technology* (pp. 926–931).

Fattahi, P., & Salehi, M. (2009). Sequencing the mixed-model assembly line to minimize the total utility and idle costs with variable launching interval. *International Journal of Advanced Manufacturing Technology*, *45*(9-10), 987–998. doi:10.1007/s00170-009-2020-0

Ford, H., & Crowther, S. (1922). *My life and work*. Doubleday.

Gang, C., & Zuo-Jun, M. S. (2007). Probabilistic asymptotic analysis of stochastic online scheduling problems. *IIE Transactions*, *39*(5), 525–538. doi:10.1080/07408170600941623

Ghosh, S., & Gagnon, R. J. (1989). A comprehensive literature review and analysis of the design, balancing and scheduling of assembly systems. *International Journal of Production Research*, *27*(4), 637–670. doi:10.1080/00207548908942574

Goldberg, D. E. (1989). *Genetic algorithms in search, optimization and machine learning*. Reading, MA: Addison-Wesley.

Gorji, A., Doustmohammadi, A., Menhaj, M., & Mohajerin, N. (2007). Modeling and control of flow shop production systems by fuzzy max-plus algebra. In *Proceedings of the 2007 IEEE International Conference on Control and Automation* (pp. 3065–3072).

Hall, L. A. (1998). Approximability of flow shop scheduling. *Mathematical Programming*, *82*, 175–190. doi:10.1007/BF01585870

Heidergott, B. (2006). *Max-plus linear stochastic systems and perturbation analysis*. Springer-Verlag. doi:10.1007/978-0-387-38995-0

Heidergott, B., Jan Olsder, G., & van der Woude, J. (2006a). *Max plus at work - modeling and analysis of synchronized systems: A course on max-plus algebra and its applications*. Princeton University Press.

Heidergott, B., Jan Olsder, G., & van der Woude, J. (2006b). *Max plus at work - Modeling and analysis of synchronized systems: A course on max-plus algebra and its applications*. Princeton University Press.

Hiroyuki, K., Hitoi, Y., & Nakao, Y. (2009, August). On scheduling a shipbuilding line based on max-plus system dynamic representation. In *Proceedings of the International Joint Conference of ICCAS-SICE* (pp. 1738–1741). Japan.

Honkomp, S. J., Mockus, L., & Reklaitis, G. V. (1999). A framework for schedule evaluation with processing uncertainty. *Computers & Chemical Engineering*, *23*(4), 595–609. doi:10.1016/S0098-1354(98)00296-8

Imaev, A. (2009). *Hierarchical modeling of manufacturing systems using max-plus algebra*. Ph.D. Dissertation, Ohio University 2010.

Imaev, A., & Judd, R. P. (2008). Hierarchical modeling of manufacturing systems using max-plus algebra. In *Proceedings of American Control Conference*, June 11 - 13, 2008.

Ishibuchi, H., Misaki, S., & Tanaka, H. (1995). Modified simulated annealing algorithms for the flow shop sequencing problem. *European Journal of Operational Research*, *81*, 388–398. doi:10.1016/0377-2217(93)E0235-P

Karabati, S., & Tan, B. (1998). Stochastic cyclic scheduling problem in synchronous assembly and production lines. *The Journal of the Operational Research Society*, *49*(11), 1173–1187.

Khoo, L. P., & Loi, M. Y. (2002). A tabu-enhanced genetic algorithm approach to agile manufacturing. *International Journal of Advanced Manufacturing Technology*, *20*, 692–700. doi:10.1007/s001700200208

Kouvelis, P., Daniels, R. L., & Vairaktarakis, G. (2000). Robust scheduling of a two-machine flow shop with uncertain processing times. *IIE Transactions*, *32*(5), 421–432. doi:10.1080/07408170008963918

Ladhari, T., & Haouari, M. (2005). A computational study of the permutation flow shop problem based on a tight lower bound. *Computers & Operations Research*, *32*, 1831–1847. doi:10.1016/j.cor.2003.12.001

Lee, Y.-D., & Lee, T.-E. (2005). Stochastic cyclic flow lines with blocking: Markovian models. *OR-Spektrum*, *27*(4), 551–568. doi:10.1007/s00291-005-0201-2

Lopes, G., Babuska, R., De Schutter, B., & van den Boom, A. (2009, December). Switching max-plus models for legged locomotion. In *Proceedings of the 2009 IEEE International Conference on Robotics and Biomimetics* (ROBIO) (pp. 221–226). Guilin.

Maniezzo, V., & Carbonaro, A. (2001). Ant colony optimization: an overview. In Ribeiro, C. C., & Hansen, P. (Eds.), *Essays and surveys in metaheuristics* (pp. 21–44). Kluwer.

Matsuo, H. (1990). Cyclic sequencing problems in the two-machine permutation flow shop: Complexity, worst-case, and average-case analysis. *Naval Research Logistics*, *37*, 679–690. doi:10.1002/1520-6750(199010)37:5<679::AID-NAV3220370507>3.0.CO;2-Q

McCormick, S. T., Pinedo, M., Shenker, S., & Wolf, B. (1989b). Sequencing in an assembly line with blocking to minimize cycle time. *Operations Research*, *37*(6), 925–935. doi:10.1287/opre.37.6.925

McCormick, S. T., Pinedo, M. L., Shenker, S., & Wolf, B. (1989a). Sequencing in an assembly line with blocking to minimize cycle time. *Operations Research*, *37*, 925–935. doi:10.1287/opre.37.6.925

McMullen, P. R., & Tarasewich, P. (2003). Using ant techniques to solve the assembly line balancing problem. *IIE Transactions*, *35*, 605–617. doi:10.1080/07408170304354

Mirzapour Al-E-Hashem. S., Aryanezhad, M., Malekly, H., & Sadjadi, S. (2009). Mixed model assembly line balancing problem under uncertainty. In *2009 International Conference on Computers and Industrial Engineering* (pp. 233-238).

Morton, T. E., & Pentico, D. W. (1993). *Heuristic scheduling systems with applications to production systems and project management*. John Wiley & Sons, Inc.

Nait-Sidi-Moh, A., Manier, M.-A., El Moudni, A., & Manier, H. (2002, October). A (max, plus) modelling approach for the evaluation of travelling times in a public transportation system. In *Proceedings of the 2002 IEEE International Conference on Systems, Man and Cybernetics*.

Nambiar, A. N., & Judd, R. (2010). Max-plus based mathematical formulation for cyclic permutation flow-shops. *International Journal Mathematical Modeling and Numerical Optimization*, *2*(1), 85–97. doi:10.1504/IJMMNO.2011.037201

Nambiar, A. N., & Judd, R. P. (2007). Scheduling and buffer allocation for cyclic permutation flow-shops using max-plus based formulation. In G. Bayraksan, W. Lin, Y. Son, & R. Wysk (Eds.), *Proceedings of the 2007 Industrial Engineering Research Conference* (pp. 1359–1363).

Nejad, B., Attia, S., & Raisch, J. (2009, October). Max-consensus in a max-plus algebraic setting: The case of fixed communication topologies. In *Proceedings of the 2009 International Symposium on Information, Communication and Automation Technologies* (pp. 1–7). Bosnia.

Ogbu, F. A., & Smith, D. K. (1990). The application of the simulated annealing algorithm to the solution of the n/m/cmax flowshop problem. *Computers & Operations Research*, *17*(3), 243–253. doi:10.1016/0305-0548(90)90001-N

Ogbu, F. A., & Smith, D. K. (1991). Simulated annealing for the permutation flowshop scheduling. *Omega*, *19*(1), 64–67. doi:10.1016/0305-0483(91)90036-S

Ohno, T. (1988). *Toyota production system: Beyond large-scale production*. Portand, OR: Productivity Press.

Osman, I. H., & Potts, C. N. (1989). Simulated annealing for permutation flow-shop scheduling. *Omega*, *17*(6), 551–557. doi:10.1016/0305-0483(89)90059-5

Ozcan, U. (2010). Balancing stochastic two-sided assembly lines: A chance-constrained, piecewise-linear, mixed integer program and a simulated annealing algorithm. *European Journal of Operational Research*, *205*(1), 81–97. doi:10.1016/j.ejor.2009.11.033

Ozcan, U., & Toklu, B. (2009a). Balancing of mixed-model two-sided assembly lines. *Computers & Industrial Engineering*, *57*(1), 217–227. doi:10.1016/j.cie.2008.11.012

Ozcan, U., & Toklu, B. (2009b). Multiple-criteria decision-making in two-sided assembly line balancing: A goal programming and a fuzzy goal programming models. *Computers & Operations Research*, *36*(6), 1955–1965. doi:10.1016/j.cor.2008.06.009

Pine, J. B. II. (1993). *Mass customization - The new frontier in business competition*. Boston, MA: Harvard Business School Press.

Pinedo, M. (1995). *Scheduling: Theory, algorithm, and system*. Englewood Cliffs, NJ: Prentice-Hall.

Pinedo, M. (2007). Stochastic batch scheduling and the smallest variance first rule. *Probability in the Engineering and Informational Sciences*, *21*(4), 579–595. doi:10.1017/S0269964807000332

Ponnambalam, S. G., Aravindan, P., & Naidu, G. M. (2000). A multi-objective genetic algorithm for solving assembly line balancing problem. *International Journal of Advanced Manufacturing Technology*, *16*, 341–352. doi:10.1007/s001700050166

Rajendran, C., & Chaudhuri, D. (1992). An efficient heuristic approach to the scheduling of jobs in a flow-shop. *European Journal of Operational Research*, *61*, 318–325. doi:10.1016/0377-2217(92)90361-C

Rajendran, C., & Ziegler, H. (2004). Ant-colony algorithms for permutation flow-shop scheduling to minimize makespan/total flow time of jobs. *European Journal of Operational Research*, *155*, 426–438. doi:10.1016/S0377-2217(02)00908-6

Rajendran, C., & Ziegler, H. (2005). Two ant-colony algorithms for minimizing total flow time in permutation flow shops. *Computers & Industrial Engineering*, *48*, 789–797. doi:10.1016/j.cie.2004.12.009

Rao, U. S., & Jackson, P. L. (1999). Estimating performance measures in repetitive manufacturing environments via stochastic cyclic scheduling. *IIE Transactions*, *28*(11), 929–939.

Ravindran, D., Noorul Haq, A., Selvakumar, S. J., & Sivaraman, R. (2005). Flow shop scheduling with multiple objective of minimizing makespan and total flow time. *International Journal of Advanced Manufacturing Technology*, *25*, 1007–1012. doi:10.1007/s00170-003-1926-1

Roundy, R. (1992). Cyclic schedules for job-shops with identical jobs. *Mathematics of Operations Research*, *17*, 842–865. doi:10.1287/moor.17.4.842

Sabuncuoglu, I. (1998). Scheduling with neural networks: A review of the literature and new research directions. *Production Planning and Control*, *9*, 2–12. doi:10.1080/095372898234460

Sabuncuoglu, I., Erel, E., & Alp, A. (2009). Ant colony optimization for the single model u-type assembly line balancing problem. *International Journal of Production Economics*, *120*(2), 287–300. doi:10.1016/j.ijpe.2008.11.017

Sabuncuoglu, I., Erel, E., & Tanyer, M. (2000). Assembly line balancing using genetic algorithms. *Journal of Intelligent Manufacturing*, *11*, 295–310. doi:10.1023/A:1008923410076

Sabuncuoglu, I., Gocgun, Y., & Erel, E. (2008). Backtracking and exchange of information: Methods to enhance a beam search algorithm for assembly line balancing. *European Journal of Operational Research*, *186*, 915–930. doi:10.1016/j.ejor.2007.02.024

Sabuncuoglu, I., & Gurgun, B. (1996). A neural network model for scheduling problems. *European Journal of Operational Research*, *93*, 288–299. doi:10.1016/0377-2217(96)00041-0

Scholl, A., & Becker, C. (2006). State-of-the-art exact and heuristic solution procedures for simple assembly line balancing. *European Journal of Operational Research*, *168*, 666–693. doi:10.1016/j.ejor.2004.07.022

Scholl, A., Fliedner, M., & Boysen, N. (2010). Absalom: Balancing assembly lines with assignment restrictions. *European Journal of Operational Research*, *200*(3), 688–701. doi:10.1016/j.ejor.2009.01.049

Scholl, A., & Klein, R. (1999). Balancing assembly lines effectively - A computational comparison. *European Journal of Operational Research*, *114*, 50–58. doi:10.1016/S0377-2217(98)00173-8

Shyu, S., Lin, B., & Yin, P. (2004). Application of ant colony optimization for no-wait flow-shop scheduling problem to minimize the total completion time. *Computers & Industrial Engineering*, *47*, 181–193. doi:10.1016/j.cie.2004.06.006

Sung, S. C., & Vlach, M. (2003). Single machine scheduling to minimize the number of late jobs under uncertainty. *Fuzzy Sets and Systems*, *139*(2), 421–431. doi:10.1016/S0165-0114(02)00505-5

Sviridenko, M. I. (2004). A note on permutation flow shop problem. *Annals of Operations Research*, *129*, 247–252. doi:10.1023/B:ANOR.0000030691.65576.28

Takahashi, H., Goto, H., & Kasahara, M. (2009). Application of a critical chain project management based framework on max-plus linear systems. In *Proceedings of the 2009 International Conference on Complex, Intelligent and Software Intensive Systems* (pp. 898–903). Fukuoda, Japan.

Tang, L., & Liu, J. (2002). A modified genetic algorithm for the flow shop sequencing problem to minimize mean flow time. *Journal of Intelligent Manufacturing*, *13*, 61–67. doi:10.1023/A:1013681029869

Tang, L., Liu, J., & Liu, W. (2005). A neural network model and algorithm for the hybrid flow shop scheduling problem in a dynamic environment. *Journal of Intelligent Manufacturing*, *16*, 361–370. doi:10.1007/s10845-005-7029-0

Temiz, I., & Erol, S. (2004). Fuzzy branch and bound algorithm for flow shop scheduling. *Journal of Intelligent Manufacturing*, *15*, 449–454. doi:10.1023/B:JIMS.0000034107.72423.b6

Toksari, M. D., Isleyen, S. K., Guner, E., & Baykoc, O. F. (2010). Assembly line balancing problem with deterioration tasks and learning effect. *Expert Systems with Applications*, *37*(2), 1223–1228. doi:10.1016/j.eswa.2009.06.005

van den Boom, T., & De Schutter, B. (2001a, December). Model predictive control for perturbed max¬plus-linear systems: A stochastic approach. In *Proceedings of the 40th IEEE Conference on Decision and Control* (pp. 4535–4540). Orlando, Florida.

van den Boom, T., & De Schutter, B. (2001b, September). MPC for perturbed max-plus-linear systems. In *Proceedings of the European Control Conference 2001* (ECC'01) (pp. 3783–3788). Porto, Portugal.

Venkatesh, J., & Dabade, B. (2008). Evaluation of performance measures for representing operational objectives of a mixed model assembly line balancing problem. *International Journal of Production Research*, *46*(22), 6367–6388. doi:10.1080/00207540701383164

Wang, L., Zhang, L., & Zheng, D.-Z. (2005). A class of hypothesis-test based genetic algorithm for flow shop scheduling with stochastic processing times. *International Journal of Advanced Manufacturing Technology*, *25*(11), 1157–1163. doi:10.1007/s00170-003-1961-y

Wang, L., Zhang, L., & Zheng, D.-Z. (2006). Genetic ordinal optimisation for stochastic flow shop scheduling. *International Journal of Advanced Manufacturing Technology*, *27*(1), 166–173. doi:10.1007/s00170-004-2154-z

Weida, X., & Xiao, T. (2008). Mixed model assembly line balancing problem with fuzzy operation times and drifting operations. In *Proceedings - Winter Simulation Conference* (pp. 1752 - 1760).

Womack, J. P., & Jones, D. T. (1996). *Lean thinking: Banish waste and create wealth in your corporation*. Free Press.

Ying, B., Hongshun, Z., & Liao, Z. (2009). Mixed-model assembly line balancing using the hybrid genetic algorithm. In I*nternational Conference on Measuring Technology and Mechatronics Automation, ICMTMA 2009* (vol. 3, pp. 242-245).

Ying, K.-C., & Liao, C.-J. (2004). An ant colony system for permutation flow-shop sequencing. *Computers & Operations Research*, *31*, 791–804. doi:10.1016/S0305-0548(03)00038-8

Zhu, Q., Sheng, W., & Xi, N. (2004, October). Max-plus algebra model for on-line task scheduling of a reconfigurable manufacturing work-cell. In *Proceedings of the 2004 IEEE/RSJ International Conference on Intelligent Robots and Systems* (pp. 1245–1250).

Chapter 12
Operator Assignment Decisions in a Highly Dynamic Cellular Environment

Gürsel A. Süer
Ohio University, USA

Omar Alhawari
Royal Hashemite Court, Jordan

ABSTRACT

Operators are assigned to operations in labor-intensive manufacturing cells using two assignment strategies: Max-Min and Max. The major concern is to see how these two approaches impact operators' skill levels and makespan values in a multi-period environment. The impact is discussed under chaotic environment where sudden changes in product mix with different operation times are applied, and also under non-chaotic environment where same product mix is run period after period. In this chapter, operators' skill levels are affected by learning and forgetting rates. The Max-Min strategy improved operators' skill levels more significantly than Max in this multi-period study; particularly in chaotic environment. This eventually led to improved makespan values under Max-Min strategy.

INTRODUCTION

Cellular manufacturing is considered as a collection of manufacturing cells that is dedicated to manufacture part families or assembly cells that are dedicated to process product families (see Askin & Standridge, 1993). The cellular manufacturing systems can be either machine-intensive or labor-intensive. In labor-intensive cells, it is easier to reconfigure cells when a product is ready to be processed. Moreover, moving equipment is much easier than it is in machine-intensive cells. Basically, in labor-intensive cells, most of the operations require light-weight, and small machines as well as equipment that require continuous operator attendance and involvement (Süer & Tummaluri,

DOI: 10.4018/978-1-61350-047-7.ch012

2008). Labor-intensive manufacturing cells have been observed in apparel, jewelry manufacturing, electromechanical assembly, sewing, shoe manufacturing, medical devices, and car seat manufacturing industries. The operator's role in machine-intensive cells is limited due to the presence of automatic machines. On the other hand, the operator has a key role in labor-intensive cells, and the number of operators and their assignment to operations has a great impact on the cell's production rate. In some cases, the number of operations is less than the number of operators. This creates the possibility that multiple operators are assigned to perform the same operation. It is important to control operator assignments; however, when the number of cells and the number of operators increase, keeping track of operator assignment becomes difficult.

In this chapter, concepts such as learning and forgetting rates are discussed to show how operator skill level varies from time to time; thus, the assignment decision is affected. Forgetting and learning rates affect the operator's skills and they are affected by their current skills. Learning takes place when the operator performs an operation continuously for a period of time, consequently, the operator will be more familiar with performing an operation. On the other hand, forgetting happens when the operator does not perform an operation in a number of consecutive periods. This chapter addresses both operator assignment and cell loading decisions. Operator assignment determines which operators are assigned to perform each task and cell loading identifies the products to be run in each cell.

The work undertaken in this chapter is an extension of work by Süer and Tummaluri (2008). The operator assignment can be made by using two different strategies; 1) Max, 2) Max-Min. Max considers only the current state of the operator skills for operator assignment to maximize output rate. On the other hand, Max-Min considers long-term effect of assignment decisions and attempts to develop more homogeneous work force without sacrificing output rate. This homogeneous work force may be more effective in dealing with drastic variations in demand and product mix in the long-term.

The objective of this chapter is to propose better mathematical models for operator assignment and also compare the performance of two major strategies, Max and Max-Min, in highly dynamic cellular environments. The main hypothesis is that Max-Min is a better strategy in operator assignment in the long-run. We want to show that long-term planning may help companies to better prepare their workforce for long-term operation than short-sided approach where only the immediate periods are considered. This approach is especially important in highly fluctuating demand environments and also in companies where product mix can quickly change. It is easier to implement such a strategy in companies where workforce is stable with low turnaround rate.

BACKGROUND

In the literature, some researchers addressed areas related to this subject such as cell loading, operator assignment, skills, learning and forgetting rate and product sequencing. Süer (1996) discussed, in his paper, the subject of optimal operator assignment and cell loading in labor-intensive manufacturing cells. He stated that the operator assignment to cells influences production rate that each cell can produce. He proposed a two-phase methodology. In phase 1, he generated operator assignments for alternative manpower levels by using a mixed integer mathematical model. In phase 2, he found the optimal manpower levels for each cell and optimal product assignment to cells.

Nembhard (2001) discussed a heuristic approach for assigning workers to task based on individual learning rate. Basically, he ran experiments based on two conditions: a long production run and a short production run. Results were interpreted and showed that the heuristic approach have an impact on overall productivity. Best results were found when workers learn more gradually. Nembhard and Osothslip (2002) discussed, in their paper, the operator behavior in terms of learning and forgetting rates; particularly, in the case of performing complex tasks. A study was conducted at a textile manufacturing plant, where different manual sewing tasks were available. Data was collected by studying the behavior of each worker over a period of one year. They used a model of individual learning and forgetting rate, which was introduced first by Nembhard and Uzumeri (2000) in order to measure the productivity rate. This model was applied to each operator, and operator learning and forgetting parameters were considered as dependent variables, whereas task complexity was considered as independent variable. Results were captured and then statistical analysis was done to find if there is a relationship between the variability of learning and forgetting rates with task complexity. Results indicated task complexity significantly affects the variance of worker learning and forgetting rates. For higher task complexities, workers are more varied in their learning and forgetting rate than they are at lower task complexities. The impact of task complexities on worker learning and forgetting affects worker assignment and productivity. Slomp et al. (2005) discussed cross-training decisions in a cellular manufacturing environment. They wanted to minimize the load of the bottleneck worker. In their study, they presented an integer programming model to calculate which workers have to be trained for which machines. Based on this model, they discussed the trade-off between the operating costs of the manufacturing cell, the costs of cross-training, and the workload among workers, they showed that the connection between workers and machines is really important to form chaining and this produces an efficient cross-training situation. In this case, workload can be shifted from heavier loaded worker to less loaded worker. Labor flexibility is needed in these environments. Unbalanced load may give feelings of unfairness in a team. Bidanda et al. (2003) presented the importance of focusing not only on technical issues, but also human issues in cellular manufacturing environments. Technical issues include cell formation and cell design, whereas human issues involve such as worker assignment strategies, skill identification, training, communication, autonomy, reward system, conflict management, and teamwork. They conducted a survey to show the importance of human issues in cellular manufacturing. The number of participants in the survey was 40, and consists of workers, managers and academicians. They were asked to rank the human issues. Their response was analyzed. The results showed that three major human issues in cellular manufacturing are communication, teamwork and training. The degree of autonomy was found the least important among all. The reward system was in the middle. The assignment strategies were found significant among academicians. The skill identification was found significant among managers and academicians whereas conflict management is significant among workers.

Shirase et al. (2001) developed a system of distributed production system which consists of some cell groups. They discussed a dynamic operator assignment method. In this cooperative method, they considered that whenever any cell in a group of cells is unable to meet the due date of certain part, it has the option to ask for one operator as a support from other groups. Eventually, cooperation is taking place between cell groups until all due dates are met. They generalized the idea in which some disturbances in a production system can be treated by cooperation between subsystems. Fan and Gassmann (1997) discussed allocation functions between worker and machines could influence the performance of manufacturing cells over a long period of time considered as 15 months. They concluded that skills development and knowledge are really important for keeping long-term competitiveness. Allocation of functions has an impact on the long term performance, in which the long period will give more vision to make the work smooth through absorbing the complexity of the nature of manufacturing environment.

Wirojanagud et al. (2005) discussed a strategic way to model worker differences for workforce planning. Impact of individual differences on management decisions is considered and then discussed. A problem of job shop environment has been formulated in a form of a mixed integer programming model. The major concern was to identify number of workers who will be hired, fired and cross-trained at a minimum cost. Experiments are run and then results are analyzed. Workers differences are playing a major role in making decisions in manufacturing system environment.

Suksawat et al. (2005) discussed the concept of evaluating the skill levels of workers. They developed a skilled worker–based scheduling method based on the skill evaluation and genetic algorithm application. They focused on the objective of improving the production rate by considering workers' skill levels. Süer and Dagli (2005) discussed manpower decisions in their paper. They considered two issues in their paper; product sequencing in a cell and cell loading. For the first issue, they target to minimize intra-cell manpower transfers. A three-phase methodology is proposed, in which optimal manpower level for each operation is found by using a mathematical model, a matrix for manpower transfers between products is formed, and traveling salesman problem is solved. For the second issue, they aim to find the optimal assignment of products to cells. Their objective is to minimize makespan and number of machines. Cesani and Steudel (2005) presented a research concerning labor assignment strategies, and their impact on the cell performance in cellular manufacturing environment. The term labor flexibility was discussed and referred to the movement of operators between cells and inside the same cell. Labor assignments, such as dedicated assignment, in which an operator is assigned to one or more machines, shared assignment in which two operators or more are assigned to one or more machines, and combined in which an operator is assigned as dedicated and shared together. They made their discussions based on workload sharing, workload balancing and bottleneck operations. Experiments and simulation models were implemented and discussed. They concluded based on results, that the balance in the operators' workload and the level of machine sharing are important factors to determine cell performance and behavior. They also referred to the importance of cross-training issue in improving cell performance.

Mahdavia et al. (2010) developed an integer mathematical model to design cellular manufacturing systems. They consider a dynamic environment as well. Their model deals with worker assignment as well as dynamic configuration of the cellular system. The overall objective is to minimize the total cost of inventory holding and backorder costs, inter-cell material handling cost, machine and reconfiguration costs and hiring, firing and salary costs of workers.

Süer, Arikan, Babayigit (2008) and Süer, Arikan, Babayigit (2009) developed fuzzy mathematical models for cell loading in labor-intensive cells subject to manpower restrictions. Süer, Cosner and Patten (2009) developed various mathematical models for cell loading and product sequencing in manufacturing cells. Süer, Subramanian and Huang (2009) developed several heuristic procedures and mathematical models for cell loading and product sequencing in a shoe manufacturing company.

Süer and Tummaluri (2008) discussed the problem of operator assignment to operations in labor-intensive cells. The operators are assigned to operations in multi-period context considering their skill levels, forgetting and learning rates. They developed a three-phase hierarchical methodology to solve this problem. The first phase is generating alternative operator levels for each product using operation standard times. The second phase is determining cell loads and cell sizes using standard times. The third phase is assigning operators to operations. A mixed integer mathematical model is used in all phases. Two different strategies (Max and MaxMin) are proposed for solving the operator assignment in the third phase. The results showed that when using Max Strategy, lower makespan allocation values are obtained, whereas Max-Min improved the skill levels more regularly. The work undertaken in this paper is an extention of their work. They found that Max-Min is superior to Max Strategy in terms of improving operator skill levels; however, they could not show that Max-Min Strategy is better in minimizing makespan as a result of improved skill levels. Their work assumed a static cellular environment, in which there are no new products entering the system and no product is leaving the system (i.e. product mix remained the same throughout the study). They have classified labor skills into nine categories following normal distribution. Their work also showed that some non-bottleneck operations became bottleneck after assigning operators. The reason for that is that they have done initial operator assignment based on standard times. When they reflected the effect of skills on processing times, some non-bottleneck operations became bottleneck and adversely affected output rates. In some cases, operators had to be re-assigned to fix the problem.

PROBLEM STATEMENT

This chapter introduces several improvements to work by Süer and Tummaluri (2008) where, 1) Max and Max-Min strategies are compared in a highly dynamic environment where product mix changes. i.e., new products enter the system and some products leave the system. It is believed that this will show the benefits of Max-Min strategy better in terms of minimizing makespan, 2) number of skill levels are reduced to seven. It is believed that this is more practical and realistic approach than having nine skill levels, 3) skill-level based processing times are used directly during operator assignment process. This helps to avoid re-computing of bottleneck operation, output rate and re-assignment of operators.

Methodology

In this section, the methodology used is described in detail.

General Methodology

This study is carried out in a multi-period environment. The number of periods included in this study is 16 and each period represents a week. This allows us to see the impact of the strategies in the long-term. Based on the assignments made in the previous periods, an operator's skill level may be adjusted. Figure 1 includes the multi-period methodology in which the proposed approach is implemented and the results are captured. These results may affect the operator skill levels; hence, they need to be revised each period. Once all periods are considered for the first strategy, then the same procedure is applied for the second strategy, and finally the results are compared.

Overview of Strategies

The major phases of the Max strategy as shown in Figure 2a are: (1) find optimal operator assignment using an integer mathematical model. In this phase, the number of workers needed for each operation is determined for each product so that production rate is maximized with the available manpower level. (2) determine cell loads to minimize makespan by using an integer mathematical model. In this phase, decision about what cell to use to produce each product is made. (3) determine product sequence in each cell by using a simple scheduling rule, SPT (Shortest Processing Time). In this phase, the products in each cell are sequenced in the increasing order of processing time to minimize average flow time as well. The first three phases of the Max-Min

Figure 1. Multi-period methodology

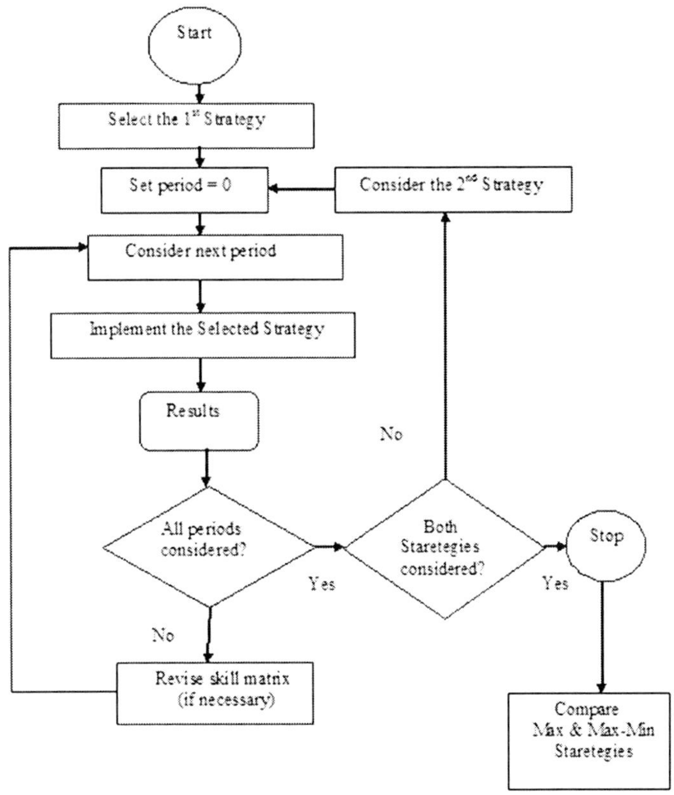

Strategy is similar to the Max Strategy. However, two more phases are included in the Max-Min Strategy (see Figure 2b), and they are: (4) identify bottleneck and non-bottleneck operations for each product in the cell. The slowest operation (lowest output) for each product is identified as bottleneck operation. (5) re-assign low-skilled operators to non-bottleneck operations using the Min-skill principle. In this phase, the focus is on non-bottleneck operations and workers are assigned to operations where their skill level is not very high as long as it does not adversely affect the production rate of the cell. By doing this, we expect that operators with low skills will get a chance to perform these operations and eventually improve their skill levels.

Performance Measures Used

The performance measures used for each task are summarized in Table 1. Production rate measures the number of units manufactured per unit time. Makespan is defined as the maximum completion time of all jobs (Equation 1) and flowtime measures how long a job remains in the system (Equation 2).

$$MS=C_{max} \quad (1)$$

$$f_i=c_i-r_i \quad (2)$$

where

c_i completion time of job i
f_i flowtime of job i
r_i ready time of job i

Figure 2. The general overview of the strategies

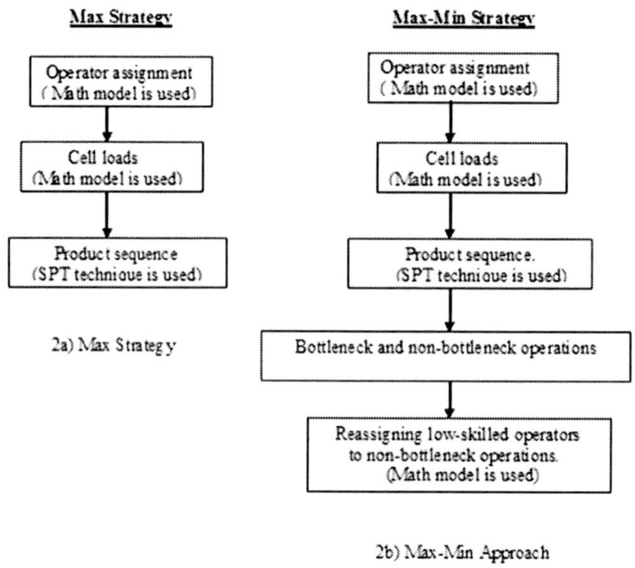

Table 1. Performance measures used for each task

Task	Max Strategy	Max-Min Strategy
Operator Assignment	Maximize Production Rate	Maximize Production Rate
Loading Cells	Minimize Makespan	Minimize Makespan
Product Sequencing	Minimize Average Flow Time	Minimize Average Flow Time
Re-assigning low-skilled operators	-----	Minimize Total Skills Without Violating Original Production Rate

Skill Levels

In this study, each operator is assumed to have a skill level for each operation he performs. These skill levels follow the normal distribution (as shown in Figure 3), in which μ represents the mean value and σ represents the standard deviation. The skill levels are divided into seven categories and their corresponding probabilities are shown in Table 2. Level 4 represents the average skill, level 7 represents the best and level 1 is the worst. This is an assumption used in this study. Süer and Tummaluri (2008) also used a similar classification except they used 9 skills as opposed to 7 suggested here.

Operation Times

Operation times are calculated according to the operator skill levels. The standard processing time for each operation is considered to be the average time, hence, the operator with skill level 4 is considered to have the average operation time. Other skills below or over will follow the normal distribution. Table 3 provides an example for different skills of an operation with a standard deviation of 5% of the mean. The σ for operation 1 for product X1 is 0.0035 (=.07*.05). As the skill level decreases by 1 level ($4 \rightarrow 3$), the operator skilled-based time increases to 0.0735 ($=.07 + \sigma$).

Figure 3. The normal distribution curve for skill levels

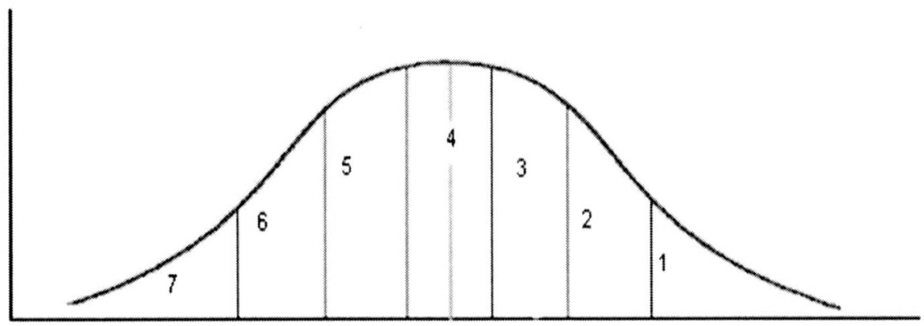

Table 2. The skill levels and probability

Skill Level	Time	Probability
1	$\mu+3\sigma$	0.0062
2	$\mu+2\sigma$	0.0606
3	$\mu+\sigma$	0.2417
4	μ	0.383
5	$\mu-\sigma$	0.2417
6	$\mu-2\sigma$	0.0606
7	$\mu-3\sigma$	0.0062

Learning and Forgetting Rates

In this study, a skill level is affected by learning and forgetting rates. An operator's skill level increases when he performs an operation for many consecutive periods. In the same manner, the length of interruption interval affects the skill levels adversely; hence, if an operator does not perform a certain operation for a number of consecutive periods, his skill level decreases. Table 4 shows the assumed number of required periods for improving or lowering skill level, and it also shows the probability for that skill to change.

The probability values follow the notion that an operator who has been performing an operation for a long time will become more experienced operator, and it will take longer for that skill to deteriorate. As shown in the same table, a skill level can be improved from 1 to 2 with a probability of 0.7, if an operator keeps performing the same operation for 3 periods consecutively. On the other hand, improving skill level from 6 to 7 requires an operator to perform the operation for 6 periods consecutively (with a probability of 0.3). Meanwhile, an operator's skill level can deteriorate from 2 to 1 with a probability of 0.7, if he does not perform the operation for 4 periods consecutively. Similarly, if he does not perform the operation for 7 periods consecutively, his skill will decrease from 7 to 6 with a probability of 0.3. No empirical study has been done to validate these assumptions due to time restrictions. However, it is believed that these numbers reflect the relations between learning and forgetting rates of workers at different skill levels along with associated probabilities reasonably well. The operator skill matrix is revised on learning and forgetting rates at the end of every period.

Table 3. The operation times for different skill levels with σ = 5% of the mean

		Skills						
Mean	Std. Dev.	7	6	5	4	3	2	1
0.07	0.035	.0595	0.063	0.0665	0.07	0.0735	0.077	0.0805

Table 4. The learning and forgetting rates

Learning				Forgetting			
Skill level		Periods	Prob.	Skill level		Periods	Prob.
From	To			From	To		
1	2	3	0.70	2	1	4	0.70
2	3	3	0.65	3	2	4	0.65
3	4	4	0.6	4	3	5	0.6
4	5	4	0.5	5	4	5	0.5
5	6	5	0.4	6	5	6	0.4
6	7	6	0.3	7	6	7	0.3

PROPOSED MATHEMATICAL MODELS

In this section, the proposed models are introduced.

Max Strategy

The Max Strategy uses the skill-based times to assign operators into operations such that maximum output is achieved. An integer mathematical model is used to assign operators. The mathematical model is formulated with the objective function of maximizing the production rate as shown in Equation (3). Equation (4) determines which operators have to be assigned to each operation. Equation (5) ensures that each operator is assigned to only one operation within a cell. Equation (6) shows that y_{kj} is a binary variable. The mathematical model is given below:

Objective Function:

$$MaxZ=R \quad (3)$$

Subject to:

$$\sum_{k\in fj}(a_{kj}y_{kj}) - R \geq 0 \; j=1,2,3,...,m \quad (4)$$

$$\sum_{j\in fk} y_{kj} = 1 \; k=1,2,3,..,n \quad (5)$$

$$y_{kj}\in(0,1) \quad (6)$$

Indices:

- k Operator index
- j Operation index

Parameters:

- a_{kj} number of units operator k can process if assigned to operation j
- m number of operations in the cell
- f_k set of operations that operator k can perform
- f_j set of operators who can perform operation j
- n Number of operators

Decision variables:

- R production rate
- y_{kj} 1 if operator k is assigned to operation j, 0 otherwise

This assignment model is run for each product by using ILOG OPL software.

Max-Min Strategy

The Max-Min strategy also uses the skill-based times to find the optimal operator assignment to maximize the output. In this Strategy, first the slowest operation, bottleneck operation, is identified. The same integer mathematical model is used to assign operators and it is solved by using ILOG/OPL. However, another constraint is added to determine production rate for each operation as given in equation (7).

$$R_j = \sum_{k \in fj} (a_{kj} y_{kj}) \text{ j=1,2,3..m} \tag{7}$$

Obviously, the lowest production rate operation is identified as the bottleneck operation as shown in equation (8)

$$R_b = \min(R_j / j = 1,2,3,\ldots,m) \tag{8}$$

The Max-Min Strategy keeps the operators assigned to the bottleneck operation the same; however, it re-assigns other operators to non-bottleneck operations to minimize total skills such that the optimal output rate is maintained. A mathematical model is used where the objective function is to minimize the total skills for the remaining operators as given in equation (9). Equation (10) guarantees that the original production rate (optimal) is not violated. Equation (11) shows a constraint in which each operator is assigned to one operation within each cell. Equation (12) shows that y_{kj} is a binary variable.

The math model used is given below as:

Objective function:

$$MinZ = \sum_{k \in fj} \sum_{j \in fk} (s_{kj} y_{kj}) \tag{9}$$

Subject to:

$$\sum_{k \in fj} (a_{kj} y_{kj}) \geq R_b \text{ j=1,2,3,....m} \tag{10}$$

$$\sum_{j \in fk} y_{kj} = 1 \text{ k=1,2,3,...n} \tag{11}$$

$$y_{kj} \in (0,1) \tag{12}$$

where,

Parameters:

s_{kj} Skill level of operator k for operation j

R_b Production rate of bottleneck operation

The difference between these two strategies is illustrated in Figure 4 using a hypothetical case. Assume; that operation 2 is the bottleneck operation with operators 3 and 7 assigned to it as shown in Figure 4a and the output rate is 80 units/hr. The Max-Min Strategy shown in Figure 4b keeps the same operators in the bottleneck operation but re-assigns other operators to minimize skills without sacrificing the optimal output rate.

Cell Loading and Product Sequencing

Cell loading is the process of assigning products to cells. In this paper, a mathematical model is used to assign products to cells and the primary performance measure is to minimize makespan. Equation (13) shows the objective function, minimizing makespan. Equation (14) shows that the total processing time in each cell should be equal to or greater than the makespan. Equation (15) ensures that each product is assigned to a cell. Equation (16) shows the sign restriction.

Objective function

$$MinZ = MS \tag{13}$$

Subject to:

$$MS - \sum_{i=1}^{n} p_{ij} x_{ij} \geq 0 \text{ j=1,2,3,...,c} \tag{14}$$

$$\sum_{j=1}^{c} x_{ij} = 1 \text{ i=1,2,3,...,n} \tag{15}$$

Figure 4. Manpower assignment using both strategies

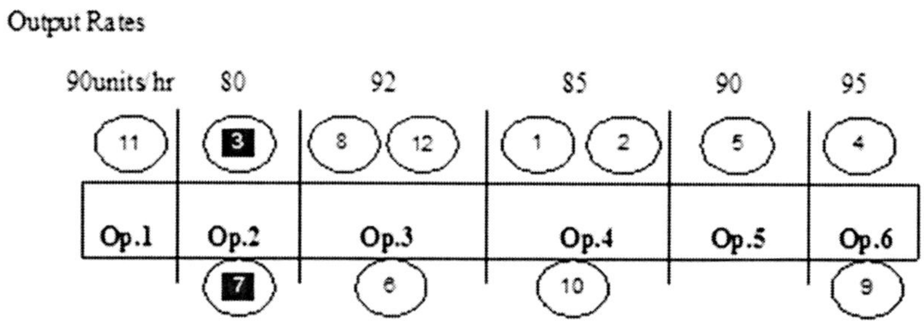

4a. The optimal operator assignment using the MAX Strategy

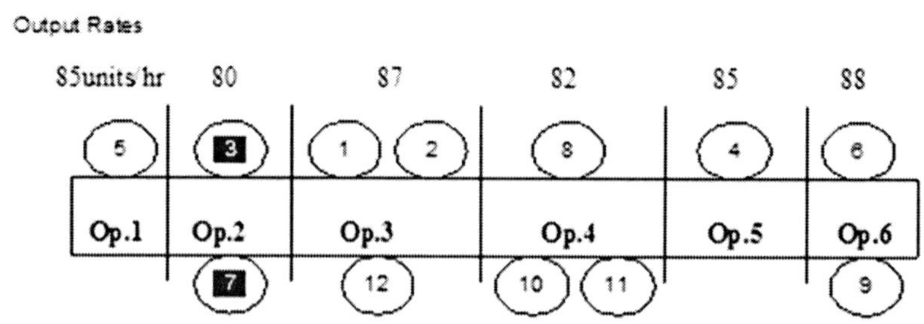

4b. The optimal operator assignment using the Max-Min Strategy

$$x_{ij} \in (0,1) \qquad (16)$$

Where,

Indices:

c Number of cells

n Number of products

Parameter:

p_{ij} Processing time of product *i* in cell *j*

Decision Variable:

x_{ij} 1 if product *i* is assigned to cell *j*, 0 otherwise

Product sequencing usually comes after cell loading in which products are arranged in such a way, that a selected performance measure is completed. In this chapter, average flow time is considered as a secondary measure and it is minimized by using the shortest processing time technique (SPT). SPT rule orders jobs in the increasing order of processing times.

DATA USED IN EXPERIMENTS

In this section, the data used in experiments are given.

Standard Times

The standard times for operations correspond to a skill level of 4. The standard times for product groups X and Y are randomly generated from a random uniform distribution in the intervals shown in Table 5. These two product groups are used to create the dynamic environment mentioned earlier.

Product Demand

The product demand for all products is randomly generated from the uniform distribution in the interval of [2200, 8500] for all periods. Table 6 shows these demand values.

Table 5. Uniform distribution intervals for standard times for product groups X and Y

Product	Operations					
	Op.1	Op.2	Op.3	Op.4	Op.5	Op.6
X	[0.04-0.09]	[0.28-0.45]	[0.37-1.18]	[0.47-0.88]	[0.18-0.45]	[0.20-0.80]
Y	[1.10-1.30]	[1.20-1.40]	[0.09-0.15]	[0.04-0.09]	[1.00-1.20]	[0.04-0.08]

Table 6. Product demand and total demand for each period

Period	Product Demand										Total Demand
	1	2	3	4	5	6	7	8	9	10	
1	3500	7500	3400	2700	2200	4000	4500	2200	2300	3000	35300
2	3500	7500	3700	2900	2200	4300	4600	2200	2500	3000	36400
3	3700	4200	3700	3000	2400	4300	4600	2400	2500	3100	37200
4	3100	6500	3700	2750	2150	3500	4400	1900	2500	3100	33600
5	3100	3750	3700	2750	2400	3900	4200	1900	2500	3300	34500
6	4200	8000	3700	3000	2500	4300	4800	2400	2500	3100	38500
7	2800	6400	3700	2250	2150	3200	4400	1900	2300	3100	32200
8	2800	6300	3700	2250	2150	3200	4400	1900	2300	3100	31800
9	3100	7750	3700	2750	2400	3900	4200	1900	2500	3300	35500
10	3300	8450	3700	3050	2600	3900	4400	2000	2500	3500	37400
11	2900	6400	3700	2300	2150	3250	4400	1900	2300	3200	32500
12	3500	7600	3400	3200	2200	3500	4500	2200	2300	3000	35400
13	4100	7700	3500	2500	2400	4800	4000	2400	2500	3400	37300
14	2800	6800	3450	2500	2150	3200	4200	2100	2300	3100	32600
15	3600	6750	3700	3000	2400	3900	4200	1900	2750	3500	35700
16	3400	6900	3700	2900	2200	4300	4600	2200	2500	3600	36300

Operator Skills

The total number of operators included in the study is 30. Each operator is assumed capable of performing 3 operations. These operators are divided into two cells with 12 operators in cell 1 and 18 operators in cell 2. The initial skill matrix is established randomly by following probabilities given in Table 2. The initial skill matrix for all operators is given in Table 7.

EXPERIMENTS

Several experiments are conducted using 10 products of type X and 10 products of type Y. Each product requires 6 operations. The experiments performed are listed below:

1. Periods 1-14, no chaos
2. Periods 1-9, no chaos; Periods 10-11, chaos
3. Periods 1-14, no chaos; Periods 15-16 chaos
4. Periods 1-14, no chaos; Period 15 chaos with new standard deviation

Table 7. Initial operator skill matrix

Operator	Operations					
	Op1.	Op.2	Op.3	Op.4	Op.5	Op.6
1	5		5	4		
2		3	4	4		
3	3	5	3			
4				4	3	4
5	2	3			5	
6			5	6		4
7	4				4	3
8	4		4	4		
9		4			4	4
10	4		3	4		
1	4			1		3
12		3	4		5	
13	4	5		4		
1		5	5	5		
1	3		3		5	
16		3	3		5	
17	5			4		3
18	5	5		5		
1		5		5		4
20	5		3		6	
21	4			3		2
22		3	5		3	
23	3		4			5
24	4			5	3	
25			4	3		6
26		3	1			7
27	3			3	5	
28		4	2		4	
29	5		4			4
30	5				4	5

Experiment 1

This experiment includes runs using Max and Max-Min strategies starting from period 1 to period 14. The results are analyzed below:

Impact on Operator Skill Levels

It was found that Max-Min improves operator skill levels more significantly than Max did. Benefits come from the Max-Min assignment strategy in which it finds the optimal operator assignment to maximize the output. Later, this strategy keeps the operators assigned to the bottleneck operation the same but re-assigns the low skilled operators to non-bottleneck operations. This strategy allows an operator to perform an operation that he or she is not skilled at; hence his or her skill does not deteriorate and certainly improves. Tables 8 and 9 show the comparison of these two strategies at the end of period 14 in terms of average operator skill levels for cells 1 and 2, respectively. In cell 1, it was found that not only the average operator skill levels is greater than the initial average skill levels when Max-Min was used, but also is greater than operators average skill levels when Max was used. On the other hand, when Max was used, 5 operators had an average skill levels greater than their initial ones, 3 operators kept the same average and 4 operators were found having lower average skill levels.

In cell 2, it was found that the average operator skill level is greater than the initial average skill levels when Max-Min was used. However, 15 operators had greater average skill levels than when Max was used, but 2 operators had the same average skill levels and 1 operator had lower average skill levels. As to Max strategy, 14 operators had an average skill levels greater than their initial ones, 1 operator kept the same average and 3 operators were found having lower average skill levels than their initial ones.

Impact on Operations

Max-Min Strategy improved operator skill levels on each operation more apparently than Max did. Table 10 shows average skill levels on each operation. It was found that operation 1 deteriorated when Max was used; however average operator skills on all operations were greater when Max-Min was used.

Impact on Makespan

The results have shown that both strategies were tied in terms of makespan for the first 11 periods. Starting period 12, Max-Min performed better in minimizing makespan by 0.2%, 0.2% and 0.1% compared to Max approach in periods 12, 13 and 14, respectively. Table 11 shows the comparison of these two strategies in terms of makespan from period 1 to period 14.

Table 8. Comparison of two strategies in cell 1 at the end of period 14

Cell 1		Average operator skill levels (at the end of Period 14)	
operator	Initial average skill level	Max-Min Strategy	Max Strategy
1	4.67	5.67	4.67
2	3.67	5.33	4.67
3	3.67	4.67	3
4	3.67	6	3.67
5	3.33	5.33	3.67
6	5	6.33	4.67
7	3.67	5.33	3.37
8	4	5	4.33
9	4	5.67	4.33
10	3.67	5.33	3.67
11	2.5	5.67	3.67
12	4	5.33	3.67

Table 9. Comparison of two strategies in cell 2 at the end of period 14

Cell 2		Average operator skill levels (at the end of Period 14)	
operator	Initial average skill level	Max-Min Strategy	Max Strategy
13	4.33	5.33	4.33
14	5	7	6
15	3.67	6.33	5.67
16	3.67	6.33	5.33
17	4	5.33	4.67
18	5	6	5.33
19	4.67	5.67	5
20	4.67	6	5
21	3	5.67	5.67
22	3.67	5.33	3.33
23	4	6	5
24	4	5	3.67
25	4.33	6.33	5.33
26	3.67	4	3
27	3.67	6	5.33
28	3.33	5.33	5.33
29	4.33	5.33	4.67
30	4.67	6	6.67

Table 10. Average skill levels on each operation

		Average operator skill levels (at the end of period 14)	
Operation	Initial average skill level	Max-Min Strategy	Max Strategy
Op.1	4	4.39	3.44
Op.2	3.92	5.23	4.23
Op.3	3.65	6	4.89
Op.4	4	6.25	5.63
Op.5	4.31	6.15	4.62
Op.6	4.25	5.92	4.62

Experiment 2

In this section, chaos is applied in periods 10 and 11 to create a big shock in the system to see how both strategies behave in terms of makespan. This is accomplished by introducing new products with very different processing times. The results have shown that when 5 products of type Y entered the system and 5 products of type X left the system, Max-Min gave better results (with 0.4% reduction

Table 11. The makespan from period 1 to period 14

period	Max	Max-Min	Difference(%)
1	58.07	58.07	0.0%
2	59.84	59.84	0.0%
3	61	61	0.0%
4	53.13	53.13	0.0%
5	55.92	55.92	0.0%
6	60.07	60.07	0.0%
7	50.23	50.23	0.0%
8	49.684	49.684	0.0%
9	55.76	55.76	0.0%
10	57.9	57.9	0.0%
11	50.64	50.64	0.0%
12	53.19	53.1	0.2%
13	56.25	56.15	0.2%
14	49.04	48.99	0.1%

in makespan) in period 10. In period 11, the entire set of products of type Y is manufactured (no product X in the system). Max-Min still worked better (with 0.4% improvement in makespan). Table 12 shows the results of makespan using these two strategies in periods 10 and 11. Table 13 shows another chaotic scenario by entering all products of type Y and releasing all products of type X in period 10. In this case, the improvement with Max-Min over Max was 0.7%.

Experiment 3

The chaos was applied in periods 15 and 16 as we did in experiment 2. The results have shown that when 5 products of type Y entered the system and 5 products of type X left the system, Max-Min gave better results (with 1.7% reduction in makespan) in period 15. In period 16, the entire set of products of type Y is manufactured (no product X in the system). Max-Min still worked better (with 1.2% improvement in makespan). Table 14 shows the results of makespan using

Table 12. Impact on makespan under chaos in periods 10 and 11

	Max-Min	Max	Difference (%)
Period 10 (5 products)	71.15	71.43	0.4%
Period 11 (5+5 products)	81.68	82.02	0.4%

Table 13. Impact on makespan under chaos in period 10

	Max-Min	Max	Difference (%)
Period 10 (10 products)	94.3	94.92	0.7%

Table 14. Impact on makespan under chaos in periods 15 and 16

	Max-Min	Max	Difference (%)
period 15 (5 products)	66.48	67.6	1.7%
period 16 (5+5 products)	88.6	89.7	1.2%

Table 15. Impact on makespan under chaos in period 15

	Max-Min	Max	Difference (%)
Period 15 (10 products)	86.81	89.01	2.2%

these two strategies in periods 15 and 16. Table 15 shows another chaotic scenario by entering all products of type Y and releasing all products of type X in period 15. In this case, the improvement with Max-Min over Max was 2.2%.

Experiment 4

In this phase, we wanted to show that the improvement in operators' skill levels should have better impact than what happened in previous experiments. We changed the standard deviation from (0.05μ) to (0.2μ) to extend the gap of operator-operation times, and then we applied these new times in period 15 by entering all products of type Y in the system. Results have shown that there was better gain in terms of minimizing makespan and total time. Max-Min performed better by 5% than Max approach. Table 16 show these results.

CONCLUSION AND FUTURE WORK

In this chapter, operators were assigned to operations to maximize the production rate using two assignment strategies: Max-Min and Max. The major concern is to see how these two approaches impact operators' skill levels, as well as their impact on makespan values. The impact is discussed under a chaotic environment where sudden changes in product mix with different operation times are applied and under a non-chaotic environment where the same product mix is run period after period. In this study, a skill level is affected by learning and forgetting rates. An operator's skill level increases when he performs an operation for many consecutive periods; on the other hand, if an operator does not perform a certain operation for a number of consecutive periods, his skill level decreases. Max-Min did improve operators' overall skill levels more significantly than Max in multi-periods. This is due to the fact that Max-Min does not only assign operators to maximize production rate, but also it re-assigns operators to

Table 16. Impact on makespan under chaos in period 15

	Max-Min	Max	Difference (%)
period 15 (10 products) (SD = 0.2)	55.23	58.02	5%

operations where they are not very skilled; thus, operator's skill levels continue to improve. On the other hand, Max is only assigning operators to operations to maximize production rate. Moreover, the previous work assumed a stable cellular environment, in which they assumed that there are no new products entering the system. Thus, in this paper, we introduced a highly dynamic cellular environment, in which new products with different processing times entered the system and some of the existing products left the system. Max-Min acts well under chaotic environment because it increases operators' skill levels well enough to face the shock applied, where the shock contains products with new processing times and this requires different manpower allocation. We also concluded that the standard deviation used in operator time matrix is an important factor for helping Max-Min approach to expand the gain in terms of minimizing makespan and total time. A standard deviation of 5% of the mean is used in operator matrix for experiments 1 and 2. In period 15, when we replaced the whole set of products of type X with products of type Y, we found the highest gain that shows Max-Min is better in minimizing makespan among all periods. This gain was captured using a standard deviation of (0.05μ); however, when we used a standard deviation of (0.2μ), we found that the gain is higher than using (0.05μ). A possible extension to this work is to take manpower level decision for each cell as a variable as opposed to using fixed manpower levels. A further expansion would be to allow operators shift from one cell to the next as opposed to fixing them to the same cell all the time. Obviously, this increases flexibility in assigning operators to cells. However, this increases computational complexity as well.

REFERENCES

Askin, R., & Standridge, C. (1993). *Modeling and analysis of manufacturing systems*. New York, NY: John Wiley & Sons, Inc.

Bhaskar, K., & Srinvasan, G. (1997). Static and dynamic operator allocation problems in cellular manufacturing systems. *International Journal of Production Research*, *35*(12), 3467–3481. doi:10.1080/002075497194192

Bidanda, B., Ariyawongrat, P., Needy, K. L., Norman, B. A., & Tharmmaphornphilas, W. (2005). Human related issues in manufacturing cell design, implementation, and operation: A review and survey. *Computers & Industrial Engineering*, *48*(3), 507–523. doi:10.1016/j.cie.2003.03.002

Cesani, V. I., & Steudel, H. J. (2005). A study of labor assignment flexibility in cellular manufacturing systems. *Computers & Industrial Engineering*, *48*(3), 571–591. doi:10.1016/j.cie.2003.04.001

Fan, S., & Gassmann, R. (1997). The affects of allocation of functions on the long-term performance of manufacturing cells—A case study. *Human Factors and Ergonomics in Manufacturing*, *7*(2), 125–152. doi:10.1002/(SICI)1520-6564(199721)7:2<125::AID-HFM5>3.0.CO;2-5

Hyer, N. L. (1984). Management's guide to group technology. In Hyer, N. L. (Ed.), *Group technology at work* (pp. 26–27). Dearborn, MI: SME.

Mahdavia, I., Aalaei, A., Paydar, M. M., & Solimanpur, M. (2010). Designing a mathematical model for dynamic cellular manufacturing systems considering production planning and worker assignment. *Computers & Mathematics with Applications (Oxford, England)*, *60*(4), 1014–1025. doi:10.1016/j.camwa.2010.03.044

Nembhard, D. A. (2001). Heuristic approach for assigning workers to tasks based on individual learning rates. *International Journal of Production Research*, *39*, 1955–1968. doi:10.1080/00207540110036696

Nembhard, D. A., & Osothsilp, N. (2002). Task complexity effects on between-individual learning/forgetting variability. *International Journal of Industrial Ergonomics*, *29*(5), 297–306. doi:10.1016/S0169-8141(01)00070-1

Nembhard, D. A., & Uzumeri, M. V. (2000). Experiential learning and forgetting for manual and cognitive tasks. *International Journal of Industrial Ergonomics*, *25*, 315–326. doi:10.1016/S0169-8141(99)00021-9

Shirase, K., Wakamatsu, H., Tsumaya, A., & Arai, E. (2001). *Dynamic co-operative scheduling based on HLA.* Retrieved September 10, 2007, from http://www.springerlink.com/content/r116462352k35gt0/fulltext.pdf

Slomp, J., Bokhorst, J. A., & Molleman, E. (2005). Cross-training in a cellular manufacturing environment. *Computers & Industrial Engineering*, *48*(3), 609–624. doi:10.1016/j.cie.2003.03.004

Süer, G. A. (1996). Optimal operator assignment and cell loading in labor-intensive manufacturing cells. *Computers & Industrial Engineering*, *31*(1-2), 155–158. doi:10.1016/0360-8352(96)00101-5

Süer, G. A., Arikan, F., & Babayigit, C. (2008). Bi-objective cell loading problem with non-zero setup times with fuzzy aspiration levels in labour-intensive manufacturing cells. *International Journal of Production Research*, *46*(2), 371–404. doi:10.1080/00207540601138460

Süer, G. A., Arikan, F., & Babayigit, C. (2009). Effects of different fuzzy operators on fuzzy bi-objective cell loading problem in labor-intensive manufacturing cells. *Computers & Industrial Engineering*, *56*, 476–488. doi:10.1016/j.cie.2008.02.001

Süer, G. A., Cosner, J., & Patten, A. (2009). Models for cell loading and product sequencing in labor-intensive cells. *Computers & Industrial Engineering*, *56*, 97–105. doi:10.1016/j.cie.2008.04.002

Süer, G. A., & Dagli, C. (2005). Intra-cell manpower transfers and cell loading in labor-intensive manufacturing cells. *Computers & Industrial Engineering*, *48*(3), 643–655. doi:10.1016/j.cie.2003.03.006

Süer, G. A., Subramanian, A., & Huang, J. (2009). Heuristic procedures and mathematical models for cell loading and scheduling in a shoe manufacturing company. *Computers & Industrial Engineering*, *56*, 462–475. doi:10.1016/j.cie.2008.10.008

Süer, G. A., & Tummaluri, R. (2008). Multi period operator assignment considering skills, learning and forgetting in labour-intensive cells. *International Journal of Production Research*, *2*, 1–25.

Suksawat, B., Hilraokai, H., & Ihara, T. (2005). *A new approach manufacturing cell scheduling based on skill-based manufacturing integrated to genetic algorithm.* Retrieved October 15, 2007, from http://www.springerlink.com/content/q885171535723r5q/fulltext.pdf

Wirojanagud, P., Gel, S., Fowler, J., & Cardy, J. (2005). *Modeling inherent worker differences for workforce planning.* Retrieved September 10, 2007, from http://ie.fulton.asu.edu/files/shared/workingpapers/Wirojanagud_Gel_Fowler.pdf.

Chapter 13
Alternative Heuristic Algorithm for Flow Shop Scheduling Problem

Vladimír Modrák
Technical University of Kosice, Slovakia

R. Sudhakra Pandian
Kalasalingam University, India

Pavol Semančo
Technical University of Kosice, Slovakia

ABSTRACT

In this chapter an alternative heuristic algorithm is proposed that is assumed for a deterministic flow shop scheduling problem. The algorithm is addressed to an m-machine and n-job permutation flow shop scheduling problem for the objective of minimizing the make-span when idle time is allowed on machines. This chapter is composed in a way that the different scheduling approaches to solve flow shop scheduling problems are benchmarked. In order to compare the proposed algorithm against the benchmarked, selected heuristic techniques and genetic algorithm have been used. In realistic situation, the proposed algorithm can be used as it is without any modification and come out with acceptable results.

INTRODUCTION

The main problems in scheduling of jobs in a manufacturing cell are, according to Wight (1984), "priorities" and "capacity". Hejazi and Saghafian (2005) characterize scheduling problem as an effort to specify the order and timing of the processing of the jobs on machines, with an objective or objectives respecting above-mentioned assumptions". Henry Gantt, as the inventor of the now well-known Gantt chart, and Frederick Taylor, with his theories of Scientific Management, give the first scientific consideration to production scheduling. Computer-based production scheduling systems that emerged later were mostly connected to the shop floor tracking systems and used dispatching rules to sequence the work (Herrmann, 2006). Such computer aided

DOI: 10.4018/978-1-61350-047-7.ch013

scheduling systems are now being integrated in manufacturing execution systems. Similar solutions of scheduling systems are now part of ERP systems that was performed in the early 1990s. Manufacturing execution systems besides their typical functions were developed and used also as the interface between ERP and process control (Modrák, 2009).

The scheduling that is related to cellular manufacturing systems is known as operation scheduling or shop scheduling. According to Cox et al. (1992), this type of scheduling is aimed to find "the actual assignment of starting and/or completion dates to operations or groups of operations to show when these must be done if the manufacturing order is to be completed on time." The scheduling of cellular manufacturing systems (CMSs) is also known as "group scheduling". Its importance has long been recognized as critical the successful implementation of Group Technology (Mosier, et al., 1984, Ruben and Mahmoodi, 1999). Existing research efforts in group scheduling can be classified in two groups: those that consider a single cell and those that consider multiple cells (Hendizadeh et al., 2008). This chapter is concerned with multi machine Flow Shop Scheduling Problems (FSPs) that present a class of Group Shop Scheduling Problems in which the operations of every job have to be processed on m machines in this same order. Johnson (1954) has shown that, in a 2-machine flow shop, an optimal sequence can be constructed. It was demonstrated later that m-machine flow shop scheduling problem (FSP) is strongly NP-hard for $m \geq 3$ (Garey et al., 1976, Lenstra et al., 1977). The criterion of optimality in a flow shop sequencing problem is usually specified as minimization of make-span that is defined as the total time to ensure that all jobs are completed on all machines. If there are no release times for the jobs then the total completion time equals the total flow time. In some cases for calculating the completion times specific constraints are assumed. For example, such a situation in the FSP arises when no idle time is allowed at machines. This constraint creates an important practical situation that arises when expensive machinery is employed (Chakraborty, 2009). The general scheduling problem for a classical shop flow gives rise to $(n!)^m$ possible schedules. With aim to reduce the number of possible schedules it is reasonable to make assumption that all machines process jobs in the same order (Gupta 1975). In the classical flow-shop scheduling problem, queues of jobs are allowed at any of m machines in processing sequence based on assumption that jobs may wait on or between the machines (Allahverdi et al., 1999). The proposed alternative algorithm for minimizing completion time is assumed for a static-deterministic permutation flow shop scheduling problem (PFSP) with n jobs and m machines. In this chapter, the objective function for the PFSS problem corresponds to the minimization of the make-span when idle time is allowed on machines.

LITERATURE REVIEW

One of the important factors that are quite frequently discussed in FSPs is the setup time (see, for instance, Allahverdi et al., 2008). The setup time represents the time required to shift from one job to another on the given machine. In the cellular manufacturing environment, "usually a negligible or minor setup is needed to change from one job to another within a family and hence can be included in the processing times of each job" (Hendizadeh et al., 2007). While this assumption considerably simplifies the problem's solution, it harmfully affects the solution quality for many applications (Allahverdi et al., 1999).

The flow-shop problem with make-span (c_{max}) criterion under above mentioned assumption can be denoted as either $n/m/F/c_{max}$ or $F//c_{max}$, where both are related to an n-job and m-machine. This notation was firstly suggested by Conway et al. (1967) and until now is handy. Pinedo (2008) introduced a term permutation flow-shop problem in which the processing sequence on the first machine is maintained throughout the remaining machines. Accordingly, the make-span criterion is denoted as $F/prmu/C_{max}$.

Solution methods for flow shop scheduling range from heuristics developed, for example, by Palmer (1965), Campbell et al. (1970), Dannenbring (1977) to more complex techniques such as Branch and Bound (Brucker, 1994), Tabu Search (Gendreau, 1998), Genetic Algorithm (Murata et al., 1996), Shifting Bottleneck procedure (Balas andVazacopoulos, 1998), Ant Colony Algorithm (Blum and Sampels, 2004) and others.

The flow shop sequencing problem is one of the most well-known classic production scheduling problems. Focusing on the PFSP with C_{max} criterion function, first classical heuristics was proposed by Page (1961). Palmer (1965) adopted his idea and proposed the slope index to be utilized for the m-machine *n*-job permutation flow shop sequencing problem. A simple heuristic extension of Johnson's rule to m-machine flow shop problem has been proposed by Campbell et al. (1970). This extension is known in the literature as the CDS (Campbell, Dudek, and Smith) heuristic. Its principle relies on constructing at most (*m*-1) different sequences from which the best sequence is chosen. Each sequence corresponds to the application of Johnson's rule on a new 2-machine problem. Another method to obtain a minimum idle time based on the optimization of idle time last machine presented Gupta (1972). A significant approach to solving the FSP proposed Nawaz et al. (1983), in which they point out that a job with larger total processing time should have higher priority in the sequence.

Modern approaches designated for larger instances are known as meta-heuristics. Approaches that combine different concepts or components of more than one meta-heuristic are named as hybrid meta-heuristic algorithms (Zobolas et al., 2009). Heuristic methods for make-span minimization have been applied, for example, by Ogbu et al. (1990) using Simulated Annealing (SA) and by Taillard (1990) applying Tabu Search (TS) algorithm. Nagar et al. (1996) proposed a combined Branch-and-Bound (BB) and Genetic Algorithm (GA) based procedure for a flow shop scheduling problem with objectives of mean flow time and make-span minimization. Similarly, Neppalli et al. (1996) were used genetic algorithms in their approach to solve the 2-machine flow shop problem with objectives of minimizing make-span and total flow time. An atypical method based on an Artificial Immune System (AIS) approach, which was inspired from vertebrate immune system, has been presented by Engin and Doyen (2004). They used the proposed method for solving the hybrid flow shop scheduling problem with minimizing C_{max}. Obviously, there are plenty of other related approaches to this problem that are identified in survey studies, such as that of Ribas et al. (2010).

THE PROPOSED APPROACH TO MULTI STAGE FLOW SHOP SEQUENCING

In the multi stage sequencing problem, the following assumptions are made.

- There are '*n*' number of jobs (*J*) and '*m*' number of machines (*M*).
- The order of sequence of operations in all machines is the same.
- The setup time is not considered for calculating make-span time.

The proposed approach works with simple steps as given in the following section 'The Algorithm Description'. The optimum sequence is identified in step 7 that adopts the method of Johnson's algorithm (Johnson, 1954), which is used to find out minimum make-span while 2-machine production schedules are included.

The Algorithm Description

Step 1: Find out the sum of processing time of n jobs in machine M1. Repeat Step 1 for machines j=1,2,3,...,m.

Step 2: Make two groups from m machines in such a way that:

$$\sum_{j=1}^{x} T_i \sim \sum_{j=x+1}^{m} T_i \rightarrow \text{minimum} \quad (1)$$

Step 3: Find out the total machines in each group. Let the number of machines in Group I = a and the number of machines in Group II = b.

Step 4: Calculate the total operational time T of jobs in each group using the formula

for the Group I and Job (J1):

$$T^{I}_{J1} = (a.t_{11})+[(a-1).t_{12}]+[(a-2).t_{13}]+\ldots+(1.t_{1a}) \quad (2)$$

Similarly calculate these values for jobs J2, J3,…,Jn.

for the Group II and Job (J1):

$$T^{II}_{J1} = (b.t_{1m})+[(b-1).t_{1m-1}]+[(b-2).t_{1m-2}]+\ldots+(1.t_{1a+1}) \quad (3)$$

Similarly calculate these values for jobs J2, J3,…,Jn.

Step 5: Tabulate these values in two rows.

Step 6: Apply a final step of Johnson's rule to find out the best sequence.

Step 7: Calculate the make-span time for the sequence obtained in Step 6.

Step 8: Store the results k.

The Algorithm Illustration

To evaluate the proposed algorithm the following 6-jobs and 5-machines problem from a real life has been used. Input values for a calculation of total operational time T of jobs in each group are shown in Table 1.

In Table 1 each row represents machine 'j' and each column represents job 'i'. The processing time of an operation of the jobs is mentioned in each cell and denoted as 't_{ij}'.

The sum of processing time of all 5 jobs in each machine is calculated in the column 'T_i' as shown in Tables 2 and 3. Two groups are formed based on the formula as given below.

$$\sum_{j=1}^{a} T_i \sim \sum_{j=a+1}^{m} T_i \rightarrow \text{minimum} \quad (4)$$

(a = the arbitrary value from 1 to 5)

Table 1. Input data for the PFSP problem of size 6 machines and 5 jobs

j\i	J1	J2	J3	J4	J5
M1	1	1,5	1,5	1	1
M2	0,5	0,75	0,75	0,5	0,5
M3	0,5	1	0,5	0,5	0,5
M4	0,5	1	0,5	0,5	0,5
M5	0,1	0,5	0,2	0,1	0,1
M6	0,2	0,3	0,3	0,1	0,1

Table 2. Group I - pseudo problem of size 5/2

j\i	J1	J2	J3	J4	J5	T_i	∑Ti
M1	1	1,5	1,5	1	1	6	9
M2	0,5	0,75	0,75	0,5	0,5	3	

Table 3. Group II - pseudo problem of size 5/4

j\i	J1	J2	J3	J4	J5	Ti	∑Ti
M3	0,5	1	0,5	0,5	0,5	3	8
M4	0,5	1	0,5	0,5	0,5	3	
M5	0,1	0,5	0,2	0,1	0,1	1	
M6	0,2	0,3	0,3	0,1	0,1	1	

$$\sum_{j=1}^{2} T_i - \sum_{j=3}^{6} T_i = 9 - 8 \quad (5)$$

Thus, the total number of machines in each group is identified.

The number of machines in Group I (Table 2), $a = 2$ (M1 and M2 are in Group-I, noted as I).

The number of machines in Group II (Table 3), $b = 4$ (M3, M4, M5 and M6 are in Group-II, noted as II).

Subsequently, for the identified groups I and II the values of T^{I}_{Ji} and T^{II}_{Ji} (for i=1 to n) are calculated for all five jobs (Table 4).

The T^{I}_{Ji} and T^{II}_{Ji} values are tabulated as shown in Table 5.

As per the step 6 of the algorithm, the best sequences obtained in this method are **J1-J2-J3-J5-J4** (or) **J1-J2-J3-J4-J5**.

When idle time is allowed on machines the make-span is calculated for the sequence J1-J2-J3-J5-J4 (see Table 6) since both sequences in given case brings identical scheduling results.

Table 4. Identified groups calculated for all five jobs

T^{I}_{J1} = (2x1,0) + 0.50 = 2,5	T^{II}_{J1} = (4 x 0.2) + (3 x 0.1) + (2 x 0.5) + 0,5 = 2,6
T^{I}_{J2} = (2x1.5) + 0,75 = 3,75	T^{II}_{J2} = (4 x 0.3) + (3 x 0.5) + (2 x 1.0) + 1,0 = 5,7
T^{I}_{J3} = (2x1.5) + 0,75 = 3,75	T^{II}_{J3} = (4 x 0.3) + (3 x 0.2) + (2 x 0.5) + 0,5 = 3,3
T^{I}_{J4} = (2x1,0) + 0,50 = 2,5	T^{II}_{J4} = (4 x 0.1) + (3 x 0.1) + (2 x 0.5) + 0,5 = 2,2
T^{I}_{J5} = (2x1,0) + 0,50 = 2,5	T^{II}_{J5} = (4 x 0.1) + (3 x 0.1) + (2 x 0.5) + 0,5 = 2,2

Table 5. The sum values of two groups

Groups\ jobs	J1	J2	J3	J4	J5
T^{I}_{J}	2,5	3,75	3,75	2,5	2,5
T^{II}_{J}	2,6	5,7	3,3	2,2	2,2

With aim to combine criterion for calculating the minimum make-span schedules when idle time is allowed on machines along with criterion for minimum process interruptions it is possible to create job schedules by manner shown in Gant chart on Figure 1.

Testing of Proposed Algorithm

Proposed algorithm is tested for nine additional problems (see Table 7) that are randomly generated with varied sizes from 4 x 4 to 30 x 25. Input data for the nine problems are depicted in Tables 18-26 (see in Appendix of this chapter). While testing this algorithm on the big-size problems, it can be found that it is a time consuming procedure of finding out the difference between two groups (see Step 2 in the algorithm description section). So it is necessary to exploit a sub-algorithm to find out two groups with minimum difference. The algorithms steps are as follows:

1. Find out the difference between processing time of first machine and the sum of processing time of remaining machines.
2. Increment with the next machine difference between first two machines with remaining machines.
3. Repeat Step 2 until the minimum difference between two groups is reached.

Based on the calculation according to steps 1-6 from the above mentioned steps described in the section 'The Algorithm Description', were generated optimal sequences for the different problems that are shown in Table 8.

CONCURRENT ALGORITHMS FOR FLOW SHOP SEQUENCING

The currently reported approximation algorithms can be categorized into two types: constructive

Table 6. Proposed method J1-J2-J3-J5-J4

J	M1		M2		M3		M4		M5		M6	
I	In	Out	in	Out	In	out	In	out	In	Out	in	Out
J1	0	1	1	1,5	1,5	2	2	2,5	2,5	2,6	2,6	2,8
J2	1	2,5	2,5	3,25	3,25	4,25	4,25	5,25	5,25	5,75	5,75	6,05
J3	2,5	4	4	4,75	4,75	5,25	5,25	5,75	5,75	5,95	6,05	6,35
J5	4	5	5	5,5	5,5	6	6	6,5	6,5	6,6	6,6	6,7
J4	5	6	6	6,5	6,5	7	7	7,5	7,5	7,6	7,6	7,7

methods or improvement methods. Constructive methods include Slope index based heuristics, CDS heuristics and others. Most of improvement approaches are based on modern meta-heuristics, such as SA, TS and GA (Chakraborty, 2009). Mentioned modern meta-heuristic algorithms can be easily applied to various FSPs. Moreover, many papers showed that by them can be obtained better solution than by constructive methods. Results obtained by Kalczynski and Kamburowski (2005) showed that many meta-heuristic algorithms are not better than the simple NEH algorithm. It conforms to the No Free Lunch Theorem (Wolpert and Macready, 1997) that states that all algorithms equal to the randomly blind search if no problem information is known, or simply said, no algorithm is better to solve all the problems. That is why the comparison of concurrent algorithms with different sizes will be in the next sections analyzed and will be still important.

Gupta's Method

Gupta (1971) was argued that the sequencing problem is a problem of sorting *n* items so as to minimize make-span. He was proposed algorithm to schedule sequence of jobs for more than two machines in a flow shop. Given a set of n independent jobs, each having m ($m>2$) tasks that must be executed in the same sequence on m machines (P1, P2,…, Pm). Output is a schedule with a minimum completion time of the last job. This algorithm is stated as follows:

Figure 1. Gant chart for the criteria minimum make-span and minimum process interruptions

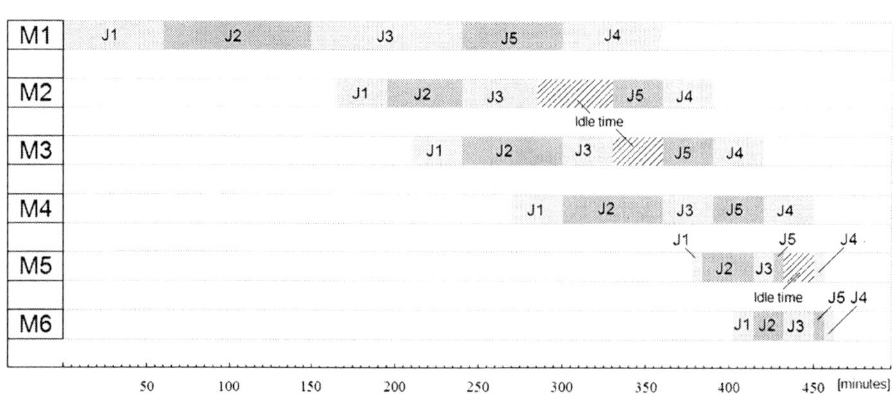

Step 1. Calculate the value of the function associated with job i, f(i), as follows:

$$f(i) = \frac{A}{\min\left(t_{ij} + t_{i,j+1}\right)}$$
$$for\ j = 1,2,\ldots,\left(M-1\right); \quad (6)$$
$$A\begin{cases} 1; \ if\ t_{iM} \le t_{i1}, \\ -1; \ otherwise. \end{cases}$$

Step 2. Arrange N jobs in ascending order of f(i) and in a favour of the job with the least sum of process times on all M machines.
Step 3. Calculate the make-span of the predetermined schedule via the recursive relation:

$$T_{ij}^{k} = \max\left[T_{ij}^{k-1}, T_{i,j-1}^{k}\right] + t_{ij} \quad (7)$$

Where T_{ij}^{k} is the cumulative processing time up to the kth order for the i job and j machine.

For the problems considered in Table 7, the Gupta's method works well and the best sequences for the 10 various sizes are given in Table 9.

Table 7. Sizes of tested scheduling problems

No.	Describing a scheduling problem	Size of scheduling problem
1	$F4 \mid prmu \mid C_{max}$	4x4
2	$F5 \mid prmu \mid C_{max}$	5x4
3	$F6 \mid prmu \mid C_{max}$	6x5
4	$F7 \mid prmu \mid C_{max}$	7x7
5	$F8 \mid prmu \mid C_{max}$	8x7
6	$F10 \mid prmu \mid C_{max}$	10x12
7	$F12 \mid prmu \mid C_{max}$	12x12
8	$F15 \mid prmu \mid C_{max}$	15x18
9	$F23 \mid prmu \mid C_{max}$	23x25
10	$F30 \mid prmu \mid C_{max}$	30x25

CDS Algorithm

The CDS algorithm, developed by Campbell, Dudek, and Smith (1970), has been used in many studies as a standard algorithm to compare with newly developed algorithms Ho and Chang (1991). The algorithm first constructs m – 1 two-machine problems. Then Johnson's 2-machine algorithm is applied. The formal steps for the algorithm execution are:

Step 1. Form number of auxiliary N-job and M-machine problems based on variable p, where p≤ M-1.
Step 2. Set k to 1 for the first auxiliary problem.
Step 3. Compute the total processing time for each job (*i*) on pseudo machine 1 (MC1), T_{i1}^{k} and pseudo machine 2 (MC2), T_{i2}^{k}:

$$T_{i1}^{k} = \sum_{i=1}^{k} t_{ij} \quad (8)$$

$$T_{i2}^{k} = \sum_{i=M+1-k}^{k} t_{ij} \quad (9)$$

Step 4. Apply Johnson's rules to N-job and 2-machine problem. Select the smallest processing time in the two-column processing time matrix. If the minimal processing time is $MC1_g$, do the gth job first. If it is $MC2_h$ do the hth job last.
Step 5. Set k to k+1 and repeat until k = p.
Step 6. Select the minimal total processing time sequence as the best sequence.

As it was above outlined, the CDS heuristics algorithm is basically an extension of Johnson's algorithm. The objective of the heuristic is the minimization of make-span in a deterministic flow shop problem. CDS heuristic forms in a simple manner a set of an m-1 artificial 2-machine subproblems for the original m-machine problem by

Table 8. Optimal sequences of the proposed algorithm (PA)

Problem size	Optimal Sequences for different problems
4x4	J4 → J1 → J2 → J3
5x4	J3 → J2 → J1 → J4
6x5	J1 → J2 → J3 → J5 → J4
7x7	J5 → J2 → J1 → J3 → J4 → J7 → J6
8x7	J2 → J4 → J7 → J1 → J3 → J6 → J5
10x12	J10 → J5 → J2 → J3 → J7 → J1 → J11 → J8 → J9 → J4 → J6 → J12
12x12	J10 → J12 → J6 → J7 → J5 → J1 → J4 → J9 → J2 → J3 → J8 → J11
15x18	J12 → J10 → J8 → J6 → J1 → J7 → J5 → J3 → J16 → J15 → J2 → J4 → J14 → J11 → J17 → J13 → J9 → J18
23x25	J18 → J13 → J9 → J24 → J17 → J16 → J15 → J2 → J5 → J7 → J23 → J21 → J14 → J11 → J1 → J3 → J4 → J19 → J22 → J6 → J20 → J8 → J12 → J25 → J10
30x25	J12 → J9 → J18 → J10 → J11 → J25 → J24 → J8 → J4 → J20 → J1 → J16 → J17 → J2 → J15 → J22 → J7 → J5 → J14 → J23 → J21 → J6 → J3 → J13 → J19

summing the processing times in a manner that combines $M_1, M_2, ..., M_{m-1}$ to pseudo machine 1 and $M_2, M_3, ..., M_m$ to pseudo machine 2. Finally, each of the 2-machine sub-problems is then solved using the Johnson's 2-machine algorithm. The best of the sequence is selected as the solution to the original m-machine problem. Let us understand from the following illustration.

In the example considered in the section 'The Algorithm Illustration' processing time for *M1* to *M5* is given as is shown in Figure 2a. The sum of processing time for all 5 machines is calculated and recorded in the last row of Table 10. This row is considered as pseudo machine 1.

Similarly, the processing time for M2 to M6 is given in Figure 2b where the sum of processing time for all 5 machines is calculated in the last row of Table 10. This row is considered as pseudo machine 2.

Let us view only the pseudo machine 1(ps_1) and 2 (ps_2) as can be seen from Table 10.

Table 9. Optimal sequences of Gupta algorithm

Problem size	Optimal Sequence for different problems
4x4	J4 → J1 → J3 → J2
5x4	J1 → J3 → J2 → J4
6x5	J2 → J3 → J1 → J5 → J4
7x7	J5 → J2 → J1 → J4 → J3 → J7 → J6
8x7	J3 → J1 → J7 → J6 → J5 → J4 → J2
10x12	J10 → J5 → J3 → J2 → J11 → J7 → J4 → J1 → J8 → J6 → J12 → J9
12x12	J12 → J9 → J10 → J6 → J7 → J5 → J3 → J1 → J4 → J2 → J11 → J8
15x18	J12 → J11 → J10 → J17 → J16 → J15 → J6 → J8 → J1 → J7 → J3 → J14 → J13 → J5 → J4 → J2 → J18 → J9
23x25	J25 → J10 → J20 → J18 → J16 → J13 → J7 → J5 → J2 → J17 → J3 → J24 → J23 → J8 → J22 → J21 → J19 → J15 → J14 → J6 → J4 → J1 → J12 → J11 → J9
30x25	J12 → J22 → J21 → J18 → J15 → J5 → J1 → J24 → J17 → J3 → J23 → J8 → J20 → J19 → J16 → J14 → J13 → J7 → J6 → J4 → J2 → J25 → J11 → J10 → J9

Figure 2. Illustration example of the CDS algorithm (a) pseudo machine 1 (b) pseudo machine 2

	J1	J2	J3	J4	J5
M1	1,0	1,5	1,5	1,0	1,0
M2	0,5	0,8	0,8	0,5	0,5
M3	0,5	1,0	0,5	0,5	0,5
M4	0,5	1,0	0,5	0,5	0,5
M5	0,1	0,5	0,2	0,1	0,1
Σ	2,6	4,8	3,5	2,6	2,6

a)

	J1	J2	J3	J4	J5
M2	0,5	0,8	0,8	0,5	0,5
M3	0,5	1,0	0,5	0,5	0,5
M4	0,5	1,0	0,5	0,5	0,5
M5	0,1	0,5	0,2	0,1	0,1
M6	0,2	0,3	0,3	0,1	0,1
Σ	1,8	3,6	2,3	1,7	1,7

b)

At this stage, the problem is considered as an n-job, 2-machine problem. The 2-machines sub-problem is now solved using the Johnson's 2-machines algorithm. The best of the sequence is selected as the solution to the original m-machine problem. Based on this procedure, the best sequences for the 10 particular sizes are given in Table 11.

Slope Index Method

The heuristic has been developed in an effort to use Johnson's rule for m≥3, since for m=2. This algorithm is slightly different from Johnson's algorithm. The idea of Slope Index method is to give priority to jobs so that jobs with processing times that tend to increase from machine to machine will receive higher priority, while jobs with processing times that tend to decrease from machine to machine will receive lower priority.

The slope index (SI) for job 'i' is calculated as:

$$SI_i = \sum_{j=1}^{m}(2j - m - 1)t_{ij}, i = 1,2,...,n. \qquad (10)$$

Table 10. Processing time of each pseudo machine

	J1	J2	J3	J4	J5
ps_1	2,6	4,75	3,45	2,6	2,6
ps_2	1,8	3,55	2,25	1,7	1,7

For illustration, the calculated values of Slope Indices for the particular 6-jobs and 5-machines problem are depicted in Figure 3.

Then a permutation sequence is determined by ordering the jobs in decreasing order of SI_i such as:

$$SI_{i1} \geq SI_{i2} \geq .... \geq SI_{in} \qquad (11)$$

By applying expression 5 for values in Figure 3, the **J1-J4-J5-J2-J3** sequence was obtained. The make-span calculation for the sequence is displayed in the Table 12.

Based on this procedure, the best sequences for the 10 particular sizes shown in Table 13 have been obtained.

Genetic Algorithm (GA)

Genetic Algorithm is a computerized search and optimization algorithm based on the mechanics of natural genetics and natural selection. GA is a search technique for global optimization in a search space. As the term suggests, they employ the concepts of natural selection and genetics using past information for directing the search with expected improved performance to achieve fairly consistent and reliable results. The traditional methods of optimization and search do not work well over a broad spectrum of problem domains. The GA attempts to mimic the biological evolution process for discovering good solutions. They are

Table 11. Optimal sequence of jobs based on CDS algorithm

Problem size	Optimal Sequence CDS
4x4	J4 → J1 → J2 → J3
5x4	J1 → J2 → J3 → J4
6x5	J2 → J3 → J1 → J5 → J4
7x7	J5 → J2 → J1 → J3 → J4 → J6 → J7
8x7	J4 → J7 → J6 → J1 → J5 → J3 → J2
10x12	J10 → J7 → J11 → J5 → J1 → J3 → J2 → J4 → J6 → J8 → J9 → J12
12x12	J10 → J12 J9 → J7 → J6 → J2 → J3 → J1 → J5 → J4 → J8 → J11
15x18	J10 → J12 → J8 → J11 → J17 → J1 → J7 → J14 → J15 → J16 → J6 → J3 → J5 → J13 → J4 → J2 → J18 → J9
23x25	J10 → J25 → J18 → J13 → J20 → J17 → J14 → J7 → J23 → J2 → J16 → J15 → J3 → J21 → J5 → J6 → J22 → J1 → J4 → J24 → J11 → J8 → J19 → J9 → J12
30x25	J12 → J9 → J18 → J24 → J17 → J16 → J22 → J1 → J4 → J15 → J5 → J23 → J21 → J6 → J3 → J7 → J2 → J14 → J11 → J20 → J8 → J25 → J19 → J13 → J10

Figure 3. An example of slope indices calculation

J1	J2	J3	J4	J5
-5	-7,5	-7,5	-5	-5
-1,5	-2,3	-2,3	-1,5	-1,5
-0,5	-1	-0,5	-0,5	-0,5
0,5	1	0,5	0,5	0,5
0,3	1,5	0,6	0,3	0,3
1	1,5	1,5	0,5	0,5
-5,2	-6,8	-7,7	-5,7	-5,7
S1	S2	S3	S4	S5

based on a direct analogy to Darwinian natural selection and mutations in biological reproduction and belong to a category of heuristics known as randomized heuristics that employ randomized choice operators in their search strategy and do not depend on complete a priori knowledge of the features of domain. These operators have been conceived through abstractions of natural genetic mechanisms such as crossover and mutation and have been cast into algorithmic forms. Holland (1976) was envisaged the concept of these algorithms in the mid-sixties and it has been applied in diverse areas such as music generation, genetic synthesis, fault diagnosis, strategy planning and also to address business problems such as Traveling Salesman Problem, production planning and scheduling problem, facility location problem, transportation problems, telecommunications and network problems, engineering design problems

Table 12. Make-span by the Slope index method

J	M1		M2		M3		M4		M5		M6	
I	In	out	In	out	In	out	in	Out	In	Out	in	Out
J1	0	1	1	1,5	1,5	2	2	2,5	2,5	2,6	2,6	2,8
J4	1	1,5	1,5	2	2	2,5	2,5	3	3	3,1	3,1	3,2
J5	1,5	2,5	2,5	3	3	3,5	3,5	4	4	4,1	4,1	4,2
J2	2,5	4	4	4,8	4,8	5,8	5,8	6,8	6,8	7,3	7,3	7,6
J3	4	5,5	5,5	6,3	6,3	6,8	6,8	7,3	7,3	7,5	7,6	7,9

Table 13. Optimal sequence using slope index approach

Problem Size	Optimal Sequence
4x4	J4 → J1 → J3 → J2
5x4	J3 → J2 → J1 → J4
6x5	J1 → J4 → J5 → J2 → J3
7x7	J2 → J5 → J1 → J3 → J6 → J7 → J4
8x7	J7 → J1 → J6 → J2 → J3 → J4 → J5
10x12	J2 → J5 → J10 → J3 → J8 → J9 → J1 → J7 → J11 → J6 → J4 → J12
12x12	J6 → J12 → J10 → J7 → J5 → J9 → J1 → J4 → J8 → J11 → J3 → J2
15x18	J6 → J1 → J10 → J8 → J7 → J12 → J4 → J5 → J3 → J11 → J16 → J2 → J9 → J17 → J14 → J15 → J13 → J18
23x25	J17 → J18 → J9 → J16 → J24 → J15 → J13 → J5 → J14 → J2 → J19 → J12 → J8 → J1J11 → J23 → J20 → J7 → J21 → J10 → J4 → J25 → J22 → J6 → J3
30x25	J1 → J6 → J5 → J10 → J4 → J18 → J21 → J3 → J2 → J11 → J8 → J22 → J17 → J23 → J12 → J25 → J24 → J7 → J15 → J9 → J20 → J16 → J14 → J13 → J19

and image processing and cell design problems. The GA differs from traditional optimization and search techniques in the following ways. It works with a coding of parameters; not with parameter themselves. The GA searches from population of points; not from a single point. It uses probabilistic rules rather than deterministic rules. In the GA, the solution is represented in terms of specific coding, for which the number of generations or iterations needs to be generated (see Figure 4). The best solution will be searched in a solution space and narrowed down as per the requirement.

In this chapter the GA is used to search for solution of make-span minimization and the results obtained from the GA are compared with the results of proposed algorithm and benchmarked algorithms as shown in Table 15. From this table it is evident that used GA method gives the best results. The following steps shown in Figure 5 introduce the GA code used for this purpose.

Representation

Representation is made in the form of solution string (t). In this problem considered, each gene represents Job number and the chromosome represents the sequence of various jobs (i.e. J1 J4 J5 J2 J3).

Figure 4. Pseudo code of genetic algorithm

- Generate and evaluate the initial population P(t), t = 0.
- Repeat the following steps until stopping condition is satisfied.
  - Selection of chromosomes from the current population
  - Apply the crossover over the parent chromosomes and produce two offsprings.
  - Apply the mutation operator over the offspring.
  - Copy the offspring to population P(t + 1).
  - Evaluate P(t + 1).
  - Replace the worst chromosome of P(t + 1) by the best chromosome found so far.
- Set t = t + 1.
- Return the best chromosome found.

Reproduction

The objective value (make-span) is computed for each string in the population and the objective is to find a string with the minimum value. The advantage is that the worst string will never be reproduced into the next generation.

Table 14. GA constraints

Parameter	Value
Population size	15
Generation Number	500
Crossover Probability	0.5
Mutation Probability	0.1

Crossover and Mutation

The crossover operator is carried out with a probability known as crossover probability. Crossover is exchange of a portion of strings at a point called crossover site. The two strings, which take part in the crossover operation, are also selected at random.

Mutation is also done randomly for each gene and it depends upon another parameter called mutation probability. In this method inversion mutation is adopted where one gene is selected at random, and exchanged with another gene mutually.

Table 15. Best sequences from the GA algorithm results

Problem size	Best Sequence Genetic Algorithm
4x4	J1 → J4 → J2 → J3
5x4	J1 → J3 → J2 → J4
6x5	J1 → J2 → J3 → J4 → J5
7x7	J5 → J1 → J2 → J3 → J4 → J7 → J6
8x7	J7 → J1 → J6 → J5 → J3 → J4 → J2
10x12	J5 → J9 → J8 → J7 → J10 → J11 → J12 → J1 → J2 → J3 → J6 → J4
12x12	J5 → J7 → J12 → J6 → J9 → J10 → J11 → J1 → J2 → J4 → J3 → J8
15x18	J4 → J6 → J3 → J5 → J2 → J7 → J1 → J9 → J10 → J11 → J12 → J13 → J14 → J15 → J16 → J18 → J8 → J17
23x25	J14 → J12 → J13 → J11 → J10 → J15 → J25 → J17 → J18 → J19 → J20 → J21 → J23 → J24 → J16 → J1 → J2 → J3 → J4 → J5 → J7 → J8 → J9 → J6 → J22
30x25	J19 → J8 → J22 → J23 → J20 → J1 → J2 → J3 → J4 → J5 → J6 → J7 → J21 → J9 → J10 → J11 → J12 → J13 → J14 → J15 → J16 → J17 → J18 → J24 → J25

Figure 5. The genetic algorithm for minimization of make-span

Step 1. Generate initial population of strings = 20 (job sequence).
Step 2. Find out the fitness value (make-span value)
Step 3. Reproduction of the strings with better make-span values
Step 4. Apply cross over with crossover probability $P_c = 0.5$
Step 5. Apply mutation with mutation probability $P_m = 0.1$
Step 6. Find out the fitness value.
Step 7. Store the best values.
Step 8. Go to step 3 and iterate till the generation value (gen=500).

For purpose of testing the genetic algorithm on 10 particular problems, the following GA constraints shown in Table 14 have been used.

The best sequences generated through GA for various sizes are given in the Table 15.

COMPARISON WITH BENCHMARKED ALGORITHMS

In order to compare the proposed algorithm (PA) against the benchmarked, the algorithms described above have been used. The make-spans for heuristic and meta-heuristic algorithms are calculated and displayed in the Table 16. Sequences obtained by using GA for the same PFS problems mostly equals to the sequences calculated by PA.

The values in Table 16 were evaluated based on percentage deviations from the best results obtained through genetic algorithm technique. As it is evident, the Genetic Algorithm is ideal for finding optimal solutions among the concurrent algorithms. In order to compare the performance of the algorithm and decide which algorithm is better to solve the particular problems average deviations values were subsequently calculated for each technique. The average values of percentage deviations are graphically displayed in Figure 6.

From the graph, it can be seen that proposed algorithm gives the closest result (2% average deviation from the optimal values) to the solution generated by meta-heuristic technique. Other heuristic algorithms are in the range of 3,14% to 5,24%.

Let us now consider the total number of permutations of all four benchmarked heuristic algorithms. Genetic algorithm is not compared to them since it is known that this meta-heuristic generates the maximum number of permutations for a given size of matrix elements. The comparison of benchmarked algorithms based on total number of permutations for a given size of matrix elements is shown in Table 17.

Based on the results in this table, we have the following conjecture. CDS, Slope index procedure and proposed algorithm generate broadly the same number of permutations. But numbers of permutations generated by Gupta's Method are in contrast to the much smaller numbers of permutations from CDS, SI and PA, especially, starting from the size 10x12.

Table 16. Comparative results of make-span

No.	Size	Optimal	Gupta	Dev [%]	CDS	Dev [%]	Slope	Dev [%]	GA	Dev [%]	PA	Dev [%]
1.	4x4	156,0	157,0	0,64	156,0	0,00	157,0	0,64	156,0	0,00	157,0	0,64
2.	5x4	51,0	51,0	0,00	51,0	0,00	53,0	3,92	51,0	0,00	54,0	5,88
3.	6x5	7,7	7,7	0,00	7,7	0,00	8,35	8,44	7,7	0,00	7,7	0,00
4.	7x7	65,0	65,0	0,00	67,0	3,08	75,0	15,38	65,0	0,00	65,0	0,00
5.	8x7	66,0	69,0	4,55	66,0	0,00	70,0	6,06	66,0	0,00	71,0	7,58
6.	10x12	96,0	106,0	10,42	104,0	8,33	104,0	8,33	96,0	0,00	96,0	0,00
7.	12x12	110,0	111,0	0,91	114,0	3,64	115,0	4,55	110,0	0,00	111,0	0,91
8.	15x18	145,0	163,0	12,41	153,0	5,52	146,0	0,69	145,0	0,00	145,0	0,00
9.	23x25	236,0	264,0	11,86	259,0	9,75	241,0	2,12	236,0	0,00	239,0	1,27
10.	30x25	268,0	285,0	6,34	271,0	1,12	274,0	2,24	268,0	0,00	278,0	3,73
Average				4,71		3,14		5,24		0,00		2,00

Figure 6. Graph of average deviations

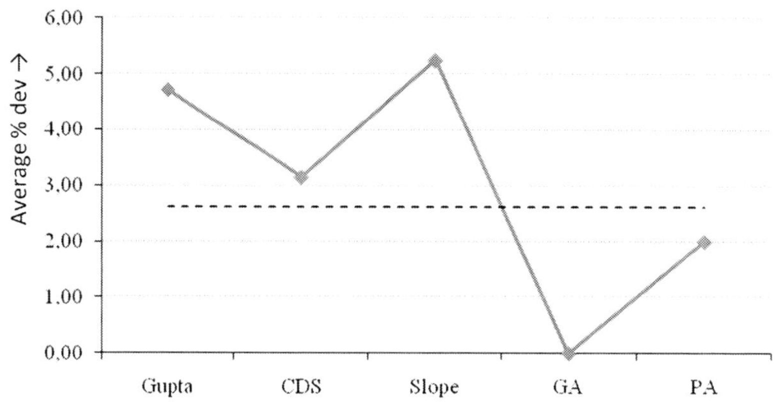

CONCLUSION

In the present study, the scheduling problem with sequence-dependent operations is dealt. The main idea is to minimize the make-span time and thereby reducing the idle time of both jobs and machines since these criteria are often applied for operational decision-making in scheduling. Based on the tested problems the proposed approach is producing at least comparable results than the benchmarked algorithms as shown in Table 16.

Many heuristics and meta-heuristics can find fast and feasible solutions to such sequencing problems that involve multiple jobs and machines and sequence-dependent operations. Taking into an account the viewpoints of the algorithm simplicity and obtained results, then the proposed method seems to be more effective than the benchmarked methods. In realistic situation, the proposed algorithm can be used as it is without any modification and come out with acceptable results.

Table 17. Number of permutations for heuristic algorithms

No.	Size	Gupta's method	CDS algorithm	Slope Index	Proposed algorithm
		No. of Perm	No. of Perm	No. of Perm	No. of Perm
1.	4x4	4	1	1	1
2.	5x4	2	1	1	1
3.	6x5	2	2	2	2
4.	7x7	4	1	1	1
5.	8x7	4	2	2	1
6.	10x12	12	3	4	1
7.	12x12	384	1	1	2
8.	15x18	69120	3	2	2
9.	23x25	14631321600	3	1	2
10.	30x25	6270566400	3	1	2

REFERENCES

Allahverdi, A., Gupta, J. N. D., & Aldowaisan, T. (1999). A review of scheduling research involving set-up considerations. *Omega*, *27*, 219–239. doi:10.1016/S0305-0483(98)00042-5

Allahverdi, A., Ng, C. T., Cheng, T. C. E., & Kovalyov, M. Y. (2008). A survey of scheduling problems with setup times or costs. *European Journal of Operational Research*, *187*, 985–1032. doi:10.1016/j.ejor.2006.06.060

Balas, E., & Vazacopoulos, A. (1998). Guided local search with shifting bottleneck for job shop scheduling. *Management Science*, *44*(2), 262–275. doi:10.1287/mnsc.44.2.262

Blum, C., & Sampels, M. (2004). An ant colony optimization algorithm for shop scheduling problems. *Journal of Mathematical Modelling and Algorithms*, *3*(3), 285–308. doi:10.1023/B:JMMA.0000038614.39977.6f

Brucker, P., Jurisch, B., & Sievers, B. (1994). A branch and bound algorithm for the job shop scheduling problem. *Discrete Applied Mathematics*, *49*(1), 109–127. doi:10.1016/0166-218X(94)90204-6

Campbell, H. G., Dudek, R. A., & Smith, M. L. (1970). A heuristic algorithm for the n job, m machine sequencing problem. *Management Science*, *16*(10), 630–637. doi:10.1287/mnsc.16.10.B630

Chakraborty, U. K. (2009). *Computational intelligence in flow shop and job shop scheduling - Studies in computational intelligence*. Springer.

Conway, R. W., Maxwell, W. L., & Miller, L. W. (1967). *Theory of scheduling*. Reading, MA: Addison-Wesley.

Cox, J. F., Blackstone, J. H. Jr, & Spencer, M. S. (Eds.). (1992). *American production and inventory, control society*. Falls Church, VA: APICS.

Dannenbring, D. G. (1977). An evaluation of flow shop sequencing heuristics. *Management Science*, *23*(11), 1174–1182. doi:10.1287/mnsc.23.11.1174

Engin, O., & Doyen, A. (2004). A new approach to solve hybrid flow shop scheduling problems by artificial immune system. *Future Generation Computer Systems*, *20*, 1083–1095. doi:10.1016/j.future.2004.03.014

Garey, M. R. D., Johnson, D. S., & Sethi, R. (1976). The complexity of flowshop and jobshop scheduling. *Mathematics of Operations Research*, *1*, 117–129. doi:10.1287/moor.1.2.117

Gendreau, M., Laporte, G., & Semet, F. (1998). A tabu search heuristic for the undirected selective travelling salesman problem. [Elsevier.]. *European Journal of Operational Research*, *106*(2-3), 539–545. doi:10.1016/S0377-2217(97)00289-0

Gupta, J. (1971). A functional heuristic algorithm for the flowshop scheduling problem. *Operational Research Quarterly*, *22*, 39–48.

Gupta, J. N. D. (1972). Heuristic algorithms for multistage flowshop scheduling problem. *AIIE Transactions*, *4*(1), 11–18.

Gupta, J. N. D. (1975). *Analysis of combinatorial approach to flowshop scheduling problems.*

Hejazi, S. R., & Saghafian, S. (2005). Flowshop scheduling problems with makespan criterion: A review. *International Journal of Production Research*, *43*(14), 2895–2929. doi:10.1080/0020754050056417

Hendizadeh, S. H., ElMekkawy, T. Y., & Wang, G. G. (2007). Bi-criteria scheduling of a flowshop manufacturing cell with sequence dependent setup time. *European Journal of Industrial Engineering*, *1*, 391–413. doi:10.1504/EJIE.2007.015388

Hendizadeh, S. H., Faramrzi, H., Mansouri, S. A., Gupta, J. N. D., & ElMekkawy, T. (2008). Metaheuristics for scheduling a flowline manufacturing cell with sequence dependent family setup times. *International Journal of Production Economics*, *111*(2), 593–60. doi:10.1016/j.ijpe.2007.02.031

Herrmann, J. W. (2006). *Handbook of production scheduling*. New York, NY: Springer.

Ho, J., & Chang, Y. (1991). A new heuristic for the n-job, m-machine flowshop problem. *European Journal of Operational Research*, *52*, 194–202. doi:10.1016/0377-2217(91)90080-F

Holland, J. H. (1976). *Adaptation in nature and artificial systems*. Ann Arbor, MI: University of Michigan Press.

Johnson, S. M. (1954). Optimal two and three stage production schedules with set-up times. *Naval Research Logistics Quarterly*, *1*, 61–68. doi:10.1002/nav.3800010110

Kalczynski, P. J., & Kamburowski, J. (2005). On the NEH heuristic for minimising the makespan in permutation flowshops. [Omega.]. *The International Journal of Management Science*, *35*(1), 53–60.

Lenstra, J. K., Rinnooy Kan, A. H. G., & Brucker, P. (1977). Complexity of machine scheduling problems. *Ann Discrete Math*, *1*, 343–362. doi:10.1016/S0167-5060(08)70743-X

Modrák, V. (2008). Integration of MES and ERP. In Koshrow-Pour, M. (Ed.), *Encyclopedia of Information Science and Technology* (2nd ed., pp. 2103–2106). Hershey, PA: Idea Group. doi:10.4018/978-1-60566-026-4.ch330

Mosier, C. T., Elvers, D. A., & Kelly, D. (1984). Analysis of group technology scheduling heuristics. *International Journal of Production Research*, *22*, 857–875. doi:10.1080/00207548408942505

Murata, T., Ishibuchi, H., & Tanaka, H. (1996). Genetic algorithms for flowshop scheduling problems. *Computers & Industrial Engineering*, *30*(4), 1061–1071. doi:10.1016/0360-8352(96)00053-8

Nagar, A., Heragu, S. S., & Haddock, J. (1996). A combined branch-and-bound and genetic algorithm based approach for a flowshop-scheduling problem. *Annals of Operations Research*, *63*, 397–414. doi:10.1007/BF02125405

Nawaz, M. E., Enscore, I., & Ham, I. (1983). A heuristic algorithm for the m machine, n job flow shop sequence problem. *Omega*, *11*(1), 91–95. doi:10.1016/0305-0483(83)90088-9

Neppalli, V. R., Chen, C. L., & Gupta, J. N. D. (1996). Genetic algorithms for the two-stage bicriteria flowshop problem. *European Journal of Operational Research*, *95*, 356–373. doi:10.1016/0377-2217(95)00275-8

Ogbu, F. A., & Smith, D. K. (1990). The application of the simulated annealing algorithm to the solution of the n/m/Cmax owshop problem. *Computers & Operations Research*, *17*, 3243–3253. doi:10.1016/0305-0548(90)90001-N

Page, E. S. (1961). An approach to scheduling of jobs on the machines. *Journal of the Royal Statistical Society. Series A (General)*, *23*, 484–492.

Palmer, D. S. (1965). Sequencing jobs through a multi-stage process in the minimum total time - a quick method of obtaining a near optimum. *Operational Research Quarterly*, *16*, 101–107. doi:10.1057/jors.1965.8

Pinedo, M. (2008). *Scheduling: Theory, algorithms and systems. Prentice Hall*. NJ: Springer.

Ribas, R., & Leisten, J. M. (2010). Review and classification of hybrid flow shop scheduling problems from a production system and a solutions procedure perspective. *Computers & Operations Research*, *37*(8), 1439–1454. doi:10.1016/j.cor.2009.11.001

Ruben, R. A., & Mahmoodi, F. (1999). Scheduling cellular manufacturing systems. In Irani, S. A. (Ed.), *Handbook of cellular manufacturing systems*. USA: John Willey and Sons. doi:10.1002/9780470172476.ch5

Taillard, E. (1990). Some efficient heuristic methods for the flow shop sequencing problem. *European Journal of Operational Research*, *47*(1), 65–74. doi:10.1016/0377-2217(90)90090-X

Wight, O. W. (1984). *Production and inventory management in the computer age*. New York, NY: Van Nostrand Reinhold Company, Inc.

Wolpert, D. H., & Macready, W. G. (1997). No free lunch theorems for optimization. [IEEE Press.]. *IEEE Transactions on Evolutionary Computation*, *1*(1), 67–82. doi:10.1109/4235.585893

Zobolas, G. I., Tarantilis, C. D., & Ioannou, G. (2009). Minimizing makespan in permutation flow shop scheduling problems using a hybrid metaheuristic algorithm. *Computers & Operations Research*, *36*(4), 1249–1267. doi:10.1016/j.cor.2008.01.007

APPENDIX

Table 18. Illustration: for the problem of size 4 machines x 4 jobs

	J1	J2	J3	J4
M1	24	61	22	21
M2	7	9	8	6
M3	7	5	6	8
M4	29	15	14	32

Table 19. Illustration: for the problem of size 5 machines x 4 jobs

	J1	J2	J3	J4
M1	7	6	5	8
M2	5	6	4	3
M3	2	4	5	3
M4	3	5	6	2
M5	9	10	8	6

Table 20. Illustration: for the problem of size 7 machines x 7 jobs

	J1	J2	J3	J4	J5	J6	J7
M1	3	2	4	5	1	3	5
M2	5	5	8	7	2	5	2
M3	7	8	1	6	8	4	8
M4	1	1	6	1	4	6	4
M5	6	6	7	8	6	8	6
M6	9	7	9	4	7	1	3
M7	4	9	1	3	4	2	2

Table 21. Illustration: for the problem of size 8 machines x 7 jobs

	J1	J2	J3	J4	J5	J6	J7
M1	5	2	4	2	5	2	1
M2	1	1	5	1	4	8	5
M3	4	2	2	5	8	5	4
M4	5	5	3	4	9	5	2
M5	8	4	5	6	2	6	4
M6	2	5	8	3	4	5	5
M7	4	3	2	2	8	8	2
M8	8	2	5	2	5	5	8

Table 22. Illustration: for the problem of size 10 machines x 12 jobs

	J1	J2	J3	J4	J5	J6	J7	J8	J9	J10	J11	J12
M1	3	2	4	5	1	3	5	4	3	2	4	6
M2	5	5	8	7	2	5	2	2	5	1	6	2
M3	7	8	1	6	8	4	8	2	4	1	6	4
M4	1	1	6	1	4	6	4	5	1	6	2	4
M5	6	6	7	8	6	8	6	7	1	2	4	3
M6	9	7	9	4	7	1	3	2	3	1	5	2
M7	4	9	1	3	4	2	2	6	7	2	5	4
M8	5	6	4	3	7	3	6	1	4	2	3	1
M9	2	4	5	3	2	1	2	4	3	1	5	1
M10	3	5	6	2	3	2	5	3	1	4	2	1

Table 23. Illustration: for the problem of size 12 machines x 12 jobs

	J1	J2	J3	J4	J5	J6	J7	J8	J9	J10	J11	J12
M1	6	6	7	8	6	8	6	7	1	2	4	3
M2	9	7	9	4	7	1	3	2	3	1	5	2
M3	4	9	1	3	4	2	2	6	7	2	5	4
M4	5	6	4	3	7	3	6	1	4	2	3	1
M5	2	4	5	3	2	1	2	4	3	1	5	1
M6	3	5	6	2	3	2	5	3	1	4	2	1
M7	5	2	4	2	5	2	1	1	1	2	1	2
M8	1	1	5	1	4	8	5	7	8	3	1	3
M9	4	2	2	5	8	5	4	1	1	1	3	2
M10	5	5	3	4	9	5	2	1	2	4	1	4
M11	8	4	5	6	2	6	4	2	2	1	4	4
M12	6	3	1	3	6	9	6	4	5	3	1	4

Table 24. Illustration: for the problem of size 15 machines x 18 jobs

	J1	J2	J3	J4	J5	J6	J7	J8	J9	J10	J11	J12	J13	J14	J15	J16	J17	J18
M1	5	6	4	3	7	3	6	1	4	2	3	1	3	2	4	5	1	3
M2	2	4	5	3	2	1	2	4	3	1	5	1	5	5	8	7	2	5
M3	3	5	6	2	3	2	5	3	1	4	2	1	7	8	1	6	8	4
M4	5	2	4	2	5	2	1	1	1	2	1	2	1	1	6	1	4	6
M5	1	1	5	1	4	8	5	7	8	3	1	3	6	6	7	8	6	8
M6	4	2	2	5	8	5	4	1	1	1	3	2	9	7	9	4	7	1
M7	3	2	4	5	1	3	5	4	3	2	4	6	4	9	1	3	4	2
M8	5	5	8	7	2	5	2	2	5	1	6	2	4	2	5	2	1	1
M9	7	8	1	6	8	4	8	2	4	1	6	4	5	1	4	8	5	7
M10	1	1	6	1	4	6	4	5	1	6	2	4	2	5	8	5	4	1
M11	6	6	7	8	6	8	6	7	1	2	4	3	4	5	1	3	5	4
M12	8	4	5	6	2	6	4	2	1	3	5	1	2	1	2	1	2	3
M13	2	5	8	3	4	5	5	2	4	1	3	2	1	5	1	2	4	2
M14	4	3	2	2	8	8	2	3	2	5	1	2	3	4	5	7	2	1
M15	8	2	5	2	5	5	8	6	2	5	4	3	2	2	5	8	2	1

Table 25. Illustration: for the problem of size 23 machines x 25 jobs

	J1	J2	J3	J4	J5	J6	J7	J8	J9	J10	J11	J12	J13	J14	J15	J16	J17	J18	J19	J20	J21	J22	J23	J24	J25
M1	5	5	8	7	2	5	2	2	5	1	6	2	4	2	5	2	1	1	9	3	3	2	5	3	1
M2	7	8	1	6	8	4	8	2	4	1	6	4	5	1	4	8	5	7	1	4	5	2	1	1	1
M3	1	1	6	1	4	6	4	5	1	6	2	4	2	5	8	5	4	1	2	5	4	8	5	7	8
M4	5	6	4	3	7	3	6	1	4	2	3	1	2	3	5	4	1	2	1	1	8	5	4	1	1
M5	2	4	5	3	2	1	2	4	3	1	5	1	2	5	4	5	6	2	3	2	1	3	5	4	3
M6	3	5	6	2	3	2	5	3	1	4	2	1	2	5	4	5	2	1	4	6	2	5	2	2	5
M7	2	5	8	3	4	5	5	2	4	1	3	2	1	5	1	2	4	2	1	4	8	4	8	2	4
M8	1	1	6	1	4	6	4	5	1	6	2	4	2	5	8	5	4	1	3	1	4	6	4	5	1
M9	6	6	7	8	6	8	6	7	1	2	4	3	4	5	1	3	5	4	1	4	7	3	6	1	4
M10	8	4	5	6	2	6	4	2	1	3	5	1	2	1	2	1	2	3	2	5	2	1	2	4	3
M11	2	5	8	3	4	5	5	2	4	1	3	2	1	5	1	2	4	2	1	2	3	2	5	3	1
M12	4	3	2	2	8	8	2	3	2	5	1	2	3	4	5	7	2	1	2	1	4	5	5	2	4
M13	5	2	4	2	5	2	1	1	1	2	1	2	1	1	6	1	4	6	3	3	8	8	2	3	2
M14	1	1	5	1	4	8	5	7	8	3	1	3	6	6	7	8	6	8	7	8	5	5	8	6	2
M15	4	2	2	5	8	5	4	1	1	1	3	2	9	7	9	4	7	1	7	3	6	8	6	7	1
M16	3	2	4	5	1	3	5	4	3	2	4	6	4	9	1	3	4	2	8	2	7	1	3	2	3
M17	5	5	8	7	2	5	2	2	5	1	6	2	4	2	5	2	1	1	9	3	4	2	2	6	7
M18	7	8	1	6	8	4	8	2	4	1	6	4	5	1	4	8	5	7	1	4	7	3	6	1	4
M19	1	1	6	1	4	6	4	5	1	6	2	4	2	5	8	5	4	1	2	5	2	1	2	4	3
M20	5	6	4	3	7	3	6	1	4	2	3	1	2	3	5	4	1	2	1	1	4	2	2	5	1
M21	2	4	5	3	2	1	2	4	3	1	5	1	2	5	4	5	6	2	3	2	2	5	4	3	1
M22	4	9	1	3	4	2	2	6	7	2	5	4	1	5	8	7	9	5	1	2	5	7	5	6	1
M23	5	6	4	3	7	3	6	1	4	2	3	1	5	2	4	6	8	7	4	5	2	1	5	2	3

Table 26. Illustration: for the problem of size 30 machines x 25 jobs

	J1	J2	J3	J4	J5	J6	J7	J8	J9	J10	J11	J12	J13	J14	J15	J16	J17	J18	J19	J20	J21	J22	J23	J24	J25
M1	3	5	6	2	3	2	5	3	1	4	2	1	2	5	4	5	2	1	4	6	2	5	2	2	5
M2	2	5	8	3	4	5	5	2	4	1	3	2	1	5	1	2	4	2	1	4	8	4	8	2	4
M3	1	1	6	1	4	6	4	5	1	6	2	4	2	5	8	5	4	1	3	1	4	6	4	5	1
M4	6	6	7	8	6	8	6	7	1	2	4	3	4	5	1	3	5	4	1	4	7	3	6	1	4
M5	8	4	5	6	2	6	4	2	1	3	5	1	2	1	2	1	2	3	2	5	2	1	2	4	3
M6	2	5	8	3	4	5	5	2	4	1	3	2	1	5	1	2	4	2	1	2	3	2	5	3	1
M7	4	3	2	2	8	8	2	3	2	5	1	2	3	4	5	7	2	1	2	1	4	5	5	2	4
M8	5	2	4	2	5	2	1	1	1	2	1	2	1	1	6	1	4	6	3	3	8	8	2	3	2
M9	1	1	5	1	4	8	5	7	8	3	1	3	6	6	7	8	6	8	7	8	5	5	8	6	2
M10	4	2	2	5	8	5	4	1	1	1	3	2	9	7	9	4	7	1	7	3	6	8	6	7	1
M11	3	2	4	5	1	3	5	4	3	2	4	6	4	9	1	3	4	2	8	2	7	1	3	2	3
M12	5	5	8	7	2	5	2	2	5	1	6	2	4	2	5	2	1	1	9	3	4	2	2	6	7
M13	7	8	1	6	8	4	8	2	4	1	6	4	5	1	4	8	5	7	1	4	7	3	6	1	4
M14	5	5	8	7	2	5	2	2	5	1	6	2	4	2	5	2	1	1	9	3	3	2	5	3	1
M15	7	8	1	6	8	4	8	2	4	1	6	4	5	1	4	8	5	7	1	4	5	2	1	1	1
M16	1	1	6	1	4	6	4	5	1	6	2	4	2	5	8	5	4	1	2	5	4	8	5	7	8
M17	5	6	4	3	7	3	6	1	4	2	3	1	2	3	5	4	1	2	1	1	8	5	4	1	1
M18	2	4	5	3	2	1	2	4	3	1	5	1	2	5	4	5	6	2	3	2	1	3	5	4	3
M19	3	5	6	2	3	2	5	3	1	4	2	1	2	5	4	5	2	1	4	6	2	5	2	2	5
M20	2	5	8	3	4	5	5	2	4	1	3	2	1	5	1	2	4	2	1	4	8	4	8	2	4
M21	1	1	6	1	4	6	4	5	1	6	2	4	2	5	8	5	4	1	3	1	4	6	4	5	1
M22	6	6	7	8	6	8	6	7	1	2	4	3	4	5	1	3	5	4	1	4	7	3	6	1	4
M23	8	4	5	6	2	6	4	2	1	3	5	1	2	1	2	1	2	3	2	5	2	1	2	4	3
M24	2	5	8	3	4	5	5	2	4	1	3	2	1	5	1	2	4	2	1	4	8	4	8	2	4
M25	1	1	6	1	4	6	4	5	1	6	2	4	2	5	8	5	4	1	3	1	4	6	4	5	1
M26	6	6	7	8	6	8	6	7	1	2	4	3	4	5	1	3	5	4	1	4	7	3	6	1	4
M27	8	4	5	6	2	6	4	2	1	3	5	1	2	1	2	1	2	3	2	5	2	1	2	4	3
M28	2	5	8	3	4	5	5	2	4	1	3	2	1	5	1	2	4	2	1	2	3	2	5	3	1
M29	4	3	2	2	8	8	2	3	2	5	1	2	3	4	5	7	2	1	2	1	4	5	5	2	4
M30	5	2	4	2	5	2	1	1	1	2	1	2	1	1	6	1	4	6	3	3	8	8	2	3	2

Chapter 14
Optimization and Mathematical Programming to Design and Planning Issues in Cellular Manufacturing Systems under Uncertain Situations

Vahidreza Ghezavati
Islamic Azad University, Iran

Mohammad Saidi-Mehrabad
University of Science and Technology, Iran

Mohammad Saeed Jabal-Ameli
University of Science and Technology, Iran

Ahmad Makui
University of Science and Technology, Iran

Seyed Jafar Sadjadi
University of Science and Technology, Iran

ABSTRACT

In practice, demands, costs, processing times, set-up times, routings, and other inputs to classical cellular manufacturing systems (CMS) problems may be highly uncertain, which can have a major impact on characteristics of manufacturing system. So, development models for cell formation (CF) problem under uncertainty can be a suitable area for researchers and belongs to a relatively new class of CMS problems that not researched well in the literature. In this way, random parameters can be either continuous or described by discrete scenarios. If probability information is known, uncertainty is described using a (discrete or continuous) probability distribution on the parameters, otherwise, continuous parameters are normally limited to lie in some pre-determined intervals. This chapter introduces basic concepts about uncertainty themes associated with cellular manufacturing systems and briefly reviews literature survey for this type of problem. The chapter also discusses the characteristics of different mathematical models in the context of cellular manufacturing.

DOI: 10.4018/978-1-61350-047-7.ch014

INTRODUCTION

During the past few decades, there have been various types of optimization techniques and mathematical programming approaches for cellular manufacturing systems under different random situations. In a cell manufacturing, once work cells and scheduling of parts in each cell are determined, it may be possible that cycle time in a specific cell be more than the other cells which creates a bottleneck in a manufacturing system. In this way, there are two different approaches in order to decrease cycle time in bottleneck cell: duplicating bottleneck machines or outsourcing exceptional parts which are known as group scheduling (GS) in the literature. Selecting each approach to balance cycle times among all cells can lead to variations in machines layout characteristics by changes in type and number of machines. Finally, formations of cells are also changed according to the changes in scheduling decisions. Thus, scheduling problem is one of the operational issues which must be addressed in design stage concurrently in an integrated problem so that the best performance of cells would be achieved. It is noted that scheduling problem includes many tactical parameters with random and uncertain characteristics. In addition, uncertainty or fluctuations in input parameters leads to fluctuations in scheduling decisions which could reduce the effects of cell formation decisions. Figure 1 indicates transmission of uncertainty from tactical parameters to the CMS problem.

Thus, in order to intensify effectiveness of the solution, integrated problem in uncertain conditions must be studied so that final solution will be robust and immune against the fluctuations in input parameters.

In the concerned problem, uncertain parameters can be listed as follows:

- Demand,
- Processing time,
- Routings or machine-part matrix,
- Machines' availability,
- Failure rate of machines,
- Capacities,
- Lead times,
- Set-up considerations,
- Market aspects,
- ...,

where the impact of each factor is discussed in the following sections.

PROBLEM BACKGROUND

Group technology (GT) is a management theory that aims to group products with similar processes or manufacturing characteristics, or both. Cellular manufacturing system (CMS) is a manufacturing concept to group products into part families based on their similarities is manufacturing processing. Machines are also grouped into machine cells

Figure 1. Illustration of uncertainty transmission to the CMS decision

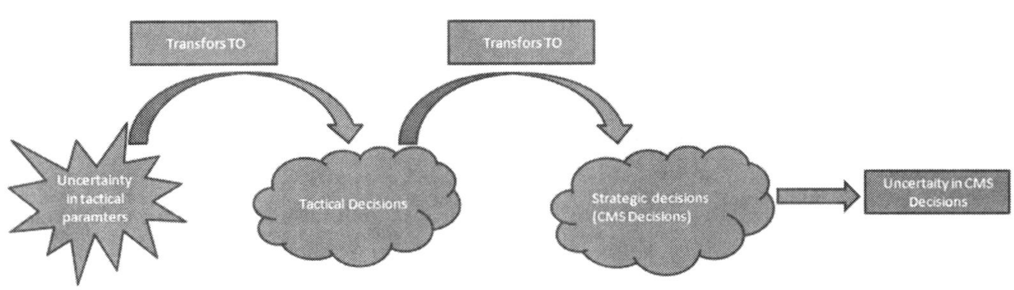

based on the parts which are supposed to be manufactured by these machines. CMS framework is a important application of group technology (GT) philosophy. The basic purpose of CM is to identify machine cells and part families concurrently, and to assign part families to machine cells in order to minimize the intercellular and intracellular costs of parts. Some real-world limitations in CF are:

- Available capacity of machines must not be exceeded,
- Safety and technological necessities must be met,
- The number of machines in a cell and the number of cells have not be exceeded an upper bound,
- Intercellular and intracellular costs of handling material between machines must be minimized,
- Machines must be utilized in effect (Heragu, 1997).

Aggregating traditional considerations with newly ones such as scheduling, stochastic approaches, processing time, variable demand, sequencing, and layout consideration can be more practical. This survey highlights studies that are relevant to the uncertainty planning of CMS problems; however, a survey of certain conditions will also be presented.

Cellular manufacturing decisions are strategic decisions which can be affected by operational decisions such as scheduling, production planning, layout consideration, utilities, productivity and etc. Thus, in order to effecting decision making related to cell formation design, it is necessary to integrate strategic decisions and operational decisions in a single problem. Recently, researchers have had some efforts in order to integrate two types of decisions. But the lack of literature is that most of them are studied in certain situations while in real-world most of the operational parameters are uncertain; and thus, integrated problems must be more studied in uncertain situations.

In the literature correspondence to CMS problems, uncertainty has been considered under different circumstances. We have classified previous researches into different groups which are discussed next.

Group 1: Uncertainty could appear either in demand or in products' mix. In this group, there are two approaches of fuzzy theory and stochastic optimization to handle uncertainty. In some of them stochastic demand is aggregated with tactical aspects such as production planning (Hurley and Whybark 1999), layout problem (Song and Hitomi1996) or dynamic and multi period conditions (Balakrishnan and Cheng 2007). Also, in other studies, uncertainty in products' demand has been resolved by fuzzy approach (Safaei et. al. 2008).

Group 2: Researchers formulated and analyzed CMS problem considering fuzzy coefficients in the objective function and constraints (Papaioannou and Wilson 2009).

Group 3: Processing times of products are assumed to be uncertain where mathematical programming and fuzzy approaches are implemented to obtain the results which are immune against the perturbation on the uncertainty. Also, some studies such as Sun and Yih (1996) and Andres et. al. (2007) attempted to achieve solutions by heuristic procedures. Some studies have formulated the problem as a queue network and then analyzed it by queuing theory (Yang and Deane 1993).

Group 4: Uncertainty normally appears due to fluctuations in design aspects during production process. Since, fluctuations in design aspects are not certain events, so uncertainty can be formulated by a set of future scenarios. In this way, some studies applied interval coefficient to resolve uncertainty (Shanker and Vrat 1998).

Group 5: In some explorations, uncertainty has been considered in the availability of resources for production equipments. In this way, some works have formulated CMS problem applying probability theory (Kuroda and Tomita (2005) and Hosseini 2000). In addition, some of them considered multi processing routes to be substituted once a machine encounters with failure (Siemiatkowski and Przybylski (2007) and Asgharpour and Javadian 2004).

Group 6: Uncertainty has been recognized in similarity coefficients. For example, a new similarity coefficient has been introduced where applied fuzzy theory and then transformed it to a binary matrix (Ravichandran and Chandra Sekhara Rao 2001).

Group 7: Capacity level of machines is considered to be uncertain. Since this critical parameter has an important role to determine bottleneck machine, thus it is vital to make flexible decisions under any realization of this parameter (Szwarc et al 1997).

Group 8: Finally, uncertainty in CMS problem has been detected in products arrival time to cells. Classical models assume that all products are available at the beginning of the production planning while in real application it may be occurred that products arrive to cell with unknown time. In this way, researchers modeled CMS problem as queue network to resolve uncertainty (Yang and Deane 1993).

Literature survey classifications can be described as follows. There exist many researches in certain situations for designing CMS in different areas such as cell formation integrated with scheduling (Solimanpur et al. (2004), Aryanezhad and Aliabadi et al 2011), considering exceptional elements in CF (Tsai et al. (1997), Mahdavi et al. 2007), some works apply meta-heuristics and heuristics methods to solve large scale problems are more practical and appealing real-case problems (Xiaodan Wu et al (2006), Venkataramanaiah 2007).

OPTIMIZATION APPROACHES IN UNCERTAIN SITUATIONS

Rosenhead et al (1972) divided decision environments into three groups of deterministic, risk and uncertain. In deterministic situations, all problem parameters are considered to be given. In risk problems parameters have probability distribution function where it is known for decision maker while in uncertain situations there is no information about probabilities.

The problems which are classified into the risk are named stochastic and the primary objective is to optimize expected value of system outcome. Also, the uncertain problems are known as robust and the primary objective is mainly to optimize performance of the system in the worst case conditions.

The aim of both stochastic and robust optimization methods is to find solution with a suitable performance in realization of any value for uncertain parameter.

Random parameters can be either continues or explained by discrete scenarios. If probability information are known, uncertainty will be explained by continues or discrete distribution functions. But if no information is available, parameters are assumed to be in predefined intervals. Scenario planning is a method in which decision makers achieve uncertainty by indicating a number of possible future states. In such conditions, the goal is to find solutions which perform well under all scenarios. In some cases, scenario planning replaces predicting as a way to assess trends and potential modifications in the industry environment (Mobasheri et al 1989). Decisions makers can thus develop strategic responses to a range of environmental adjustments, more adequately preparing themselves for the uncertain future. Under such conditions, scenarios are qualitative descriptions of possible future states, consequences from the present state with consideration of potential key industry events. In other cases, scenario planning is used as a tool for modeling and solving specific operational problems (Mulvey 1996). While scenarios here also depict a range of future states, they do so through quantitative descriptions of the various values that problem input parameters may resolve. Scenario based planning has two main negative aspects. The first is that identifying scenarios and assigning probabilities to them are a difficult task. The second is that we are unable to increase the number of scenarios since due to limitation on computation time which consequently limits the future correspondence situations for decision making. This approach has the advantageous that provides statistical correlation between parameters (Snyder 2006).

DECISION MAKING APPROACHES IN UNCERTAIN SITUATIONS

There are different approaches which can be applied in modeling process based on the problem characteristics: Stochastic Optimization (SO), Robust Optimization (RO) and Queuing Theory (QT) with defined decision tree as follows.

- Stochastic Optimization
- Discrete Planning - Set of Scenario
- Continues Optimization
- Mean Value model: the most popular objective in any SO problem is to optimize expected value of the system outcome. For example, minimizing expected cost or maximizing expected income.
- Mean – Variance Model: in some studies variance and expected of system performance are considered simultaneously in optimization problem.
- Probability Approaches
- Max Probability Optimization: Maximizing the probability of a random event that solution performs good under each realization of random parameter.
- Chance Constrained Programming: a probability event located in problem constraint sets such as service level constraint.
- Queuing Theory & Markov Chain: It is a well-known approach.
- Robust Optimization

The objective in any stochastic optimization problem mainly focuses on optimizing the expected value of system outcome such as maximizing expected profit or minimizing total expected cost.

In any stochastic programming we must determine which variables are considered in the first stage (design variable) and which are considered in the second stage (control variable). In other words, which variables must be determined first and which of them must be determined after uncertainty is resolved. In modeling process for cellular manufacturing problem, cell formation decisions are the first and operational and tactical decisions are the second variables. If both decisions are made in a single stage, the model is reduced to a certain problem in which uncertainty of parameters are replaced by mean of variables.

Mean-Variance Models

The mean value models discuss only the expected performance of the system without reflecting on the fluctuations in performance and the decision maker's risk aversion limitations. However, a portion of literature incorporates the company's level of risk aversion into the decision-making process, classically by applying a mean–variance objective function.

$$\text{Min} = E(\text{Cost}) + \lambda \text{Var}(\text{Cost})$$

Probabilistic Approaches

The mean-variance models consider only the expected value or variance of the stochastic objective function, there is an extensive portion of literature which considers probabilistic information about the performance of the system; for example, maximizing the probability that the performance is good or minimizing the probability that it is bad, under suitable and predefined explanations of "good" and "bad". We introduce two such approaches: (1) max-probability problems; (2) chance-constrained programming;

Queuing Theory for CMS Problem

Queuing theory can be applied to any manufacturing or service systems (also, in cellular manufacturing systems). For example, in a machine shop, jobs wait to be machined; (Heragu 1997b). In a queuing system, customers arrive by some arrival process and wait in a queue for the next available server. In the manufacturing framework, customers can be assumed as parts and servers may be machines or working cells. The input process shows how parts arrive at a queue in a cell. An arrival process is commonly identified by the probability distribution of the number of arrivals in any time interval. The service process is usually described by a probability distribution. The service rate is the number of parts served per unit time. The arrival rate of a queuing system is usually given as the number of parts arriving per unit time. Thus, measurements of a queue system such as maximization the probability that each server is busy (utilization factor), minimization waiting time in queues (that leads to minimization work in process in cells) and etc can be optimized and cells will be formed optimality.

Robust Optimization

Once there is no probability information about the uncertain parameters, the expected cost and other objectives discussed in previous section are inappropriate. Many measurements of robustness have been introduced for this condition. The two most common are mini-max cost and mini-max regret, which are directly related to one another. Just like the stochastic optimization case, uncertain parameters in robust optimization problems may be considered as being either discrete or continuous. Discrete parameters are formulated applying the scenario based planning. Continuous parameters are normally assumed to lie in some predefined interval, because it is often impossible to consider a "worst case scenario" when parameter values are unbounded. This type of uncertainty is described as "interval uncertainty".

The two most common robustness measurements consider the regret of a solution, which is the difference (absolute or percentage) between the cost of a solution in a given scenario and the cost of the optimal solution for that scenario. Regret is sometimes described as opportunity loss: the difference between the quality of a given strategy and the quality of the strategy that would have been chosen had one known what the future held (Snyder 2006).

As it was already described, the performance of a cellular manufacturing system heavily influenced by tactical and operational decisions such as scheduling, production planning, layout and etc. Notable point is that the tactical decisions and operational parameters are dependent on many uncertainties that affect the system. As a result, the tactical and operational decisions are suffering from uncertainty. This causes to transfer uncertainty into the decisions related cell formation. Therefore, it is essential for researchers to recognize different types of uncertainty in the problem and make decisions regarded to their impact into the problem.

The most important parameters with uncertainty in manufacturing cell formation problem considered as below:

- Demand
- Processing time

- Routings or machine-part matrix
- Machine's failure rate
- Capacities

One of the factors causing uncertainty in the problem associated with product design changes during the course of production. Moreover, changes in product design with many features of the product are altered. Design changes can occur based on a variety of reasons such as changes in customer expectations, short-term life products, and competing products to market entry.

Under such circumstances, many characteristics of products such as demand and time will find a process of change.

Note that the reasons of changes expressed are not certain events in the future and thus they have to be predicated as some discrete scenarios. In such case analytical space problem is discrete and can be optimized by discrete optimization.

As it was discussed earlier, one of the product features which can be changed due to changes in product design is product routings. In this way, sequence of machines in which product has to visit them may be changed and therefore part – machine index may be changed. In such cases, the values within the part – machine matrix unlike classical models that were only zero or one can be a probabilistic value between zero and one. In such problems, discrete optimization can be applied to formulation.

Another factor with uncertainty is the rate of access to machines based on their failure. Since failure and machine downtime are not certain events, the machine accessibility for the decision maker at the time of manufacturing cells with defined uncertainty is also under uncertainty.

Another parameter that is uncertain and can affect formation of work cells is features of capacity. These factors include different items: the capacity of processing machinery on parts as well as physical capacities for manufacturing framework. Such variations must be predicted at the beginning planning horizon.

The summary of above discussions can be found Table 1.

MATHEMATICAL MODELLING

In this section, different mathematical models with different optimization approaches which include two new models and one published model are discussed. The selected approaches are stochastic optimization and queuing theory.

Table 1. Summary uncertainty developments in CMS problem

No.	Uncertain parameter	Optimization Approach	Decision space
1	demand	Stochastic	Continuous & Discrete
2	Processing time	Stochastic	Continuous & Discrete
3	Processing time	Robust	Continuous & Discrete
4	Processing time	Queuing Theory	Continuous
5	Routing	Stochastic	Discrete
6	Routing	Queuing Theory	Discrete
7	Capacity	Stochastic	Discrete
8	Machines' Availability	Queuing Theory	Continuous & Discrete
9	Machines' Availability	Stochastic	Continuous & Discrete
10	Lead times	Stochastic & Robust	Continuous & Discrete

Model 1

In this section, a bi-objective mathematical model to form manufacturing cells is presented where uncertainty is accessed in part – machine matrix. As discussed earlier, due to changes in design characteristics of products, several factors are subject to changes such as the processing routings of parts. Thus, according to the forecasting based on the scenario planning, forecasting different routing processes for a part in uncertain situation is possible. In this condition, each part can have different routing process for each scenario. Therefore, in order to design cellular configuration efficiently, all planning conditions must be considered. In current problem the factor with uncertainty is part – machine matrix. In classical models, only zero-one elements are used in part – machine matrix while in the presented problem each element can be a continuous value between zero and one. Each array denotes the probability that part i visits machine j with regard to all scenarios. For example, if there are two scenarios in which the probability of the first scenario is 0.4 and for the second one is 0.6, we have:

$p_1 = 0.4 \Rightarrow$ Routing in scenario 1 for part 1: Machine 1 → Machine 2 → Machine 3 → Finish
$p_2 = 0.6 \Rightarrow$ Routine in scenario 2 for part 1: Machine 1 → Machine 2 → Machine 3 → Finish

$$a_{[ij]} = \begin{bmatrix} M.1 & M.2 & M.3 & M.4 \\ 1 & 0.4 & 1 & 0.6 \end{bmatrix}$$

Where element [ij] indicates the probability that part i processed on machine j.

Since, in both scenarios, machines 1 and 3 are the same in processing routing, so part 1 has to visit them surely (or with probability 1) to do operation process. But, based on the first scenario this part has to visit machine 2 with probability 0.4 and also machine 4 with probability 0.6. As it can be seen, in introduced part – machine matrix each array can have a value between zero and one based on the probability occurrence for scenarios.

In a mathematical model which is presented in this section, the first objective function minimizes the costs associated with the under utilization in a manufacturing system. Also, the second objective function is optimizing a random event in manufacturing system unlike the classical models which optimized only certain events. As it was discussed in definitions of a cellular manufacturing system, one of the most important objectives is to minimize the number of inter cellular transportation. In this problem, since processing rout for parts is uncertain, therefore the number of inter cellular transportation is uncertain too. A random event which is considered for optimization is to "minimizing the probability that the number of inter cellular transportation exceeds the upper bound limitations". For computing above objective the following notations are defined:

Parameters

$$a_{ijs} = \begin{cases} 1 \\ 0 \end{cases}$$

1 if part *i* needs to be processed on machine *j* in scenario *s*
0 otherwise
p_s: Probability of occurring scenario s
N: Maximum number of intercellular transportation allowed in each scenario

Decision Variables

n_s: Number of intercellular transportation in scenario *s*.

$$e_s = \begin{cases} 1 \\ 0 \end{cases}$$

1 if no. of intercellular transportation in scenario s configuration exceeded up bound N
0 otherwise

or

$$e_s = \begin{cases} 1 \\ 0 \end{cases}$$

1 if $n_s \geq N$
0 if $n_s < N$
z_s: Integer additional variable for each scenario.

$$x_{ik} = \begin{cases} 1 \\ 0 \end{cases}$$

1 if part i is assigned to cell k
0 otherwise

$$y_{jk} = \begin{cases} 1 \\ 0 \end{cases}$$

1 if machine j is assigned to cell k
0 otherwise

In order to minimizing under utilization costs in the first objective function, the following function is defined:

$$MinZ_1 = \sum_s \sum_i \sum_j p_s \times (1 - a_{ijs}) \times x_{ik} \times y_{jk} \quad (1)$$

Also, based on the above definitions, an attractive random event for minimizing the second function can be defined as follows:

p (no. of intercellular transportation in each condition $\geq N$)

The above random event must be optimized by minimizing the probability of occurrence that leads to maximum utility for decision maker in final solution. In other words, above probability transforms to the following function:

$$MinZ_2 = \sum_s e_s \times p_s \quad (2)$$

Since there is s scenarios in the proposed problem which are similar to s independent random events, thus probability of total events will be equal to summation of probability of each event. In other words, assuming $s_1, s_2, \ldots, s_n$ as n independent random events, we have:

$$P\left(s_1 \cup s_2 \cup \ldots \cup s_n\right) = P\left(s_1\right) + P\left(s_2\right) + \ldots + P\left(s_n\right)$$

As a result, in above function if in scenario s the number of inter cellular transportation exceeds the upper bound limitation then we can assume that inter cellular transportation may be occurred with the probability of p_s. Finally, the summation of the probability of scenarios with unsatisfied inter cellular transportation restriction denotes the final probability of the problem.

In this model, the objective functions and also, the following constraints are effective:

$$MinZ_1 = \sum_s \sum_i \sum_j p_s \times (1 - a_{ijs}) \times x_{ik} \times y_{jk}$$
$$MinZ_2 = \sum_s e_s \times p_s$$

Constraints

$$\sum_k x_{ik} = 1 \qquad \forall i \quad (3)$$

$$\sum_k y_{jk} = 1 \qquad \forall j \quad (4)$$

$$n_s - \sum_i \sum_j a_{ijs} \times x_{ik} \times (1 - y_{jk}) = 0 \qquad (5)$$

$$z_s - \left\lceil \frac{n_s}{N} \right\rceil = 0 \qquad (6)$$

$$z_s \leq M \times e_s \qquad (7)$$

$$x_{ik}, y_{jk}, e_s \in \{0,1\} \quad z_s \text{ integer} \geq 0 \quad n_s \geq 0$$

The first objective minimizes total expected cost associated with the utilization computed when a part do not need to be processed on a machine placed together in a same cell. The second objective minimizes the probability that number of inter cellular transportation exceeds the maximum transportation. Set constraint (3) says that each part must be assigned to a single cell. Set constraint (4) states that each machine can be assigned only to one cell. Set constraint (5) computes total number of inter transportation in each scenario. In set constraint (6) additional variable z_s will be zero if the number of inter transportations in scenario s is less than the maximum limit and it is an integer value greater than 1 else. Set constraint (7) guarantees that if $n_s \geq N$ then e_s will be 1. Otherwise, e_s will be 0.

Model 2

Applying Queuing Theory to CMS Problem

In this section, we formulate a CMS problem as a queue system. Also, assume a birth-death process with constant arrival (birth) and service completion (death) rates. The role of the birth-death process in automated manufacturing systems is described in detail in Viswanadham and Narahari (Viswanadham and Narahari 1992). Specifically, let λ and μ be the arrival and service rate of parts, respectively, per unit time. If arrival rate is greater than the service rate, the queue will grow infinitely. The ratio of λ to μ is named utilization factor or the probability that a machine is busy and is defined as $\rho = \lambda \mid \mu$. Therefore, for a system in steady state, this ratio must be less than one. In this research, we assume M/M/1 queue system for each machine in CMS where each part arrives to cells with rate λi an_d parts served by machines. In this condition, due to operate different parts (or different customers) on each machine and each part has different arrival rate, so for each machine (server) ρ is computed using the following property. Figure 2 illustrates modeling of cellular manufacturing system by queuing theory approach.

Figure 2. A CMS problem and queuing theory framework (Ghezavati and Saidi-Mehrabad 2011)

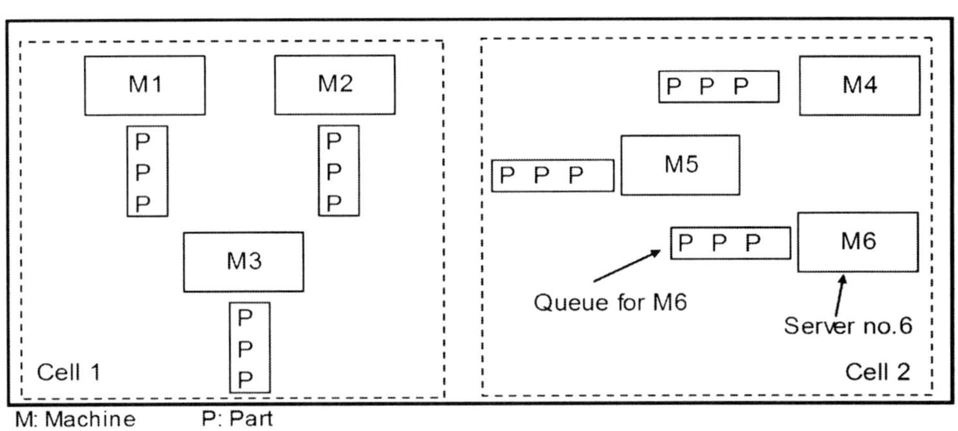

Property 1 the minimum of independent exponential random variables is also, exponential. Let F1, *F2*, ..., *Fn* be independent random variables with parameters $\lambda 1, \lambda 2, \ldots, \lambda n$. Let Fmin = min{F1, F2, ..., Fn}. Then for any t ≥ 0,

$$P(F_{\min} > t) = P(F_1 > t) \times P(F_2 > t) \times \ldots \times P(F_n > t)$$
$$= e^{-\lambda_1 t} e^{-\lambda_2 t} \ldots e^{-\lambda_n t} = e^{-[\lambda_1 + \lambda_2 + \ldots + \lambda_n] t}$$

An interesting implication of this property to inter-arrival times is discussed in Hillier and Lieberman (Hillier and Lieberman 1995). Suppose there are n types of customers, with the ith type of customer having an exponential inter-arrival time distribution with parameter λi, arrive at a queue system. Let us assume that an arrival has just taken place. Then from a no-memory property of exponential distribution, it follows that the time remaining until the next arrival is also exponential. Using mentioned property, we can see that the inter-arrival time for entire queue system or efficient arrival rate (which is the minimum among all inter-arrival times) has an exponential distribution with parameter:

$$\lambda_{eff} = \sum_{i=1}^{N} \lambda_i$$

Hence, utilization factor or the probability that each machine (*j*) is busy is as follow (efficient arrival rate divided by service rate):

$$\rho_j = \frac{\lambda_{eff}}{\mu_j} = \frac{\sum_{i=1}^{N} \lambda_i}{\mu_j} \tag{8}$$

Chance Constrained Programming

Since, both arrival time and service time are uncertain so the amount of time in which each customer spends in server will be uncertain, too. In order to prevent long waiting time for each customer, a chance constraint must be considered in the formulation. Note that distribution function denoting total time for each customer in a M/M/1 system is as follows:

$$P(W_s \geq t) = e^{-\mu(1-\rho)t} \tag{9}$$

Proof: Assume that there are *N* customers in a system once a new customer is arrived. Thus, based on the conditional probability theory:

$$P(W_s \geq t) = \sum_{n=0}^{\infty} P(W_s \geq t \mid N = n) \times P(N = n) \tag{10}$$

On the other side, total time in which a new customer has to wait is equal to:

$$W_q = F_1 + F_2 + \ldots + F_n \tag{11}$$

Where F_i denotes service time for customer *i*. So:

$$W_s = W_q + F_{n+1} \tag{12}$$

where F_{n+1} denotes service time for new arrived customer. It is obvious that sum of the *n*+1 random variables with exponential distribution with rate μ will be an Erlang random variable with parameters n+1 and μ. So:

$$P(W_s \geq t \mid N = n) = P(\sum_{i=1}^{n+1} F_i > t) = \int_t^{\infty} \mu \times e^{-\mu y} \frac{(\mu y)^n}{n!} d_y \tag{13}$$

Note that the probability of being n customers in a M/M/1 model system is:

$$p_n = \rho^n (1-\rho) \quad \text{where} \quad \rho = \frac{\lambda}{\mu} \tag{14}$$

Based on the Equations 13 and 14, Equation 10 will be computed as:

$$P(W_s \geq t) = \sum_{n=0}^{\infty} \rho^n (1-\rho) \int_t^{\infty} \mu \times e^{-\mu \cdot y} \frac{(\mu y)^n}{n!} d_y \tag{15}$$

$$= \mu(1-\rho) \int_t^{\infty} e^{-\mu \cdot y} \sum_{n=0}^{\infty} \frac{\rho^n (\mu y)^n}{n!} d_y \tag{16}$$

Also, based on the exponential series, we have:

$$\sum_{n=0}^{\infty} \frac{\rho^n (\mu y)^n}{n!} = e^{\rho \cdot \mu \cdot y} = e^{\lambda \cdot y} \tag{17}$$

If we replace Equation 17 to the Equation 16, the Equation 9 will be proven. It can be found that W_s has an exponential distribution function with parameter $\mu - \lambda$.

In order to satisfy service level this probability must be at most α. So, the chance constraint will be determined as $P(Ws \geq t) \leq \alpha$. In order to linearize this nonlinear constraint the following procedure is performed:

$$P(W_s \geq t) \leq \alpha \tag{18}$$

$$\Rightarrow e^{-\mu(1-\rho)t} \leq \alpha \tag{19}$$

$$\Rightarrow -\mu(1-\rho)t \leq Ln(\alpha) \tag{20}$$

The achieved constraint indicates that a customer will be in system more than critical time *t* with probability at most α.

Property 2. If *n* types of customers have to visit a server to receive service with different arrival rate λi then the probability of a random customer in which visits the server be *i*th type will be as follows:

$$p_i = \frac{\lambda_i}{\sum_j \lambda_j} \tag{21}$$

In concerned model, the characteristics of a Jacson service network will be applied. In a Jacson network, it is assumed that each customer has to visit multiple servers in order to complete service stages. For example, each part refers to several machines to complete operation processes. In such network, input rate for machines needed for the first operation will be equal to the arrival rate of the part to the system. But, the ratio for any machines need for the second operation input rate will be equal to the output rate from the previous server (or machine). Similarly, any machine needs for the third operation input rate will be equal to the exit rate of the second machine and this process goes on for the other machines.

In a cellular manufacturing problem formulated as a queue system, each part based on its routing process visits machines or multi cells in order to receive service. Figure 3 illustrates such process.

For each machine, effective input rate is made of two elements. The first fraction is the summation of arrival rate of parts which visit the machine in the first operation. The second fraction is the summation of input rate of parts which visit the machine after the second operation. This rate is equal to the output rate of the previous machine. Figure 3 illustrates difference between arrival rates for machines per a specific part. In this model, such procedure will be applied to compute effective input rate for each machine.

In this section, a part—machine matrix—will be applied where sequence operations of parts are determined. This can help us formulate problem as a Jacson network. Each element of this matrix is defined as follows:

Figure 3. Arrival rate for part 1 into the different machines based on the routing

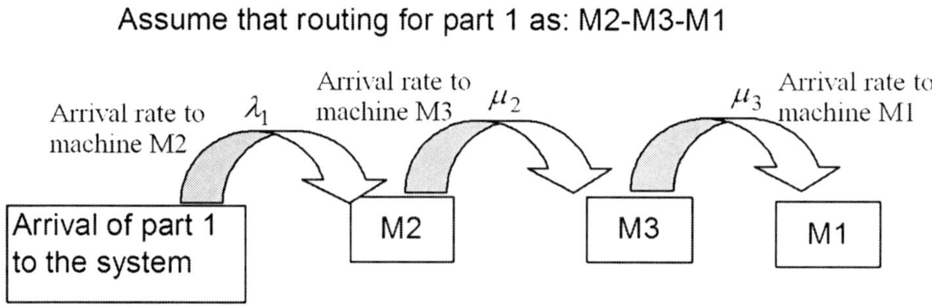

$$a_{ik} = \begin{cases} j \\ 0 \end{cases}$$

j if kth process of part i is completed by machine j
0 otherwise

$$b_{ij} = \begin{cases} k \\ 0 \end{cases}$$

k if part i refers to machine j to complete kth process
0 otherwise

Other parameters are defined as follows:

$$z_{ij} = \begin{cases} 1 \\ 0 \end{cases}$$

1 if operation on machine j is the first operation of part i
0 otherwise

$$c_{ij} = \begin{cases} 1 \\ 0 \end{cases}$$

1 if part i needs to be processed on machine j
0 otherwise

λI= Arrival rate of part i to manufacturing system.
μj= service rate of machine j (or $[1/\mu j]$ denotes average operation time on machine j).
pi_j= The probability that a random part type i leaves machine j.
β = Penalty rate multiplied to arrival process if intercellular movement occurs.

It is assumed that if an operation of a part has to transfer to the other cell (or inter cellular movement) then arrival rate will be multiplied by β in which included transfer time and also waiting time between cells.

λj_e^{ff}= Effective arrival rate for machine j.

Based on the above definitions λj_e^{ff} will be computed by the following equation.

$$\lambda_j^{eff} = \sum_{i=1}^{m} \left(z_{ij} \times \lambda_i \right) + \sum_{i=1}^{m} (1 - z_{ij}) \times c_{ij} \times \lambda_{a_{i,b_{ij}-1}}^{eff} \times p_{i,a_{i,b_{ij}-1}}$$

In above equation, in order to compute effective input rate for each machine two fractions are considered: the first fraction is the summation of the arrival rate for parts which visit machine j in the first operation. The second term is the summation of input rate for parts which visit the machine after the second operation. Number of operation which completed by machine j is b_{ij} based on the defined parameters. Thus, number of previous operation is $b_{ij} - 1$. Finally, according to the definition of a_{ik} (the machine completes kth operation of part i), the machine which completes previous operation of part i will be $a_{i,b\mu-1.}$ Therefore, the second term of above equation effective arrival rate for parts visit machine j after the second operation are computed as follows: effective arrival rate of a machine needs before machine j multiplied by the probability of leaving for part i from previous machine.

For example, assume that customers arrive to a book store with Poisson distribution with rate 10 per hour where 60 percent is man and 40 percent is women. Hence, the number of men arrives to the store will be Poisson with rate 15×0.6 per hour and also, number of women arrives to the store will be Poisson with rate 15×0.4 per hour.

Note that if operation j of part i needs inter cellular transportation, machine j is penalized by increasing arrival rate of part i and the rate is multiplied by β. Finally, the model must determine whether each operation needs inter cellular transportation or not. It must be mentioned that operation j of part i needs inter cellular transportation when machine j and part i are not located in the same cell. Based on the above description λ^{eff} is computed as follows:

$$\lambda_j^{eff} = \sum_{i=1}^{m}\left[\left(x_{ik} \times y_{jk}\right) \times \lambda_i + x_{ik} \times \left(1 - y_{jk}\right) \times \lambda_i \times \beta\right] \times z_{ij} + \sum_{i=1}^{m} p_{i,a_{i,b_{ij}-1}} \times \left[\left(x_{ik} \times y_{jk}\right) \times \lambda^{eff}_{a_{i,b_{ij}-1}} + x_{ik} \times \left(1 - y_{jk}\right) \times \lambda^{eff}_{a_{i,b_{ij}-1}} \times \beta\right] \times c_{ij} \times (1 - z_{ij}) \tag{22}$$

Mathematical Model

In this section, a mathematical model optimizes cell formation decisions based on the queuing theory will be proposed. The objective function is to minimize total cost included under utilization cost. Also, a chance constraint will be considered in order to prevent additional waiting time of parts in a queue line in front of each machine. As it was discussed, assuming each machine as a M/M/1 model, the chance constraint (13) satisfies considered objective.

$$MinZ = \sum_i \sum_j (1 - a_{ij}) \times x_{ik} \times y_{jk} \tag{23}$$

Constraints

$$\sum_k x_{ik} = 1 \qquad \forall i \tag{24}$$

$$\sum_k y_{jk} = 1 \qquad \forall j \tag{25}$$

$$\lambda_j^{eff} = \sum_{i=1}^{m}\left[\left(x_{ik} \times y_{jk}\right) \times \lambda_i + x_{ik} \times \left(1 - y_{jk}\right) \times \lambda_i \times \beta\right] \times z_{ij} + \sum_{i=1}^{m} p_{i,a_{i,b_{ij}-1}} \times \left[\left(x_{ik} \times y_{jk}\right) \times \lambda^{eff}_{a_{i,b_{ij}-1}} + x_{ik} \times \left(1 - y_{jk}\right) \times \lambda^{eff}_{a_{i,b_{ij}-1}} \times \beta\right] \times c_{ij} \times (1 - z_{ij}) \tag{26}$$

$$\rho_j - \frac{\lambda_j^{eff}}{\mu_j} = 0 \tag{27}$$

$$-\mu_j \times (1 - \rho_j)t \le Ln(\alpha) \qquad \forall j \tag{28}$$

$$\rho_j \le 1 \qquad \forall j \tag{29}$$

$$p_{ij} - \frac{\lambda_i \times c_{ij}}{\sum_{r=1} \lambda_r \times c_{rj}} = 0 \qquad \forall i, j \tag{30}$$

$$x_{ik}, y_{jk} \in \{0,1\} \qquad \rho_j, p_{ij} \ge 0$$

Constraints (24) and (25) compute effective arrival rate and utilization factor for each machine, respectively. Set constraint (28) guaranties satisfaction of chance constrained for each machine where the probability that each part has to wait more than critical time t is at most a. Set constraint (29) ensures that utilization factor for each machine will be less than one. Set constraint

(30) determines the probability that a random part leaves machine *j* be type *i*.

Model 3

Recently, Ghezavati and Saidi-Mehrabad (2010) proposed a stochastic cellular manufacturing problem in where uncertainty is captured by discrete fluctuations in processing times of parts on machines. The aim of their model was to optimize scheduling cost (expected maximum tardiness cost) plus cell formation costs, concurrently. The mathematical model is represented in this part and interested readers are referred to read the paper for more details.

Parameters

$$a_{ij} = \begin{cases} 1 \\ 0 \end{cases}$$

1 if part *i* required to be process on machine *j*
0 otherwise

c_i: *Penalty* cost of subcontracting for part *i*
μi_j: Cost part i not utilizing machine j
M_{max}: Maximum number of machines permitted in a cell
C_μ: Maximum number of cells permitted
p_s: Probability of scenario *s* occurs
t_{ijs}: Processing time for part *i* on machine *j* in scenario *s*
DD_i: Due Date of part *i*
pc: Penalty cost for unit time delayed

Decision Variables

$$x_{ik} = \begin{cases} 1 \\ 0 \end{cases}$$

1 if part *i* processed in cell *k*
0 otherwise

$$y_{jk} = \begin{cases} 1 \\ 0 \end{cases}$$

1 if machine *j* assigned to cell *k*
0 otherwise

$$Z_{is[r]} = \begin{cases} 1 \\ 0 \end{cases}$$

1 if part i assigned to sequence [r] in scenario s
0 otherwise

F[r]ks: The time in which process of part with sequence [r] ends in cell k and scenario s
FD[r]ks: Due date of part with sequence [r] in cell k in scenario s
L[r]ks: Tardiness of part with sequence [r] in cell k in scenario s
MLs: Maximum Tardiness occurred in scenario s
Diks: Total processing times of part i needs to be processed in cell k and scenario s
T[r]ks: Total processing times of a part with sequence [r] assigned to cell *k* in scenario *s*

CF decisions are scenario – independent: they must be made before occurring scenarios and they are made based on their similarities in processing parts and are independent to quantity of processing time. Scheduling decisions are scenario – dependent, thus Z, D, T, FD, L, ML and F variables are indexed by scenario since they must be made after we realize scenario and where the processing time is occurred.

Mathematical Model (Ghezavati, V.R. and Saidi-Mehranad, M., 2010)

$$\text{Minimize } Z = \sum_s pc \times p_s \times ML_s + \sum_k \sum_j \sum_i c_i a_{ij} x_{ik} (1 - y_{jk}) + \sum_k \sum_j \sum_i u_{ij} (1 - a_{ij}) x_{ik} y_{jk} \quad (31)$$

Subject to:

$$\sum_k x_{ik} = 1 \qquad \forall i \qquad (32)$$

$$\sum_k y_{jk} = 1 \qquad \forall j \qquad (33)$$

$$\sum_r Z_{is[r]} = 1 \qquad \forall i, s \qquad (34)$$

$$\sum_i x_{ik} Z_{is,[r+1]} \le \sum_i X_{ik} Z_{is[r]} \qquad \forall k, s, r \qquad (35)$$

$$D_{iks} = \sum_j a_{ij} t_{ijs} x_{ik} y_{jk} \qquad \forall i, k, s \qquad (36)$$

$$\sum_i x_{ik} Z_{is[r]} \le 1 \qquad \forall r, s, k \qquad (37)$$

$$T_{[r]ks} = \sum_i Z_{is[r]} D_{iks} \qquad \forall k, s, r \qquad (38)$$

$$F_{[r]ks} = \sum_{r=1} \sum_{\alpha=1}^{r} T_{\alpha ks} \qquad \forall k, s, r \qquad (39)$$

$$FD_{[r]ks} = \sum_i x_{ik} \times Z_{is[r]} \times DD_i \qquad \forall k, s, r \qquad (40)$$

$$L_{[r]ks} = \max\left\{0, F_{[r]ks} - FD_{[r]ks}\right\} \qquad \forall k, s, r \qquad (41)$$

$$ML_s = Max\{L_{[r]ks} : k = 1, \ldots, C \ and \ [r] = 1, \ldots, P\} \quad \forall s \qquad (42)$$

$$\sum_j y_{jk} \le M_{\max} \qquad \forall k \qquad (43)$$

$$x_{ik}, y_{jk}, Z_{isr} \sim (0,1) \qquad (44)$$

$$D_{iks}, T_{rks}, F_{rks}, FD_{rks} \ge 0 \qquad (45)$$

Set constraints (32), (33) and (43) indicate cell formation constraints and set constraints (34), (35), (36), (37), (38), (39), (40), (41) and (42) perform scheduling computations and rational constraints.

Linearization Approaches

In above formulation, since there are both binary and continuous variables where are multiplied to each other, nonlinear terms are appeared in formulation process. Two common types of non-linear terms are:

Type 1: Pure 0-1 polynomial problem in which n binary variables are multiplied to each other such as $Z = x_1 \times x_2 \times \ldots \times x_n$.

Type 2: Mixed 0-1 polynomial problems which n binary variables are multiplied to each other and this term is multiplied to a continuous variable such as $Z = x_1 \times x_2 \times \ldots \times x_n \times Y$.

For linearization type 1 the following method can be applied by introducing some new auxiliary constraints:

$$Z \le x_i \qquad i = 1, 2, \ldots, n$$

$$Z \ge \sum_{i-1}^{n} x_i - (n+1)$$

Also, for linearization type 2 in a minimization problem, the following auxiliary constraints will be applied:

P1: Nonlinear problem

$$MinZ = x_1 \times x_2 \times \ldots \times x_n \times y$$

St:

$$L(X, Y)$$

P2: Linear form

$$Min \quad Z$$

St:

$$Z \geq y - |U| \times \left(n - \sum_{i=1}^{n} x_i \right)$$
$$Z \geq 0 \quad L(X,Y)$$

where U is upper bound for continuous variable y and therefore Z will be a continuous variable (Ghezavati and Saidi-Mehrabad 2011).

CONCLUSION

In summary, in this chapter basic principles of uncertainty in a cellular manufacturing system were established. Since CMS problem is affected by tactical decisions such as scheduling, production planning, layout considerations, utilization aspects and many other factors, thus each CMS problem must be aggregated with tactical decisions in order to achieve maximum efficiency. As it is known, tactical decisions are made of many uncertain parameters. Since strategic decisions are influenced by tactical decisions, therefore CMS decisions will be mixed with uncertainty. There are some popular approaches which can analysis uncertain problems such as: Stochastic Optimization, Discrete Planning - Set of Scenario, Continues Optimization, Mean Value model, Mean – Variance Model, Max Probability Optimization, Chance Constrained Programming, Queuing Theory and Markov Chain, and Robust Optimization. This chapter has proposed two sample mathematical models and also one published model [32]. It was assumed that processing routing, inter arrival and service time and also processing time to be uncertain. Stochastic optimization and queuing theory were to resolve uncertainty in formulation process. A complete survey on meta-heuristic methods to solve CMS problems can be found by Ghosh et al (2011). For future directions, the following suggested developments can be applied for researchers and readers:

- Uncertain Processing time optimized by robust approach in continuous or discrete space
- Uncertain capacities optimized by stochastic or robust approach in discrete space
- Uncertain machines' availability optimized by stochastic or queuing theory approaches in continuous or discrete space.
- Aggregating CMS problem with logistics considerations in uncertain environments.
- Aggregating CMS problem with production planning aspects in uncertain environments.
- Aggregating CMS problem with layout considerations in uncertain environments.
- Aggregating CMS problem with scheduling concerns in uncertain environments.

REFERENCES

Andres, C., Lozano, S., & Adenso-Diaz, B. (2007). Disassembly sequence planning in a disassembly cell. *Robotics and Computer-integrated Manufacturing*, *23*(6), 690–695. doi:10.1016/j.rcim.2007.02.012

Aryanezhad, M. B., & Aliabadi, J. (2011). A new approach for cell formation and scheduling with assembly operations and product structure. *International Journal of Industrial Engineering*, *2*, 533–546..doi:10.5267/j.ijiec.2010.06.002

Asgharpour, M. J., & Javadian, N. (2004). Solving a stochastic cellular manufacturing model using genetic algorithm. *International Journal of Engineering, Transactions A. Basics*, *17*(2), 145–156.

Balakrishnan, J., & Cheng, C. H. (2007). Dynamic cellular manufacturing under multi-period planning horizon. *European Journal of Operational Research*, *177*(1), 281–309. doi:10.1016/j.ejor.2005.08.027

Ghezavati, V. R., & Saidi-Mehrabad, M. (2011). An efficient hybrid self-learning method for stochastic cellular manufacturing problem: A queuing-based analysis. *Expert Systems with Applications*, *38*, 1326–1335. doi:10.1016/j.eswa.2010.07.012

Ghezavati, V. R., & Saidi-Mehranad, M. (2010). Designing integrated cellular manufacturing systems with scheduling considering stochastic processing time. *International Journal of Advanced Manufacturing Technology*, *48*(5-8), 701–717. doi:10.1007/s00170-009-2322-2

Ghezavati, V. R., & Saidi-Mehranad, M. (2011). An efficient linearization technique for mixed 0-1 polynomial problems. *Journal of Computational and Applied Mathematics*, *235*(6), 1730–1738. doi:10.1016/j.cam.2010.08.009

Ghosh, T., Sengupta, S., Chattopadhyay, M., & Dan, P. K. (2011). Meta-heuristics in cellular manufacturing: A state-of-the-art review. *International Journal of Industrial Engineering Computations*, *2*(1), 87–122. doi:10.5267/j.ijiec.2010.04.005

Heragu, S. (1997a). *Facilities design* (p. 316). Boston, MA: PWS Publishing Company.

Heragu, S. (1997b). *Facilities design,* (p. 345). Boston, MA: PWS publishing company

Hillier, F. S., & Lieberman, G. J. (1995). *Introduction to operation research* (6th ed.). New York, NY: McGraw-Hill.

Hosseini, M. M. (2000). An inspection model with minimal and major maintenance for a system with deterioration and Poisson failures. *IEEE Transactions on Reliability*, *49*(1), 88–98. doi:10.1109/24.855541

Hurley, S. F., & Clay Whybark, D. (1999). Inventory and capacity trade-off in a manufacturing cell. *International Journal of Production Economics*, *59*(1), 203–212. doi:10.1016/S0925-5273(98)00101-7

Kuroda, M., & Tomita, T. (2005). Robust design of a cellular-line production system with unreliable facilities. *Computers & Industrial Engineering*, *48*(3), 537–551. doi:10.1016/j.cie.2004.03.004

Mahdavi, I., Javadi, B., Fallah-Alipour, K., & Slomp, J. (2007). Designing a new mathematical model for cellular manufacturing system based on cell utilization. *Applied Mathematics and Computation*, *190*, 662–670. doi:10.1016/j.amc.2007.01.060

Mobasheri, F., Orren, L. H., & Sioshansi, F. P. (1989). Scenario planning at southern California Edison. *Interfaces*, *19*(5), 31–44. doi:10.1287/inte.19.5.31

Mulvey, J. M. (10996). Generating scenarios for the Towers Perrin investment system. *Interfaces*, *26*(2), 1–15. doi:10.1287/inte.26.2.1

Papaioannou, G., & Wilson, J. M. (2009). Fuzzy extensions to integer programming of cell formation problem in machine scheduling. *Annals of Operations Research*, *166*(1), 1–19. doi:10.1007/s10479-008-0423-1

Ravichandran, K. S., & Chandra Sekhara Rao, K. (2001). A new approach to fuzzy part family formation in cellular manufacturing system. *International Journal of Advanced Manufacturing Technology*, *18*(8), 591–597. doi:10.1007/s001700170036

Rosenhead, J., Elton, M., & Gupta, S. K. (1972). Robustness and optimality as criteria for strategic decisions. *Operational Research Quarterly*, *23*(4), 413–431.

Safaei, N., Saidi-Mehrabad, M., Tavakkoli-Moghaddam, R., & Sassani, F. (2008). A fuzzy programming approach for cell formation problem with dynamic & uncertain conditions. *Fuzzy Sets and Systems*, *159*(2), 215–236. doi:10.1016/j.fss.2007.06.014

Shanker, R., & Vrat, P. (1998). Post design modeling for cellular manufacturing system with cost uncertainty. *International Journal of Production Economics*, *55*(1), 97–109. doi:10.1016/S0925-5273(98)00043-7

Siemiatkowski, M., & Przybylski, W. (2007). Modeling and simulation analysis of process alternative in cellular manufacturing of axially symmetric parts. *International Journal of Advanced Manufacturing Technology*, *32*(5-6), 516–530. doi:10.1007/s00170-005-0366-5

Snyder, L. V. (2006). Facility location under uncertainty: A review. *IIE Transactions*, *38*, 537–554. doi:10.1080/07408170500216480

Solimanpur, M., Vrat, P., & Shankar, R. (2004). A heuristic to optimize makespan of cell scheduling problem. *International Journal of Production Economics*, *88*, 231–241. doi:10.1016/S0925-5273(03)00196-8

Song, S.-J., & Hitomi, K. (1996). Determining the planning horizon and group part family for flexible cellular manufacturing. *Production Planning and Control*, *7*(6), 585–593. doi:10.1080/09537289608930392

Sun, Y.-L., & Yih, Y. (1996). An intelligent controller for manufacturing cell. *International Journal of Production Research*, *34*(8), 2353–2373. doi:10.1080/00207549608905029

Szwarc, D., Rajamani, D., & Bector, C. R. (1997). Cell formation considering fuzzy demand and machine capacity. *International Journal of Advanced Manufacturing Technology*, *13*(2), 134–147. doi:10.1007/BF01225760

Tsai, C. C., Chu, C. H., & Barta, T. (1997). Analysis and modeling of a manufacturing cell formation problem with fuzzy integer programming. *IIE Transactions*, *29*(7), 533–547. doi:10.1080/07408179708966364

Venkataramanaiah, S. (2007). Scheduling in cellular manufacturing systems: An heuristic approach. *International Journal of Production Research*, *1*, 1–21.

Viswanadham, N., & Narahari, Y. (1992). *Performance modeling of automated manufacturing systems*. Englewood Cliffs, NJ: Prentice Hall.

Wu, X. D., Chu, C. H., Wang, Y. F., & Yan, W. L. (2006). Concurrent design of cellular manufacturing systems: A genetic algorithm approach. *International Journal of Production Research*, *44*(6), 1217–1241. doi:10.1080/00207540500338252

Yang, J., & Deane, R. H. (1993). Setup time reduction and competitive advantage in a closed manufacturing cell. *European Journal of Operational Research*, *69*(3), 413–423. doi:10.1016/0377-2217(93)90025-I

Chapter 15
Planning Process Families with PROGRES

Linda L. Zhang
IESEG School of Management, France

ABSTRACT

Process family planning has been well recognized as an effective means of maintaining production efficiency by exploiting process reuse and near mass production efficiency underlying product families. To support process family planning automation, this study develops a PROGRES-based approach to modeling planning data, knowledge and reasoning. The PROGRES-based process family planning models are hierarchically organized. At the top level, a meta model is defined to conceptualize process family planning in general. Based on this meta model, generic models are defined for planning process families for specific product families (i.e., specific process family planning). Finally, instance models are obtained by instantiating the generic models, representing production processes for given product family members. The proposed approach is illustrated with planning processes for a textile spindle family.

INTRODUCTION

To survive, manufacturing companies strive to timely offer a large number of customized products at affordable costs. Developing product families, instead of single products, has been well accepted as an effective means to accommodate the increasingly individualized customer expectations while leveraging cost of delivering the resulting variety (Meyer and Utterback, 1993). A product family refers to a set of customized products that assume some common structures and yet possess specific features and functionalities to meet particular customer requirements. Many approaches and methodologies have been introduced to accommodate product family development (e.g. Agard and Kusiak, 2004; Anderson, 1997; Hsiao and Liu, 2005). With focus on design, these methods

DOI: 10.4018/978-1-61350-047-7.ch015

can help companies reduce design costs and time and reuse proven design knowledge, as evidenced by some successful industrial cases, including Sony Walkmans (Sanderson and Uzumeri, 1997), Compaq personal computers (Meyer, 1997) and Lutron lighting systems (Pessina and Renner, 1998). However, they are not able to facilitate other issues of product family development (e.g., production). Authors have pointed out successfully developing product families hinges on efficiency of both design and production (do Carmo-Silva and Alves, 2006; Wiendahi et al., 2007)

Due to the finite manufacturing resources existing on shop floors and the short delivery lead times, product families lead to difficulties in production process planning, and further in production (Wortmann et al., 1997). This is because the production optimality of single products may conflict that of a product family (Jiao et al., 2007). In response to the inefficiency of the traditional approaches to planning, planning production processes for product families rather than single products (i.e., process family planning) has been put forward as an effective means for companies to obtain production efficiency of product families (Martinez et al., 2000; Schierholt, 2001; Zhang and Rodrigues, 2009). This is accomplished by exploiting process reuse and near mass production efficiency underlying product families. The rationale of process family planning lies in anchoring the planning of production processes for product family members to a common platform so as to reconfigure the existing processes and capitals (Azab et al., 2008).

A process family refers to the set of production processes to produce the set of product variants in a family. In this regard, a production process is to produce a complete product, which consists of both component parts and component assemblies. It is formed by a number of operations, operations precedence and manufacturing resources. Further, in accordance with the hierarchy of a product in consideration, operations and other associated process elements in a production process can be grouped as several subprocesses. Each such subprocess is to produce the corresponding component part or assembly in the product hierarchy. In this regard, dealing with part manufacturing, cellular manufacturing compliments process family planning. With operations and manufacturing resources determined in process family planning, the appropriate methods reported in cellular manufacturing can be used to, e.g., group parts with similar routings.

Despite the similar product structures and the same item types, product variants in a family differ from one another in specific item variants, be they parts or assemblies. Thus, the associated diverse designs of product variants and component items impose requirements for different production processes. As a result, process family planning involves large volumes and different types of data pertaining to a product family and the process family to be planned. In this regard, how to organize these data as a common platform becomes a major issue in process family planning. Due to the different component parts, assemblies involved in a product family and the limited manufacturing resources available on shop floors, there are numerous constraints that must be satisfied when planning production processes for product variants (Zhang and Xu, 2010). These constraints complicate the reasoning behind process family planning, and further the planning process. In this regard, how to model the dynamics (i.e., the planning process) and the planning reasoning presents itself as another major issue.

To support process family planning automation, in this chapter, we focus on the above two issues. We approach the first issue by addressing the static representation of data involved in process family planning and the second one by handling the dynamic modeling of the planning process. With the static representation and dynamic modeling developed in this study, we expect to shed light on 1) the organization of the large volumes of product and process family data in a logical way, and 2) the reasoning behind process family planning.

Many modeling tools/languages (e.g., Petri nets, simulation, mathematical programming, data diagrams, flow charts) have been reported in the literature. However, they are designed to either represent a system from the static aspect (i.e., structure representation involving constituent elements and their relationships) or model a system from the dynamic aspect (i.e., dynamic modeling), but not both. In response to this limitation, PROGRES (PROgrammed Graph Rewriting System) is developed to model systems by capturing both system's static structures and dynamic behavior (Schurr et al., 1995). Its applications have been seen in a wide range of areas, such as data structure specification, process modeling and configuration management (Szuba, 2005). In addition, PROGRES excels in handling derived fact, providing parametric rewriting rule specification, and supporting data and system consistency (Schurr et al., 1995; 1998). The readers may refer to (Schurr et al., 1995; 1998) for details about PROGRES.

Recognizing its modeling capabilities, we adopt PROGRES to model process families and their planning. By modeling a process family, the attempt is to show how the large volumes of data involved in process family planning should be organized; by modeling the planning process, the aim is to clarify the reasoning behind planning production processes for product families.

In section 2, we present a review of the literature relevant to process family planning. An overview of PROGRES-based process family planning modeling is given in Section 3. Also discussed are the guidelines of PROGRES-based modeling. Sections 4 and 5 present the graph schema and graph transformations and generic models for process family planning modeling. The results of a case study are presented in Section 6 to demonstrate the potential and feasibility of the proposal. In Section 7, this chapter is ended with conclusions, limitations and potential avenues for future research.

RELATED WORK

In view of the importance of production efficiency in successfully developing products and the limitations of traditional approaches to planning, more and more efforts have been put in developing new planning methods. The hope is that with these new methods, companies are able to maintain production of product families as stable as possible by eliminating the unnecessary process variations (Lu and Botha, 2006; Pisano, 1997). Schierholt (2001) puts forward a concept of process configuration for configuring process plans to manufacture part families. Similarly, Williams et al. (2007) introduce process parameter platforms for deriving process parameters to be included in machining processes for parts. To facilitate process plan generation based on process configuration, Zheng et al. (2008) present a systematic model of hierarchical, historical and case-based manufacturing process knowledge. While the above work addresses planning processes for part families, other reported work presents methods for assembly planning. Gupta and Krishnan (1998) introduce a methodology to design assembly sequence for a product family. Gottipolu and Ghosh (2003) develop an approach for generation, representation and selection of assembly plan alternatives by translating geometric and mobility constraints into contact and translational functions. Marian et al. (2003) first formalize and solve the assembly sequence planning problem; subsequently, they discuss an approach based on genetic algorithm to optimize the generated assembly plan alternatives (Marian et al., 2006). In summary, based on given inputs, such as parts, operations and machines, methods in the above work attempt to specify detailed process parameters (e.g., cutting speed, feed rate, collision path) for manufacturing parts or producing assemblies. To ensure the accuracy of these planning activities by providing the correct/optimal inputs, production processes for final products should be planned (do Carmo-Silva and Alves, 2006; Wiendahi et al., 2007).

A concept of process family planning is introduced in (Zhang and Rodrigues, 2009), in attempting to determine the processes to achieve optimal production in terms of e.g. lead time, resource utilization for the product family as a whole. While the authors put forward the concept of process family planning, they focus on the construction of a generic routing structure underpinning the product and process families from data existing in companies' databases. In attempting to facilitate constraint construction, Jiao et al. (2008) discuss an approach based on association rule mining to identify the mapping relationships between product and process families. In a recent work (Zhang and Xu, 2010), process family planning is formulated as a constraint satisfaction problem (CSP) and solved based on the techniques existing in the CSP literature. While these studies approach process family planning from different perspectives by formulating and solving the problems concerned, they leave data organization and planning reasoning untouched. In view of their importance in process family planning automation, this chapter thus focuses on data organization and planning reasoning.

OVERVIEW OF PROGRES-BASED PROCESS FAMILY PLANNING

Process Family Planning

For a given product family, a generic routing structure can be constructed by integrating a generic product structure with a generic process structure (Zhang and Rodrigues, 2009). While the generic product structure models all design data pertaining to the product family, the generic process structure organizes all process data describing the corresponding process family. With such a generic routing structure, process family planning generally entails activities from two views, including the design view and the production view. From the design view, process family planning is to specify product variants based on given customer requirements; from the production view, it determines the corresponding production processes. Thus, in the design view, process family planning is characterized by the generic product structure involving a set of design parameters, compatible constraints among these parameters, product items and relationships among them; in the production view, process family planning is characterized by processes, sequence relationships, operations and operations precedence organized in the generic process structure. The processes are associated with component items that can be located in the generic product structure. More specifically, a process is to produce a parent item by taking several child items as input. These processes are connected by sequence relationships, and detailed by operations and operations precedence. Unlike the output of processes, the output of some operations are pseudo items, which cannot be found in the generic product structure, and are formed by child product items and/or child pseudo items.

Consistent with the above understanding, planning a specific production process involves two phases: product variant specification and production process determination. With given customer requirements, a user (e.g., a designer) assigns values to parameters, resulting in a list of compatible parameter value pairs which define an end-product. These parameter values then propagate along the hierarchy of the generic product structure, determining parameter values of items. Each item, be it a secondary or a primary, is instantiated according to the parameter values obtained from parameter propagation. Item instantiation leads to a specific hierarchy pertaining to the product variant. With the item variants and the goes-into relationships (i.e., parent-child relationships) among them, the processes, sequence relationships, operations and operations precedence in the generic process structure are instantiated. Such instantiation is accommodated by the interconnections between the generic product and process structures and the conditions to include an operation or process. The instantiation results a production process including several groups of ordered operations for the product variant, with each group producing one product item.

Overview of PROGRES-Based Process Family Planning

By following PROGRES language constructs, this chapter develops a PROGRES model of process family planning. In the model, design parameters, compatible constraints, items, processes, operations, etc. are represented as nodes; relationships among them are represented as edges; and manipulations of items, processes and operations are modeled as productions. The graph representing product family elements that construct the starting point of process family planning is adopted as a starting graph. The graph of a production process is transformed based on the starting graph by invoking proper productions. In addition to productions, graph generation is accomplished with control structures, which define the execution order of productions.

To achieve the above PROGRES-based process family planning modeling, a graph schema is designed to model all involved objects. It consists of a set of graph entities common to graphs of production processes of the family. In addition, it describes all the necessary types of nodes and edges, as well as their associated attributes. Furthermore, the productions and control structures are programmed to manipulate graphs, more specifically nodes and edges, by reasoning about planning production processes for given product variants.

In accordance with the stratified character of PROGRES, models at three levels of abstraction are developed in the PROGRES-based modeling of process family planning. They include meta models at the top (or meta) level, generic models at the family level and instance models at the variant level. A meta model captures the abstraction of objects/concepts and their relationships common to planning of production processes for different product families. The abstraction is specified by defining the corresponding node classes and edge types. In addition, to generalize the graph manipulations that are common to all process families, the related graph transformations are defined at this meta level. The generic model represents the unified generic data structure of a process family, where family related elements, such as items, processes, operations and design parameters, are specified using node classes. The relationships among these elements are specified using edge types. The graphical representation of a generic model is called a family graph. For a process family of a particular product family, the corresponding generic model can be defined by adapting relevant entities of the meta model to specific characteristics of the product and process families. A family graph can be transformed to variant graphs, representing operations and operations precedence of a specific production process. Essentially, these variant graphs are instance models.

The meta model and graph transformations are defined by the classes and productions in the PROGRES formalism. To adapt the meta model to a family specific generic model, node types of the process family are specified from two different views: the design view and the production view. By emerging these specific types with classes and productions, a complete PROGRES specification of the particular process family is obtained. The starting graph for design consists of design parameters, products, items, compatible constraints, goes into relationships, selection constraints and the relationships among them, thus named as the design view family graph. The starting graph for production consists of processes, sequence relationships, operations, precedence relationships, selection constraints and the relationships among them, thus named as the production view family graph.

Based on the family specific generic model and the given customer requirements, users input the values for design parameters. The design view family graph is then rewritten according to the pre-defined control structures. The result is a variant graph – graphical representation of a product variant that can satisfy the given customer requirements, more specifically its product hierarchy. The product hierarchy is then transferred to the production view in the form of items and their goes into relationships. Taking these items and goes into relationships as input, the production view family graph starts to transform according to the pre-defined control structure. The resulted variant graph represents the production process to produce the product variant.

META MODELS OF PROCESS FAMILY PLANNING

Applicable to planning of process families for different product families, the meta model includes a class level graph schema (as shown in Figure 1) and graph transformations. The class level graph schema generalizes the entities common to different process families and models them as node classes and edge types. By involving a number of productions, transactions and control structures, graph transformations model the dynamic behavior of process family planning.

Graph Schema

The class level of PROGRES graph schema contains all common entities of process families, and defines all nodes and edge classes occurring in process family planning. As shown in Figure 1, PFP_OBJECT acts as the root of the class hierarchy. To model the two views of process family planning (i.e., the design view and the production view), two subclasses – DSGN_OBJECT and PROD_OBJECT – are defined. These two subclasses are further elaborated in the design view and the production view meta models, respectively.

As shown, DSGN_OBJECT is a superclass encompassing all entities occurring in the design view meta model. A defines edge between PRODUCT and PARAMETER indicates that design parameters define products. The edge, valueOf, between PARAMETER and VALUE models the fact that a parameter can assume a number of values, the determination of which defines specific products (i.e., product variants). An affects edge models the fact that a parameter (the affecting parameter) whose value is the antecedent of the value of another parameter (the affected parameter) must be assigned a value before the value assignment of the latter parameter. (See below the meaning of antecedent.) Assigned is a derived attribute of PARAMETER, indicating whether or not the value of this parameter has been selected (true) or not (false). Its default value is false. When a suitable value is selected, it becomes true.

To handle compatible constraints between parameter values, a node class, COMPATIBLE_CONSTR-AINT, is defined. Moreover, a COMPATIBLE_CONSTRAINT node connects the antecedent and consequent of the modeled compatible constraint with the edges antecedent and consequent, respectively. The derived attribute, AnteSelected, of COMPATIBLE _CONSTRAINT is determined by the Selected attribute of its antecedent. A COMPATIBLE _CONSTRAINT can be a REQUIRE_CONSTRAINT or an EXCLUDE_CONSTRAINT. A REQUIRE_CONSTRAINT is defined to model that: if a specific value, A^*_x, of parameter A (the antecedent) is selected and assigned to A, then a value, B^*_y, of parameter B (the consequent) must be selected for B. An EXCLUDE_CONSTRAINT captures that: if A^*_x is selected for A, then B^*_y must not be selected for B.

To model product items and the relationships among them, a GOES INTO_RELATIONSHIP node class and an ITEM node class are defined. An

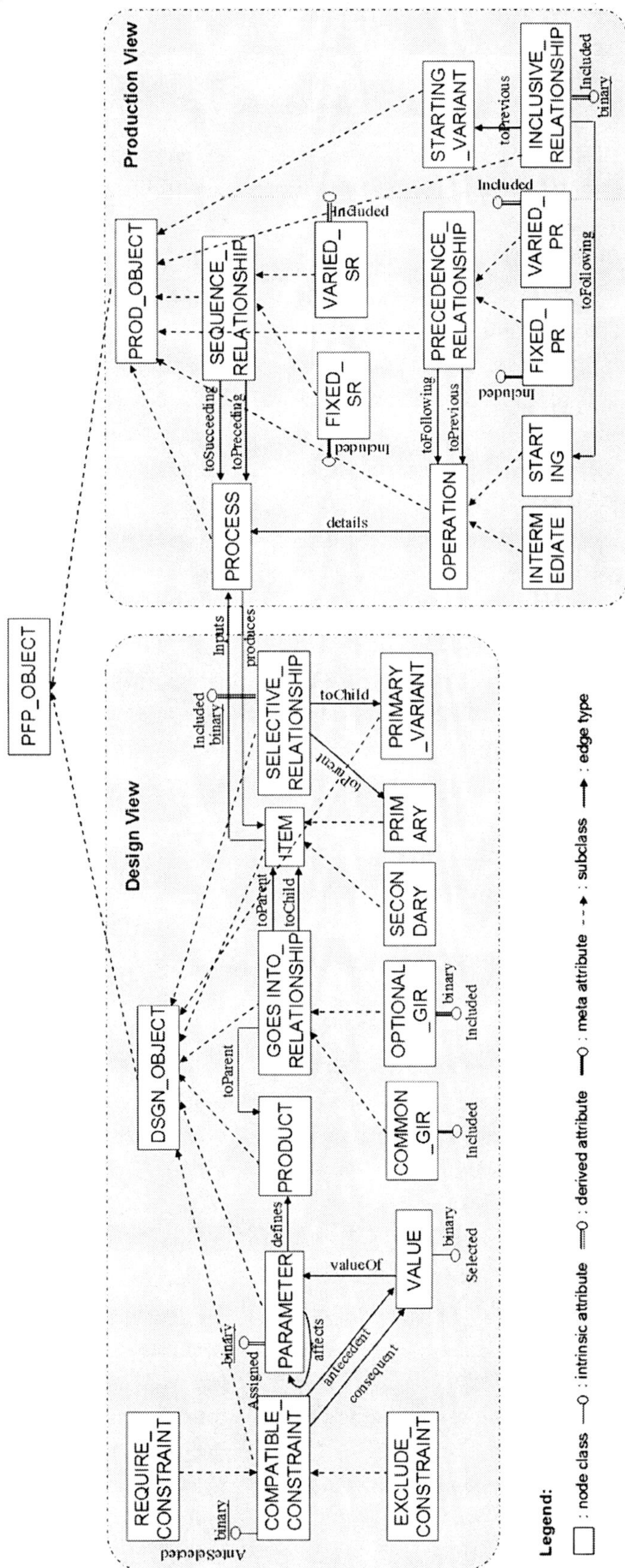

Figure 1. The graph schema of PROGRES-based process family planning

ITEM can be a PRIMARY or SECONDARY one. A primary item cannot be further decomposed and, is represented by one or more PRIMARY_VARIANT. A secondary item is the parent item of lower level child items that may be primary or secondary. PRODUCT contains items of both types.

A GOES INTO_RELATIONSHIP can be a COMMON_GIR or OPTIONAL_GIR. While the COMMON _GIR models those relationships between parent and child items that are common to all product variants in a family, OPTIONAL_GIR captures these parent-child item relationships that appear in several, but not all, product variants. To indicate a COMMON_GIR, an attribute, included, is introduced as a meta attribute whose value is always true. The derived attribute, included, of OPTIONAL_GIR is determined by the parameters of parent items, which are ultimately determined by these of the products. The node class, SELECTIVE_RELATIONSHIP, defines the relationships between primary items and their variants. The included attribute of a SELECTIVE_RELATIONSHIP is instantiated according to the parameters of the associated primary item. A primary item variant is included in the product's structure when included = true.

To model the relationships among node classes, such as PRODUCT and GOES INTO_ RELATION-SHIP, two edge classes, namely toParent and toChild, are defined in the schema. A toParent edge links a GOES INTO_RELATIONSHIP to the associated parent item; and a toChild edge connects a GOES INTO _RELATIONSHIP with the associated child item. In addition, a toParent edge links a SELECTIVE_ RELATIONSHIP to the primary item; and a toChild edge connects a SELECTIVE_RELATIONSHIP with a primary item variant.

As shown in Figure 1, PROD_OBJECT is a superclass covering all entities occurring in the production view meta model. A node class, PROCESS, is defined to model processes to produce parent items while taking child items as input. In this regard, two edge classes, produces and inputs, connect the ITEM node in the design view with the PROCESS node in the production view. A SEQUENCE_RELATIONSHIP node class models the relationships between items' processes. In accordance with the common and optional goes into relationships in the design view meta model, a SEQUENCE_RELATIONSHIP can be either a FIXED_SR or VARIED_SR. While FIXED_SR models those sequence relationships common to all production processes of product variants, VARIED_SR captures sequence variations, meaning the relevant sequence relationships are only assumed by production processes of several product variants. Similarly, a meta attribute, included, with a value true is introduced to FIXED_SR, indicating the modeled sequence relationship appears in production processes of all product variants. The derived attribute, included, of VARIED_SR is determined by the inclusion of the items to be produced. Two edge classes, toSucceeding and toPreceding, link SEQUENCE_RELATIONSHIP with PROCESS. A production process of a product variant thus contains all processes to produce product items and the sequence relationships among them.

An OPERATION node class is further defined to model the operations detailing processes to produce items. An operation can be a STARTING or an INTERMEDIATE one. A starting operation is the first one involved in a process to produce an item, whereas an intermediate one can be any other operation including the last one. PRECEDENCE_RELATIONSHIP models the precedence relationships between operations. It can be a FIXED_PR modeling precedence relationships common to all processes in relation to all variants of an item family or VARIED_PR indicating operations variations. To model a FIXED_PR, the meta attribute, included, with value true is introduced; and to indicate a VARIED_PR, the derived attribute, included, is determined by the previous operation. Two edge classes, toFollowing and toPrevious, connect PRECEDENCE_RELATIONSHIP with OPERATION, modeling the previous operation and the following operation in a precedence relationship. Another node class, INCLUSIVE_RELATIONSHIP, models the inclusion of STARTING_VARIANT (modeling variants of starting operations) in a process. The included attribute of an INCLUSIVE_RELATIONSHIP is determined by parameters of the item to be produced. Included = true indicates that a starting operation variant is included in the process.

Graph Transformations

While the class level schema addresses the static part of PROGRES-based process family planning, graph transformations deal with the operational behavior. The basic operations involved in process family planning are modeled as productions; and complex operations (defined as graph transactions) thus involve a number of productions to be executed by following control structures. In this regard, graph transformations associate with transactions, productions and control structures.

To support process family planning modeling, some operations are necessary at the meta level, thus being independent of process families of particular product families. First, there should be operations to allow designers to assign values to design parameters characterizing a product family. These operations are modeled by the production, AssignValue, as shown in Figure 2. The dashed rectangles above and below the separator ::= define the LHS (left hand side) and RHS (right hand side) of the production, respectively. The rule can be applied only if all conditions are fulfilled. The first statement in the condition part is used to check whether the parameter has been assigned a value; the second statement ensures that there is no affecting PARAMETER to which a value should be assigned before the values is assigned to this parameter or the affecting parameters have been assigned values; finally, the third one is to check that the affected parameters have not been assigned values. If all condition statements hold true, the elements of the LHS in the family graph are replaced by the elements of the RHS. Those unselected values are thus removed from the graph and the node attribute receives its new values according to the transfer function.

If a parameter value, which is the antecedent of a compatible constraint, is selected, there should be operations to process consequent according to whether it is an exclude constraint or a require constraint. In case of an exclude constraint, the consequent of the constraint should be deleted from those possible values of the affected parameter. In case of a require constraint, the consequent of the constraint should be deleted from those possible values of the affected parameter. The two productions in Figure 3 are designed to model these two types of operations. While the productions in Figures 2 and 3 are performed on the design view meta model, the productions modeling the below operations are preformed on the design and production views family graphs.

There should also be operations to delete items, primary item variants and processes that are not included in the graphs of a product variant and the production process. The productions in Figure 4 are designed to model these operations; they

Figure 2. Production modeling assigning values to a parameter

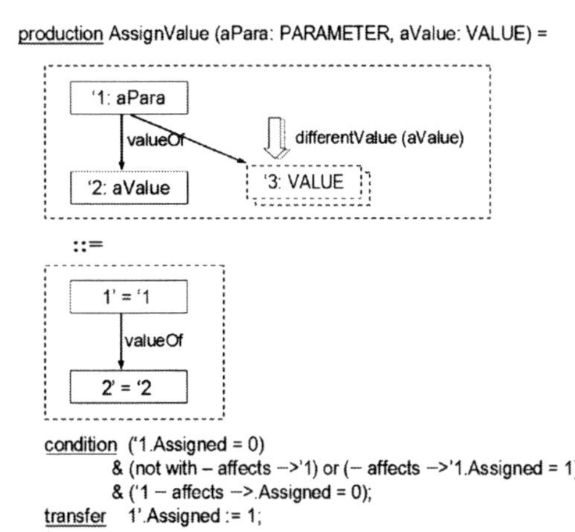

Figure 3. Productions modeling constraint handling for the design view meta model

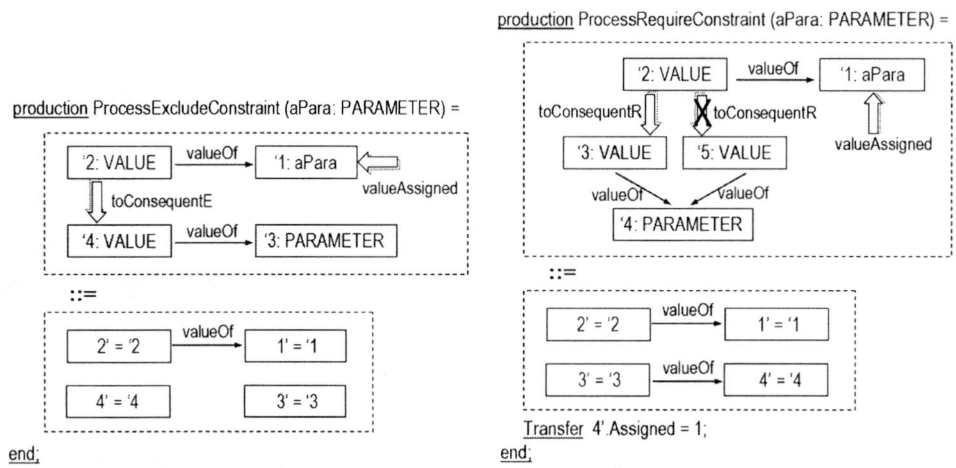

(a) Processing an exclude constraint (b) Processing a require constraint

are performed on the design view family graph. At last, there should be operations to 1) determine operations to be included in an item's process, and 2) delete operations and starting operations variants that are not in included in the graph of the desired production processes. The productions in Figure 5 are designed to model these operations; they are performed on the production view family graph.

Control Structures in Production Process Derivation

Due to the complexity involved, transforming a family graph into a variant graph necessitates the execution of more than one production. In this regard, imperative control structures are a key to enforce certain orders of production application (Schurr, 1990). The control structure is designed to manage the execution of productions for deriving production processes. It consists of two parts with the first part deriving a product variant (see Appendix A), and the second part deriving the corresponding production process (see Appendix B).

First, a graph test on the design view family graph is performed (see Appendix A). A parameter, whose value has not been assigned, is thus selected. If there is no value-unassigned parameter preceding this selected parameter, the selected parameter and its allowable values are listed to users for them to choose. After obtaining the users' input (i.e., a selected value), a production, AssignValue, is applied to the family graph. If the selected value is the antecedent of certain constraints, these constraints will be processed. It then goes back to process another value-unassigned parameters till every parameter obtains a value. With all the

Figure 4. Productions removing items and item variants

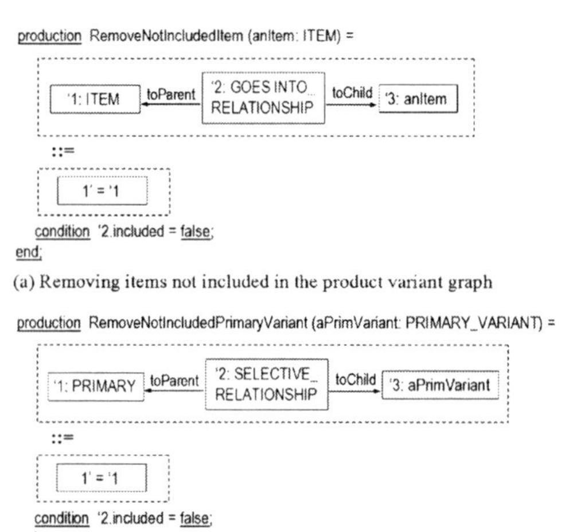

(a) Removing items not included in the product variant graph

(b) Removing primary item variants not included in the product variant graph

Figure 5. Productions removing processes, operations and operations variants

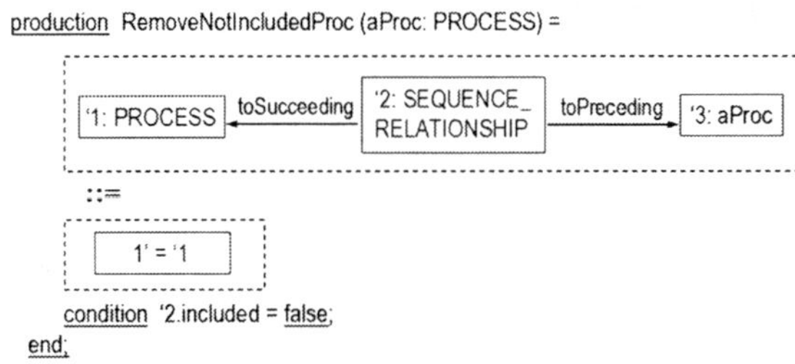

(a) Removing processes not included in the production process graph

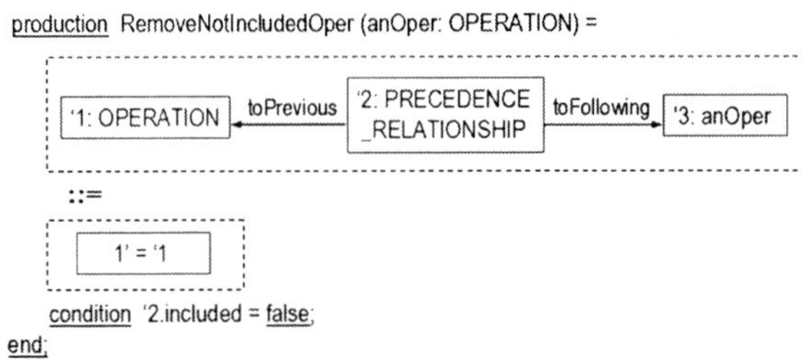

(b) Removing operations not included in the production process graph

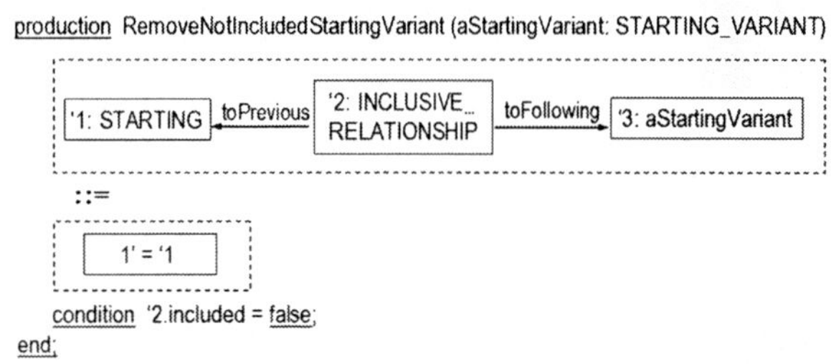

(c) Removing starting operations variants not included

parameter values assigned, the control structure starts to determine the product items and their relationships to be included in the product variant. This operation is automatic as both parameter propagation from parent items to child items and include conditions have been modeled as derived attributes (see the following section). Derived attributes are evaluated; and all those un-included items and primary item variants are removed. The desired product variant graph is thus obtained.

As with the graph test on the design view family graph, a graph test is conducted on the production view family graph first (see Appendix B). An item whose process has not been determined is thus selected. If the process of its parent item has been determined, then a process will be selected to produce this item. Once all the processes together with their sequence relationships have been determined for the corresponding items, the control structure will specify, for each process, the operations and starting operation variants to be removed from the production view family graph. Similarly, the operation to determine processes and operations is automatic since include conditions have been modeled as derived attributes according to the goes into relationships of items and the parameter values of items, respectively. The production process graph is finally derived.

GENERIC MODELS OF PROCESS FAMILY PLANNING

While a meta model acts as a general pattern of process family planning, a generic model functions as the fundamental mechanism to support a specific process family planning (i.e., planning a process family for a particular product family). Hence, meta models need to be transformed to generic models so as to enable the planning of specific process families. To do so, all family specific parameters, generic items, generic processes, generic operations, etc. in the generic routing structure of the process family are specified as node types. A node type declaration defines the label of a group of nodes (i.e., node instances) and the node class to which it belongs. It determines the static properties of node instances. The declaration is accomplished by defining three kinds of attributes: the intrinsic, derived and meta attributes. Some examples of node type definitions are given for AssyProc4Chair, AssyProc4Armrest and MOp-4Wheel (see Appendix C). AssyProc4Chair is an abbreviation for *assembly process for chair* modeling a process family (i.e., a generic assembly process) to assemble a chair family (i.e., a generic chair) from immediate child item families (i.e., generic child items); AssyProc4Armrest captures an assembly process family for an armrest family; and finally, MOp4Wheel models a machining operations family involved in manufacturing a wheel family. (Note an armrest is one of the immediate child items of a chair.)

Intrinsic attributes have the values that are directly assigned and, do not depend on the values of any other attribute. For example, in the assembly process family of the chair family, such attributes as *item_ toProduce*, *item#1_Input*, *proc#1_Preceding* and *proc#n_Preceding* are intrinsic attributes (see Appendix C). An intrinsic attribute has a type-dependent initial value, which may be changed by performing a graph transformation. If an item is modeled as an intrinsic attribute, its default value can be set to be the initial value of the attribute. Unlike intrinsic attributes, meta attributes are the attributes that posses constant type-dependent values. Thus, those attributes whose values are common among family members can be assigned as meta attributes. This enables the handling of those node properties having the same value for all instances of a given node type. For example, for the assembly process family of the chair family, a statement that the value of a meta attribute, included, is true means that production processes of all chair variants in the family assume a chair assembly process (see Appendix C).

In determining product items, parameter propagation from a parent node (representing a parent item) to child nodes (representing child items) can be modeled by derived attributes that have node instance specific values and change their values as a result of graph transformations performed. Similarly, with derived attributes, the corresponding processes can be determined to be included in production processes of product variants. For example, for the assembly process family of the armrest family, whether or not an armrest assembly process is included depends on the fact whether or not an armrest is involved in the chair variant (see Appendix C). In this regard, the derived attributes of processes are associated with the relevant product items. Unlike determining items to be included in products, which is of top-down, specifying operations involved in processes follows a bottom-up approach (i.e., the inclusion of operations is determined based on that of the previous operations). This is modeled by the derived attributes as well. For example, in MOp4Wheel, whether or not the machining operation to be included in manufacturing a wheel depends on the inclusion of its previous operation – FOp4Wheel (fabrication operation for wheel).

Node types together with associated edges comprise generic models of product and process families, which can be represented as family graphs. The design view family graph consists of family specific node types for parameters, values, compatible constraints, goes into relationships, items, primary item variants and selective relationships. The production view family graph includes family specific node types for processes, operations, sequence relationships, starting operations variants, precedence relationships and inclusive relationships.

CASE STUDY

The proposed PROGRES-based process family planning is applied to textile spindle production in an Indian company. (Due to the confidential issue, the company's name is not revealed; and the original data is modified without losing the capability to highlight the characteristics of this study.) On one hand, compared with other products (e.g., airplanes, semiconductor manufacturing equipment), textile spindles are not complicated products; on the other hand, the degree of textile spindles' product complexity not only allows illustrative simplicity but also enables the case application to be representative enough. The fact that meta models in Section 4 are common to process family planning of any product families highlights the importance in addressing generic models and instance models in application cases. Hence, in this case study, the focus is on the generic models of process family planning of a textile spindle family and how production processes of textile spindle variants are derived based on family graphs in the generic models and graph transformations defined in the meta models.

Family Graphs

From the design view, there are four design parameters characterizing the textile spindle product family, as shown in Table 1, including length, diameter, thread pitch and chamfer. Also shown are the possible values of these parameters and the compatible constraints among them. A textile spindle is assembled from two immediate child items, shaftassy (shaft assembly) and rockerarmassy (rockerarm assembly), which are secondary items. The secondary and primary items involved in the textile spindle family are given in Table 1 as well. The determination of the parameter values of these item variants is based on these defining textile spindle variants in Table 1.

Based on the available product-related data and the company's designers' domain knowledge, the generic model in the design view is constructed, including node type specifications in the textual form (see Appendix D) and the corresponding family graph in Figure 6.

From the production view, planning production processes for the textile spindle family involves a number of processes in accordance with the child items at different levels of the product hierarchies. Each of these processes is further detailed by operations together with operations precedence. Similarly, the production view generic model is constructed, capturing all process-

Table 1. Parameters, values, compatible constraints and secondary and primary items

Elements in PFP for the textile spindle family	
Parameter	**Value**
Length	45mm, 50mm
Diameter	10mm, 12mm
Thread pitch	1mm, 2mm
Chamfer	30°, 45°
Compatible constraint	If Diameter = 12mm, Then Thread pitch = 1mm;
	If Thread pitch = 2mm, Then chamfer = 30°;
	If Length = 45mm, Then Diameter ≠ 12mm.
Secondary items	Textile spindle, shaftassy, rockerarmassy
Primary items	Shaft, needle, rockerarm, sleeve.
Common items	Shaft, needle, rockerarm, sleeve.

related elements and their relationships involved in process family planning for the textile spindle family. While Appendix E shows this generic model in the textual form (i.e., the node type specifications), Figure 7 shows the graphic representation of this generic model (i.e., the family graph). For illustrative simplicity, only the operations details for Proc4St (process for shaft) and for Proc4RA (process for rockerarmassy) are shown in Figure 7.

Production Process Derivation

Essentially, the above family graphs are starting graphs of process family planning for the textile spindle family. While family graphs concern all node labels, node attributes and edge labels from the static structural perspective, other elements of process family planning such as productions and control structures, after being adapted to the generic models, enable graph transformations from the dynamic behavior perspective. Therefore, production process derivation entails a series of graph

Figure 6. The family graph in the design view

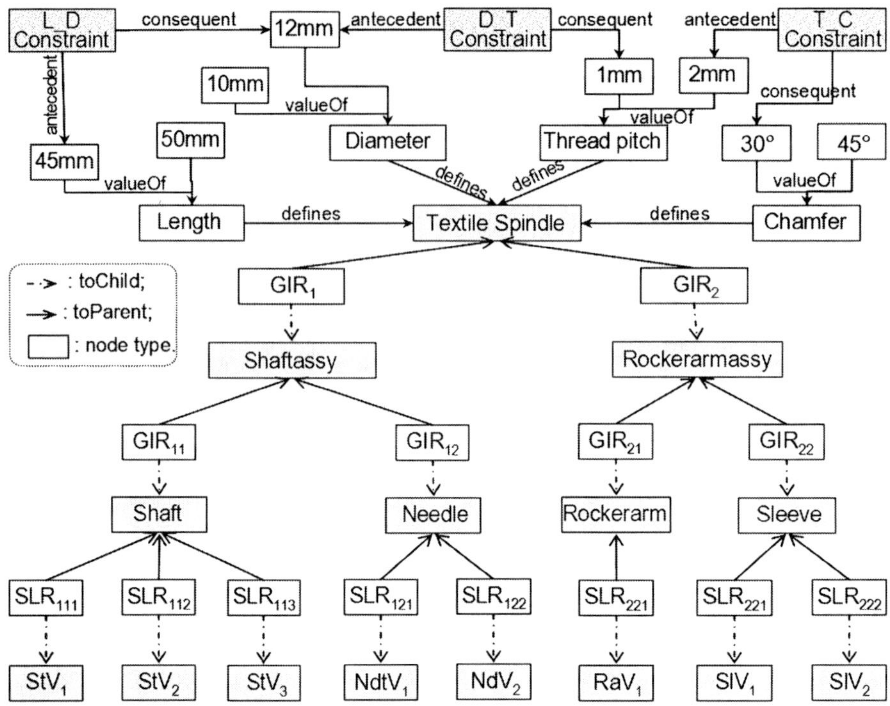

Figure 7. The family graph in the production view

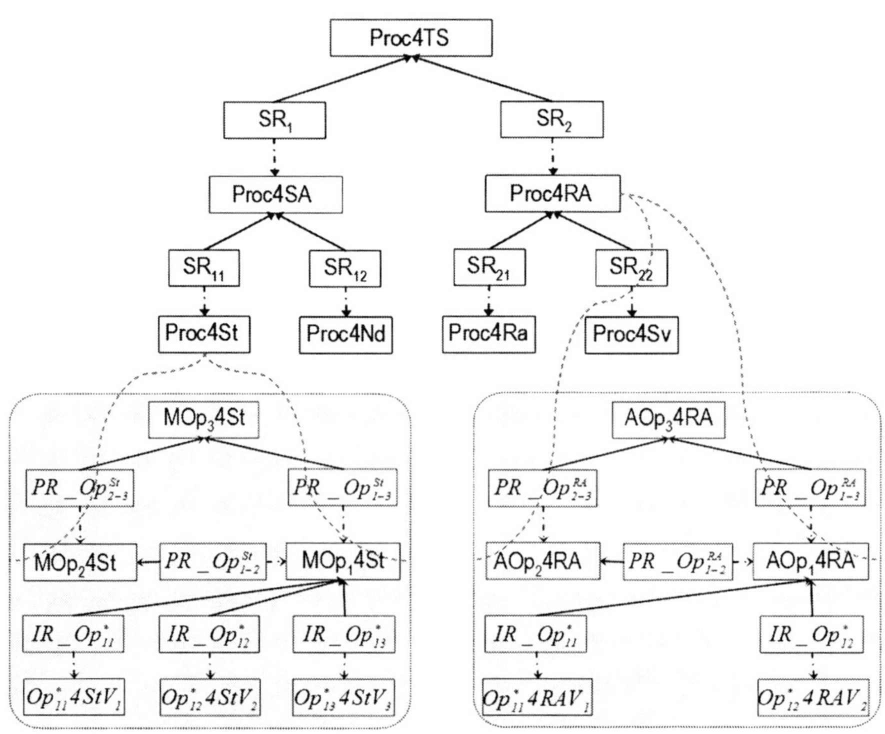

SR: Sequence relationship; PR: Precedence relationship; IR: Includisve relationship

→ : toSucceeding; - · → : toProceeding; - · → : toPrevious; → : toFollowing; ☐ : node type.

Figure 8. Variant graphs derived by graph transformation

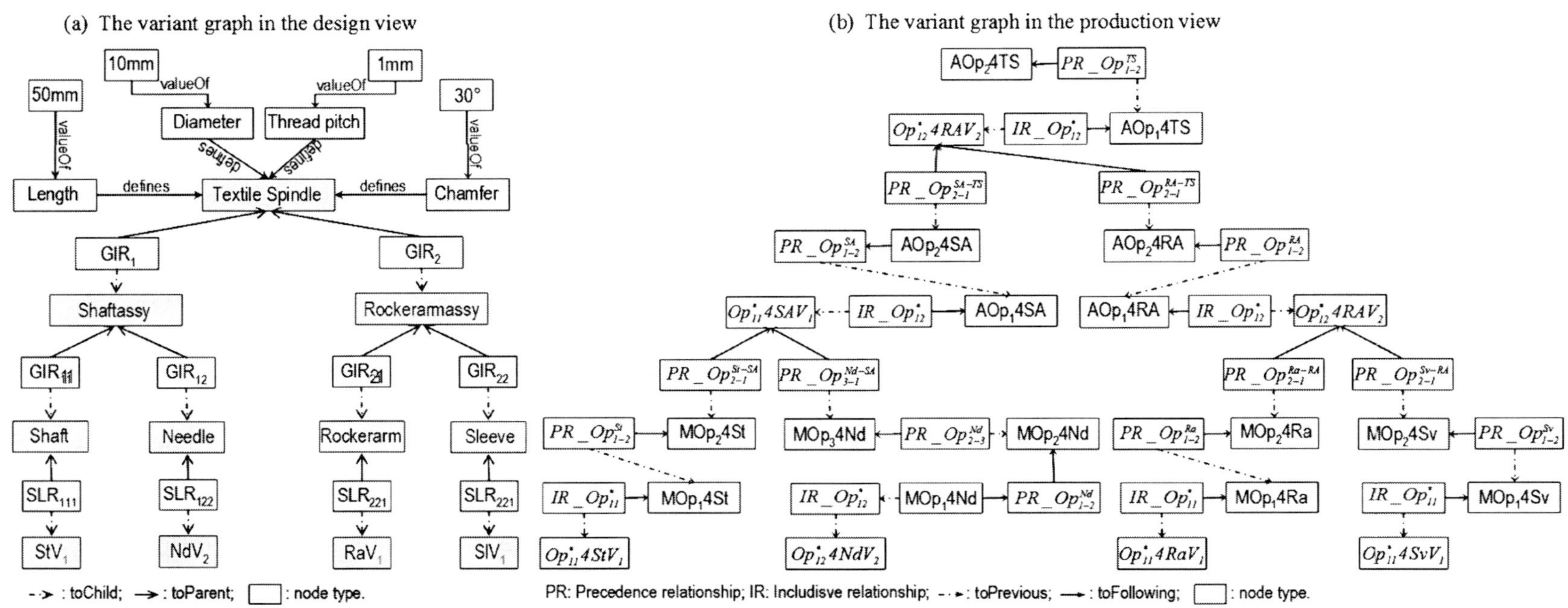

transformations. All production processes that can be obtained by graph rewriting form the process family to produce the textile spindle family.

Suppose a designer decides the following parameter values when defining a textile spindle variant: length = 50mm, diameter = 10mm, thread pitch = 1mm, chamfer = 30°. With these input values, the PFP system can first generate the spindle variant in terms of product items together with their relationships and, finally generate the production process desired. Figure 8(a) shows the spindle variant in the design view. It results from removing all unselected items and item variants from the design view family graph in Figure 6. The transformation of Figure 6) to Figure 8(a) thus demonstrates the graph rewriting process from a starting graph to a variant graph. The production process graph, as shown in Figure 8(b), is derived by transforming the production view family graph in Figure 7 based on the control structure in Appendix B. To better represent the textile spindle variant and its production process, the graphs in Figure 8 can be transformed into the corresponding BOM (bill of materials) -like graph in Figure 9(a) and the tree-like graph in Figure 9(b).

CONCLUSION

Process family planning appears to be a promising approach to planning production processes for product families in order to achieve product family production efficiency. To support process family planning automation, this chapters develops a PROGRES-based process family planning model. The modeling addresses both the static structure of a process family and the dynamic process of its planning. It is approached from two views: design and production. In the design view, process family planning is represented as a design view family graph. The nodes of the graph describe diverse product family elements, such as design parameters, values and items. The edges model the relationships among these product family elements. In the production view, process family planning is modeled as a production view family graph. This graph consists of a number of nodes representing processes, sequence relationships, operations and operations precedence, etc. along with edges denoting the relationships among them. These family graphs act as the starting graphs for graph transformations, through which

Figure 9. The BOM structure and the production process tree

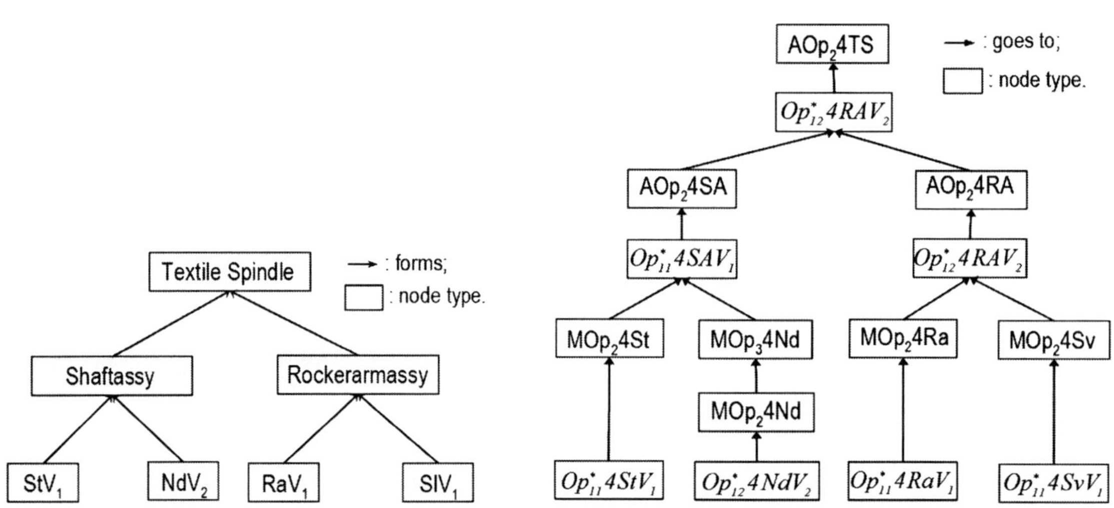

(a) The BOM-like graph after graph transformations (b) The tree-like graph after graph transformations

production process graphs can be derived by executing productions according to pre-defined control structures. Each production process graph represents the production process to produce a product variant.

In line with the unique features of PROGRES (e.g., distinguishing among node classes, node types and node instances), we define 1) meta models for family graphs by generalizing process family planning at a higher level, and 2) generic models modeling all specific elements pertaining to particular process families, and obtain data structures describing production processes for product variants. Attribute dependency in PROGRES is used to model parameter propagation in the generic product structure and the include conditions in the generic process structure. The hierarchical graph schema supports multiple inheritances of graph elements. Parametric rewriting rules support controlled use of formal node type parameters within generic subgraph tests and graph transformations.

The application of the proposed PROGRES-based process family planning to textile spindling production process planning demonstrates that it accommodates documenting the knowledge from existing design and planning practice. Besides, the PROGRES-based process family planning further accommodates knowledge reuse in an interactive environment, where users can select parameter values to define product variants. Consequently, it facilitates process family planning automation. Nevertheless, the PROGRES-based process family planning is developed based on one implicit assumption, that is, the product family has been designed. This allows the organization of data pertaining to the product and process families as the generic routing structure. Hence, for a new product family design, the proposed model might not be sufficient to facilitate planning its process family. Accordingly, efforts may be geared to develop such models that allow planning process families for new product families. In accordance with the link between process family planning and cellular manufacturing (i.e., the former provides inputs, such as parts, operations and resources to the latter), further research might be directed to rigorously formulate this link in order to establish interconnection between production of complete products and manufacturing of parts. With such interconnection, the complexities in the low-volume, high mix production might be reduced.

REFERENCES

Agard, B., & Kusiak, A. (2004). Data-mining based methodology for the design of product families. *International Journal of Production Research*, *42*, 2955–2969. doi:10.1080/00207540410001691929

Anderson, D. M. (1997). *Agile product development for mass customization*. Chicago, IL: Irwin Professional Publishing.

Azab, A., Elmaraghy, H. A., & Samy, S. N. (2008). Reconfigurable process plans: A new approach to minimize change. In Elmaraghy, H. A. (Ed.), *Changeable and reconfigurable manufacturing systems* (pp. 179–194). London, UK: Springer-Verlag.

do Carmo-Silva, S., & Alves, A. C. (2006). Detailed design of product oriented manufacturing systems. In J. Riezebos & J. Slomp (Eds.), *Proceedings of the 3rd International Conference on Group Technology/Cellular Manufacturing* (pp. 1-10). Groningen, The Netherlands.

Gottipolu, R. G., & Ghosh, K. (2003). A simplified and efficient representation for evaluation and selection of assembly sequences. *Computers in Industry*, *50*, 251–264. doi:10.1016/S0166-3615(03)00015-0

Gupta, S., & Krishnan, V. (1998). Product family-based assembly sequence design methodology. *IIE Transactions*, *30*, 933–945. doi:10.1080/07408179808966547

Hsiao, S. W., & Liu, E. (2005). A structural component-based approach for designing product family. *Computers in Industry*, *56*, 13–28. doi:10.1016/j.compind.2004.10.002

Jiao, J., Zhang, L., & Pokharel, S. (2007). Process platform planning for variety coordination from design to production in mass customization manufacturing. *IEEE Transactions on Engineering Management*, *54*(1), 112–129. doi:10.1109/TEM.2006.889071

Jiao, J., Zhang, L., Zhang, Y., & Pokharel, S. (2008). Association rule mining for product and process variety mapping. *International Journal of Computer Integrated Manufacturing*, *21*(2), 111–124.

Lu, Q., & Botha, B. (2006). Process development: A theoretical framework. *International Journal of Production Research*, *44*(15), 2977–2996. doi:10.1080/00207540600547430

Marian, R. M., Kuong, L. H. S., & Abhary, K. (2003). Assembly sequence planning and optimization using genetic algorithms Part I: Automatic generation of feasible assembly sequences. *Applied Soft Computing*, *2*(3F), 223–253. doi:10.1016/S1568-4946(02)00064-9

Marian, R. M., Kuong, L. H. S., & Abhary, K. (2006). A genetic algorithm for the optimization of assembly sequences. *Computers & Industrial Engineering*, *50*, 503–527. doi:10.1016/j.cie.2005.07.007

Martinez, M. T., Favrel, J., & Ghodous, P. (2000). Product family manufacturing plan generation and classification. *Concurrent Engineering: Research and Applications*, *8*(1), 12–22.

Meyer, M. H. (1997). Revitalize your product lines through continuous platform renewal. *Research Technology Management, 40*(2).

Meyer, M. H., & Lehnerd, A. P. (1997). *The power of product platform-building value and cost leadership*. New York, NY: Free Press.

Meyer, M. H., & Utterback, J. (1993). The product family and the dynamics of core capability. *Sloan Management Review*, (Spring): 29–47.

Pessina, M. W., & Renner, J. R. (1998). Mass customization at Lutron Electronics – A total company process. *Agility & Global Competition*, *2*(2), 50–57.

Pisano, G. P. (1997). *The development factory: Unlocking the potential of process innovation*. Boston, MA: Harvard Business School Press.

Sanderson, S. W., & Uzumeri, M. (1997). *Managing product families*. Chicago, IL: Irwin.

Schierholt, K. (2001). Process configuration: Combining the principles of product configuration and process planning. *Artificial Intelligence for Engineering Design, Analysis and Manufacturing*, *15*(5), 411–424.

Schürr, A. (1990). PROGRES: A VHL-language based on graph grammars. *Proceedings of the 4th International Workshop on Graph Grammars and Their Application to Computer Science, LNCS 532* (pp. 641-659). Bremen, Germany: Springer-Verlag.

Schürr, A., Winter, A., & Zündorf, A. (1995). Graph grammar engineering with PROGRES. *Proceedings of the 5th European Software Engineering Conference, LNCS 989* (pp. 219-234). Barcelona, Spain: Springer-Verlag.

Schürr, A., Winter, A., & Zündorf, A. (1998). PROGRES: Language and environment. In Rozenberg, G. (Ed.), *Handbook on graph grammars: Applications* (*Vol. 2*, pp. 1–61). Singapore: World Scientific.

Szuba, J. (2005). *Graphs and graph transformations in design in engineering*. Ph.D. dissertation, Polish Academy of Sciences, Warsaw, Poland.

Wiendahl, H. P., El Maraghy, H. A., Nyhuis, P., Zah, M. F., Wiendahl, H. H., Duffie, N., & Brieke, M. (2007). Changeable manufacturing: Classification, design and operation. *Annals of the CIRP*, *56*(2), 783–809. doi:10.1016/j.cirp.2007.10.003

Williams, C. B., Allen, J. K., Rosen, D. W., & Mistree, F. (2007). Designing platforms for customizable products and processes in markets of non-uniform demand. *Concurrent Engineering: Research and Applications*, *15*(2), 201–216. doi:10.1177/1063293X07079328

Wortmann, J. C., Muntslag, D. R., & Timmermans, P. J. M. (1997). *Customer-driven manufacturing*. London, UK: Chapman and Hall.

Zhang, L., & Rodrigues, B. (2009). A tree unification approach to constructing generic processes. *IIE Transactions*, *41*(10), 916–929. doi:10.1080/07408170903026049

Zhang, L., & Xu, Q. (2010). *Solving production configuration based on constraint satisfaction*. The 2010 IEEE International Conferences on Industrial Engineering and Engineering Management (IEEM), December 07 – 11, 2010, Macau, China.

Zheng, L. Y., Dong, H. F., Vichare, P., Nassehi, A., & Newman, S. T. (2008). Systematic modeling and reusing of process knowledge for rapid process configuration. *Robotics and Computer-integrated Manufacturing*, *24*, 763–772. doi:10.1016/j.rcim.2008.03.006

APPENDIX A: THE CONTROL STRUCTURE SUPPORTING PROCESS FAMILY PLANNING

Figure 10. Deriving product variants

```
transaction DefineProduct [1:1]
      use aPara: PARA do
          loop
              FindUnassignedPara (out aPara);
            & ConfirmNoUnassignedAffectingPara (aPara);
            & write (aPara, allowableValue);
            & read (aValue in allowableValue);
            & AssignValue (aPara, aValue);
            & use aConstraint: EXCLUDE_CONSTRAINT do
                  loop
                      ProcessExcludeConstraint (aPara, aValue);
                  end
                end [0:n];
            & use aConstraint: REQUIRE_CONSTRAINT do
                  loop
                      ProcessRequireConstraint (aPara, aValue);
                  end
                end [0:n];
          end
      end
    & use anItem: ITEM do
          loop
              RemoveNotIncludedItem (anItem);
          end
      end;
    & use aPrimaryVariant: PRIMARY_VARIANT do
            loop
                RemoveNotIncludedPrimaryVariant (aPrimaryVariant);
            end
        end;
    & CloseDefine (aProduct);
end.
```

APPENDIX B: THE CONTROL STRUCTURE SUPPORTING PROCESS FAMILY PLANNING

Figure 11. Deriving production processes

```
transaction PlanProdProc [1:1]
        use anItem: ITEM do
             loop
                    FindProcUndeterminedItem (out anItem);
                 & ConfirmNoSucceedingUndeterminedItem (anItem);
                 & DetermineItemProc (aProc: PROCESS);
                 & use aProc: PROCESS do
                        loop
                              use anOperation: OPERATION do
                                    loop
                                          RemoreNotIncludedOperation (anOperation);
                                    end
                              end;
                           &  use aStartingVariant: STARTING_VARIANT do
                                    loop
                                          RemoreNotIncludedStartingVariant (aStartingVariant);
                                    end
                              end;
                        end
                   end
           end
        & RemoveNotIncludedProc;
        & ClosePlan (aProdProc);
end.
```

APPENDIX C

Figure 12. Declaration examples of intrinsic, derived, and meta attributes

```
node type AssyProc4Chair : PROCESS
        intrinsic  item_toProduce := [undefined];  /* item to be produced
                   item#1_Input := [undefined]; /* the 1st input child item
                     :
                   item#n_Input := [undefined]; /* the nth input child item
                   proc#1_Preceding := [undefined]; /* the 1st preceding process of an input child item
                     :
                   proc#n_Preceding := [undefined]; /* the nth preceding process of an input child item
         meta   proc_Succeeding := nil;  /* no succeding process
                included := true;  /* this process is assumed by the entire chair family
end
node type AssyProc4Armrest : PROCESS
        intrinsic  item_toProduce := [undefined];  /* item to be produced
                   item#1_Input := [undefined]; /* the 1st input child item
                     :
                   item#n_Input := [undefined]; /* the nth input child item
                   proc#1_Preceding := [undefined]; /* the 1st preceding process of an input child item
                     :
                   proc#n_Preceding := [undefined]; /* the nth preceding process of an input child item
       derived  included := [/(self. -produces-> Armrest)| false];
       meta   proc_Succeeding := AssyProc4Chair;  /*  the generic chair assembly process follows all the
                                                      armrest assembly process variants
              machine #1 := Assembly machine #1;  /* all variants of this assembly process require Assembly
                                                      machine #1
              machine #2 := Assembly machine #2;  /* all variants of this assembly process require Assembly
                                                      machine #2
              operator := 2;  /* all variants of this assembly process require 2 operators

end
node type MOp4Wheel : OPERATION
        intrinsic  item_toProduce := [undefined];  /* item to be produced
                   item_Input := [undefined]; /* the input item
                   op_Previous := [undefined]; /* the previous operation
                   op_Following := [undefined]; /* the following operation
         derived  included := [/(self. <-precede- FOp4Wheel)| false];
         meta   machine := CNC machine #1;  /* all variants of this operation is performed by this machine
                operator := 1;  /* all variants of this operation require 1 operator
end
```

APPENDIX D

Figure 13. Node type specifications of the generic model in the design view

```
node_type   Textile Spindle : PRODUCT
     derived   defined := false;
end;
node_type   Length : PARAMETER
     derived      assigned := [if(self.<-valueOf-VALUE.Selected)|false];
end;
node_type   L_D Constraint : EXCLUDE_CONSTRAINT
end;
node_type   D_T Constraint, T_C Constraint : REQUIRE_CONSTRAINT
end;
node_type   GIR1, GIR2 : COMMON_GIR
     derived      included := true;
end;
node_type   Shaftassy : SECONDARY
     derived      Length := [if(self.-toParent->Textile Spindle.Length)|false];
                  Diameter := [if(self.-toParent->Textile Spindle.Diameter)|false];
end;
node_type   Rockerarmassy : SECONDARY
     derived      Length := [if(self.-toParent->Textile Spindle.Length)|false];
                  Diameter := [if(self.-toParent->Textile Spindle.Diameter)|false];
                  Thread pitch := [if(self.-toParent->Textile Spindle.Thread pitch)|false];
                  Chamfer := [if(self.-toParent->Textile Spindle.Chamfer)|false];
end;
node_type   GIR11, GIR12, GIR21, GIR22 : COMMON_GIR
     derived      included := true;
end;
node_type   Shaft : PRIMARY
     derived      Length := [if(self.-toParent->Shaftassy.Length)|false];
                  Diameter := [if(self.-toParent->Shaftassy.Diameter)|false];
end;
...
node_type   StV1, StV2, StV3, NdV1, NdV2, RaV1, SlV1, SlV2 : PRIMARY_VARIANT
end;
node_type   SLR111, SLR112, SLR113 : OPTIONAL_SLR
     derived   included :=[(if(self.-toParent->Shaft.Length)&if(self.-toParent->Shaft.Diameter))|false];
end;
...
node_type   SLR221  : COMMON_SLR
     meta   included :=true;
end;
```

APPENDIX E

Figure 14. Node type specifications of the generic model in the production view

```
node_type   Proc4TS : PROCESS
        instrinsic   item_toProduce := [undefined];
                     item#1_Input := [undefined];
                     item#2_Input := [undefined];
         meta   Proc_Succeeding := nil;
                included := true;
end;
node_type   SR1, SR2 : FIXED_SR
        derived   included := true;
end;
node_type   Proc4SA : PROCESS
        instrinsic   item_toProduce := [undefined];
                     item#1_Input := [undefined];
                     item#2_Input := [undefined];
         meta   Proc_Succeeding := Proc4ST;
                included := true;
                #ofOperator := 2;
end;
...
node_type   SR11, SR12, SR21, SR22: FIXED_SR
        derived   included := true;
end;
node_type   Proc4St : PROCESS
        instrinsic   item_toProduce := [undefined];
                     item#1_Input := [undefined];
        meta   included := true;
               #ofOperator := 3;
end;
...
node_type   MOp34St : INTERMEDIATE
        instrinsic   item_toProduce := [undefined];
                     item_Input := [undefined];
        meta   included := true;
               #ofOperator := 1;
end;
```

```
node_type   MOp24St : INTERMEDIATE
        instrinsic   item_toProduce := [undefined];
                     item_Input := [undefined];
        derived   included := [f(PR_Op^s_{1-2}.included)|false];
end;
node_type   MOp14St : STARTING
        instrinsic   item_toProduce := [undefined];
                     item_Input := [undefined];
        derived   included := true;
end;
node_type   PR_Op^s_{1-1} : VARIED_PR
        derived   included := [f(IR_Op^*_{11}.included)|false];
end;
node_type   PR_Op^s_{2-2} : VARIED_PR
        derived   included := [f(self -toPrevious->MOp24St.included )
                               |false];
end;
node_type   PR_Op^s_{1-2} : VARIED_PR
        derived   included := [f( Op^*_{12}4StV_2.included ) OR
                               f( Op^*_{13}4StV_3.included)|false];
end;
node_type   IR_Op^s_{13} : VARIED_IR
        derived   included := [f(Op^*_{13}4StV_3.included )|false];
end;
...
node_type   Op^*_{13}4StV_3 : STARTING_VARIANT
        intrinsic   item_toProduce := [underined];
                    item_Input := [undefined];
        derived   included := [f( StV_3 )|false]
end;
```

Section 3
Related Issues to Cellular Manufacturing Systems

Chapter 16
Lean Thinking Based Investment Planning at Design Stage of Cellular/Hybrid Manufacturing Systems

M. Bulent Durmusoglu
Istanbul Technical University, Turkey

Goksu Kaya
Istanbul Technical University, Turkey

ABSTRACT

This chapter focuses on providing a methodology for lean thinking based investment planning from the perspective of cellular or hybrid manufacturing systems. The chapter has been divided into three parts. First part provides a general explanation of why lean thinking is so beneficial for managing manufacturing processes and obtaining sustained improvement. This part then moves to the aim of cell formation, and then uses value stream mapping to map current state for visualizing material-information flow and to design a desired future state for examining economic aspects of new machine investment decisions aligned with lean manufacturing principles. The purpose of second part is to explore axiomatic design approach; it provides an overall view of what to do. The third part presents the actual use of the methodology with implementation of hybrid system at a furniture factory; it helps to see application results of this methodology as part of a lean manufacturing program.

DOI: 10.4018/978-1-61350-047-7.ch016

INTRODUCTION

Traditional manufacturing systems are built on a functional layout or an assembly line with the principle of economies of scale. This point of view causes much capital investments in high-volume operations and large work-in-process inventories. As an alternative to traditional manufacturing, the principles of the Toyota Production System (TPS) have been widely adopted in recent years. Application of TPS principles have led to lean manufacturing (Sullivan *et al.*, 2002). Womack & Jones (1996) used the term lean thinking as the thinking process of Taiichi Ohno and the set of methods describing the Toyota Production System (Womack & Jones, 1996; Monden, 1993). Lean manufacturing emerged as a global approach that uses different tools to focus on waste elimination and to manufacture products that meet customer's needs (Hines & Taylor, 2000). Lean manufacturing has been increasingly adopted as a potential solution for many organizations, particularly within the automotive industries (Womack, *et al.*, 1990; Day, 1998; Jones, 1999) and aerospace industries (Abbett *et al.*, 1999; Womack & Fitzpatrick, 1999).

Lean production requires the analysis of the "value stream". A value stream is defined as all the value-added and non-value-added operations required manufacturing specific products and services to a customer (Womack & Jones, 1996; Rother & Shook, 1998). Value Stream Map is an enterprise improvement technique to visualize entire production process, representing information and material flows, to improve the production process by identifying waste and its sources (Rother & Shook, 1998).

Lean manufacturing focuses on the waste elimination and produces products that meet customer expectations. Lean production uses production and assembly cells consisting of product focused resources. The aims of the cell formation are smoothing work flow with flexible operation across a wide variety of low cost and high quality products by means of waste elimination. Economic benefits of lean manufacturing include smaller floor space requirements, lower work-in process, reduced lead-times and higher throughput (Sullivan, *et al.*, 2002). Lean production focuses on value pulled from the next upstream activity as customer. As value is specified, value streams are identified eliminating steps that do not create value, so the product will flow smoothly toward the customer. A value stream mapping is an enterprise improvement technique to visualize an entire production process by identifying waste (Braglia *et al.*, 2006).

Cellular manufacturing is an important technique in the planning and controlling of manufacturing system. Cellular manufacturing offers three groups of benefits. These benefits are: human related factors facilitated by empowerment in smaller cells; improved flow and supervisory control in cells to deal with smaller number of parts and facilities; improved operational efficiency, obtainable due to similarity; setup reduction; batch size reduction; improvement in performance related to productivity, quality and agility (Babu *et al.*, 2000). In practice, it is usually hard to partition all machines into independent cells. So, a functional layout generally becomes necessary. Because of this fact, hybrid cellular manufacturing systems (HMS) are required (Suresh 1991).

Hybrid manufacturing system (HMS) is the system where manufacturing cells and functional layout coexist (Shambu & Suresh, 2000), and also it has an advantage of more product flexibility (Satoglu *et al.*, 2009) and less capital investment for machines. Utilization of alternative machines in the HMS reduces additional machine purchasing requirements, and therefore it is beneficial (Satoglu *et al.*, 2009).Empirical evidences also show that hybrid manufacturing system is common for practice (Marsh *et al.*, 1999).

Lean manufacturing tools and techniques provide economical basis to managers for investment planning decisions. Value stream mapping creates a common language about a production process, enabling more purposeful decisions to improve the production system. Value stream map of the system should be taken into account for the design of future state to examine the economic aspects of new machine investment decisions (Sullivan, *et al.*, 2002).This chapter attempts to provide insight as to the choice and use of appropriate tools for designing a successful lean manufacturing system. Although it does not cover every lean manufacturing aspect, it does offer a road map that can guide a company for effective new machine investment decisions toward the development of a lean manufacturing environment.

Investment planning is the determination of suitable machines for manufacturing of part families within the cell. Investment planning for cellular manufacturing can enhance the operating characteristics of the system. After equipment purchases are decided, it is necessary to measure the effectiveness of this decision by evaluating the effects of the new equipment on the system (Gosh, 1989). In existing factories, investment decisions for purchasing new machines in favor of cellular or hybrid manufacturing systems can be troublesome to managers. Many of investment decisions in manufacturing industry are not beneficial and economical (Sullivan, *et al.*, 2002). For this reason, this chapter aims to be a guideline to managers for effective new machine investment decisions in implementation of cellular or hybrid manufacturing layout. This chapter focuses on providing a methodology for lean thinking based investment planning in restructuring manufacturing environment from the perspective of cellular or hybrid manufacturing layout. To do this, we used axiomatic design methodology for creating a systematic perspective.

Axiomatic design (AD) is a design theory that was created by Professor Nam Pyo Suh of the Massachusetts Institute of Technology (Suh 1990). The goal of AD is to establish a scientific basis for design and to improve design activities by providing logical and rational processes and tools to designer. In accomplishing this goal, AD provides a systematic search process through the design space to minimize the random search process and determines the best design solution among many alternatives (Kulak *et al.*, 2005).

Some of algorithmic methods can be effective if the design has to satisfy only one functional requirement, but when many functional requirements must be satisfied at the same time, they are less effective. Axioms provide the boundaries within which these algorithms are valid, in addition to providing the general principles. AD includes four domains: customer domain, functional domain, physical domain and process domain. In the customer domain, there are customers' needs or attributes (CAs) from a product, service or system. The functional domain includes customer needs transformed into functional requirements (FRs). To answer FRs, physical domain has design parameters (DPs). Finally, process domain is characterized by the process variables (PVs) to develop a process for production (Suh, 2001). These functional domain and physical domain and mapping between them are illustrated in Figure 1.

The most important concept in axiomatic design is the existence of the design axioms. These axioms are (Suh, 2001):

1. **Axiom:** Independence axiom: The FRs are defined as the minimum set of independent requirements that the design must satisfy. A set of FRs is the description of design goals. The independence axiom states that

Figure 1. Hierarchical structure of axiomatic design (Durmusoglu & Kulak, 2008)

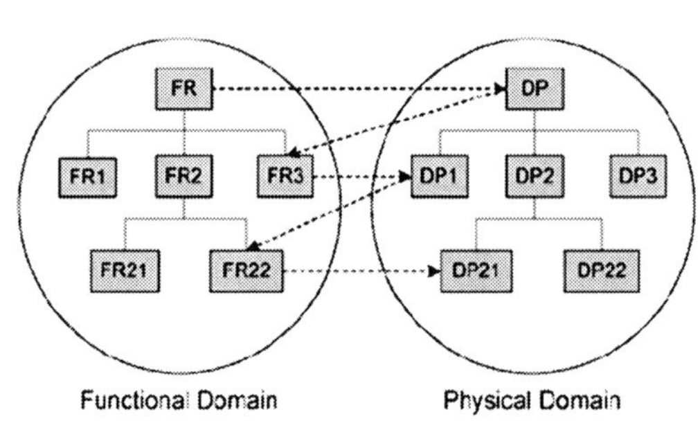

when there are two or more FRs, the design solution must be such that each one of the FRs can be satisfied without affecting the other FRs. That means we have to choose a correct set of DPs to be able to satisfy the FRs and maintain their independence (Suh, 2001).

2. **Axiom:** Information axiom: The design that has the smallest information content is the best design. The information content is defined in terms of probability; this axiom also states that the design that has the highest probability of success is the best design (Suh, 2001).

$$\{FR\} = [A]\ \{DP\} \quad (1)$$

Here, {FR} is the functional requirement vector, {DP} is the design parameter vector, and [A] is the design matrix that characterizes the design. The structure of [A] matrix defines the type of design being considered. In order to satisfy the independence axiom, [A] matrix should be an uncoupled (a diagonal matrix) or decoupled design (a triangular matrix) (Suh, 2001).

The design of an ideal manufacturing system depends on the selection of functional requirements (FRs) the system must satisfy within a given set of constraints (Cs). Therefore, an ideal manufacturing system design is not time invariant. It changes with the selection of specific sets of FRs and Cs. An efficient manufacturing system must utilize things, people and information in a rational manner, consistent with basic principles (Suh, 2001).

BACKGROUND

There have been many publications on lean manufacturing and cellular manufacturing systems. Some subsequent steps need for a complete cell formation as Durmusoglu and Nomak (2005) observed in their CMS design implementation. Nancy and Wemmerlöv (2002), published a paper for defining part families and related machine groups, and Sarher and Mondal (1999) studied the evaluation of grouping efficiency measures. Aneke and Carrie (1986) studied on designing of multi product flow lines on the basis of one piece flow.

Many AD applications in designing products, systems, organizations and software have appeared in the literature so far. AD theory and principles have been introduced first time by Suh (1990). Suh (1997) described a conceptual approach for designing of the systems using AD methodology. Suh *et al.* (1998) provided an AD-based model for an ideal production system based on lean principles. Cochran, Eversheim *et al.* (2000) used lean principles to structure smaller production segments by AD. Houshmand & Jamshidnezhad (2002) also provided a lean manufacturing based production system design model using AD approach. In this model, organizational capabilities, technological capabilities and value stream analysis are used as the basis. Kulak *et al.* (2005) published a paper for a complete cellular manufacturing system design methodology based on axiomatic design principles.

Sullivan *et al.* (2002) illustrated an equipment replacement decision problem within the context of lean manufacturing implementation. In particular, they demonstrated how the value stream mapping (VSM) tools can be used to map the current state of a production line and design a desired future state. Further, they provided a roadmap for how VSM can provide necessary information for analysis of equipment replacement decision problems encountered in lean manufacturing implementation.

Considering the literature mentioned above, a road map including a systematic design model for equipment investments in cellular/hybrid manufacturing system is not found. In this study, a methodology is developed using AD principles in order to fill this gap.

LEAN THINKING BASED INVESTMENT PLANNING BY AXIOMATIC DESIGN FOR CMS OR HMS

Main focus of the chapter is giving a systematic methodology by axiomatic design principles to design a desired future state aligned with lean manufacturing principles by using value stream mapping and its associated tools and to provide a roadmap for investment decisions in cellular/hybrid manufacturing.

The first step of axiomatic design is defining the functional requirement at the highest level of the system hierarchy of functional domain (Suh *et al.*, 1998).

Step 1. Defining Functional Requirements (FR1): The design goal of the production system (functional requirement at the highest level) was defined as;

FR1= Improve the system performance

The companies, focusing on lean, aim eliminating wastes and improving system for responding quickly to customer needs.

Step 2. Mapping of FRs in the Physical Domain (DP1): Design parameter (DP), which satisfies the FRs established in the previous step, is selected as below.

DP1= Design of a lean manufacturing system

Step 3. Decomposition of FR (FR1): If the DPs can not be implemented without further clarification, the AD principles recommend returning to the functional domain for decomposing the FRs into their lower functional requirement set (Suh, 2001). The following functional requirements are defined for decomposing the FR determined in the first step.

- FR11= Define customer requirements and expectations
- FR12= Make experts and employees conscious on lean thinking
- FR13= Divide production system into sub-systems for simplification of the system
- FR14= Visualize wastes
- FR15= Apply the proposed plan to eliminate waste in production processes and offices

Step 4. Defining the Corresponding DPs of FRs (DP1): We move from the functional domain to the physical domain. The following DPs in response to satisfy the five FR1ns defined in step 3 are listed below.

- DP11= Market research
- DP12= Training procedure on lean manufacturing
- DP13= Selection procedure of product families
- DP14= Current state value stream mapping
- DP15= Future state value stream management system

Step 5. Structuring of Design Matrix (FR1-DP1): The FR-DP sets are defined in Step 3 and Step 4, and the corresponding design matrix (DM) providing the relationship between the FR and DP elements is structured. In the design matrix, a symbol X represents a strong relationship between the corresponding FR-DP pair. It is important to ensure that this DM satisfies the Independence Axiom (IA) of the AD principles. If the DM matrix is uncoupled or decoupled, then it satisfies the Independence Axiom of AD principles (Suh, 2001). The design equation and the DM corresponding to the FR-DP sets are as follows and depicted in Figure 2. Equation 2 is a decoupled design, and satisfies the IA.

$$\begin{bmatrix} FR11 \\ FR12 \\ FR13 \\ FR14 \\ FR15 \end{bmatrix} = \begin{bmatrix} X & 0 & 0 & 0 & 0 \\ X & X & 0 & 0 & 0 \\ 0 & X & X & 0 & 0 \\ 0 & X & X & X & 0 \\ 0 & X & X & X & X \end{bmatrix} * \begin{bmatrix} DP11 \\ DP12 \\ DP13 \\ DP14 \\ DP15 \end{bmatrix} \tag{2}$$

Defining Customer Requirements and Expectations (FR11-DP11): Making market research is one of the effective working criteria for companies. The companies not giving importance to the research face various risks. Market research is collecting and analyzing of systematic and objective information about products, market and consumers (Megep, 2008). Market research is very important for defining customer demands, sales forecasting in production planning and new product developments.

Making Experts and Employees Conscious on Lean Thinking (FR12-DP12): Employee training is the most efficient method for using human resource effectively in a company. Employee training directly affects company's profitability and decreasing employee turnover by increasing employee satisfaction. Applying of lean thinking and making it company's culture are possible with by training program which makes employees conscious about lean.

Dividing Production System into Sub-systems for Simplification of the System (FR13-DP13): The products should be classified up to the similarities based on production system's characteristics. So, the first step consists of identification of product families and in the selection of one with its production sub-system as the pilot application for improvement. This step continues until all of the product families have been selected.

Pareto Analysis (ABC analysis) is used for separating the vital few from trivial many. Pareto analysis is essential for visualizing volume of the product families in total production. By doing this analysis, it is possible to see how the customer demand is distributed among different product types (Durmusoglu & Kulak 2008). High volume products are responsible for the largest amount of waste's costs (work in process, material handling, other operational costs, etc.). Focusing on these high volume products affects overall performance of the company.

Product family could be a group of high volume products, which pass through the similar operational steps at common machines. Another example of product families is to divide products into catalogued products and specific-project type products. Therefore determining product families depends on production system's characteristics. In order to determine product family, the relationship matrix, including products, production volume, processes, production functions and customers is needed (Durmusoglu & Kulak 2008).

Visualizing Wastes (FR14-DP14): Value stream mapping (VSM) visualizes value and waste resources through the production processes for a defined product family. Therefore, it is used for understanding how the process flow must be, and then, it combines lean tools.

Value stream mapping is a tool, which was created for redesigning production system (Rother & Shook 1998; Womack & Jones, 2002; Pavnaskar *et al.*, 2003). To find causes of waste, it is useful to show parameters for each production processes in detail (Braglia *et al.*, 2006).

Applying the Proposed Plan to Eliminate Waste in Manufacturing Processes and Offices (FR15-DP15): The last step of the proposed methodology is the future state value stream management. For future state, applied for developing the system in redesign stage, the initial aim must be performing a plan that is proposed without any investment. Before the application stage, several scenarios can be tried for deciding new machine investment(s) is needed or not, and finally the cost analysis must be performed based on lead time for decision making. Initial aim of the axiomatic design model is designing cells for products families processed in current machine resources on hand. As seen in the industry, new machine investments to solve bottlenecks of the system are dominated point of view as the key solution. Whereas the majority of new machinery purchasing decisions may create new unnecessary waste resources. This systematic study underlines the necessity of value stream mapping for new investment decisions.

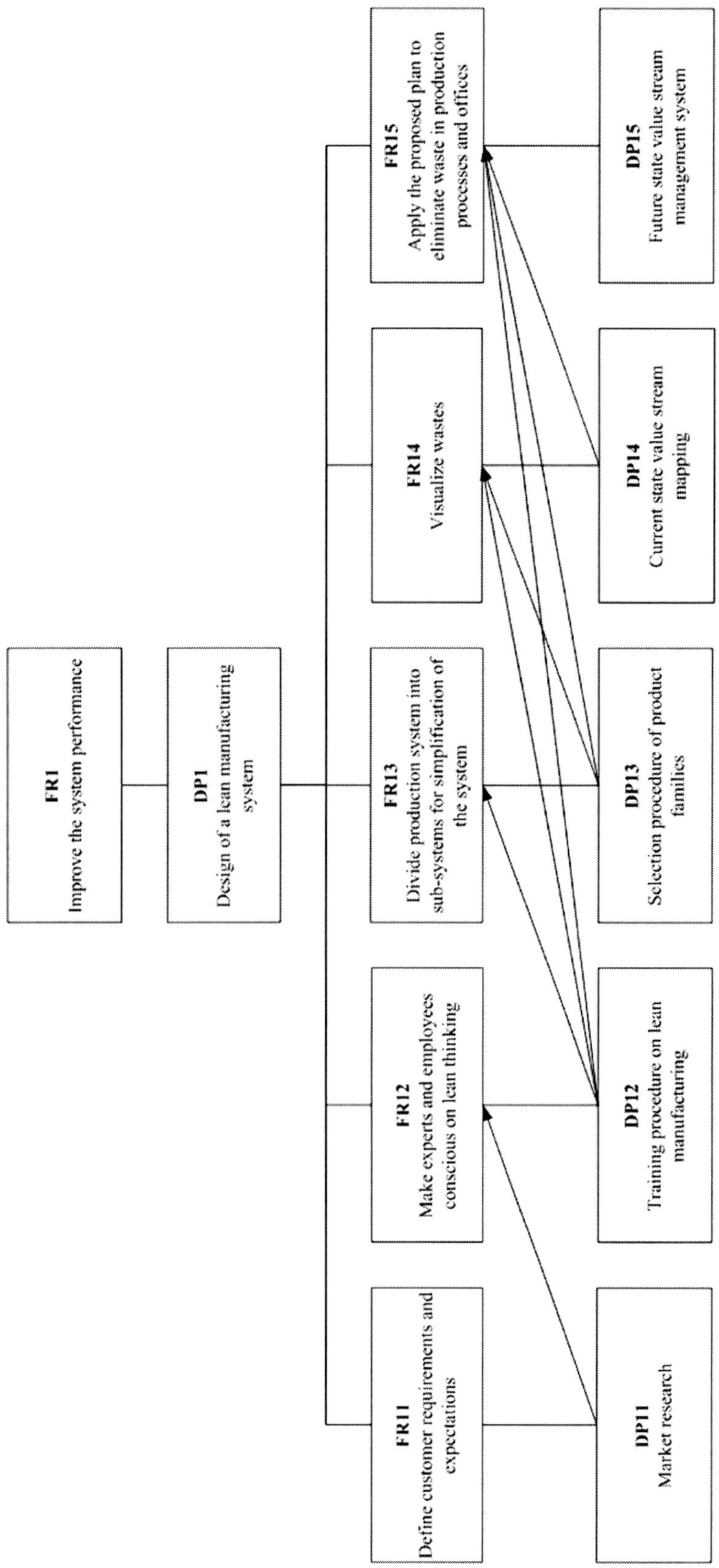

Figure 2. The decomposition FR1-DP1

For future state value stream management plan, cell design, kaizen plans, poka-yoke applications, total productive maintenance, heijunka boxes for leveling of production and pull/hybrid system design must be done as lean manufacturing tools with using current resources. Supporting lean manufacturing practices to increase overall performance is possible by lean transition within the whole organization. Therefore, lean office applications have been planned for other units that support production. After the determination of the pilot study group for lean office applications, designing new processes and office cells, and applying visual management practices are aimed to be strengthening with the help of value stream maps and new process flow charts.

Step 6. Decomposition of FR12: FR12 (Make experts and employees conscious on lean thinking) and DP12 (Training procedure on lean manufacturing) are decomposed below:

- FR121= Teach about lean manufacturing
- FR122= Provide the basic information of lean manufacturing to participants
- FR123= Prepare a training program by using current resources effectively
- FR124= Ensure continuity of lean manufacturing training activities
- FR125= Ensure and increase the effectiveness of lean manufacturing training

Design parameters satisfying the five FRs defined above are as follows;

- DP121= Choice procedure of qualified trainer
- DP122= Conceptualization of determined training program
- DP123= Planned training schedule
- DP124= Determination of participants procedure
- DP125= Training performance measurement and evaluation procedure

The design equation and the DM corresponding to the FR-DP sets are as followed and depicted in Figure 3. This is a decoupled design, and satisfies the IA.

$$\begin{bmatrix} FR121 \\ FR122 \\ FR123 \\ FR124 \\ FR125 \end{bmatrix} = \begin{bmatrix} X & 0 & 0 & 0 & 0 \\ X & X & 0 & 0 & 0 \\ X & X & X & 0 & 0 \\ X & X & X & X & 0 \\ X & X & X & X & X \end{bmatrix} * \begin{bmatrix} DP121 \\ DP122 \\ DP123 \\ DP124 \\ DP125 \end{bmatrix} \quad (3)$$

Teaching about Lean Manufacturing (FR121-DP121): To increase the efficiency of lean manufacturing and to ensure continuity, employees should be educated on this issue. Planned training program for this purpose is required to be updated continuously to ensure the continuity. Therefore, the experienced instructors from outside or inside the company must be determined for contributing to the preparation and implementation of training programs.

Providing the Basic Information of Lean Manufacturing to Participants (FR122-DP122): Lean manufacturing system, considered to be implemented, brings all the principles of new business culture understanding. To achieve the success of the designed system, employee resistance to new ideas must be eliminated, and then, the new philosophy must be adapted to all employees. Training needs of all employees must be determined and after this, preparation of multi-purpose training programs that meet these needs is necessary.

Preparing a Training Program by Using Current Resources Effectively (FR123-DP123): A timetable for trainers and participants should be identified in parallel with the implementation

Figure 3. The decomposition FR12-DP12

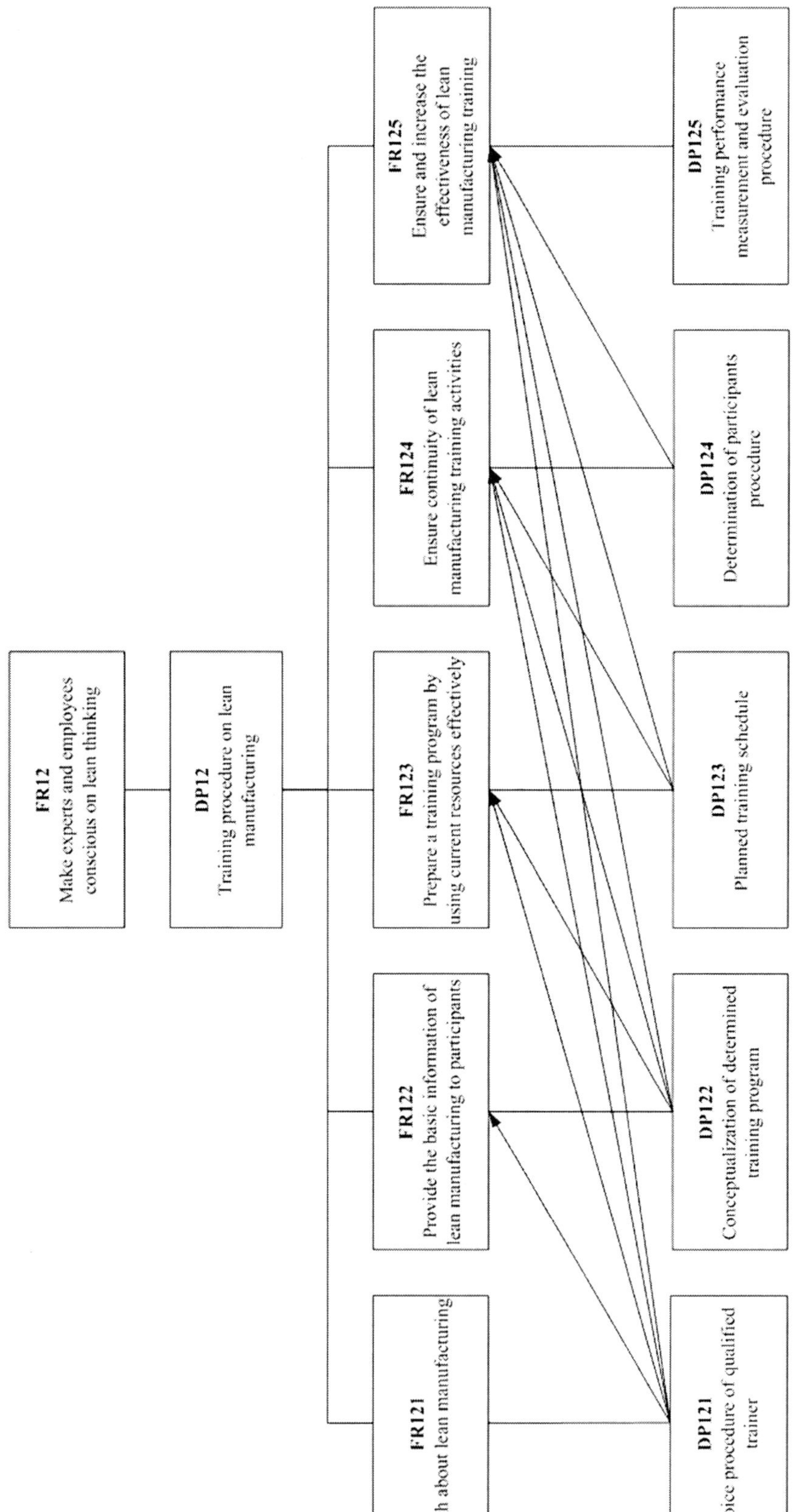

steps of designed system, and all staff should be informed about lean tools by training programs.

Ensuring Continuity of Lean Manufacturing Training Activities (FR124-DP124): Participants have to be determined according to the company's planned training program and schedule. Selected instructor(s) should determine the training periods and the number of employees to participate in training with a balanced planning in order to ensure the effectiveness of training.

Ensuring and Increasing the Effectiveness of Lean Manufacturing Training (FR125-DP125): At the end of the training, questionnaires should be distributed to evaluate the content of training program. Performances of instructors and training program's content should be reviewed according to the evaluation results obtained from questionnaires. Finally, the scope of further training programs should be revised according to the evaluation results.

Step 7. Decomposition of FR13: FR13 (Divide production system into sub-systems for simplification of the system) and DP13 (Selection procedure of product families) are decomposed as below:

- FR131= Classify products based on customer demand
- FR132= Determine product families based on production system's characteristics

The corresponding DPs satisfying FRs at this step are stated as:

- DP131= Pareto Analysis
- DP132= Products- production volume-processes-production functions-customers' relation matrix

The design matrix for the above set of FRs and DPs are as followed and depicted in Figure 4. This is a decoupled design, and satisfies the IA.

$$\begin{bmatrix} FR131 \\ FR132 \end{bmatrix} = \begin{bmatrix} X & 0 \\ X & X \end{bmatrix} * \begin{bmatrix} DP131 \\ DP132 \end{bmatrix} \quad (4)$$

Classify Products Based on Customer Demand (FR132-DP132): In order to simplify the production system for analysis, products should be classified based on customer demand. Pareto (ABC) analysis is used for this classification and it is essential for visualizing volume of the product families. High volume products are responsible for the largest amount of waste's costs (work in process, material handling, other operational costs, etc.). Focusing on these high volume products affects overall performance of the company. So, pareto analysis is used for determining which product families should be selected first for pilot applications.

Determine Product Families Based on Production System's Characteristics (FR131-DP131): Products based on production systems characteristics are grouped as families. For example, products belonging to "A" class, which pass through the similar operations, are grouped as a family. In order to determine product families, constructing a relationship matrix, including products, production volume, processes, production functions and customers, is needed.

Step 8. Decomposition of FR14: FR14 (Visualize wastes) and DP14 (Current state value stream mapping) are decomposed as below:

- FR141= Define process parameters in proposed processes and collect information in response to these parameters
- FR142= Define relations between operations and knowledge processes

The corresponding DP14s satisfying FR14s at this step are stated as followed:

- DP141= Information collecting procedure
- DP142= Value stream mapping procedure

Figure 4. The decomposition FR13-DP13

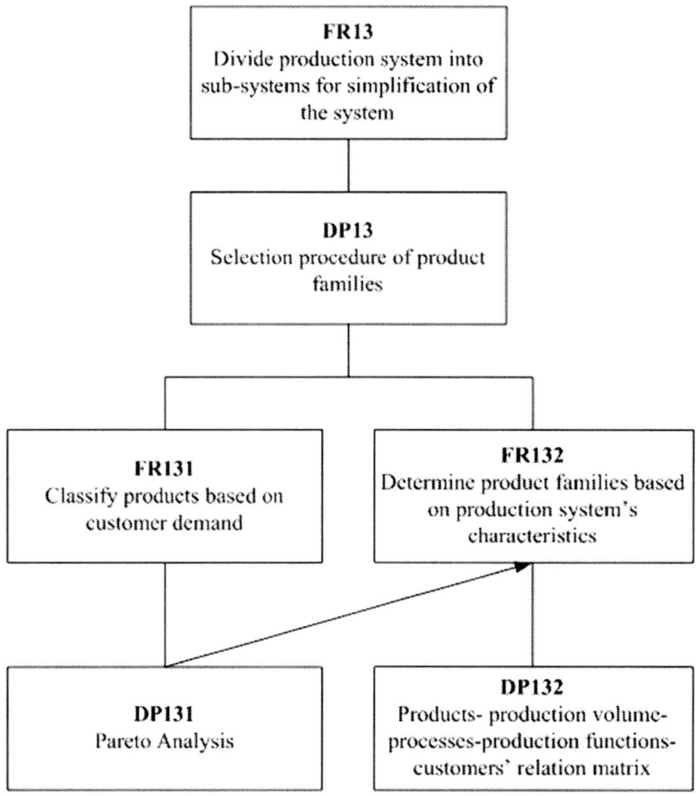

The design equation and the DM corresponding to the FR14-DP14 sets are as followed and depicted in Figure 5. This is a decoupled design, and satisfies the IA.

$$\begin{bmatrix} FR141 \\ FR142 \end{bmatrix} = \begin{bmatrix} X & 0 \\ X & X \end{bmatrix} * \begin{bmatrix} DP141 \\ DP142 \end{bmatrix} \tag{5}$$

Defining Process Parameters in Proposed Processes and Collect Information in Response to These Parameters (FR141-DP141): The first step for the implementation of lean manufacturing tools is visualization of wastes. Before drawing the current state value stream mapping, necessary information such as the process routes, the daily product demands, supplier delivery schedules, available production time, cycle times, setup times, uptime, scrap rates, number of employees, number of shifts, inventory locations and quantities, times between processes, should be collected, and then, should be transferred to value stream maps.

Defining Relations between Operations and Knowledge Processes (FR142-DP142): Drawing current state value stream map helps to visualize the whole system through operation steps by the representation of information and material flow. It also helps to eliminate the non-value added activities. By the use of value stream map, purposeful decisions can be made for the selection of necessary lean tools to eliminate wastes in the system.

Value stream mapping has two parts: big picture mapping and detailed mapping. Before starting detailed mapping of any core process, it is useful to develop an overview of the key features of that entire process. This will (Hines & Taylor, 2000):

Figure 5. The decomposition FR14-DP14

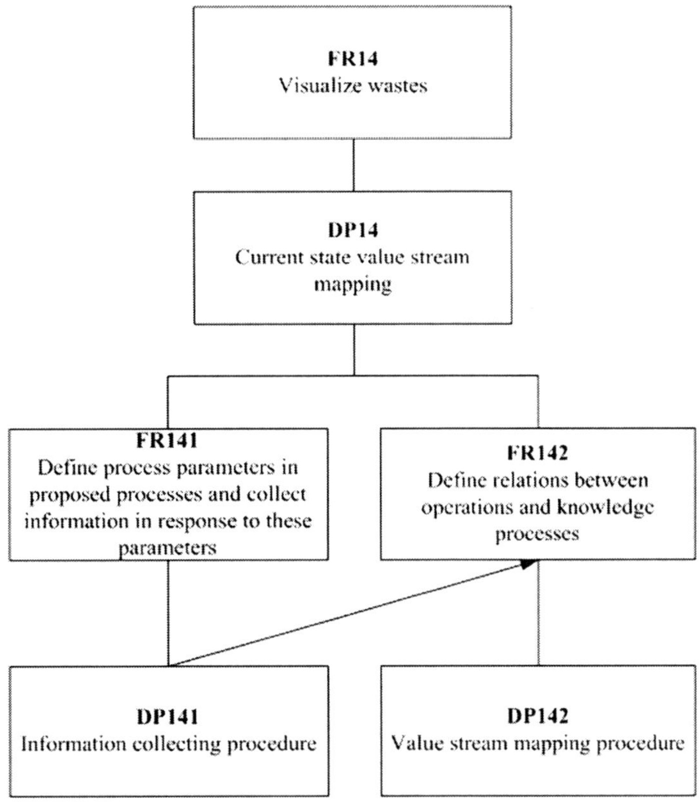

- help you to visualize the flows,
- help you to see where waste is,
- pull together the lean thinking principles,
- help you to decide who should be in the implementation teams,
- show relationships between information and physical flows.

Big picture mapping makes easier to understand current state of the system. Big picture mapping consists of five basic steps (Hines & Taylor, 2000):

Phase 1: Record customer requirements
Phase 2: Add information flows
Phase 3: Add physical flows
Phase 4: Linking physical and information flows
Phase 5: Complete mapping

After big picture mapping, detailed mapping should be done for drawing value stream map to visualize whole system. Collected information should be reflected to the map by value stream mapping symbols.

Step 9. Decomposition of FR15: FR15 (Apply the proposed plan to eliminate waste in manufacturing processes and offices) and DP15 (Future state value stream management system) are decomposed as below:

- FR151= Ensure production pace based on customer demand
- FR152= Ensure the part family flows in product families in the system
- FR153= Ensure one-piece flow
- FR154= Visualize the planned system
- FR155= Increase performance on the systems supporting manufacturing

- FR156= Define other parameters needed for improvement
- FR157= Eliminate the rest trouble bottlenecks
- FR158= Visualize the system with investment
- FR159= Evaluate the system performance with investment
- FR1510= Decrease the excessive volume fluctuations and variations in production
- FR1511= Provide pulling between inter cell flows

The corresponding DP15s satisfying FR15s at this step are stated as:

- DP151= Calculated takt time
- DP152= Clustered machine groups based on part families
- DP153= Cell(s) design procedure
- DP154= Future state value stream mapping procedure
- DP155= Lean office design
- DP156= Kaizen procedures correspondent to parameter
- DP157= Multi-attribute decision making procedure for machine and/or software selection
- DP158= Value stream mapping of invested future state
- DP159= Cost analysis based on lead time
- DP1510= Heijunka system
- DP1511= Kanban/hybrid system design

The design equation and the DM corresponding to the FR15-DP15 sets are as followed and depicted in Figure 6. This is a decoupled design, and satisfies the IA.

$$\begin{bmatrix} FR151 \\ FR152 \\ FR153 \\ FR154 \\ FR155 \\ FR156 \\ FR157 \\ FR158 \\ FR159 \\ FR1510 \\ FR1511 \end{bmatrix} = \begin{bmatrix} X & 0 & 0 & 0 & 0 & 0 & 0 & 0 & 0 & 0 & 0 \\ X & X & 0 & 0 & 0 & 0 & 0 & 0 & 0 & 0 & 0 \\ X & X & X & 0 & 0 & 0 & 0 & 0 & 0 & 0 & 0 \\ X & X & X & X & 0 & 0 & 0 & 0 & 0 & 0 & 0 \\ 0 & 0 & X & X & X & 0 & 0 & 0 & 0 & 0 & 0 \\ 0 & X & X & X & X & X & 0 & 0 & 0 & 0 & 0 \\ 0 & X & X & X & X & X & X & 0 & 0 & 0 & 0 \\ 0 & 0 & 0 & 0 & 0 & X & X & X & 0 & 0 & 0 \\ 0 & 0 & 0 & 0 & X & X & X & X & X & 0 & 0 \\ 0 & X & X & X & 0 & X & X & X & X & X & 0 \\ 0 & X & X & X & X & X & X & X & X & X & X \end{bmatrix} * \begin{bmatrix} DP151 \\ DP152 \\ DP153 \\ DP154 \\ DP155 \\ DP156 \\ DP157 \\ DP158 \\ DP159 \\ DP1510 \\ DP1511 \end{bmatrix} \quad (6)$$

The stage after drawing the current state value stream map is put into practice the lean principles and lean tools by focusing on waste points. Primarily aim at this point will be making planning for elimination of wastes without any investment in equipment. At this stage, if equipment investment is needed, the system should be re-planned in accordance to the new situation. The starting point is making a decision to focus on which specific areas (the price reduction request from customers, lead time reduction demand, presence of competing products on the market, product quality problems, etc.). For the purpose of waste elimination plan without any investment, workflow should be simplified, takt time should be calculated to determine the production pace according to customer demand, cell designs should be done, and later other parameters should be taken into account for developing. After configuration plan of the cells, Heijunka and pull system should be established in accordance with the lean principles for reduction of work in process inventories and performing manufacturing just in time according to the customer demand. After this planning stage, the value stream map should be drawn by visualization of the developed system without investment.

After elimination of wastes in production environment, lean office working must be performed in order to increase the performance of systems, supporting manufacturing. By the result

Figure 6. The decomposition of FR15-DP15

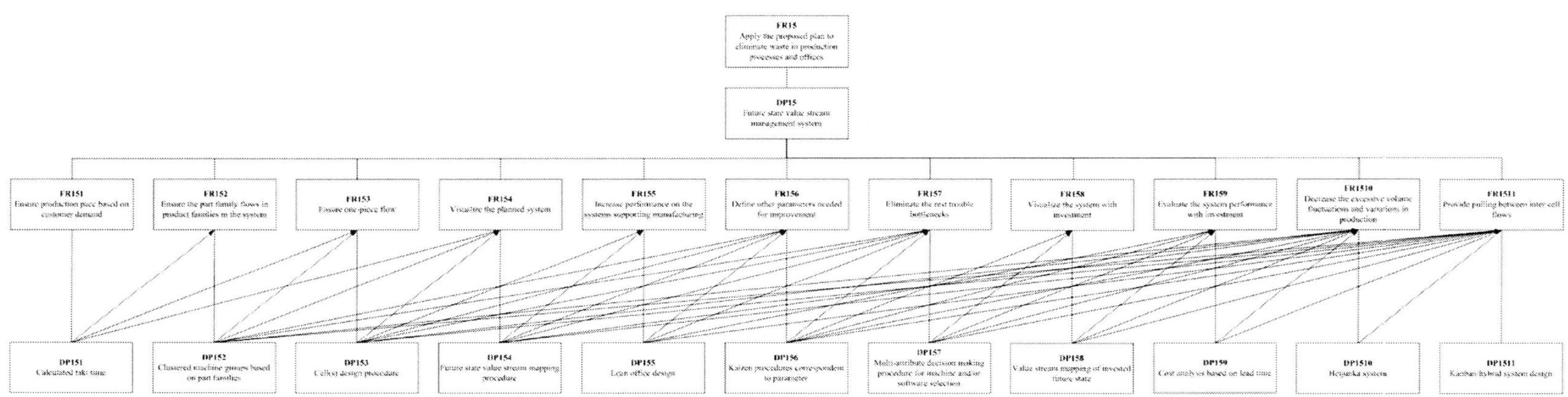

of all these activities, if there are still bottlenecks in the system, software or machinery investment decisions were taken into account and the future state value stream map must be revised. Cost analysis based on lead time is recommended for the system's performance evaluation before any investment.

Ensuring Production Pace based on Customer Demand (FR151-DP151): Rather than producing at high-speed, producing according to takt time is preferred (Byrne, 1995). In accordance with the competitiveness conditions, enterprises have to build production systems that will produce the amount that customer demands. Takt time is the cycle time that customers request. Takt time that is calculated by dividing daily operating time to daily customer demand is used to determine the pace of production.

Ensuring the Part Family Flows in Product Families in the System (FR152-DP152): In a manufacturing system for a product family, determination of parts and manufacturing processes/ machines assigned to the cell(s) is needed.

For this purpose, there are three approaches:

- Intuitive and Visual Analysis
- Classification & Coding
- Analysis of Routings, Part-Machine Clustering Procedures

If the system is complex, the clustering procedure is necessary. Using clustering method is more practical and appropriate than alternative classification and coding methods for configurations of cells (Gallagher & Knight 1973).

Ensuring One-Piece Flow (FR153-DP153): Cellular manufacturing system leads to one-piece flow (Miltenburg, 2001). The one-piece flow principle stipulates that parts in a batch travel between machines or processes as single pieces, and do not wait for the rest of the batch to be completed. In other words, part operations on different machines are overlapped and carried out in parallel, which reduces part waiting times, and therefore manufacturing lead-times (Satoglu *et al.*, 2009).

Creating a cellular layout helps to achieve the targeted one-piece flow without repeat and backtracking movements between machines and/or work stations (Satoglu *et al.*, 2009). Figure 7 shows all possible parts movements. If the workflow is complex within the cell, an algorithm, for example, developed by Aneke & Carrie, (1986) would be useful for the design of intra-cell movements.

After using of multi product flow line algorithm developed by Aneke & Carrie, (1986), one piece flow in the cells is ensured. Following the determination of relative positions of the work stations in the cell, the work station layout, preferably U-shaped layout, must be decided. While planning the layout of work stations, transport of materials within the cell and transportation distance should be minimized. Material handling costs in total production costs are ranged from 30% to 75% (Sule,1994). Also, the cells, which have high material transportation traffic between each

Figure 7. Four types of product flow (Aneke & Carrie, 1986)

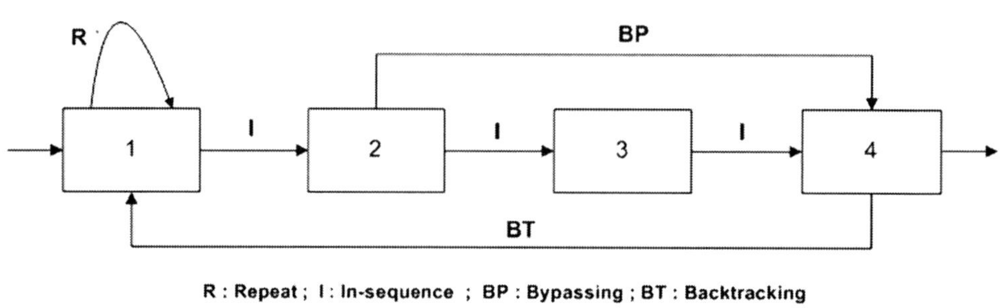

other, should be placed as close to each other for reducing the number and size of stock area.

Visualizing the Planned System (FR154-DP154): Primary goal is performing the future state plan without any equipment investment. For this purpose, implementation of improvement methods (Kaizen applications, heijunka boxes for production leveling and pull/hybrid system design, reduction of setup times, total productive maintenance, etc.) should be planned. If these improvement methods are not sufficient and there are still some bottlenecks in the production system, new machine investment should be considered. For visualization of the planned system, drawing of future state value stream map is necessary.

Drawing of future state value stream map includes customer demand, workflow and distribution of operations to machines by reducing of work in process inventory, etc.. Moving from the current state value stream map, future state is designed by using lean metrics and tools (Tapping, *et al.*, 2002).

Increasing Performance on the Systems Supporting Manufacturing (FR155-DP155): One of the reasons for long respond time to customer inquiries can be long lead time related to office operations (Suri, 1998). An important method for solving this problem should be lean office. Lean office is elimination of waste from workplace to provide better service to internal and external customers. For reducing non-value added activities and decreasing lead times, the organizational structure should be reconfigured into office cells. At the same time, office cells create a good environment for teamwork. For office cell design, product/service/project families must be determined (with clustering), accordingly the team members should be selected and relevant skill development planning should be done (team building), then the physical office layout should be configured. Moreover, an effective management model should be developed for team work within cells (Durmusoglu & Kulak, 2008).

Defining Other Parameters Needed for Improvement (FR156-DP156): To increase the success in implementation of the planned future state value stream map, Kaizen plans (continuous improvement plans) are recommended (Tapping, *et al.*, 2002). For parameters decided to develop, kaizen tools as 5S for work environment, Single Minute Exchange of Die (SMED) for setup time reduction, Poka-Yokes for reduction of defects, Total Productive Maintenance (TPM) for reduction of potential equipment breakdowns and process kaizens should be applied.

Eliminating the Rest Trouble Bottlenecks (FR157-DP157): After application of kaizen plans, there are still bottlenecks in the system, new investments can be considered. The selection of appropriate machines is very important for prevention of problems caused by production quality, delivery and cost. Many of the decision variables in selection of machines between alternatives make decision making process difficult for managers. In the literature, there are many decision making methods (analytic hierarchy process, second axiom in axiomatic design, etc.) related to machine selection (Saaty, 1990; Babic 1999). On the other side, in the previous step of the axiomatic design model for lean office operations, software investment can be made for accelerating office works, easier information sharing and keeping up to date of office operations.

Visualizing the System with Investment (FR158-DP158): Since the equipment investment decision affects the cell configurations, the future state value stream map should be revised according to the new situation.

Evaluating the System Performance with Investment (FR159-DP159): Waste elimination in the existing plants and responsibility for new equipment investments are challenging for managers (Sullivan, *et al.*, 2002). Economic analysis based on value stream map should be done in born of the need for equipment investment. Better investment decisions can be made with the use of value stream maps. Cost analysis based on lead time should be made before implementation of future state value stream maps with investment and non-investment decisions. In addition to this, the economic lives of the new equipments affect the investment decision.

Decreasing the Excessive Volume Fluctuations and Variations in Production (FR1510-DP1510): For application of pull production control system according to the lean principles, the production flow must be smoothed primarily. If the flow is leveled according to production-mix throughout the production, the work-in-process inventories and therefore the total lead time will be reduced dramatically. By the accordingly designed Heijunka boxes, the amount of production and how much time it will take to produce this amount can be determined.

Providing Pulling between Inter Cell Flows (FR1511-DP1511): Pull production control system combines the demand and production. Kanban cards are used for the pull production control system. To be successful in Kanban, smoothed production flow is required. Kanban system is a different approach for manufacturing according to the customer demand and communication in the production environment for purchasing the materials needed. Kanban assigned to the cells draws parts from the cells, which produces these parts (Suzaki, 1987).

So far, a road map was presented for lean thinking based investment planning at design stage of cellular/hybrid manufacturing system (see Figure 8). Design matrixes also showed that leaf-level design decisions are consistent.

Implementation of the Proposed Methodology

The proposed methodology was tested on a real case application. The proposed methodology was implemented step by step for transforming the existing traditional manufacturing system to cellular/hybrid manufacturing system at a furniture factory. Before implementation, the facility layout of the system was functionally organized (see Figure 9). Plans have been devised for company-wide participation. The first main step (FR11-DP11) was the collection of information about market to define customer requirements. The second main step (FR12-DP12) was making experts and employees conscious on lean thinking by training programs. Around this time, lean manufacturing training programs were prepared and nearly 90% of the employees participated to lectures and workshops about lean tools and techniques.

The third main step (FR13-DP13) was to divide production system into sub-systems for simplification of the system. At this stage, the product families were defined based on manufacturing characteristics. The fourth main step (FR14-DP14) was drawing current state value stream map for the selected product family to visualize wastes and to define which lean tools would be applied to eliminate waste resources. The last main step (FR15-DP15) for the roadmap was applying the proposed plan to eliminate waste in production processes and offices by future state value stream management. The takt time was determined for the selected product family. Determination of parts for that product family and proper allocation of machines for each cell have also been done.

By application of the multi product flow line algorithm developed by Aneke & Carrie (1986), the new layout of the facility was decided based on one piece flow. The new layout consists three newly formed cells and a functional area. Staffing of the cells was finalized regarding of the current cell configurations according to the takt time. Final acquisition in each cell was determined with the participation of cell teams. Machine locations were defined in the sequence of part movements. After this stage, to increase performance of the systems, supporting manufacturing, lean office studies were planned and started. Before making new investment, the system was tried to improve by kaizen procedures. For maintaining the 5S discipline, 5S evaluation worksheets were designed and routinely implemented. Once the standard works

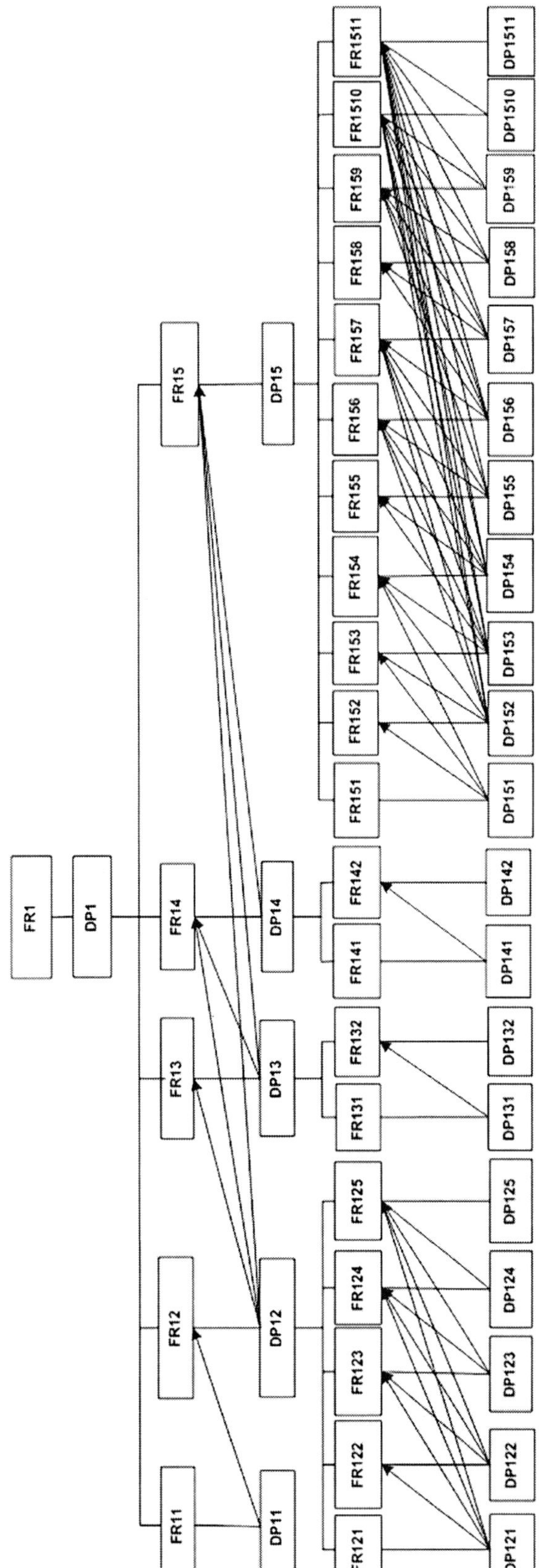

Figure 8. The decomposition of AD model

Figure 9. Spaghetti diagram before cell formation

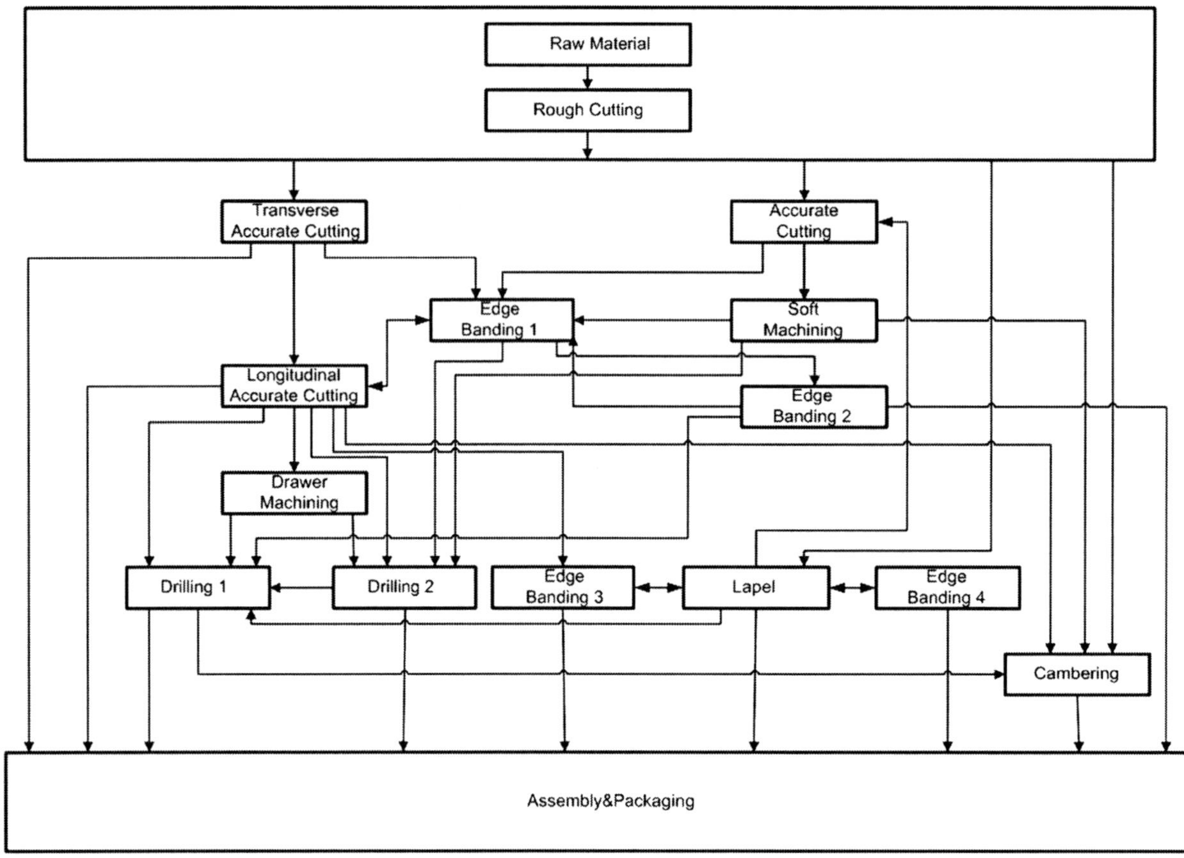

in each cell were defined, proper metrics were developed for continuous improvement monitoring. In order to maintain and improve the performance, equipment effectiveness was calculated and also, total productive maintenance applications and visual management tools were developed. Single minute exchanges of die (SMED) plans were applied to machines to reduce setup times. Beside these, process kaizens were applied continuously for the bottleneck areas.

At this stage, there were still some bottlenecks on the manufacturing system. New machines investment was needed for solving those bottlenecks. Future state value stream map was revised considering new machine investment decision. Finally, four new machines investment is needed to be installed in order to maintain smooth product flow (see Figure 10). Each machine was selected with an AHP based model for the cell and functional area. To evaluate the system performance with investment before purchasing, some economical cost analysis based on lead time was done for the cases of before and after investment. Finally, we decided to make new machine investments by considering the cost analysis results.

At this stage, new machine purchasing procedure was completed, and then, to decrease the excessive volume fluctuations and variations in production, heijunka boxes were planned and to provide continues part flow, a combination of Kanban and push production control system were planned. Based on customer demands and forecasted sales, mixed leveled production plans were established for each cell. Based on these leveled

Figure 10. Spaghetti diagram after cell formation

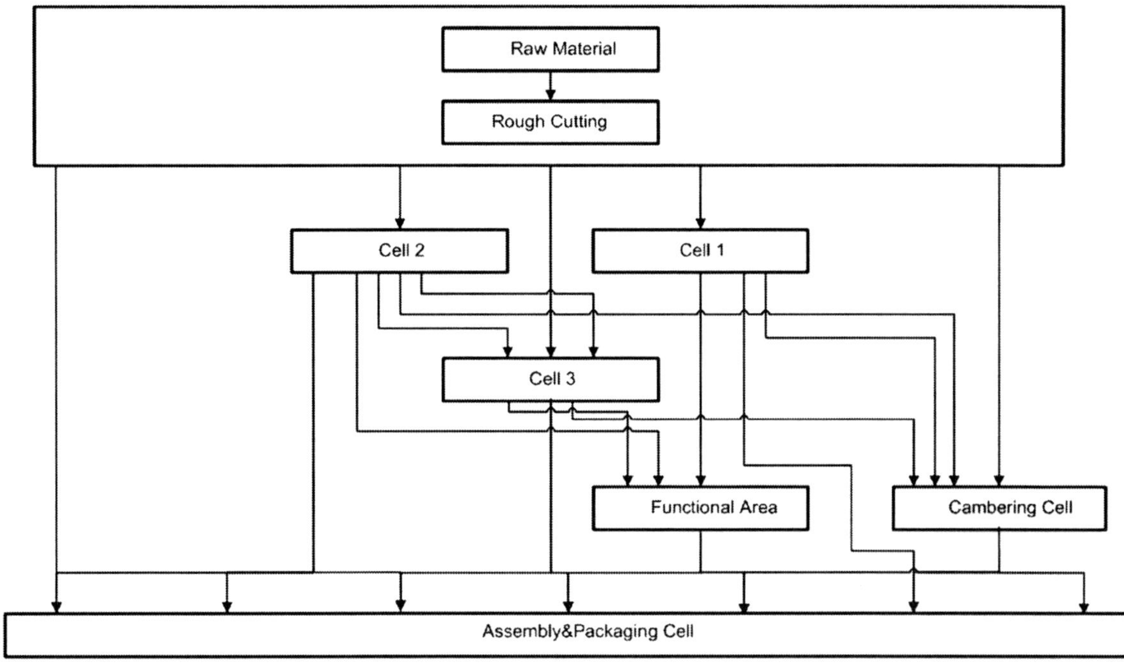

production plans, the convenient batch sizes have been determined for each product family. For Kanban card implementation, new cards were designed. The Kanban cards traffic between the cells and supermarkets were planned. Finally, it has been seen that this methodology is very practical and beneficial for production system performance improvement and economical investment decisions. The result of this implementation on this real case is listed in Figure 11.

Some of the indicators used in Figure 11 are as explained as follows:

Manufacturing lead time is called as duration from the moment the chipboards entered to the storage is ready for assembly. It is calculated as following:

$$Lead\ Time = \frac{Work\ In\ \Pr ocess\ Inventory\,(units)}{Demand\ Rate\,(units\ /\ day)} \tag{7}$$

Training time is called as multiplication of the planned duration of training for lean manufacturing applications by the number of participants. Cellular equipment effectiveness (CEE) is the

Figure 11. Comparison of business metrics

Performance Criteria	Before Cellular Manufacturing	After Cellular Manufacturing
Lead time (days)	7.5	3.5
Training time (person-hour/year)	7200	16000
Equipment effectiveness (%)	38	85
Number of Poka-Yokes	-	5
Number of Kaizen suggestions (1/year)	4154	4944
Machine breakdown (%)	9.12	5
First time through (%)	50	95
Scrap (%)	3	1
Setup time reduction (%)	0	50

ability to produce high quality and at the right speed at the machineries of cell and to use machines when necessary.

$$Equipment\ Effectiveness = Availability * Performance * Quality \quad (8)$$

$$Availability = \frac{(Total\ Time - Downtime)}{Total Time} \quad (9)$$

$$Performance\ Efficiency = \frac{Actual\ Run\ Rate}{Ideal\ Run\ Rate} \quad (10)$$

$$Ideal\ Run\ Rate = \frac{1}{Takt\ Time} \quad (11)$$

$$Quality = \frac{Total\ Quantity\ Manufactured - Number\ \ Rejected}{Total\ Quantity\ Manufactured} \quad (12)$$

Number of Poka-Yoke is the number of error preventing applications. Number of Kaizen suggestion is the annual number of suggestions for continuous improvement. First time through is called as the ratio of the number of error-free parts to total quantity manufactured. Scrap rate is the ratio of the number of scrapped parts to total quantity manufactured. Reduction of setup time is the decline in the percentage of duration between the last good part of the previous setup and the first acceptable part in the new setup.

As shown in Figure 11, all the indicators were improved significantly. In order to simplify the work flow further as a continuous improvement effort, a new facility layout has also been planned as shown in Figure 12. This facility layout is ex-

Figure 12. Spaghetti diagram for future state goal

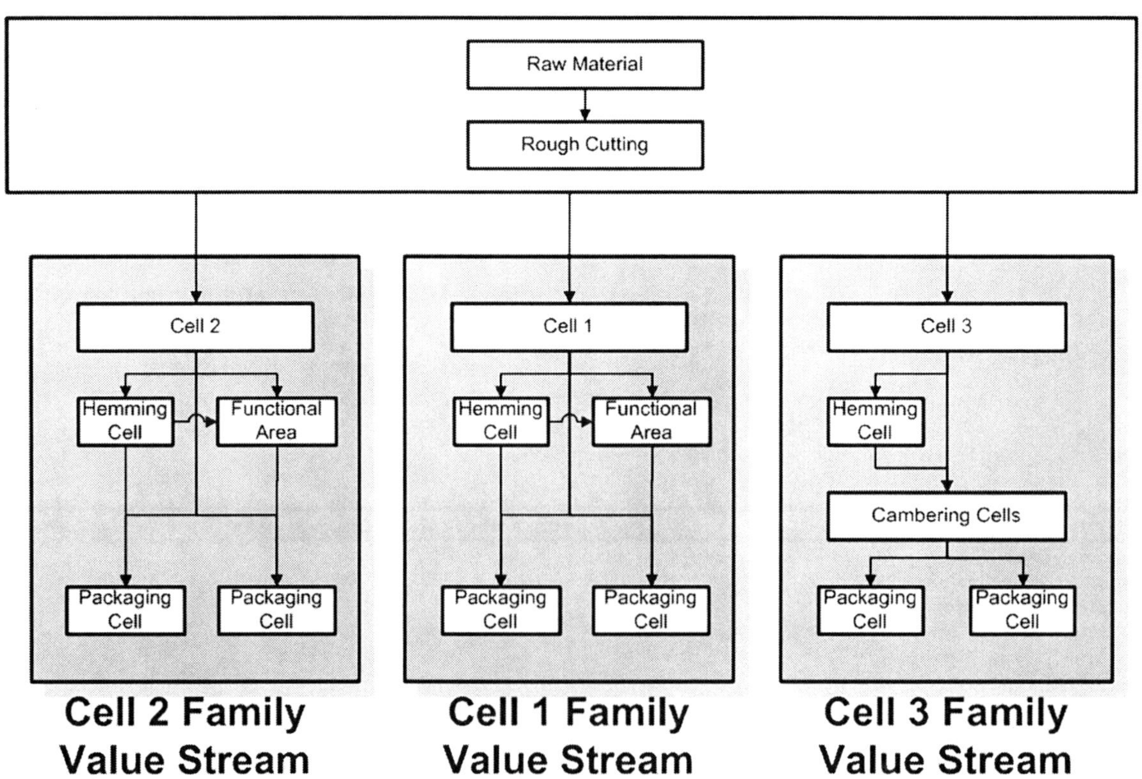

pected to decrease the current lead time by 20%. After investigation on the accidents has occurred previously, the number of Poka-Yokes was decided to enhance for the prevention of accidents resulting from misuse of machines. The types of kaizen gifts have been decided to increase. After maturation of lean practices, the number of monthly control tables was planned to increase. For the development related to the considered parameters, it was also decided to form the numerous Kaizen teams.

CONCLUSION

Lean manufacturers must use lean thinking when implementing new purchasing strategies for machines that coincide with the product value stream. Most companies are ignoring the value stream in the decision-making process. In this chapter, we have provided a methodology for transforming a process oriented manufacturing facility into a CMS or HMS with a reasonable investment decision making. The proposed methodology is based on Axiomatic Design principles. Here, a roadmap for how lean thinking based investment planning can be applied at the design stage of cellular/hybrid manufacturing system has been presented. The concept in this chapter can be applied easily to real cases. Value stream mapping is the best tool to visualize wastes and to determine which lean tools can be applied to the system. Visualizing the current state helps to plan the future state by value stream mapping.

In future studies, the proposed methodology for new investment planning can be extended to the process domain which is comprised of the process variables. Decision making when there are multiple axiomatic design alternatives is another research field for the future.

REFERENCES

Abbett, D., Payne, V., & Gulfstream, V. (1999). *Value stream tour*. Atlanta, GA: Lean Enterprise Institute.

Aneke, N. A. G., & Carrie, A. S. (1986). A design technique for the layout of multi-product flow lines. *International Journal of Production Research*, *24*(3), 471–481. doi:10.1080/00207548608919743

Babic, B. (1999). Axiomatic design of flexible manufacturing systems. *International Journal of Production Research*, *37*(5), 1159–1173. doi:10.1080/002075499191454

Babu, A. S., Nandurkar, K. N., & Thomas, A. (2000). Development of virtual cellular manufacturing systems for SMEs. *Logistics Information Management*, *13*(4), 228–242. doi:10.1108/09576050010340866

Braglia, M., Carmignani, G., & Zammori, F. (2006). A new value stream mapping approach for complex production systems. *International Journal of Production Research*, *44*(18–19), 3929–3952. doi:10.1080/00207540600690545

Byrne, A. (1995). How wire mold reinvented itself with kaizen. *Target*, *11*(1), 8–14.

Cochran, D. S., Eversheim, W., Kubin, G., & Sesterhenn, M. L. (2000). The application of ad and lean management principles in the scope of production system segmentation. *International Journal of Production Research*, *38*(6), 1377–1396. doi:10.1080/002075400188906

Day, J. C. (1998). Learning about lean systems at Freudenberg-NOK: Where continuous improvement is a way of life. In Likert, J. K. (Ed.), *Becoming lean* (pp. 179–198). Portland, OR: Productivity, Inc.

Durmusoglu, M. B., & Kulak, O. (2008). A methodology for the design of office cells using axiomatic design principles. *Omega International Journal of Management Science*, *36*, 633–652. doi:10.1016/j.omega.2005.10.007

Durmusoglu, M. B., & Nomak, A. (2005). GT cells design and implementation in a glass mould production system. *Computers & Industrial Engineering*, *48*, 525–536. doi:10.1016/j.cie.2003.03.003

Gallagher, C. C., & Knight, W. A. (1973). *Group technology*. London, UK: Butterworths.

Ghosh, B. K. (1989). Equipment investment decision analysis in cellular manufacturing. *International Journal of Operations & Production Management*, *10*(7).

Hines, P., & Taylor, D. (2000). *Going lean*. UK: Lean Enterprise Research Centre Cardiff Business School.

Houshmand, M., & Jamshidnezhad, B. (2002). Conceptual design of lean production systems through an axiomatic design. *Proceedings of ICAD2002, Second International Conference on Axiomatic Design,* Cambridge, MA (ICAD033).

Jones, D. T. (1999). *Seeing the whole: Macro value stream mapping*. Atlanta, GA: Lean Enterprise Institute.

Kulak, O., Durmusoglu, M. B., & Tufekci, S. (2005). A complete cellular manufacturing system design methodology based on axiomatic design principles. *Computers & Industrial Engineering*, *48*, 765–787. doi:10.1016/j.cie.2004.12.006

Marsh, R. F., Shafer, S. M., & Meredith, J. R. (1999). A comparison of cellular manufacturing research presumptions with practice. *International Journal of Production Research*, *37*(14), 3119–3138. doi:10.1080/002075499190202

Megep. (2008). *Research for marketing*. Project for Improving the Professional Education and Training, Ministry of National Education, Republic of Turkey (In Turkish).

Miltenburg, J. (2001). One-piece flow manufacturing on U-shaped production lines: A tutorial. *IIE Transactions*, *33*(4), 303–321. doi:10.1080/07408170108936831

Monden, Y. (1993). *Toyota production system: An integrated approach to JIT*. Georgia: Engineering Management Press.

Nancy, H., & Wemmerlöv, U. (2002). *Reorganizing the factory* (pp. 95–145). Productivity Press.

Pavnaskar, S. J., Gershenson, J. K., & Jambekar, A. B. (2003). Classification scheme for lean manufacturing tools. *International Journal of Production Research*, *41*, 3075–3090. doi:10.1080/0020754021000049817

Rother, M., & Shook, J. (1998). *Learning to see: Value stream mapping to add value and eliminate muda* (2nd ed.). Brookline, MA: The Lean Enterprise Institute, Inc.

Saaty, T. L. (1990). How to make a decision: The analytic hierarchy process. *European Journal of Operational Research*, *48*, 9–26. doi:10.1016/0377-2217(90)90057-I

Sarher, B. R., & Mondal, S. (1999). Grouping efficiency measures in cellular manufacturing: A survey and critical review. *International Journal of Production Research*, *37*(2), 285–314. doi:10.1080/002075499191779

Satoglu, S. I., Durmusoglu, M. B., & Ertay, T. (2009). A mathematical model and a heuristic approach for design of the hybrid manufacturing systems to facilitate one-piece flow. *International Journal of Production Research*, *48*(17), 5195–5220. doi:10.1080/00207540903089544

Shambu, G., & Suresh, N. C. (2000). Performance of hybrid cellular manufacturing systems: A computer simulation investigation. *European Journal of Operational Research*, *120*(2), 436–458. doi:10.1016/S0377-2217(98)00371-3

Suh, N. P. (1990). *The principles of design*. New York, NY: Oxford University Press.

Suh, N. P. (1997). Design of systems. *Annuals of the CIRP*, *46*(1), 75–80. doi:10.1016/S0007-8506(07)60779-3

Suh, N. P. (2001). *Axiomatic design: Advances and applications*. New York, NY: Oxford University Press.

Suh, N. P., Cochran, D. S., & Paulo, C. L. (1998). Manufacturing system design. *Annuals of the CIRP*, *47*(2), 627–639. doi:10.1016/S0007-8506(07)63245-4

Sule, D. R. (1994). *Manufacturing facilities: Locations, planning and design*. Boston, MA: PWS Publishing Company.

Sullivan, W. G., McDonald, T. N., & Van Aken, E. M. (2002). Equipment replacement decisions and lean manufacturing. *Robotics and Computer-integrated Manufacturing*, *18*, 255–265. doi:10.1016/S0736-5845(02)00016-9

Suresh, N. C. (1991). Partitioning work centers for group technology: Insights from an analytical model. *Decision Sciences*, *22*(4), 772–791. doi:10.1111/j.1540-5915.1991.tb00364.x

Suri, R. (1998). *Quick response manufacturing* (pp. 303–333). Portland, OR: Productivity Press.

Suzaki, K. (1987). *The new manufacturing challenge: Techniques for continuous improvement*. New York, NY: The Free Press.

Tapping, D., Luyster, T., & Shuker, T. (2002). *Value stream mapping*. Portland, OR: Productivity Press.

Womack, J. P., & Fitzpatrick, D. (1999). *Lean thinking for aerospace: the industry that can afford its future*. Atlanta, GA: Lean Enterprise Institute.

Womack, J. P., Jones, D., & Roos, D. (1990). *The machine that changed the world*. New York, NY: Rawson Associates.

Womack, J. P., & Jones, D. T. (1996). *Lean thinking: Banish waste and create wealth in your corporation*. New York, NY: Simon & Schuster.

Womack, J. P., & Jones, D. T. (2002). *Seeing the whole. Mapping the extended value stream*. Massachusetts Lean Enterprise Institute.

Chapter 17
Performance Comparison of Cellular Manufacturing Configurations in Different Demand Profiles

Paolo Renna
University of Basilicata, Italy

Michele Ambrico
University of Basilicata, Italy

ABSTRACT

Cellular manufacturing systems (CMSs) are an effective response in the economic environment characterized by high variability of market. The aim of this chapter is to compare different configurations of cellular models through the main performance. These configurations are fractal CMS (defined FCMS) and cellular systems with remainder cells (defined RCMS), compared to classical CMS used as a benchmark. FCMSs consist of a cellular system characterized by identical cells each capable of producing all types of parts. RCMSs consist of a classical CMS with an additional cell (remainder cell) that in specific conditions is able to perform all the technological operations. A simulation environment based on Rockwell ARENA® has been developed to compare different configurations assuming a constant mix of demand and different congestion levels. The simulation results show that RCMSs can be a competitive alternative to traditional cells developing opportune methodologies to control the loading of the cells.

DOI: 10.4018/978-1-61350-047-7.ch017

INTRODUCTION

Competitiveness in today's market is much more intense compared to the past decades. Considerable resources are invested on facilities planning and re-planning in order to adapt the manufacturing systems to the market changes. A well-established manufacturing philosophy is the group technology concept.

Group technology (GT) can be defined as a manufacturing philosophy identifying similar parts and grouping them together to take advantage of their similarities in manufacturing and design (Selim et al.,1998). It is the basis of so-called cellular manufacturing systems (CMSs). In current production scenario demand for products is characterized by continuous fluctuations in terms of volumes, type of product (part mix), new products introduction and the life cycle of products has significantly reduced. The planning horizon needs to be divided into smaller horizons (time bucket) and the length of each period is related to the characteristics of products. These characteristics need to be considered in design process of a manufacturing system. Introduction of Cellular Manufacturing Systems has already introduced significant improvements. They are conceived with the aim of reducing costs such as setup costs or handling costs and also to reduce lead time and work in process (WIP). They combine advantages of flow shop and job shop, but a further step can be accomplished to be competitive in the market. They allow significant improvements such as: product quality, worker satisfaction, space utilization. Benefits and disadvantages (Irani et al.,1999) are showed in Table 1. They documented that companies implementing cellular manufacturing have a very high probability of obtaining improvements in various areas.

The first column of Table 1 shows the case studies with improvements and the second column reports the percentage of improvement of the measures. Similarly, the third column shows the percentage of cases with worsening and in the fourth column is evidenced the rate of deterioration.

The demand volatility and continuous new product introduction lead to re-configure several times the cellular manufacturing systems in order to keep a high level of performance.

Table 1. Benefits and disadvantages of CMS

Measure	Percentage cases with improvements	Average percentage improvement	Percentage cases with worsening	Average percentage worsening
Tooling cost	31%	-10%	69%	+17%
Labor cost	91%	-33%	9%	+25%
Setup Time	84%	-53%	16%	+32%
Cycle Time	84%	-40%	16%	+30%
Machine utilization	53%	+33%	47%	-20%
Subcontracting	57%	-50%	43%	+10%
Product quality	90%	+31%	10%	-15%
Worker satisfaction	95%	+36%	5%	-
Space utilization	17%	-25%	83%	+40%
WIP inventory	87%	-58%	13%	+20%
Labor turnover/absenteeism	100%	-50%	0	-
Variable production cost	93%	-18%	7%	+10%

For the above reasons, new configurations have been proposed in literature such as Virtual Cell Manufacturing System (VCMS), Fractal Cell Manufacturing System (FCMS), Dynamic Cell Manufacturing System (DCMS), with the aim of keeping high flexibility of manufacturing systems.

The concept of DCMS was introduced for the first time by Rehault et al. (1995). It provides a physical reconfiguration of the cells. The reconfiguration activity can be periodic or resulting from the variation of performance parameters. Reconfigure can mean duplicating machines, relocating machines between cells, removing machines, or also subcontracting some parts to other companies. These problems must be addressed by the decision maker.

The concept of VCMS requires that the machines are dedicated to a part family but these machines are not necessarily close together in a classical cell. One machine can belong simultaneously to different cells. Hence sharing of machine makes the system more flexible. Moreover the machines are not shifted as dynamic cellular system therefore costs of reallocation are eliminated. On the other hand we must consider the increase in the movements of parts (or batches) across machines. A further problem may be the complication in the measurement of performance of the cells. This is because monitoring stations are usually located out of the cell, but in this case the cell does not exist physically.

The FCMSs are based on the constructions of identical cells and they are not built for different families. The idea comes from Skinner (1974) and that is to build factory within a factory with duplication of processes. Each cell can work all products. Working time will be greater but these configurations are very effective if there are changes in part mix and in cases of machine breakdowns. Even if for example there are flash orders.

A further idea was mentioned by Sripathy Maddisetty (2005). The author referred to so-called remainder cells and we can call them RCMSs. In addition to traditional cells refer to the product families you may create an additional cell that operates when conditions such as machine failures or overloaded machines occur. Focusing on an advanced design the RCMSs could provide interesting results in terms of competitiveness.

Our goal in this chapter is to compare the various approaches to the design of manufacturing systems, making a complete performance comparison. In particular we aimed to compare the following systems: CMSs, FCMSs and RCMSs. A simulation environment has been developed to compare the performance (WIP, Throughput Time, Tardiness, Throughput and Average Utilization) using as a benchmark the classic CMS. The aim is to evaluate the responses of different systems when market fluctuations occur in terms of arrival demand. The chapter is structured as follows. Section 2 provides an overview of the literature of various manufacturing system configurations, while in section 3 the system context is formulated. In section 4 there is a brief description of scheduling approaches. Section 5 presents the simulation environment and the case study while in section 6 are discussed simulation results. In Section 7 conclusions and future developments are discussed.

BACKGROUND

Recently, several authors have investigated the configuration of manufacturing cells in order to keep a high level of performance when the market conditions change.

Hachicha et al. (2007) proposed a simulation based methodology which takes into consideration the stochastic aspect in the CMS. They took into account the existence of exceptional elements between the parts and the effect of the correspondent inter-cell movements. They compared two strategies: permitting intercellular transfer and exceptional machine duplication. They used the simulation (Rockwell Arena) and they analyzed the following performance: mean transfer time, mean machining time, mean wait time, mean flow time. They assumed demand fixed and known for the parts. They did not consider failures of machines and maintenance policies.

A multi-objective dynamic cell formation was presented by Bajestani et al. (2007) where purpose was to minimize simultaneously total cell load variation and sum of miscellaneous costs (machine cost, inter-cell material handling cost, and machine relocation cost). Since the problem is *NP-hard* they used a scatter search approach for finding locally Pareto-optimal frontier.

Safei et al. (2007) proposed to use an approach based on fuzzy logic for the design of CMS under uncertain and dynamic conditions. They began by finding that in most of research related on DCMS input parameters were considered deterministic and certain. Therefore they introduced fuzzy logic as a tool for the expression of the uncertainty in design parameters such as part demand and available machine capacity.

Ahkioon et al. (2007) tried to investigate DCMS focusing on routing flexibility. They studied the creation of alternate contingency process routings in addition to alternate main process routings for all part types. Contingency routings had the function to provide continuity in case of exceptional events such as machine breakdowns but also flash orders. Furthermore their work provided discussions on the trade-off between the additional cost related to the formation of contingency routings and the advantages of increased flexibility. Linearized model proposed by the authors was solved with CPLEX.

Aryanezhad et al. (2008) developed a new model which simultaneously embrace dynamic cell formation and worker assignment problem. They focused on two separate components of cost: the machine based costs such as production costs, inter-cell material handling costs, machine costs and human related costs such as hiring costs, firing costs, training costs and wages. They made the comparison of two models. One considered the machine costs and the other considered both machine costs and human related costs. The model was *NP-hard* even though they did not consider learning curve.

Xiaoqing Wang et al. (2008) proposed a nonlinear multi-objective mathematical model in dynamic cells formation problem by giving weighing to three conflicting objectives: machine relocation costs, utilization rate of machine capacity, and total number of inter-cell moves over the planning horizon. A scatter search approach was developed to solve the nonlinear model. Results were compared with those obtained by CPLEX. They considered certain demand and they did not consider machine breakdowns.

Safei et al. (2009) proposed an integrated mathematical model of the multi-period cell formation and production planning in a dynamic cellular manufacturing system (DCMS). The focus was on the effect of the trade-off between production and outsourcing costs on the reconfiguration of the cells.

Balakrishnan (2005) discussed cellular manufacturing system under conditions of changing product demand. He made a conceptual comparison to virtual cell manufacturing and he discussed a case study.

Kesen et al. (2008) investigated three different types of system (cellular layout, process layout and virtual cells) by using simulation. They paid attention to the following performance: mean flow time and mean tardiness. Based on these simulations they used regression meta-models to estimate the systems behaviours. They only considered one family-based scheduling scheme and they did not consider extraordinary events such as machine failures.

Vakharia et al. (1999) proposed and validated analytical approximations for comparing the performance of virtual cells and multistage flow shops. First they used these approximations and hypothetical data to identify some key factors that influenced the implementation of virtual cells in a multistage flow shop environment. Then they concluded with an application of approximations to industrial data.

Kesen et al. (2009) examined the behaviours of VCMs, process layouts and cellular layouts. They addressed the VCMs by using family-based scheduling rule. The different systems were compared by simulation. Subsequently they developed an ant colony optimization based meta-models to reflect the system's behaviours.

Kesen et al. (2010) presented a genetic algorithm based heuristic approach for job scheduling in virtual manufacturing cells (VMCs). Cell configurations were made to optimize the scheduling objective under changing demand conditions. They considered the case with multiple jobs and different processing routes. It was considered multiple machine types with several identical machines in each type and they were located in different locations in the shop floor. The objective was to minimize the total travelling distance. To evaluate the effectiveness of the genetic algorithm heuristic they compared it with a mixed integer programming solution. Results showed that genetic algorithm was promising in finding good solutions in very shorter.

Uday Venkatadri et al.(1997) proposed a methodology for designing job shops under the fractal layout organization as an alternative to the more traditional function and product organizations. The challenge in assigning flow to workstation replicates was that flow assignment is in itself a layout dependent decision problem. They proposed an iterative algorithm that updated layouts depending on flow assignments, and flow assignments based on layouts. Their work has had the far-reaching consequence of demonstrating the validity of the fractal layout organization in manufacturing systems (FCMSs).

Montreuil (1999) developed a new fractal alternative for manufacturing job shops which allocated the total number of workstations for most processes equally across several fractal cells. He introduced fractal organization and he briefly discussed the process of implementing fractal designs. He illustrated a case example and he showed that system is characterized by great flexibility.

Maddisetty (2005) discussed the design cells in a probabilistic demand environment. He discussed idea of remainder cells (RCMS). A remainder cell is a kind of lung to cope in changes in demand. He examined the following performance: total WIP, average flow time, machine utilization. He proposed a comparison using three different approaches: mathematical, heuristic, and simulation.

Süer et al. (2010) proposed a new layered cellular manufacturing system to form dedicated, shared and remainder cells to deal with the probabilistic demand. Moreover they proposed a comparison of its performance with the classical cellular manufacturing system. Simulation and statistical analysis were performed to help identify the best design within and among both layered cellular design and classical cellular design. They observed that the average flow time and total WIP were not always the lowest when additional machines were used by the system, but the layered cellular system performed better when demand fluctuations was observed.

There are several limitations encountered in existing literature. In previous research the demand of products was usually determined at the beginning of each period and it was known. The change in part mix was rarely assumed. Frequently the bottleneck station in each cell was considered as fixed and independent of the type of the part. Almost never were held in account exceptional events such as machine failures and maintenance. Almost never flash orders was considered and similarly backorders. The concept of learning curve was rarely covered. Furthermore hardly researchers focused on a wide range of performance measures.

In this chapter the objective is to evaluate the reaction of different manufacturing systems configurations (CMSs, FCMSs and RCMSs) when there is a fluctuation in terms of arrival demand. The configurations are investigated considering the same machines for all cases; the machines are set in order to obtain the particular configuration. The analysis conducted allows to highlight the most promising configurations in terms of performance measures. Another objective of the chapter is to develop a simulation environment based on Rockwell Arena® tool in order to analyse the different configurations. The simulation allows build a model with minor simplification compared to mathematical models which require significant simplifications (linearization) in cases of complex systems. Moreover the dynamic model (demand not known a priori, unexpected events like machine breakdowns) cannot be obtained with mathematical models.

MANUFACTURING SYSTEM CONTEXT

The mentioned objective of this chapter is to compare the performance of different manufacturing systems. In particular, the configurations analyzed by using simulation tools based on the software Rockwell ARENA® are: CMS, FCMS and RCMS. Moreover another configuration has been considered changing the layout of machines and obtaining a CMS in line.

The manufacturing system consists of *M* machines general purpose that are used for each configuration. It has been considered three part families. We consider a constant mix of each part family.

We introduce the following assumptions for the model:

- the demand for each part type is unknown at priori end it is extracted randomly from an exponential distribution. Therefore, the parameter to set is the exponential parameter;
- set-up times are not simulated. When, the manufacturing cells are configured the set-up times are very low for the product family assigned to the cell;
- the due date is obtained by processing time multiplied with an index greater than or equal to 1;
- Machine breakdowns and maintenance are not considered;
- intra-cell handling times are negligible;
- it is assumed that parts moved in units;
- each configuration presents the same number of machines in order to make a comparison in the same conditions.

The performance measures used to compare the manufacturing systems are the following:

- Work in Process (WIP);
- Average utilization of the manufacturing system;
- Throughput time;
- Average throughput time;
- Tardiness (total of all the parts);
- Throughput.

Figure 1 describes the parameters and the performance analyzed in this research.

The first manufacturing system configuration considered is a classical cellular system (CMS). The scheme is showed in Figure 2.

The system manufactures *N* product families with *N* cells. Each cell is specialized to perform the technological operations required by the product family assigned (setup time is not necessary). In this chapter, it has been also considered a CMS with a different routing, as showed in Figure 3.

The second configuration considered is the FCMS. In this case, the allocation of machines to cells is performed in order to obtain N identical cells. Each cell manufactures all product families with higher processing time, because the machines

Figure 1. Manufacturing configurations analysis

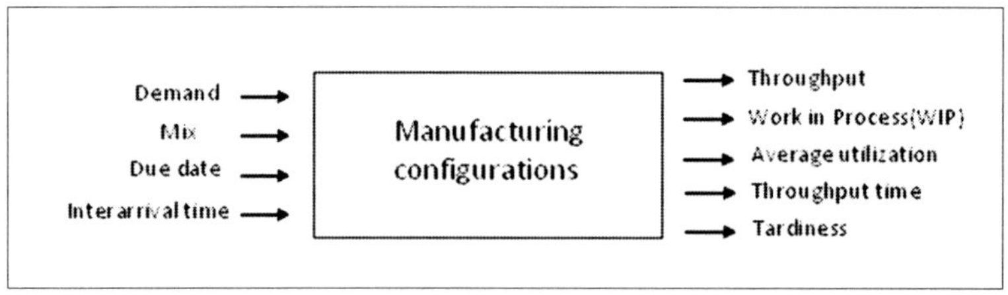

Figure 2. CMS configuration

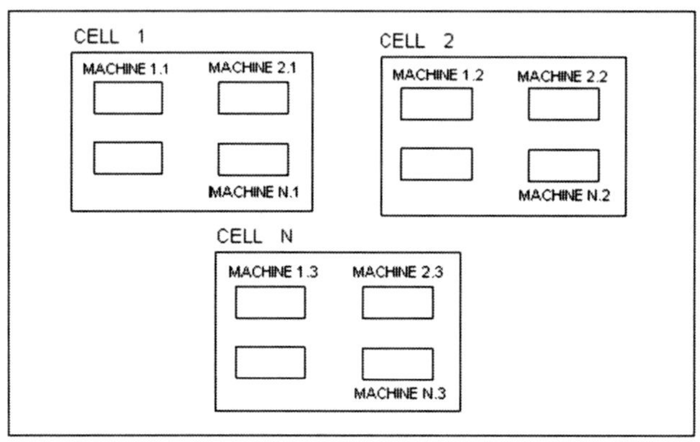

Figure 3. CMS in line configuration

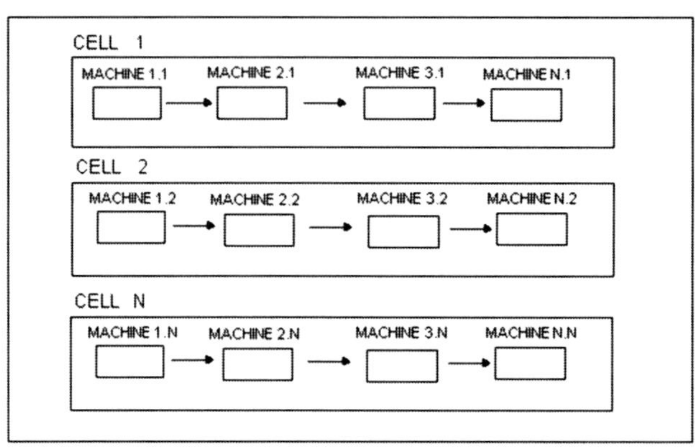

will be able to perform all the technological operations required. The scheme of FCMS is shown in Figure 4.

The third configuration considered is the RCMS. In this configuration there are N cells respectively for N product families. In addition, there is a further cell called remainder cell where all operations can be performed with higher processing times. It may be useful in case of machine failures but also in case of congestion of the system. The scheme of RCMS is showed in Figure 5.

Figure 4. FCMS configuration

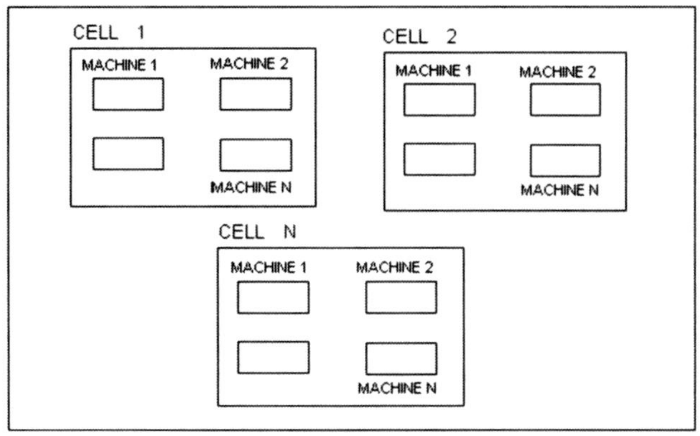

Each configuration includes the same number of machines and the time to manufacture each part is assumed the same, except for fractal cells (belonging to FCMS) and the remainder cell (belonging to RCMS) where machines can produce all kinds of part with a higher processing time (general purpose machine configuration). Therefore the processing time of machine *i-th* in fractal cell (pt_{if}) and processing time of machine *i-th* in remainder cell (pt_{ir}) is major of processing time of the machine *i-th* in the cell *j-th* in CMS (machine configured for the technological operations of a particular family) (pt_{ij}):

$$over = \frac{pt_{if}}{pt_{ij}} = \frac{pt_{ir}}{pt_{ij}}, over > 1 \qquad (1)$$

LOADING POLICY

In the previous section we have discussed the different cell configurations. Each configuration needs a loading approach policy to operate.

For classical CMS parts arrive in the system and each family has its own cell competence. In CMS we have provided two different layouts: one with parallel machines and other with machines in line, as described above.

Figure 5. RCMS configuration

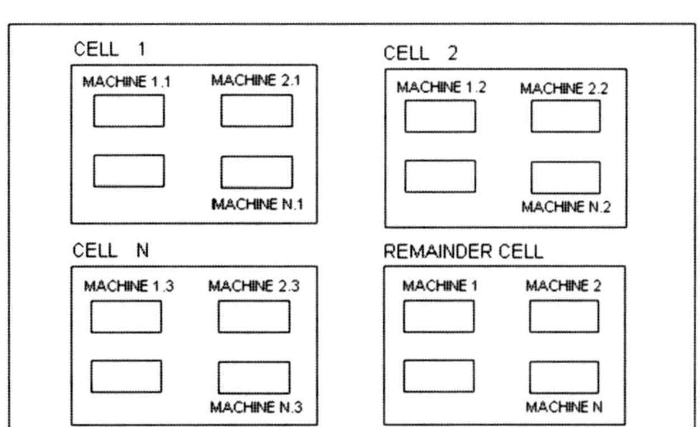

In FCMS configuration parts arrive in the system and they are routed to cells with minor workload.

The RCMS needs a specific loading policy for the use of the remainder cell. Parts arrive in the system and each cell is designed for a part family. In each cell, there is a controller that adopts the following strategy: it measures the number of parts in queue in each machine. If the measured value in cell *j-th* is greater than a maximum threshold of the cell (defined $Smax_j$) then the part is conveyed to the remainder cell. Similarly, when the measured value is minor of a minimum threshold of the cell *j-th* (defined $Smin_j$) then the part is assigned to the cell designed for the part family. The logic of controller above described is showed in the flowchart of Figure 6.

SIMULATION ENVIRONMENT

The manufacturing system consists of $M=10$ machines. All different configurations are obtained re-allocating the same number of machines available. It is considered that each machine functions for 24 hours a day. Therefore total numbers of minute that system works is considered to be 43200 minutes per month. This is the simulation horizon considered.

In order to evidence only the difference among the configurations, it is assumed that each part needs 40 minutes to complete processing. This technological time is divided by the number of machines used in the process, depending on manufacturing configuration. As above introduced it is equal for all parts except for those made in fractal cells and remainder cell where machines take more time.

We assume three product families. The product mix is as follow: Product 1 (40%), Product 2 (40%) and Product 3 (20%).

We have analyzed the performance of four different cellular systems changing one parameter: the average inter-arrival time. We have considered five different values of inter-arrival time that leads to different congestion levels of the manufacturing system (see Table 2).

These values were selected to keep the average utilization of machines in a range that goes from 0.56 (low utilization) to 0.99 (high utilization).

The demand for each part type is unknown at priori and it is extracted randomly from an exponential distribution with mean equal to the inter-arrival time reported in Table 2.

Figure 6. The logic of RCMS

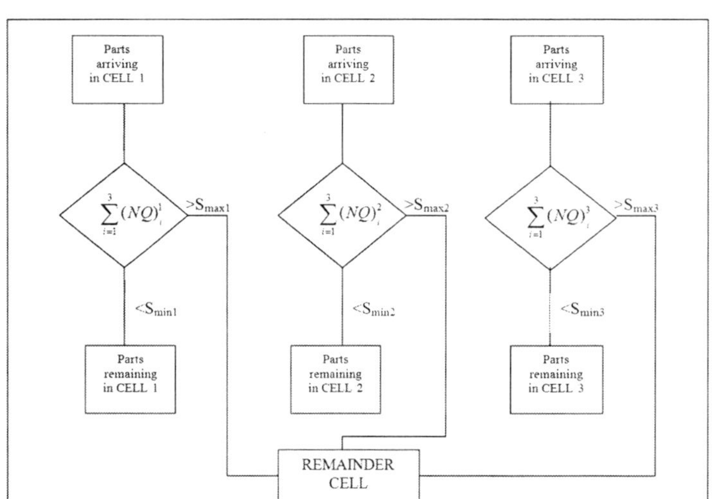

Table 2. Average inter-arrival times

Average inter-arrival times (min)
4
4.5
5
6
7

The due date is obtained from the sum of the arrival time (*tnow*) and the technological working time (*WT*) multiplied with an index (DdateINDEX), as showed in equation 2.

$$Ddate = tnow + (WT \cdot Ddate_{INDEX}) \qquad (2)$$

The *WT* is obviously equal to 40 minutes. The $Ddate_{index}$ is 1.5 for parts 1 and 2, while it is 1 for part 3. The minor index of part 3 is justified by the lower demand than other part-mix, so there is no shift of the due date. However the due dates are the same for all configurations examined. Therefore not affect the comparison, but they are included in the model for completeness.

Cellular systems analyzed are those already mentioned: CMS, CMS in line, RCMS and FCMS. The benchmark system is the CMS. The simulation environment has been developed by Rockwell Arena® tool.

Arena is characterized by a block diagram that makes it more familiar environment simulation. The arrival stations of the parts and the exit station are showed in the Figure 7.

In the first three boxes are showed the arrival stations where to each part is assigned a delivery time and a destination in the respective cell for processing; then the parts leave the arrival station.

Exit station is equal for all types of configuration: if the delivery time has been observed then the WIP is updated and the part leaves the system. Otherwise the delay is calculated.

Cellular Manufacturing System

In this case we consider three cells of production. The first two cells containing four identical machines working in pairs and in parallel. These cells are respectively for both products type 1 and type 2. The third cell contains 2 machines for products of type 3 (minor product mix). Each machine has a process time equal to 20 minutes. The scheme is showed in Figure 8.

In each rectangle is indicated the working time.

Cellular Manufacturing System in Line

In this case we also consider 3 cells of production. The first two cells containing 4 machines in

Figure 7. Arrival and exit stations

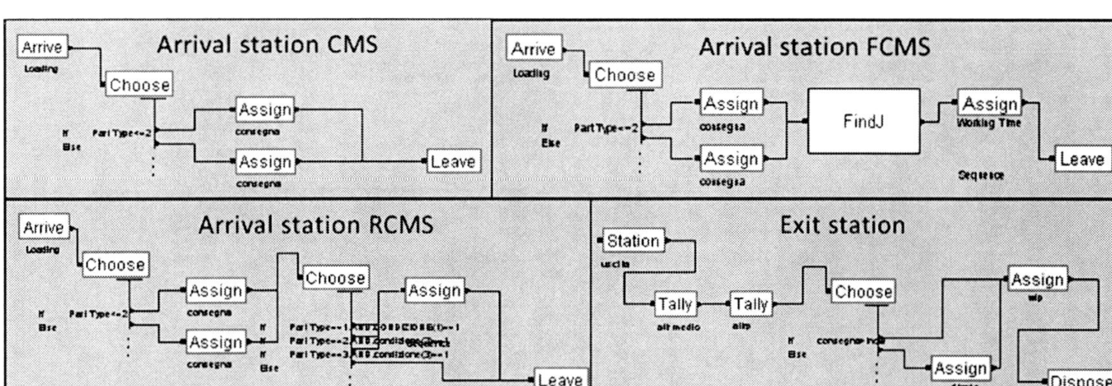

Figure 8. CMS considered in simulation

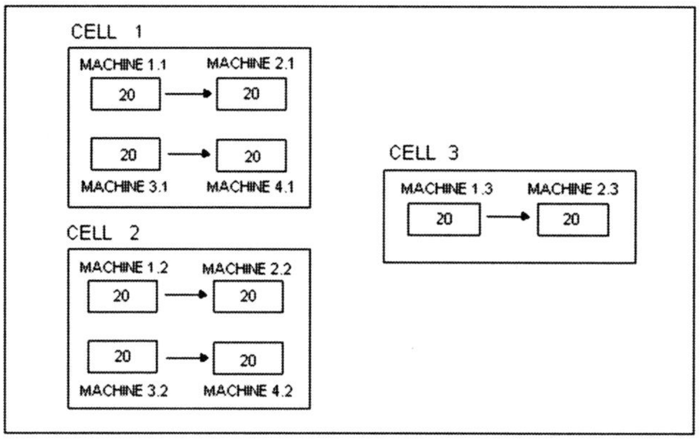

line. Each machine has a process time equal to 10 minutes. These cells are respectively for type 1 and type 2. The third cell contains 2 machines for product type 3, each machine has a process time equal to 20 minutes. The scheme is showed in Figure 9.

Fractal Cellular Manufacturing System

In this case there are 5 identical cells. Each cell contains 2 machines and each cell is able to work on all the product mix. The scheme is showed in Figure 10.

Naturally the machines perform the manufacturing operations with a major process time (see equations 1 and 2) because they are not dedicated to a part family but they are configured for all operations. In fact the process time of each machine is equal to 20 units time increased by 20% (over=1.2).

Remainder Cellular Manufacturing System

In this case there are 3 cells (one for each part type) and there is a remainder cell where is defined a loading policy based on the number of parts in queue in other cells. The scheme is showed in Figure 11.

The three machines operating in cell 1 (product type 1) has a process time equal to 13,33

Figure 9. CMS in line considered in simulation

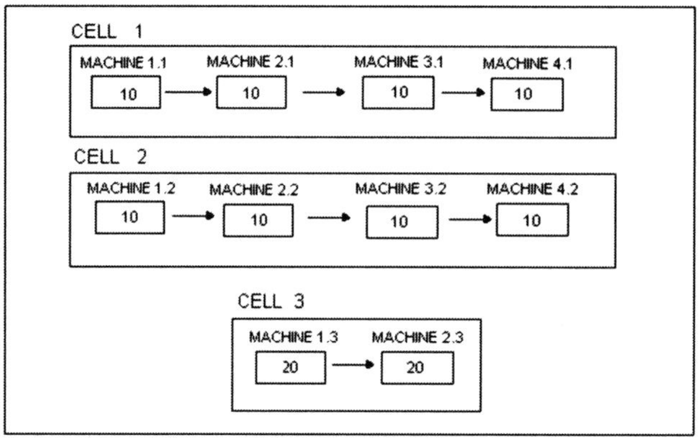

Figure 10. FCMS considered in simulation

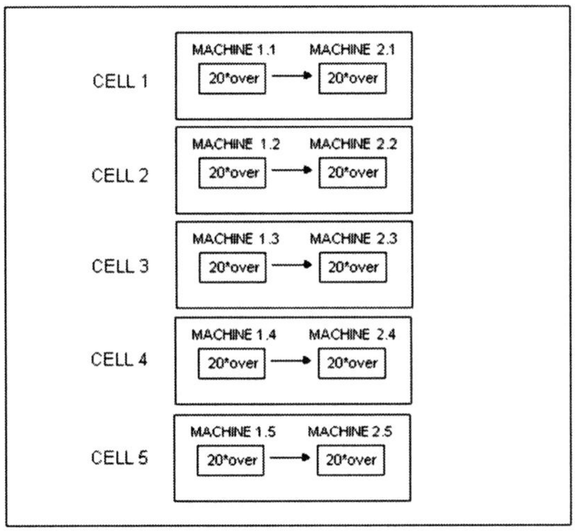

minutes. The same for the machines operating in cell 2 (product type 2). The two machines operating in cell 3 has a process time equal to 20 minutes. The machines assigned to the remainder cell perform the manufacturing operations with a major process time (see equations 1 and 2) because they are configured for all operations; the process time of each machine is equal to 20 units time increased by 20%(over=1.2).

In this work, it has been investigated different instances of the same policy loading about the use of remainder cell. Each cell has a controller that measures the number of parts in queue in each machine. Using thresholds the parts can be conveyed to the remainder cell. In ARENA the controller is showed in Figure 12.

The first *"scan"* controls the maximum threshold ($Smax_j$) and therefore assigns the part to the cell. Similarly, the second *"scan"* checks the minimum thresholds($Smin_j$).

For the values of maximum ($Smax_j$) and minimum ($Smin_j$) thresholds have been considered respectively six cases, equal for all three cells (see Table 3).

SIMULATION RESULTS

The length of each simulation is fixed to 43200 minutes. During this period the average inter-arrival time and part mix are both constant. Table 4 reports the design of simulation experiments conducted for all four configurations of the manufacturing system.

Combining the five inter-arrival times, four system configurations, and for the last configuration (RCMS) six cases regarding the thresholds, it has been obtained 45 experimental classes.

For each experiment class have been conducted a number of replications able to assure a 5% confidence interval and 95% of confidence level for each performance measure.

As previously described the performance measures investigated are the following:

Figure 11. RCMS considered in simulation

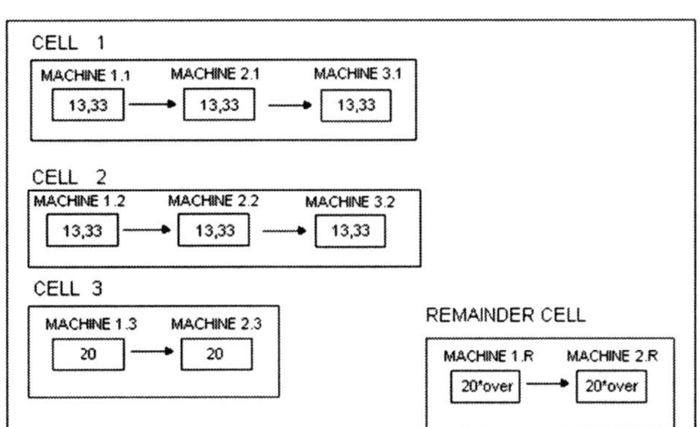

Figure 12. Control blocks cell 1

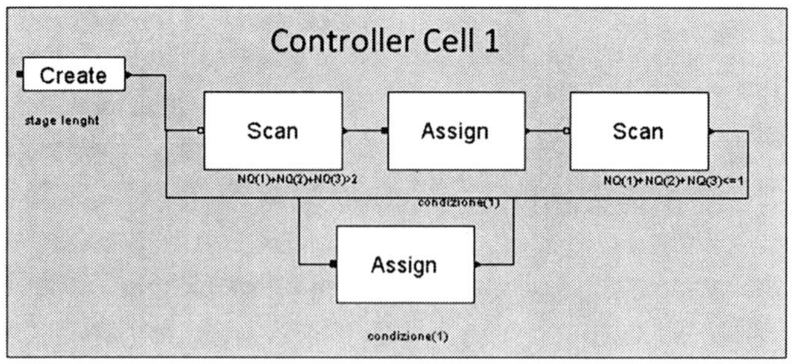

- Work in Process (WIP);
- Average utilization of the manufacturing system (*av.utilization*);
- Throughput time for each part j(*thr. Time j*);
- Average throughput time (*average thr. Time*);
- Total tardiness time of all the parts (*tardiness*);
- Throughput (*thr.*).

The objective of the analysis of simulation results is the comparison between different manufacturing configurations and classical cellular configuration (CMS, used as base for percentage computation). The aim is to use the performance parameters to highlight the behaviour of different configurations when changing the volume of demand (the variation of average inter-arrival times).

Table 5 shows the average utilizations of machines in classical CMS at different inter-arrival times.

Table 3. Threshold values

Cases	Smax	Smin
1	7	5
2	5	3
3	3	2
4	4	1
5	3	1
6	2	1

Therefore the simulations are performed for five congestion levels of the manufacturing system. It is important to emphasize that the results showed do not include machine breakdowns.

Table 6 reports the first three parameters (WIP, Tardiness and Throughput) for the different manufacturing configurations. Table 6 shows the average values over inter-arrival times with the respective standard deviations (St.dev). The standard deviation allows to highlight the variability of the results when the inter-arrival changes. The percentages refer to the comparison with the classical CMS. The positive percentages represent an increase of the respective factor while the negative percentages represent a decrease. Table 7 is the same for the throughput time of different parts and for the average throughout time.

Tables 6 and 7 show that CMS with configuration in line has almost the same behaviour of the classical CMS except for the tardiness that increments significantly.

Tables 6 and 7 also show that fractal configuration (FCMS) is the worst configuration. This is because the scheduling policy used is more simply. An opportune policy needs to be implemented for the FCMS. This is a limit of FMCS configuration, because a more complex control system has to be designed. The standard deviation shows the variability of the performance measures related to the inter-arrival changes in fact the FCMS is the configuration with the higher dependence on the inter-arrival changes. As the reader can notice,

Table 4. Experimental classes

Exp. No.	Configuration	Inter-arrival	Exp. No.	Configuration	Inter-arrival
1	CMS	4	26	RCMS(3,2)	4
2	CMS	4,5	27	RCMS(3,2)	4,5
3	CMS	5	28	RCMS(3,2)	5
4	CMS	6	29	RCMS(3,2)	6
5	CMS	7	30	RCMS(3,2)	7
6	CMS in line	4	31	RCMS(4,1)	4
7	CMS in line	4,5	32	RCMS(4,1)	4,5
8	CMS in line	5	33	RCMS(4,1)	5
9	CMS in line	6	34	RCMS(4,1)	6
10	CMS in line	7	35	RCMS(4,1)	7
11	FCMS	4	36	RCMS(3,1)	4
12	FCMS	4,5	37	RCMS(3,1)	4,5
13	FCMS	5	38	RCMS(3,1)	5
14	FCMS	6	39	RCMS(3,1)	6
15	FCMS	7	40	RCMS(3,1)	7
16	RCMS(7,5)	4	41	RCMS(2,1)	4
17	RCMS(7,5)	4,5	42	RCMS(2,1)	4,5
18	RCMS(7,5)	5	43	RCMS(2,1)	5
19	RCMS(7,5)	6	44	RCMS(2,1)	6
20	RCMS(7,5)	7	45	RCMS(2,1)	7
21	RCMS(5,3)	4			
22	RCMS(5,3)	4,5			
23	RCMS(5,3)	5			
24	RCMS(5,3)	6			
25	RCMS(5,3)	7			

the RCMS performance depends on the choose of the threshold values.

Table 8 reports the variation of performance observed in correspondence of three values of inter-arrival times (5, 6, and 7). The percentages always refer to the comparison with the classical CMS.

Table 5. Average utilizations

Configuration	Inter-arrival time	Av. utilization
CMS	4	0,99
	4,5	0,88
	5	0,80
	6	0,66
	7	0,57

Among the various configurations of RCMS is showed only one (with thresholds 2, 1) with the most interesting results (see Table 8). Except for value of tardiness (when inter-arrival time is equal to 5) the other performance converge to values close to CMS configuration with differences about 10%. The better performance of RCMS is obtained with inter-arrival time equal to 5 therefore with a medium –high average utilization of the manufacturing system (see Table 5). With high and low congestion levels the other configurations compared to CMS have very

Table 6. Simulation results

	WIP		Tardiness		Throughput	
	average	St. dev	average	St. dev	average	St. dev
CMS(in line)	2,15%	1,62%	85,97%	179,28%	0,01%	0,19%
FCMS	495,96%	699,08%	956,98%	1583,56%	-4,49%	6,74%
RCMS 7,5	62,55%	64,65%	148,50%	49,65%	-0,77%	1,66%
RCMS 5,3	76,76%	91,60%	136,93%	49,38%	-0,99%	2,10%
RCMS 3,2	107,54%	118,07%	134,58%	71,72%	18,64%	45,45%
RCMS 4,1	95,70%	133,95%	133,78%	96,78%	-1,47%	2,92%
RCMS 3,1	132,86%	170,05%	191,64%	237,27%	-1,62%	3,36%
RCMS 2,1	203,37%	265,32%	315,70%	514,08%	-2,21%	3,81%

Table 7. Simulation results

	Thr. Time 1		Thr. Time 2		Thr. Time 3		Average Thr. Time	
	average	St. dev	average	St. dev	average	St. dev	average	St. dev
CMS(in line)	3,27%	1,24%	2,21%	2,68%	0,42%	0,79%	2,17%	1,58%
FCMS	551,65%	775,44%	547,81%	770,79%	352,77%	508,21%	496,34%	699,14%
RCMS 7,5	76,62%	63,94%	75,71%	63,20%	-12,66%	13,99%	53,20%	43,86%
RCMS 5,3	88,43%	87,56%	87,40%	86,09%	-7,56%	5,58%	62,97%	63,25%
RCMS 3,2	113,21%	101,38%	112,44%	100,65%	17,07%	44,23%	87,72%	78,76%
RCMS 4,1	101,34%	117,21%	100,78%	116,89%	-0,88%	16,07%	74,25%	89,56%
RCMS 3,1	139,38%	162,54%	136,99%	159,71%	12,55%	31,48%	104,75%	125,62%
RCMS 2,1	208,51%	275,98%	205,19%	272,66%	38,91%	75,03%	161,62%	218,90%

Table 8. Simulation results: arrival comparison

	Inter-arrival time	WIP	Thr. time 1	Thr. time 2	Thr. time 3	Average Thr. Time	Tardiness	Throughput
CMS in line	5	3,20%	4,41%	3,21%	0,83%	3,08%	7,56%	0,17%
	6	3,08%	3,72%	4,00%	0,02%	2,94%	5,88%	0,22%
	7	2,56%	3,56%	3,51%	0,15%	2,77%	4,80%	-0,23%
FCMS	5	65,16%	77,98%	76,58%	29,02%	64,87%	247,06%	0,08%
	6	13,64%	19,60%	19,42%	-4,64%	13,72%	22,65%	-0,10%
	7	13,23%	17,97%	17,96%	0,26%	13,92%	22,87%	-0,58%
RCMS 2,1	5	18,39%	31,34%	30,25%	-18,17%	18,25%	63,03%	0,08%
	6	7,95%	14,47%	14,27%	-11,52%	8,18%	6,19%	-0,20%
	7	9,17%	13,43%	13,43%	-5,43%	9,14%	12,61%	0,12%

low performance level. This is confirmed in Figure 13 that shows the profile of performance at different congestion levels.

Figures 14 and 15 show the comparison of the performance measures. It is clear that FCMS configuration in all cases performs worse especially for average inter-arrival time equal to 5. The design of this configuration needs to be re-thought. For higher inter-arrival times the differences tend to decline. The behaviour of RCMS is more interesting and there is more possibility for improvement. In Figure 13 observing the curve of RCMS (2,1), it is interesting to note that the throughput time of product 3 performs better than other configurations. This is probably due to the fact that the cell 3 has lower loads (since part mix 3 is 20%) and it obtains more synergy from the remainder cell. In that configuration queues larger than 2 units (parts) are not tolerated. In this case, the remainder cell is used frequently and this is the key to a better behaviour of system configuration. The results showed indicate that a better balance of utilizations between dedicated cells and remainder cell leads to an improvement in performance.

CONCLUSION AND FUTURE DEVELOPMENT

This chapter investigates several configurations of the cellular manufacturing systems. A simulation environment is used to create equal operating conditions for different cellular systems. Each simulation includes the same number of machines. Thus the comparison between systems is normalized. Volume changes are analysed changing of inter-arrival times. It has been considered interesting to compare the performance because the economic environment is extremely turbulent. In particular our attention has focused on alternative approaches to traditional cells. A solution that looks interesting results is the remainder cellular manufacturing system (RCMS). The results of this research can be summarized as it follows:

- the classical cellular configuration with machines placed in line (CMS in line) is the best solution with static market conditions; the results are very close to the case of machines that are not in line (CMS);
- the fractal cellular configuration(FCMS) gives bad results as it is conceived in a static environment and should think a more complex logic with different loading policies;
- the cellular manufacturing system with remainder cell (RCMS) is already competitive in some cases with larger inter-arrival times; the best configuration is one that requires more stringent threshold values which imply a greater use of the remainder cell.

From this it follows that RCMS could become very competitive when the presence of a turbulent market would involve a greater use of remainder cell, and similarly the presence of noise on the manufacturing system(such as machine breakdowns or maintenance).

In literature, it is known that the FCMS and the RCMS are not very competitive against classical CMS.

But in previous studies remainder cells were often used as support cells with exclusive use in special circumstances. Our proposal is to adopt loading policies designed to achieve a strategic use of the remainder cells. Simple loading policies included in simulation models show how the remainder cell can be used to keep different performance under certain conditions. This work aims to demonstrate under certain dynamic conditions the proposed configurations can be competitive with classical CMS. Furthermore this chapter

Figure 13. Performance comparison: RCMS

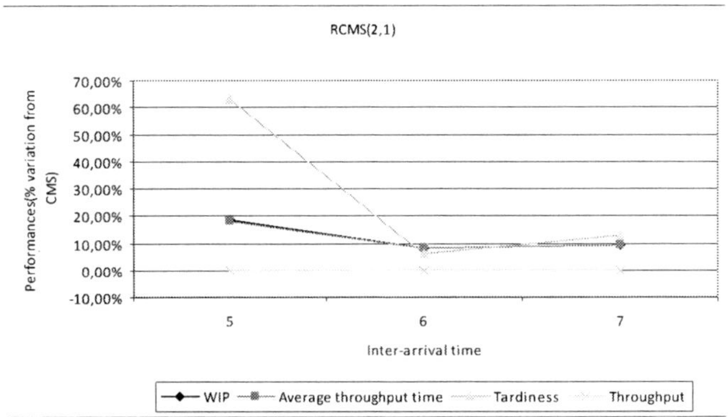

Figure 14. Performance comparison: interarrival time

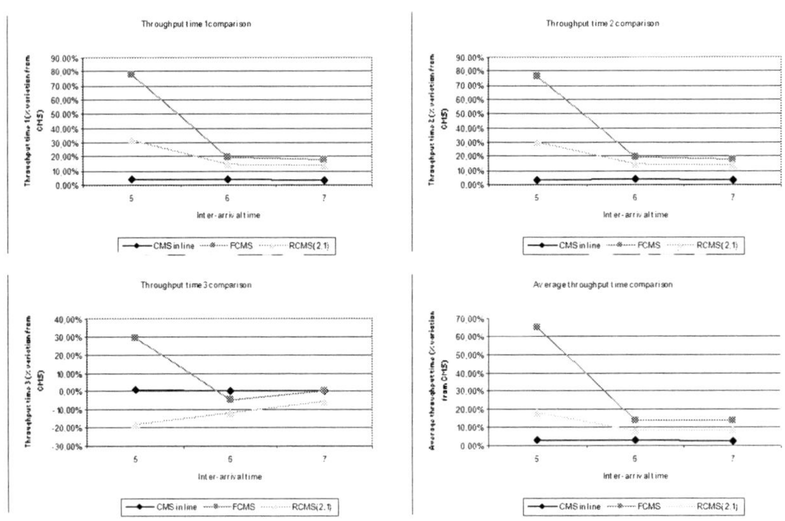

Figure 15. Performance comparison: interarrival time

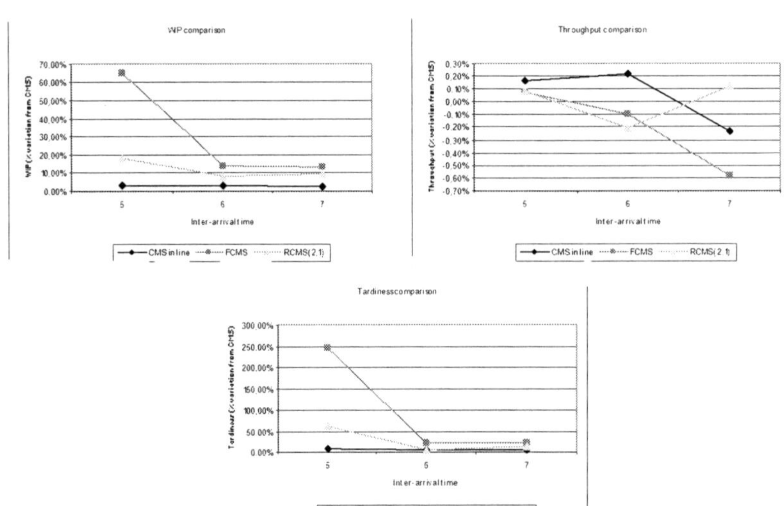

demonstrates the strong dependence of the results from the design of loading approaches, which deserve special attention.

Future research could focus on defining complex loading policies able to maintain high performance of the manufacturing system in different operating conditions and also taking into account the need for maintenance and possible failures of the machines(also for those belonging to remainder cell). These policies will certainly improve both the RCMS as for FCMS.

Moreover in the RCMS the logic of loading machines have a strong influence on the performance. Under dynamic conditions with market fluctuations these strategies using remainder cells can avoid the reconfigurations of manufacturing systems, avoiding downtimes and reducing costs.

Future works could investigate a variety of systems that integrate the configurations showed in this chapter with decision-making systems with intelligence to interpret the variability of real production scenario, moreover would also be interesting to analyze the economic aspect of different manufacturing solutions and how it may influence the choices.

REFERENCES

Ahkioon, S., Bulgak, A. A., & Bektas, T. (2009). Cellular manufacturing systems design with routing flexibility, machine procurement, production planning and dynamic system reconfiguration. *International Journal of Production Research*, *47*(6), 1573–1600. doi:10.1080/00207540701581809

Aramoon Bajestani, M., Rabbani, M., Rahimi Vahed, A. R., & Baharian Khoshkhou, G. (2009). A multiobjective scatter search for a dynamic cell formation problem. *Computers & Operations Research*, *36*, 777–794. doi:10.1016/j.cor.2007.10.026

Aryanezhad, M. B., & Deljoo, V., Mirzapour Al-e-hashem, S. M. J. (2009). Dynamic cell formation and the worker assignment problem: A new model. *International Journal of Advanced Manufacturing Technology*, *41*, 329–342. doi:10.1007/s00170-008-1479-4

Chen, C. H., & Balakrishnan, J. (2005). Dynamic cellular manufacturing under multiperiod planning horizons. *Journal of Manufacturing Technology Management*, *16*(5), 516–530. doi:10.1108/17410380510600491

Chen, C. H., & Balakrishnan, J. (2007). Multiperiod planning and uncertainty issues in cellular manufacturing: A review and future directions. *European Journal of Operational Research*, *177*, 281–309. doi:10.1016/j.ejor.2005.08.027

Hachicha, W., Masmoudi, F., & Haddar, M. (2007). An improvement of a cellular manufacturing system design using simulation analysis. *International Journal of Simulation Modelling*, *4*(6), 193–205. doi:10.2507/IJSIMM06(4)1.089

Irani, S. A., Subramanian, S., & Allam, Y. S. (1999). Introduction to cellular manufacturing system. In Irani, S. A. (Ed.), *Handbook of cellular manufacturing systems* (pp. 29–30). John Wiley & Sons. doi:10.1002/9780470172476.ch

Kelton, W. D., & Sadowski, R. P. (2009). *Simulation with Arena*. McGraw-Hill.

Kesen, S. E., Sanchoy, K., & Gungor, Z. (2010). A genetic algorithm based heuristic for scheduling of virtual manufacturing cells (VMCs). *Computers & Operations Research*, *37*, 1148–1156. doi:10.1016/j.cor.2009.10.006

Kesen, S. E., Toksari, M. D., Gungor, Z., & Guner, E. (2009). Analyzing the behaviors of virtual cells (VCs) and traditional manufacturing systems: Ant colony optimization (ACO)-based metamodels. *Computers & Operations Research*, *36*(7), 2275–2285. doi:10.1016/j.cor.2008.09.002

Maddisetty, S. (2005). *Design of shared cells in a probabilistic demand environment.* PhD Thesis. College of Engineering and Technology of Ohio University, Ohio(USA).

Montreuil, B. (1999). Fractal layout organization for job shop environments. *International Journal of Production Research*, *37*(3), 501–521. doi:10.1080/002075499191643

Rheault, M., Drolet, J., & Abdulnour, G. (1995). Physically reconfigurable virtual cells: A dynamic model for a highly dynamic environment. *Computers & Industrial Engineering*, *29*(1–4), 221–225. doi:10.1016/0360-8352(95)00075-C

Safei, N., Saidi-Mehrabad, M., & Babakhani, M. (2007). Designing cellular manufacturing systems under dynamic and uncertain conditions. *Journal of Intelligent Manufacturing*, *18*, 383–399. doi:10.1007/s10845-007-0029-5

Safei, N., & Tavakkoli-Moghaddam, R. (2009). Integrated multi-period cell formation and subcontracting production planning in dynamic cellular manufacturing systems. *International Journal of Production Economics*, *120*, 301–314. doi:10.1016/j.ijpe.2008.12.013

Selim, M. S., Askin, R. G., & Vakharia, A. J. (1998). Cell formation in group technology: Review evaluation and directions for future research. *Computers & Industrial Engineering*, *34*(1), 3–20. doi:10.1016/S0360-8352(97)00147-2

Süer, G. A., Huang, J., & Maddisetty, S. (2010). Design of dedicated, shared and remainder cells in a probabilistic demand environment. *International Journal of Production Research*, *48*(19), 5613–5646. doi:10.1080/00207540903117865

Vakharia, A. J., Moily, J., & Huang, Y. (1999). Evaluating virtual cells and multistage flow shops: An analytical approach. *International Journal of Flexible Manufacturing Systems*, *11*, 291–314. doi:10.1023/A:1008117329327

Venkatadri, U., Rardin, R. L., & Montreuil, B. (1997). A design methodology for fractal layout organization. *IIE Transactions*, *29*, 911–924. doi:10.1080/07408179708966411

Wang, X., Tang, J., & Yung, K. (2009). Optimization of the multi-objective dynamic cell formation problem using a scatter search approach. *International Journal of Advanced Manufacturing Technology*, *44*, 318–329. doi:10.1007/s00170-008-1835-4

Chapter 18
Petri Net Model Based Design and Control of Robotic Manufacturing Cells

Gen'ichi Yasuda
Nagasaki Institute of Applied Science, Japan

ABSTRACT

The methods of modeling and control of discrete event robotic manufacturing cells using Petri nets are considered, and a methodology of decomposition and coordination is presented for hierarchical and distributed control. Based on task specification, a conceptual Petri net model is transformed into the detailed Petri net model, and then decomposed into constituent local Petri net based controller tasks. The local controllers are coordinated by the coordinator through communication between the coordinator and the controllers. Simulation and implementation of the control system for a robotic workcell are described. By the proposed method, modeling, simulation, and control of large and complex manufacturing systems can be performed consistently using Petri nets.

INTRODUCTION

Manufacturing systems, where the materials which are handled are mainly composed of discrete entities, for example parts that are machined and/or assembled, are called discrete manufacturing systems. Due to its complexity, manufacturing system control is commonly decomposed into a hierarchy of abstraction levels: planning, scheduling, coordination and local control. Each level operates on a certain time horizon. The planning level determines at which time each product will be introduced in the manufacturing system. The scheduling level produces a sequence of times for the execution of each operation on each machine or a total ordering of all the operations. The

DOI: 10.4018/978-1-61350-047-7.ch018

coordination level updates the state representation of the manufacturing system in real-time, supervises it and makes real-time decisions. The local control level implements the real-time control of machines and devices etc., interacting directly with the sensors and actuators. All the emergency procedures are implemented at this level, so real-time constraints may be very hard. At each level, any modeling has to be based on the concepts of discrete events and states, where an event corresponds to a state change (Martinez, 1986), (Silva, 1990).

A flexible manufacturing system is formed of a set of flexible machines, an automatic transport system, and a sophisticated decision making system to decide at each instant what has to be done and on which machine. A manufacturing cell is an elementary manufacturing system consisting of some flexible machines (machine tools, assembly devices, or any complex devices dedicated to complex manufacturing operations), some local storage facilities for tools and parts and some handling devices such as robots in order to transfer parts and tools. Elementary manufacturing cells are called workstations. At the local control level of manufacturing cells many different kinds of machines can be controlled, and specific languages for different application domains are provided; for example, block diagrams for continuous process control and special purpose languages for CNC or robot programming. For common sequential control, special purpose real-time computers named Programmable Logic Controllers (PLCs) are used. PLCs are replacements for relays, but they incorporate many additional and complex functions, such as supervisory and alarm functions and start-up and shut-down operations, approaching the functionalities of general purpose process computers. The most frequent programming languages are based on ladder or logic diagrams and boolean algebra. However, when the local control is of greater complexity, the above kinds of languages may not be well adapted. The development of industrial techniques makes a sequential control system for manufacturing cells more large and complicated one, in which some subsystems operate concurrently and cooperatively. Conventional representation methods based on flowcharts, time diagrams, state machine diagrams, etc. cannot be used for such systems.

To realize control systems for flexible manufacturing cells, it is necessary to provide effective tools for describing process specifications and developing control algorithms in a clear and consistent manner. In the area of real-time control of discrete event manufacturing cells the main problems that the system designer has to deal with are concurrency, synchronization, and resource sharing problems. For this class of problems, Petri nets have intrinsic favorable qualities and it is very easy to model sequences, choices between alternatives, rendezvous and concurrent activities by means of Petri nets (Reisig, 1985). When using Petri nets, events are associated with transitions. Activities are associated to the firing of transitions and to the markings of places which represent the states of the system. The network model can describe the execution order of sequential and parallel tasks directly without ambiguity (Murata, et al. 1986), (Crockett, et al. 1987). Moreover, the formalism allowing a validation of the main properties of the Petri net control structure (liveness, boundedness, etc.) guarantees that the control system will not fall immediately in a deadlocked situation. In the field of flexible manufacturing cells, the last aspect is essential because the sequences of control are complex and change very often. Furthermore, a real-time implementation of the Petri net specification by software called a token player can avoid implementation errors, because the specification is directly executed by the token player and the implementation of these control sequences preserves the properties of the model (Bruno, 1986). In this approach, the Petri net model is stored in a database and the token player updates the state of the database according to the operation rules of the model. For control purposes, this solution is very well suited to the need of flexibility, because, when the control sequences change, only the database needs to be changed(Silva, et al. 1982), (Valette, et al. 1983).

In addition to its graphic representation differentiating events and states, Petri nets allows the modeling of true parallelism and the possibility of progressive modeling by using stepwise refinements or modular composition. Libraries of well-tested subnets allow components reusability leading to significant reductions in the modeling effort. The possibility of progressive modeling is absolutely necessary for flexible manufacturing cells because they are usually large and complex systems. The refinement mechanism allows the building of hierarchically structured net models.

Some techniques derived from Petri nets have been successfully introduced as an effective tool for describing control specifications and realizing the control in a uniform manner. However, in the field of flexible manufacturing cells, the network model becomes complicated and it lacks the readability and comprehensibility. Therefore, the flexibility and expandability are not satisfactory in order to deal with the specification change of the control system. Despite the advantages offered by Petri nets, the synthesis, correction, updating, etc. of the system model and programming of the controllers are not simple tasks (Desrochers, et al. 1995), (Lee, et al. 2006).

In this chapter, a Petri net based specification and real-time control method for large and complex manufacturing cells is presented. Based on the hierarchical and distributed structure of the manufacturing cell, the specification procedure is a top-down approach from the conceptual level to the detailed level such that the macro representation of the system is broken down to generate the detailed Petri nets at the local machine control level. Then the Petri nets are decomposed and assigned to the machine controllers to perform distributed control using Petri net based multitask processing. An algorithm is proposed for coordination of machine controllers. By the proposed method, modeling, simulation and control of large and complex manufacturing cells can be performed consistently using Petri nets.

MODELING OF MANUFACTURING CELLS USING MODIFIED PETRI NETS

A manufacturing process is characterized by the flow of workpieces or parts, which pass in ordered form through subsystems and receive appropriate operations. Each subsystem executes manufacturing operations, that is, physical transformations such as machining, assembling, or transfer operations such as loading and unloading.

From the viewpoint of discrete event process control, an overall manufacturing process can be decomposed into a set of distinct activities (or events) and conditions mutually interrelated in a complex form. An activity is a single operation of a manufacturing process executed by a subsystem. A condition is a state in the process such as machine operation mode.

Considering the nature of discrete event manufacturing cells which are characterized by the occurrence of events and changing conditions, the condition-event net based specification method has been investigated. The Petri net is one of the effective means to represent condition-event systems. The specification method is a graphical model used as a tool to identify types of activities, conditions, and their mutual interrelation. It describes explicitly the concept of the manufacturing process to be carried out in the discrete event manufacturing cells.

Considering not only the modeling of the systems but also the actual well-designed control, the guarantee of safeness and the capability to represent input and output signals from and to the machines are required. In the condition-event systems, deadlock occurs when the system enters into a state that is not possible for any event to occur. Further bumping occurs when, despite the holding of a condition, the preceding event occurs. This can result in the multiple holding of that condition. Therefore, the basic Petri net which is called Place/Transition-net should be modified and extended in order to represent the activity contents and control strategies for the manufacturing system control in detail.

The extended Petri net consists of the following six elements: (1) Place, (2) Transition, (3) Directed arc, (4) Token, (5) Gate arc, (6) Output signal arc (Hasegawa, et al. 1984). A place represents a condition of a system element or action. A transition represents an event of the system. A directed arc connects from a place to a transition or from a transition to a place, and its direction shows the input and output relation between them. Places and transitions are alternately connected using directed arcs. The number of directed arcs connected with places or transitions is not restricted. A token is placed in a place to indicate that the condition corresponding to the place is holding.

A gate arc connects a transition with a signal source, and depending on the signal, it either permits or inhibits the occurrence of the event which corresponds to the connected transition. Gate arcs are classified as permissive or inhibitive, and internal or external. An output signal arc sends the signal from a place to an external machine. Thus a transition is enabled if and only if it satisfies all the following conditions:

1. It does not have any output place filled with a token.
2. It does not have any empty input place.
3. It does not have any internal permissive arc signaling 0.
4. It does not have any internal inhibitive arc signaling 1.

Figure 1 shows the place and gate variables for transition firing test.

Formally, the enabling condition and the external gate condition of a transition *j* are described using the logical place and gate variables as follows:

$$t_j(k) = \bigcap_{m=1}^{M} p_{j,m}^{I}(k) \wedge \bigcap_{n=1}^{N} \overline{p_{j,n}^{O}(k)} \wedge \bigcap_{q=1}^{Q} g_{j,q}^{IP}(k) \wedge \bigcap_{r=1}^{R} \overline{g_{j,r}^{II}(k)} \quad (1)$$

$$g_j^{E}(k) = \bigcap_{u=1}^{U} g_{j,u}^{EP}(k) \wedge \bigcap_{v=1}^{V} \overline{g_{j,v}^{EI}(k)} \quad (2)$$

where,

- M: set of input places of transition j
- $p_{j,m}^{I}(k)$: state of input place m of transition j at time sequence k
- N: set of output places of transition j
- $p_{j,n}^{O}(k)$: state of output place n of transition j at time sequence k
- Q: set of internal permissive gate signals of transition j

Figure 1. Place and gate variables for transition firing test

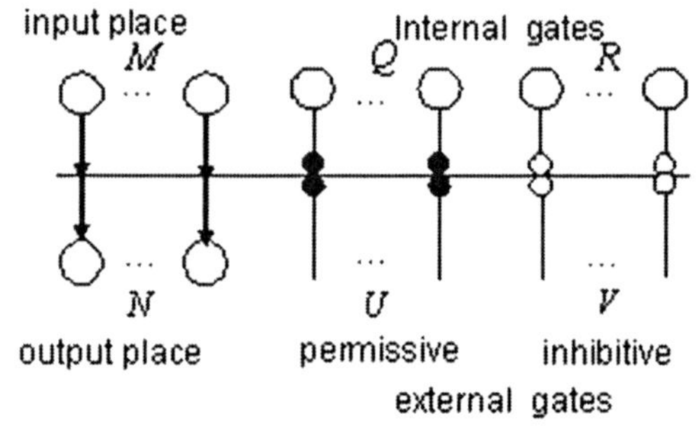

- $g_{j,q}^{IP}(k)$: internal permissive gate signal variable q of transition j at time sequence k
- R: set of internal inhibitive gate signals of transition j
- $g_{j,r}^{II}(k)$: internal inhibitive gate signal variable r of transition j at time sequence k
- U: set of external permissive gate signals of transition j
- $g_{j,u}^{EP}(k)$: external permissive gate signal variable u of transition j at time sequence k
- V: set of external inhibitive gate signals of transition j
- $g_{j,v}^{EI}(k)$: external inhibitive gate signal variable v of transition j at time sequence k

The state (marking) change, that is, the addition or removal of a token of an input or output place, is described as follows:

$$p_{j,m}^{I}(k+1) = p_{j,m}^{I}(k) \wedge \overline{(t_j(k) \wedge g_j^{E}(k))} \quad (3)$$

$$p_{j,n}^{O}(k+1) = p_{j,n}^{O}(k) \vee (t_j(k) \wedge g_j^{E}(k)) \quad (4)$$

An enabled transition may fire when it does not have any external permissive arc signaling 0 nor any external inhibitive arc signaling 1. The firing of a transition removes a token from each input place and put a token in each output place connected to it. The assignment of tokens into the places of a Petri net is called marking and it represents the system state. In any initial marking, there must not exist more than one token in a place. According to these rules, the number of tokens in a place never exceeds one, thus the Petri net is essentially a safe graph; the system is free from the bumping phenomenon.

For the actual control, the operations of each machine are broken down into a series of unit motions, which is represented by mutual connection between places and transitions. A place means a concrete unit motion of a machine. From these places, output signal arcs are connected to the machines, and external gate arcs from the machines are connected to the transitions of the Petri net when needed, for example, to synchronize and coordinate operations. When a token enters a place that represents a subtask, the machine defined by a machine code is informed to execute a specified subtask with positional data and control parameters; all the code and data are defined as the place parameters.

If a place has two or more input or output transitions, these transitions may be in conflict for firing. When two or more transitions are enabled only one transition should fire using gate arcs or some arbitration rule. The Petri net described in detail by such a procedure mentioned above can be used as a program for the system control, while features of discrete event manufacturing cells such as ordering, parallelism, asynchronism, concurrency and conflict can be concretely described through the extended Petri net.

The extended Petri net is a tool for the study of condition-event systems and used to model condition-event systems through its graphical representation. Analysis of the net reveals important information about the structure and the dynamic behavior of the modeled condition-event system. This information can then be used to evaluate the modeled condition-event system and suggest improvements or changes.

DESIGN OF HIERARCHICAL AND DISTRIBUTED CONTROL SYSTEM

A specification procedure for discrete event manufacturing cells based on Petri nets is as follows. First, the conceptual level activities of the discrete event manufacturing cells are defined through a Petri net model considering the task specification corresponding to the manufacturing process. A conceptual Petri net model describes the aggregate manufacturing process. At the level, each subtask composing the task specification is represented as a place of the Petri net, where the activity of each equipment is also represented as a place. Then, the detailed Petri nets describing the activities are deduced based on activity specification and required control strategies. The macro representation of the manufacturing process is effectively used for achieving a top-down interpretation down to the concrete lower level activities using Petri nets. Based on the hierarchical approach, the Petri net is translated into the detailed Petri net by stepwise refinements from the highest conceptual level to the lowest machine control level. This procedure is repeated up to an appropriate level corresponding to the control level of the equipment responsible for the activity execution. At each step of detailed specification, places of the Petri net are substituted by a subnet in a manner which maintains the structural properties (Miyagi, 1988). The overall procedure of the Petri net based implementation of hierarchical and distributed control for robotic manufacturing cells is summarized as shown in Figure 2.

It is natural to implement a hierarchical and distributed control system, where one controller is allocated to each control layer or block. For the robotic manufacturing cells composed of robots, machine tools, and conveyors, an example structure of hierarchical and distributed control is composed of one station controller and three machine controllers as shown in Figure 3, although each robot may be controlled by one robot controller. The detailed Petri net is decomposed into subnets, which are assigned to machine controllers.

In the decomposition procedure, a transition may be divided and distributed into different machine controllers as shown in Figure 4. The machine controllers should be coordinated so that these transitions fire simultaneously, that is, the aggregate behavior of decomposed subnets should be the same as that of the original Petri net. Decomposed transitions are called global transitions, and other transitions are called local transitions.

Figure 2. Flow chart of Petri net based implementation of hierarchical and distributed control system

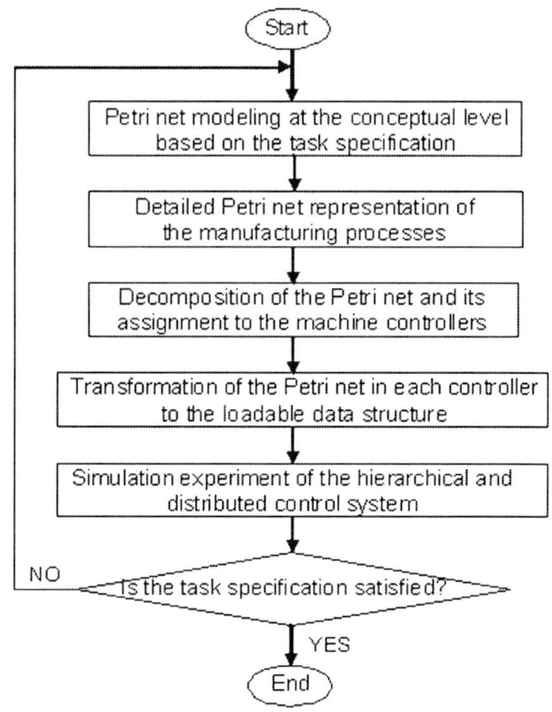

By the Petri net model, the state of the discrete event system is represented as the marking of tokens, and firing of any transition brings about change to the next state. So the firing condition and state (marking) change before decomposition should be the same as those after decomposition. If transition *j* is divided into *s* transitions *j1*, *j2*,,, *js*, as shown in Figure 5, the firability condition of the transition after decomposition is described as follows:

$$t_{jsub}(k) = \bigcap_{m=1}^{Msub} p^{I}_{jsub,m}(k) \wedge \overline{\bigcap_{n=1}^{Nsub} p^{O}_{jsub,n}(k)} \wedge \bigcap_{q=1}^{Qsub} g^{IP}_{jsub,q}(k) \wedge \overline{\bigcap_{r=1}^{Rsub} g^{II}_{jsub,r}(k)} \quad (5)$$

$$g^{E}_{jsub}(k) = \bigcap_{u=1}^{Usub} g^{EP}_{jsub,u}(k) \wedge \overline{\bigcap_{v=1}^{Vsub} g^{EI}_{jsub,v}(k)} \quad (6)$$

Figure 3. Example structure of distributed control system

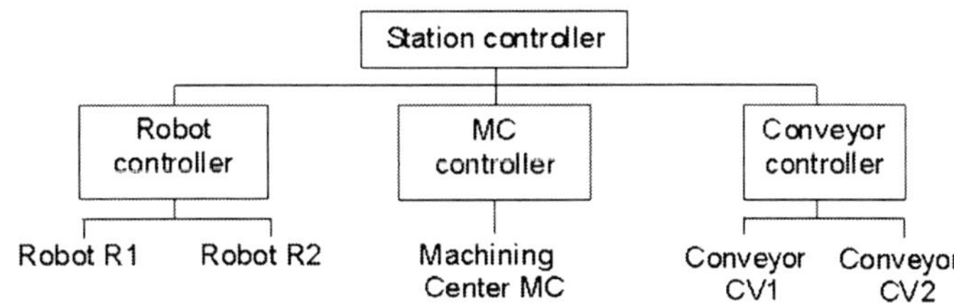

From Equation 1 and Equation 5,

$$t_j(k) = \bigcap_{sub=1}^{S} t_{jsub}(k) \tag{7}$$

From Equation 2 and Equation 6,

$$g_j^E(k) = \bigcap_{sub=1}^{S} g_{jsub}^E(k) \tag{8}$$

where,

- *S*: total number of subnets
- *Msub*: set of input places of transition *jsub* of subnet *sub*
- $p_{jsub,m}^I(k)$: state of input place *m* of transition *jsub* of subnet *sub* at time sequence *k*
- *Nsub*: set of output places of transition *jsub* of subnet *sub*
- $p_{jsub,n}^O(k)$: state of output place *n* of transition *jsub* of subnet *sub* at time sequence *k*
- *Qsub*: set of internal permissive gate signals of transition *jsub* of subnet *sub*
- *Rsub*: set of internal inhibitive gate signals of transition *jsub* of subnet *sub*
- *Usub*: set of external permissive gate signals of transition *jsub* of subnet *sub*
- *Vsub*: set of external permissive gate signals of transition *jsub* of subnet *sub*

The addition or removal of a token of a place connected to a decomposed transition is described as follows:

Figure 4. Decomposition of transitions

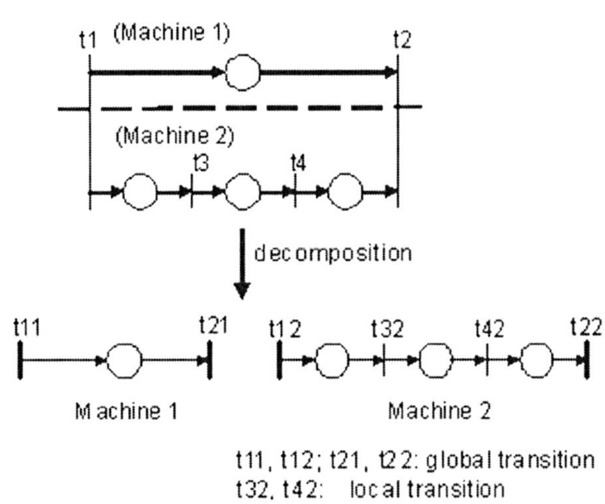

Figure 5. Place and gate variables after decomposition of transition

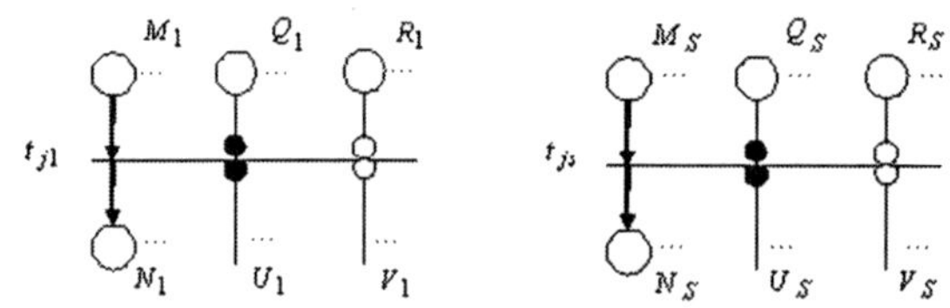

$$p^{I}_{jsub,m}(k+1) = p^{I}_{jsub,m}(k) \wedge \overline{(t_j(k) \wedge g^{E}_j(k))} \tag{9}$$

$$p^{O}_{jsub,n}(k+1) = p^{O}_{jsub,n}(k) \vee (t_j(k) \wedge g^{E}_j(k)) \tag{10}$$

Consequently it is proven that the firability condition of the original transition is equal to AND operation of firability conditions of decomposed transitions. If and only if all of the decomposed transitions are firable, then the global transitions are firable. To utilize the above results, the coordinator program has been introduced to coordinate the decomposed subnets so that the aggregate behavior of decomposed subnets is the same as that of the original Petri net.

There may exist a place which has several input transitions and/or several output transitions. This place is called a conflict place. The transitions connected to a conflict place are in conflict when some of them are firable at the same time. In this case, only one of them can be fired and the others become disabled. The selection of firing transition is done arbitrarily using an arbiter program.

In case that a transition in conflict with other transitions is decomposed as shown in Figure 6, these transitions should be coordinated by the system controller. If arbitrations of the transitions are performed independently in separate subnets, the results may be inconsistent with the original rule of arbitration. Therefore the transitions should be arbitrated together as a group. On the other hand, arbitration of local transitions in conflict is performed by local machine controllers.

The hierarchical and distributed control system composed of one station controller and several machine controllers has been implemented. The conceptual Petri net model is allocated to the Petri net based controller for management of the overall system. The detailed Petri net models are allocated to the Petri net based controllers in the machine controllers. Each machine controller directly monitors and controls the sensors and actuators of its machine.

The control of the overall system is achieved by coordinating these Petri net based controllers. Figure 7 shows the Petri net based control structure with the coordinator. System coordination is performed through communication between the

Figure 6. Decomposition of transition in conflict

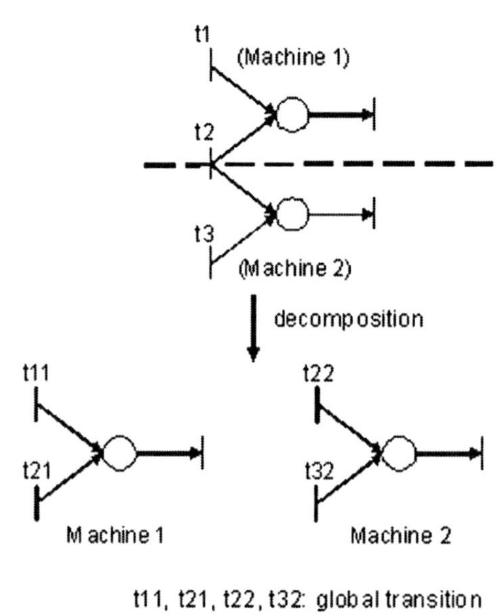

coordinator in the station controller and the Petri net based controllers in the machine controllers as the following steps.

1. When each machine controller receives the start signal from the coordinator, it tests the firability of all transitions in its own Petri net, and sends the information on the global transitions and the end signal to the coordinator.
2. The coordinator tests the firability of the global transitions, arbitrates conflicts among global transitions, and sends the names of firing global transitions and the end signal to the machine controllers.
3. Each machine controller arbitrates conflicts among local transitions using the information from the coordinator, generates a new marking, and sends the end signal to the coordinator.
4. When the coordinator receives the end signal from all the machine controllers, it sends the output command to the machine controllers.
5. Each machine controller outputs the control signals to its actuators simultaneously.

Figure 7. Petri net based control structure with coordinator

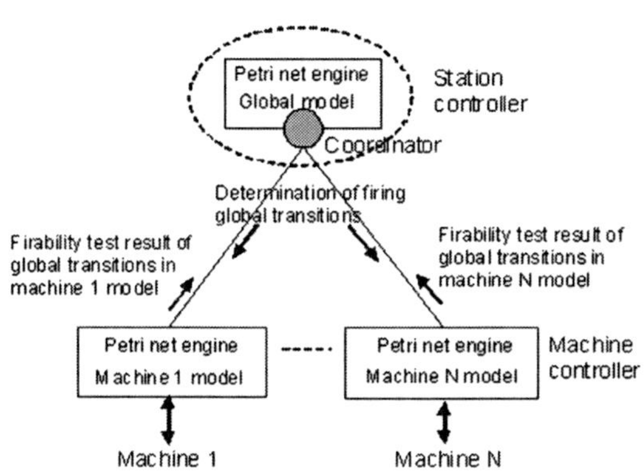

REAL-TIME CONTROL OF A MANUFACTURING CELL

The example manufacturing system has two robots, one machining center, and two conveyors, where one is for carrying in and the other is for carrying out, as shown in Figure 8. The main execution of the system is indicated as the following task specification:

1. A workpiece is carried in by the conveyor CV1.
2. The robot R1 loads the workpiece to the machining center MC.
3. The machining center MC processes the workpiece.
4. The robot R2 unloads the workpiece from the machining center and places it on the conveyor CV2.
5. The workpiece is carried out by the conveyor CV2.

A conceptual Petri net model is first chosen, which describes the aggregate manufacturing process. The places which represent the subtasks indicated as the task specification are connected by arcs via transitions in the specified order corresponding to the flow of subtasks and a workpiece. The places representing the machines are also added to connect transitions which correspond to the beginning and ending of their subtasks. Thus at the conceptual level the manufacturing process is represented as shown in Figure 9. In this step, if necessary, control conditions such as the capacity of the system between the respective subtasks must be connected to regulate the execution of the Petri net. For the cell with one robot, the place "Robot" has two input transitions and two output transitions, but these transitions are not firable at the same time, so they are not in conflict for firing. The firing of only one of these transitions is permitted using gate arcs (Yasuda, 2008). Next, each place representing a subtask at the conceptual

Figure 8. Example of robotic manufacturing system

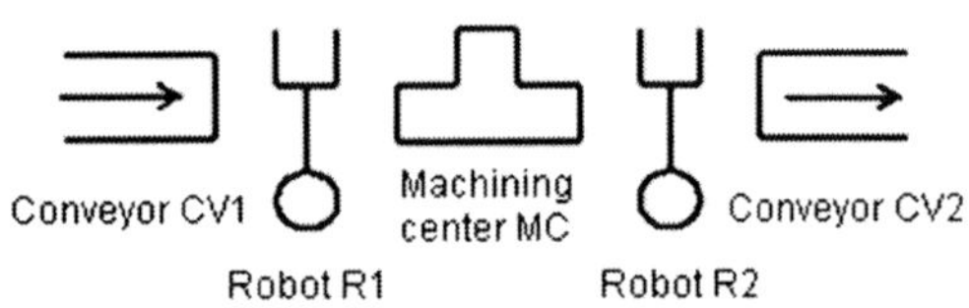

level is translated into a detailed subnet. Figure 10 shows the detailed Petri net representation of loading, processing and unloading in Figure 9.

For the example system, the hierarchical and distributed control system has been realized using a set of PCs. Each machine controller is implemented on a dedicated PC. The station controller is implemented on another PC. Communications among the controllers are performed using serial communication interfaces.

The names of global transitions and their conflict relations are loaded into the coordinator in the station controller. The connection structure of a decomposed Petri net model and conflict relations among local transitions are loaded into the Petri net based controller in the corresponding machine controller. In the connection structure, a transition of a Petri net model is defined using the names of its input places and output places; for example, t1-1=p1-1, -p1-11, where the transition no.1 (t1-1) of Robot controller (subsystem no.1) is connected to the input place no.1 and the output place no.11. For the distributed control system shown in Figure 3, the Petri net representations assigned to the machine controllers are shown in Figure 11. Structural information for the robot controller inputted to the loader is shown in Table 1.

Using the names of transitions in the subsystems, global transitions are defined; for example, G2: t0-2, t1-21, t2-22, t3-23 indicates that the global transition G2 is composed of the transition no.2 of Station controller (subsystem no.0), the transition no.21 of Robot controller, the transition no.22 of MC controller (subsystem no.2), and the transition no.23 of Conveyor controller (subsystem no.3). Then, the coordinator information for the example distributed control system is shown in Table 2.

By executing the coordinator and Petri net based controllers algorithms based on loaded information, simulation experiments have been performed. The robot controller executes robot motion control through the transmission of command. The MC controller and the conveyor controller communicate with a dedicated PLC. The Petri net simulator initiates the execution of the subtasks attached to the fired transitions through the serial interface to the robot or other external machine. When a task ends its activity, it informs the simulator to proceed with the next activations by the external permissive gate arc (Yasuda, 2010). The detailed Petri net representa-

Figure 9. Petri net representation of the example system at the conceptual level

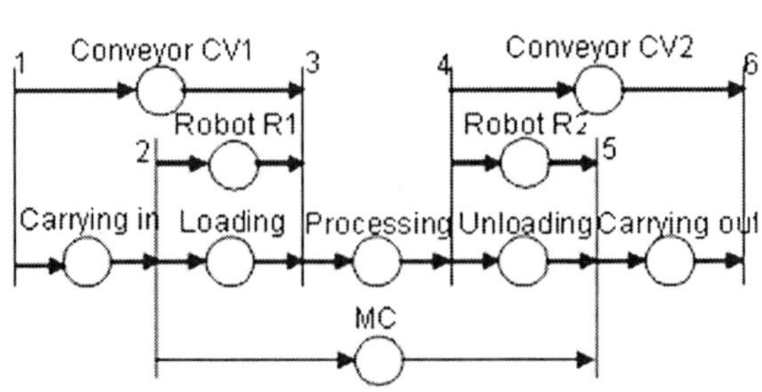

Figure 10. Detailed Petri net representation of subtasks

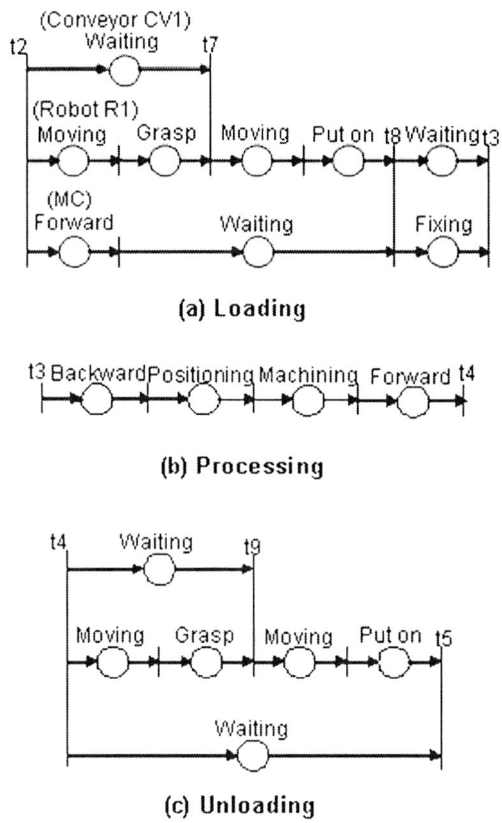

tion for real-time external machine control is shown in Figure 12. External permissive gate arcs from sensors for detecting the completion of work handling are employed as shown in Figure 13.

The machine controllers control two conveyors or robots, so control software on each PC is written using multithreaded programming. Petri net simulation and task execution program through serial interface are implemented as threads in each machine controller. Real-time task execution using multithreaded programming is shown in Figure 14.

In multithreaded system a thread is an independent flow of execution. All threads share code, data, stack, and system memory areas. So, all threads can access to all global data. Further, because each thread has a separate CPU state and its own stack memory, all local variables and function arguments are private to a specific thread. Context switching between threads involves simply saving the CPU state for the current thread and loading the CPU state for the new thread. It's easy for threads to interact with each other by way of synchronization objects and intertask communications, because they have some shared memory. Memory management data structures do not need to be changed, because the threads share the same memory areas (Grehan, et al. 1998).

In the implementation both the Petri net simulation and task execution threads access to the external gate variables as shared variables; the task execution thread writes the new values of the gate variables and the Petri net simulation thread reads them. Mutual exclusive access control is implemented, so that, while one thread accesses to the shared variables, the other thread can not access to them. Control software using multithreaded programming was written in Visual C# under OS Windows XP SP3 on a general PC. Experiments using a real industrial robot show that the Petri net simulation thread and the task execution threads proceed concurrently with even priority, and the values of external gate variables are changed successfully; after the task execution threads write the new values, the Petri net simulator thread reads them immediately.

Experimental results show that the decomposed transitions fire simultaneously as the original transition of the detailed Petri net of the whole system task. The robots cooperated with the conveyors and the machining center, and the example manufacturing cell performed the task specification successfully. Firing transitions and marking of tokens can be directly observed on the display at each time sequence using the simulator on each PC (Yasuda, 2009), as shown in Figure 15. The Petri net modeling thread is executed as the main thread to perform the modeling, drawing and modification of the overall system control net model based on task specification. When the

Figure 11. Petri net representation of machine controllers (: global transition, : local transition)

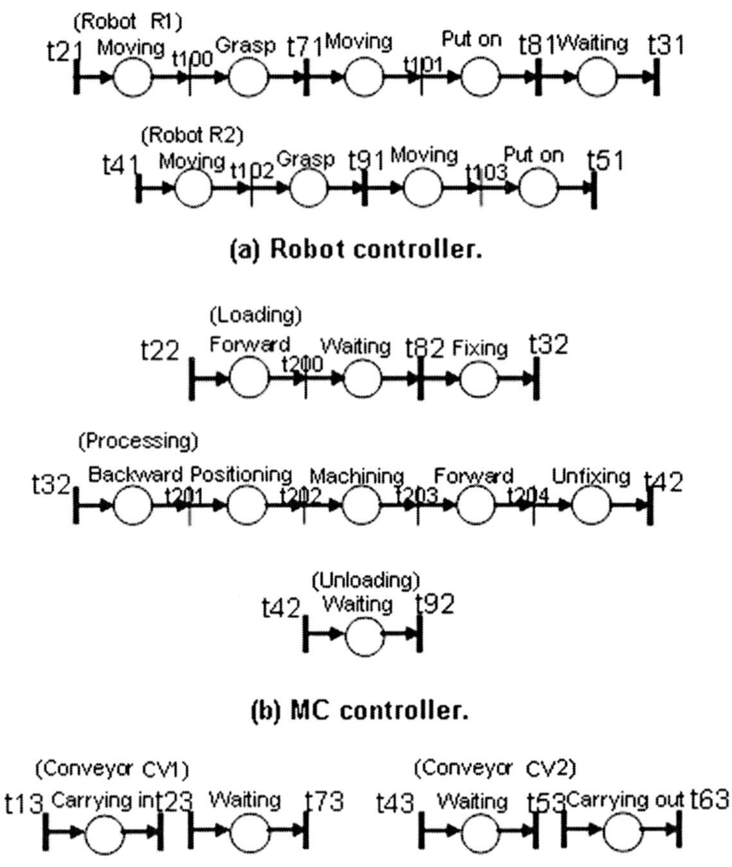

(a) Robot controller.

(b) MC controller.

(C) Conveyor controller.

Table 1. Structural information for the robot controller, as inputted to the loader

t1-21=-p1-11	t1-41=-p1-21
t1-100=p1-11, -p1-12	t1-102=p1-21, -p1-22
t1-71=p1-12, -p1-13	t1-91=p1-22, -p1-23
t1-101=p1-13, -p1-14	t1-103=p1-23, -p1-24
t1-81=p1-14, -p1-15	t1-51=p1-24
t1-31=p1-15	

Table 2. Coordinator information for the example distributed control system

G1: t0-1, t3-13	start of carrying in
G2: t0-2, t1-21, t2-22, t3-23	start of loading from CV1
G3: t1-71, t3-73	end of grasp on CV1
G4: t1-81, t2-82	end of putting on MC
G5: t0-3, t1-31, t2-32	end of loading into MC
G6: t0-4, t1-41, t2-42, t3-43	start of unloading from MC
G7: t1-91, t2-92	end of grasp on MC
G8: t0-5, t1-51, t3-53	end of putting on CV2
G9: t0-6, t3-63	end of carrying out

Figure 12. Detailed Petri net representation of real-time machine control

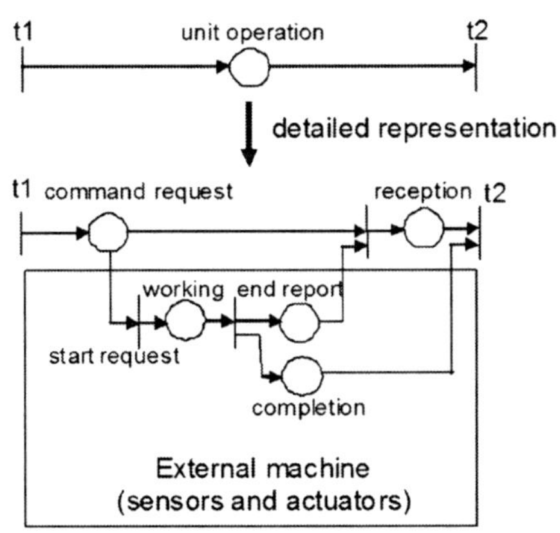

Figure 13. External permissive gate arcs from sensors for work handling

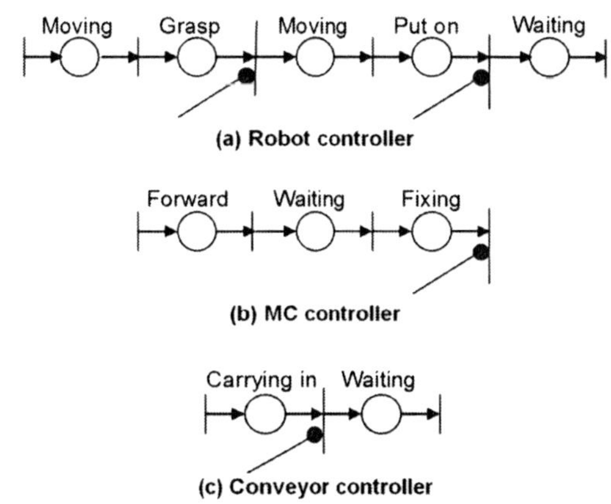

Figure 14. Real-time execution of subtasks using multithreaded programming

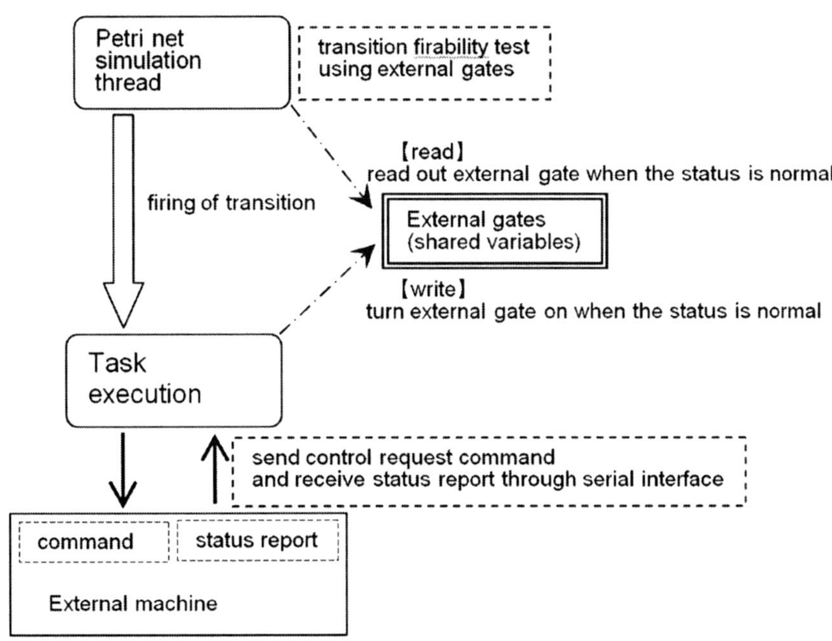

transformation of graphic data of the Petri net model into internal structural data is finished, the Petri net simulation thread starts. During simulation and task execution, liveness or deadlock is possibly decided, and the user can stop the task execution at any time.

CONCLUSION

A methodology to construct hierarchical and distributed control systems, which correspond to the hardware structure of manufacturing cells, has been proposed. By introduction of the coordinator, the Petri net based controllers are arranged

Figure 15. View of Petri net simulator

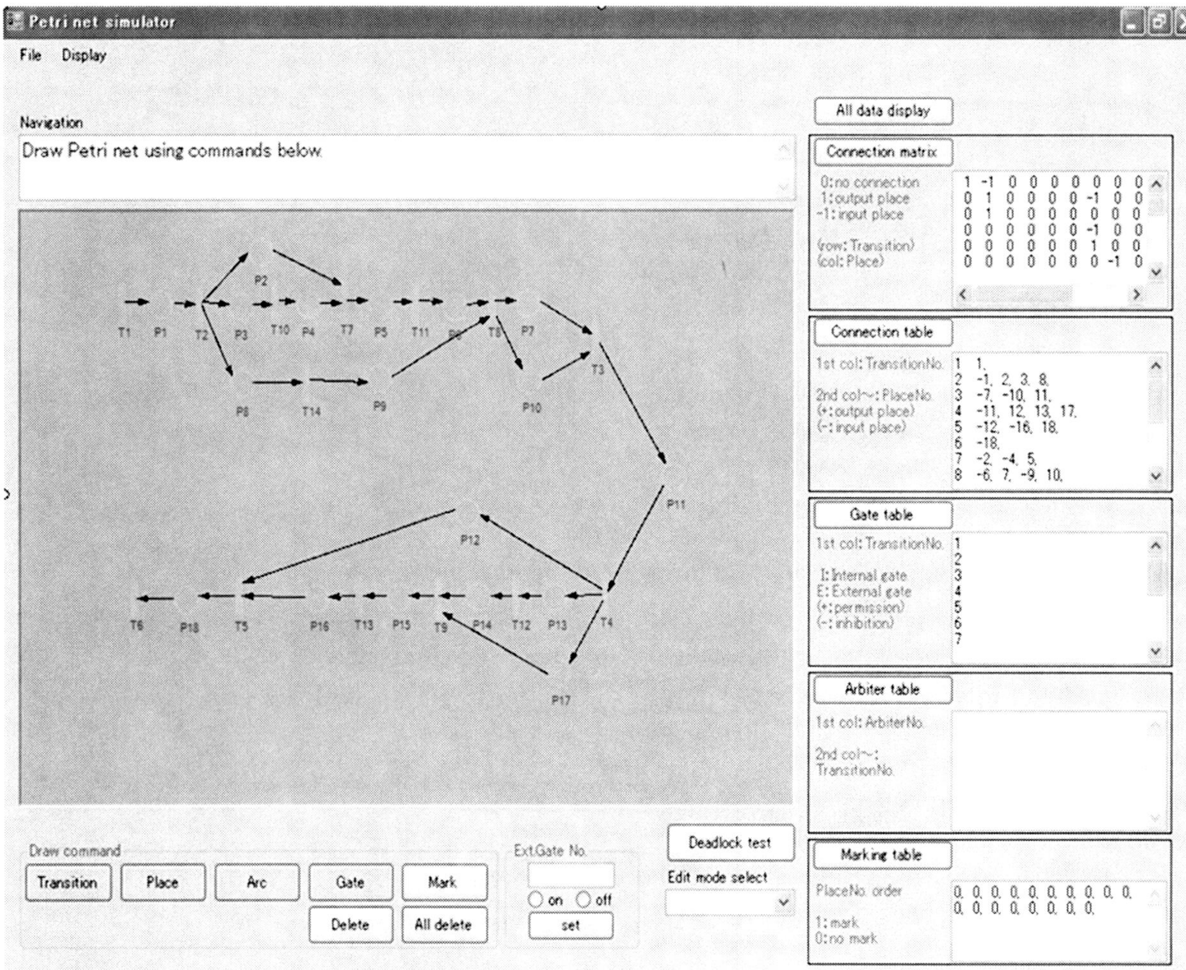

according to the hierarchical and distributed nature of the overall manufacturing system; the coordination mechanism is implemented in each layer repeatedly. The Petri net model in each Petri net based machine controller is not so large and easily manageable. The overall control structure of the example robotic manufacturing cell was implemented on a communication network of PCs using multithreaded programming. In accordance with the implementation using multithreaded programming, hierarchical and distributed implementation under a real-time operating system on a network of microcomputers connected via a serial bus is also possible, where each microcomputer is dedicated to the local Petri net model of a subsystem in the manufacturing cell. Thus, modeling, simulation and control of large and complex manufacturing systems can be performed consistently using Petri nets.

The proposed methodology has the following advantages especially for manufacturing cells composed of several parallel processes.

1. Because of decomposition of detailed Petri net model, design and implementation of lower level controllers can be performed efficiently for each subsystem. Modification or breakdown in a lower level controller is

restricted to the corresponding controller. Machine controllers can be realized using conventional relay circuits, PLCs, or microcomputers. From the practical point of view, transformation Petri nets into ladder diagrams, PLC or microcomputer programs, and inversely transformation them into Petri nets, are desirable.

2. By monitoring the flow of token in the upper level controller, the global state of the system can be conceptually understood.
3. By realizing the upper level controller using a general PLC or computer, the Petri net model (places and transitions) can be easily changed according to task specification.

From the proposed coordinator algorithm the coordination mechanism is performed with the master-slave multicomputer architecture; the upper level controller is the master and the lower level controllers are slaves. So access conflict and lock out do not occur, and both hardware and software structures can be simply realized, especially including non-homogeneous system. Although in the master-slave architecture of operating systems the performance of the upper level controller is required to be sufficiently high, there is no problem because robotic arms, conveyors and machine tools are very slow in comparison with the controller. In case of an accident or breakdown of the upper level controller, manual operation modes must be provided for the lower level controllers. In order to improve the Petri net based control system for manufacturing cells, future works include the following:

1. The number of token is increased and some kinds of tokens are provided in a place.
2. The time required for transition firing condition in a place is provided.

The above extensions of the Petri net is necessary to optimize the process control in a manufacturing cell using simulations without real machines. It is also possible to analyze deadlock phenomena in Petri net models based on detailed simulations.

REFERENCES

Bruno, G., & Marchetto, G. (1986). Process-translatable Petri nets for the rapid prototyping of process control systems. *IEEE Transactions on Software Engineering*, *SE-12*(2), 346–357.

Crockett, D., Desrochers, A., Dicesare, F., & Ward, T. (1987). Implementation of a Petri net controller for a machine workstation. *Proceedings of 1987 IEEE International Conference on Robotics and Automation*, (pp.1861-1867).

Desrochers, A. D., & Al-Jaar, R. Y. (1995). *Applications of Petri Nets in manufacturing systems: Modeling, control and performance analysis*. IEEE Press.

Grehan, R., Moote, R., & Cyliax, I. (1998). *Real-time programming A guide to 32-bit embedded development*. Reading, Massachusetts: Addison-Wesley.

Hasegawa, K., Takahashi, K., Masuda, R., & Ohno, H. (1984). Proposal of mark flow graph for discrete system control. *Trans. of SICE*, *20*(2), 122–129.

Lee, E. J., Togueni, A., & Dangoumau, N. (2006). A Petri net based decentralized synthesis approach for the control of flexible manufacturing systems. *Proceedings of the IMACS Multiconference Computational Engineering in Systems Applications*.

Martinez, J., Alla, H., & Silva, M. (1986). Petri nets for the specification of flexible manufacturing systems. In *Modeling and Design of Flexible Manufacturing Systems* (pp. 389–406). New York: Elsevier Science Publishing.

Miyagi, P. E., Hasegawa, K., & Takahashi, K. (1988). A programming language for discrete event production systems based on production flow schema and mark flow graph. *Trans. of SICE*, *24*(2), 183–190.

Murata, T., Komoda, N., Matsumoto, K., & Haruna, K. (1986). A Petri net based controller for flexible and maintainable sequence control and its application in factory automation. *IEEE Transactions on Industrial Electronics*, *33*(1), 1–8. doi:10.1109/TIE.1986.351700

Reisig, W. (1985). *Petri Nets*. Berlin, Germany: Springer-Verlag.

Silva, M. (1990). Petri nets and flexible manufacturing. In Rozenberg, G. (Ed.), *Advances in Petri Nets LNCS 424* (pp. 374–417). Berlin, Germany: Springer-Verlag.

Silva, M., & Velilla, S. (1982). Programmable logic controllers and Petri nets: A comparative study. In Ferrate, G., & Puente, E. A. (Eds.), *IFAC Software for computer control* (pp. 83–88). Oxford, UK: Pergamon.

Valette, R., Courvoisier, M., Begou, J. M., & Albukerque, J. (1983). Petri net based programmable logic controllers. *Proceedings of 1st International IFIP Conference on Computer Applications in Production and Engineering*, (pp. 103-116).

Yasuda, G. (2008). Hierarchical and distributed control of robotic manufacturing processes based on Petri nets. *Proceedings of the IEEE International Conference on Automation and Logistics*, (pp.221-226).

Yasuda, G. (2009). Design and distributed control of discrete event robotic manufacturing systems using Petri nets. *Proceedings of the 2009 IEEE International Conference on Automation and Logistics*, (pp.1861-1866).

Yasuda, G. (2010). Petri net model based specification and distributed control of robotic manufacturing systems. *Proceedings of International Conference on Advanced Mechatronics 2010* (ICAM 2010), (pp. 410-415).

Chapter 19
Equipment Replacement Decisions Models with the Context of Flexible Manufacturing Cells

Ioan Constantin Dima
Valahia University of Târgovişte, Romania

Janusz Grabara
Częstochowa University of Technology, Poland

Mária Nowicka-Skowron
Częstochowa University of Technology, Poland

ABSTRACT

The main problem of establishing equipment replacement decisions rules under specific conditions is to find decision variables that minimize total incurred costs over a planning horizon. Basically, the rules differ depending on what type of production type is used. For a batch production organization the suitable criterion is built on the principle of economies of scale. Proposed econometric models in this chapter are focused on a multiple machine replacement problem in flexible manufacturing cells with several machines for parts' processing, and industrial robots for manipulation and transportation of manufactured objects. Firstly, models for a simple case multiple machine replacement problems are presented. Subsequently, the more complicated case is considered where technological improvement is taken into account.

DOI: 10.4018/978-1-61350-047-7.ch019

INTRODUCTION

Historically, development of production processes has passed from production structures in automatic rigid flow lines, efficient for mass and wide-range production, to flexible structures, especially efficient in low and medium-range production. Because manufacturing firms has to be flexible towards new market requirements, flexible production forms are increasingly seen as one of the most important manufacturing concepts. Currently, the trend in flexible manufacturing systems is toward small flexible manufacturing structures, called flexible manufacturing cells (FMC). In this sense, two or more CNC machines are considered a flexible cell and two or more cells are considered a flexible manufacturing system (Groover, 2001). Flexible manufacturing cell, in general, allows the processing of pieces which are different in terms of shape and dimensions, in a determined range. This creates prerequisites for the accomplishment of variable products, under high yield conditions. Considerable savings are made because the utilizations increases, the processing time is shortened, the handling distances are reduced, intermediate storage expenses decrease, the area required for production is reduced, the process may be systematized, proper conditions for continuous work are created and direct expenses are reduced. However, the real occurrence of failures during the exploitation stage can markedly modify the FMC performances (Corbaa et al, 1997). For this reason, downtime of FMC has to be analyzed and its influence on the processing cost has to be pondered over. In addition, to ensure that manufacturing process is held to be competitive, upgrading or replacing of equipment due to rapid innovations in technology also has to be considered. In this context often encountered issues in production planning are: Should this equipment be replaced? If not now, then when? Usually, written equipment replacement policy, in which units are scheduled for replacement based on age and expected condition, contains answers on such questions. In this chapter we wish to show several econometric models that could inspire managers to develop their own specific tools in building of an effective equipment replacement policy.

THE PROBLEM STATEMENT

Theoretically, any equipment replacement decision would be made based on thorough modeling equipment deterioration and projected remaining life. Practical approaches to equipment replacement decisions are mostly based on subjective appraisal. But it is generally accepted that tools for equipment replacement decision create important element of repair/replacement policy. Such a policy provides guidance to production and economic manager regarding when to replace existing equipment or its part; how to conduct the acquisition process; and what should be done with the equipment being replaced. Then, the main importance of developing of equipment replacement decision models in production planning consist in establishing rules for the replacement of old equipment or its part(s) by new. The main problem of establishing the rules is to find decision variables that minimize total incurred costs over a planning horizon (Dehayem Nodem et al 2009). Basically, the rules differ depending on what type of production type is used. For batch production organizations suitable criteria are built on the principle of economies of scale, where the large fixed costs of production are depreciation-intensive because of huge capital investments made in high-volume operations and are spread over large production batch sizes in an effort to minimize the total unit costs of owning and operating the manufacturing system (Sullivan, 2002). When solving equipment/parts replacement problem within a flexible manufacturing cell, it is necessary to consider the impacts of the replacement decisions on all of the components of the system. Therefore, possible equipment stoppages due to wrong decision results at least in diminishing capacity or stopping the operations in a manufacturing cell. Accordingly, proposed methods are dedicated for a multiple machine replacement problem that is also characterized as a flexible flow shop problem. A parallel flow shop production concept is consisting of a number of production lines. Jobs in such work shop may be composed of a series of works, each requiring several machines (Jianhua and Fujimoto, 2003).

Firstly we will model a simple case multiple machine replacement problem that is characterized for a parallel flexible flow shop environment, in which no technological improvement in equipment is in concern. Then we will consider the more complicated case where we also take account of technological improvement.

RELATED WORK

Equipment replacement, as a specific field of knowledge and practice, has been extensively studied in the professional literature from the third decade of the 20th century (Castro et al 2009). Operations research approaches utilized in this domain are classified based on methods used to solve replacement problems, such as: integer programming (Hritonenko and Yatsenko, 2007), dynamic programming (Flynn and Chung, 2004), simulation techniques (Freeman, 1996) and Markov decision problems (Love et al, 2000).

Equipment replacement decision approaches related to this work can be divided into two basic types: parallel and series. The difference between these two categories is that in parallel models, the capacity of the system is simply the sum of the capacities of the individual assets and in the series – flow shop models, the minimum capacity assets in the series defines the capacity of the system (Hartman and Ban, 2002). The literature on parallel models is relatively rich. Among the many papers published on this topic, the interested reader may refer to Bean et al. (1994). As regards to series models, there is limited work. For instance, Tanchoco and Leung (1987), Suresh (1991, 1992) and Stinson and Khumawala (1987) present various approaches in which machines operate in series.

Equipment Replacements models can be grouped as: simple and complex ones. By simple models are meant those with a small number of unknown parameters. An instance, where only a small number of observations of time to fail are required to determine a near-optimal value of the critical age for preventive replacement may be an example for age-based replacement models (Baker and Scarf, 1995). The second group of models with a large number of parameters is characterized by high correlations between parameter estimates. This indicates that the available data is insufficient to distinguish between equally plausible parameter combinations (Scarf, 1997).

Recent discussions in econometric models are highlighted the question when to use a capital replacement modeling or an economic life modeling. By comparison of these two approaches, capital replacement modeling methods are evidently more application-oriented than second ones (Christer and Scarf, 1994; Scarf and Bouamra, 1995).

Pioneering approaches of equipment replacement studies are mainly addressed to the replacement of single machines or systems. Developed methods stated to multi-machine systems have mostly assumed linear production flows, with limited operational flexibility. A multi-period replacement model for flexible automated systems was developed by Lotfi and Suresh (1994). Their model was formulated as a nonlinear integer programming problem and was intended to serve as an analytical approximation along with closed queuing networks.

Obviously, it would be possible to mention more similar works from different authors on that topic, but this was comprehensively provided, for instance, by Fine and Freund (1990) or Cheevaprawatdomrong and Smith (2003).

MODELS DEVELOPMENT

The further presented equipment replacement decisions models are econometric-based methods. Econometric methods are in generally concerned with using relevant data for modeling relations between economic and business variables. In these methods one problem is the fact that the selection of variables is somewhat subjective. Their role in decision support for the equipment management and replacement consists of finding the adequate moment to change machine-tool in use or its part(s), based on a specified criterion. In the next subparagraphs several mathematical methods regarding the equipment replacement decision are described.

The OEC and ORC Dependence Based Method

The problem is to choose an optimal replacement policy such that sum of operating equipment cost (OEC) and replacement equipment cost (REC) per unit time is minimized.

In general, the calculation of operating costs (OC) requires the examination of various influencing parameters. Moreover, there is some difference of opinion about whether the wages of equipment operators should be included in the operating equipment cost (Sears et al 2008). In this method the wages are included to this cost. Because we are looking at all costs from cash flow perspective equipment, thus a replacement cost in our approach deals with present value analysis.

The instrumental assumptions of this method imply that at the beginning of every year, data regarding operation and replacement costs of a certain machine are collected. The data usually shows an increase in the operation cost, because of the damages in certain components of the machine. Some of these components may be replaced, thus the equipment operating cost are reduced. The replacement thereof implies costs with the materials and salaries and, hence, such costs have to be compensated through the savings which may be obtained pursuant to the reduction of operating costs. Thus, we want to determine an optimal replacement policy, able to minimize the sum of operating and replacement expenses during the period between two successive data collections.

Let us consider $c(t)$, the operating cost per time unit at the moment t, after replacement and c_r, the cost of a replacement. Then, the relation between the operating cost, replacement cost and time is shown in Figure 1.

The replacement policy is presented in Figure 2, with the following notations: $[0,T]$, the time interval regarding the collection of data on the machine and t_r, intervals when n replacements shall occur.

The assumed goal is the determination of the optimal interval between successive replacements, so that the sum of the operating and replacement cost $C(t_r)$ is minimal.

Then, $C(t_r)$ present the replacement cost during the period $[0,T]$, plus operating cost during the period $[0,T]$.

Replacement cost by period $[0,T]$ is calculated by

$$\sum C_r = n \cdot C_r \tag{1}$$

Thus, n are the number of replacements by period $[0,T]$ and C_r, the cost of one replacement.

The total cost per time unit $C(t_r)$, for the replacement performed at the moment t_r is: $C(t_r)$ i.e. total cost in the interval $(0,t_r)$ related to the length of the interval.

The total cost in the interval $(0,t_r)$ is the operating cost plus replacement cost.

$$C(t_r) = \int_0^{t_r} c(t)\,dt + C_r \approx \left[\frac{\int_0^{t_r} c(t)\,dt + C_r}{t_r}\right] \tag{2}$$

Figure 1. Relation between operation cost and replacement cost

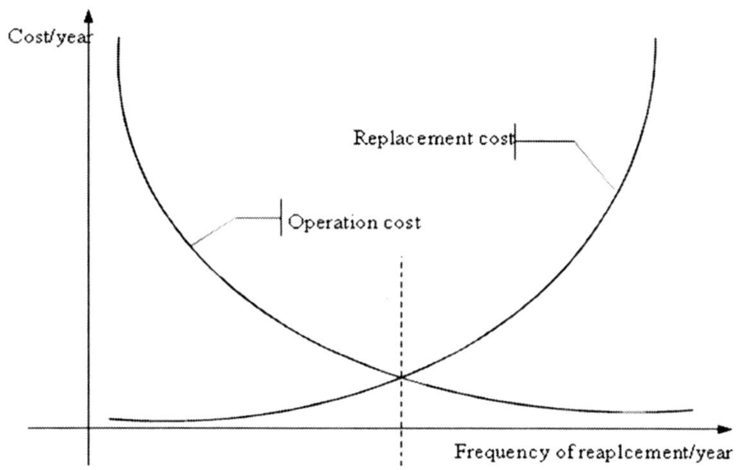

Figure 2. Graphical representation of the replacement policy

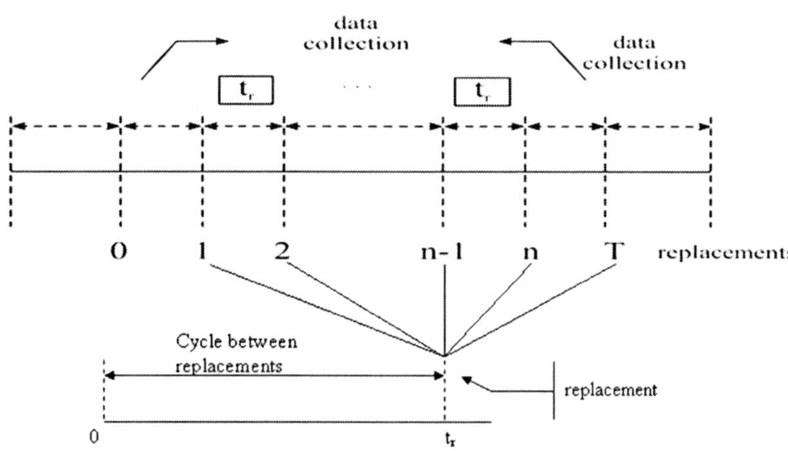

As we may see, the two different cost calculation procedures are similar, because the minimization of $C(t_r)$ is desired, depending on t_r.

Neither of the two procedures considers the time T_r (see Figure 3) required for performing a replacement.

If the time required for performing a replacement is considered, Equation 2 becomes:

$$C(t_r) = \frac{\int_0^{t_r} c(t)\,dt + C_r}{t_r + T_r} \quad (3)$$

Even though this method is also applicable for job shop problems, the given procedures meet essential requirements of the Total Productive Maintenance concept that have a very important role to effective use the automated production systems like FMC and flexible manufacturing systems.

Figure 3. Time structure of cycle between two replacements

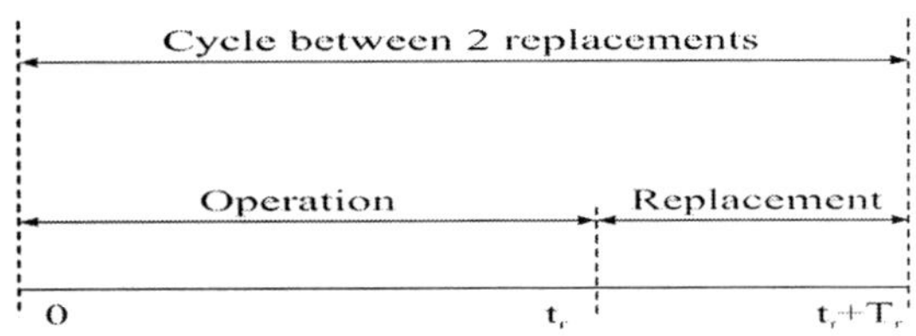

The Method of Replacing the Equipment at a Certain Age

We consider that the machine shall be used for a certain number of years. Further, we assume that a certain machine fabricates certain products, according to the production plan. In this case the goal, in order to minimize the total operating and replacement cost for a fixed time period, consists in the determination of the replacement policy establishing whether: at a certain age of the machine; the latter should be replaced or left to operate continuously.

Let us use *I* to denote the age of the equipment (from the last replacement, with *n* plan periods of proper operation, until the end of the production plan); *c(a)* to represent the cost of operating the equipment for a plan period, when the equipment has age *a*; *J* to represent the age of the equipment from the moment of the last replacement, having (*n-1*) operating time periods until the end of the production plan; C_r, replacement cost; *C(I,J)*, total cost during the period when the equipment develops from age *I* to age *J*. The proposed goal consists in the determination of a replacement policy, so that the cost of operating and replacing the machine $C_n(i)$, along the following *n* time periods is minimal. When $C_n(i)$ has a minimal value, the smallest cost is defined as $f_n(i)$.

Ten weeks before the end of the production plan, two decisions may be made: continuous use or replacement of the machine. If it is decided that the machine should operate further, the equipment shall have age *4* when a new decision may be made (Figure 4).

The total operating cost for the period (10, 9) is:

$$C\left(3,4\right) = C\left(3\right) \tag{4}$$

If the decision to replace the machine is made, then the total cost for the period (10, 9) shall be:

$$C\left(3,1\right) = C_r + C\left(0\right) \tag{5}$$

Thus, C_r is the replacement cost, and *C(0)*, the operating cost for a period, when the machine has age *0*.

The optimal replacement policy is graphically presented in Figure 5.

The mathematical model used for identifying this optimal policy has the following form:

Figure 4. Replacement policy for the machine with age "4"

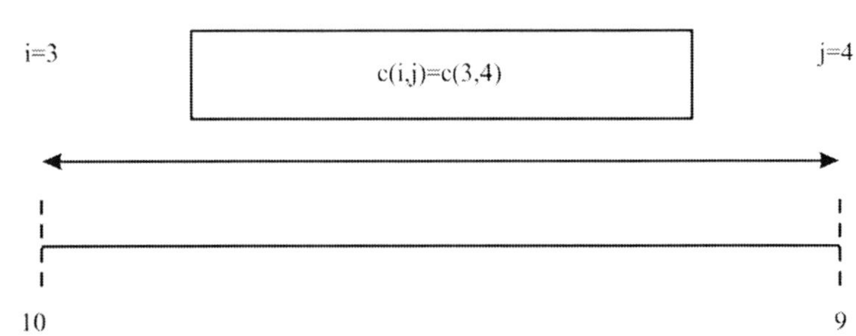

Figure 5. Optimal replacement policy

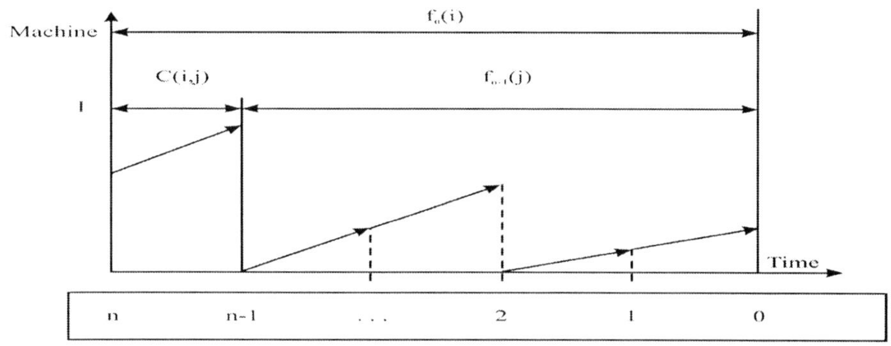

Consider: $f_n(i)$, the minimal cost resulting from taking the best decision at the beginning of the period n plus the cost of the best decision taken on the remaining periods ($n-1$); $C(i,j)$ represent the cost resulting from taking the decision at the beginning of the period n; $f_{n-1}(j)$, minimal cost by periods ($n-1$) remaining at the moment when the machine has the age J.

The cost by the n periods is:

$$C\left(i,j\right)+f_{n-1}\left(j\right), \tag{6}$$

so:

$$f_n\left(i\right)=\min\left[C\left(i,j\right)+f_{n-1}\left(j\right)\right] \tag{7}$$

with:

$$f_0\left(i\right)=0$$
$$j=i+1 \quad \text{or } 0$$

The equation (7) that may be solved by means of dynamic programming can be used to determine the replacement policy under a given specifications formulated in a section 'The problem statement'.

The Method of Replacement Based on the Existence of Equipment in Standby

It implies the replacement of assets provided that the manufacturing flow contains a spare asset, and the operating cost increases with the use of the asset existing in production.

In this case, an optimal replacement policy must be determined, combined for the two assets, which shall minimize the total replacement and operating cost for a fixed time period.

The state of the production system at the beginning of a period shall be noted with I, where I is equivalent to the pair of numbers (x, y), where x refers to the asset (A or B) which is generally used, and y, to the age of the asset.

Consider: $C_x(y)$, the operating cost for a period; j, the state of the production system at the end of a period, where j is equivalent to (x,y); C_r = replacement cost, considered equal for both assets $C(i,j,)$, total cost of the system between the states of the system i and j. The time required for the replacement of an asset is a period when the replacement decision is made, and then the *stand by* asset becomes operative. The proposed goal is the determination of an optimal combined policy for replacement/operation, so that the operating

Figure 6. Replacement policy given the existence of the standby asset

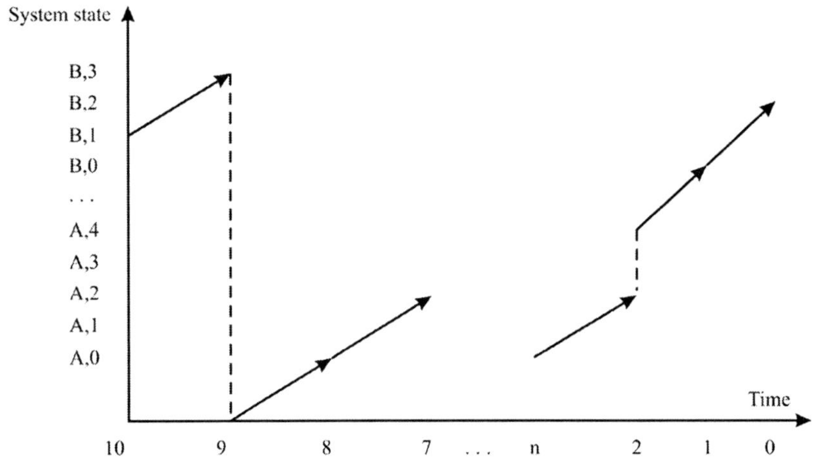

and replacement cost for the following n time periods is minimal. Figure 6 shows such a policy, where $n=10$, the system is in state $I=(B,2)$. At the beginning of period 10, a decision is made to go on with asset B.

At the beginning of period 9, a decision is made to replace asset B etc. The total minimal cost for replacement and operation, for the n periods is $f_n(i)$.

The cost of the first decision taken at the beginning of the period n is $C(i,j)$. At the end of this period, the system is in state j, having $(n-1)$ operating periods. Then the minimal cost for the remaining period $f_{n-1}(j)$ is:

$$Total\cos t = C\left(i,j\right) + f_{n-1}\left(j\right) \tag{8}$$

and

$$f_n\left(i\right) = \min\left[C\left(i,j\right) + f_{n-1}\left(j\right)\right] \tag{9}$$

In the next section we will describe the replacing policy model based on considering technological changes.

The Method of Replacing the Equipments Based on Technological Improvement in Finished Time Horizon

Considers that the replacement of an old machine by a new one not always is an exact copy of the old one, but that the latter is better, so that operating and maintenance costs are smaller, efficiency is higher etc. The following model aims at determining the way how the new available machines may be used with a successful purpose, considering that the time period is fixed and finite.

Consider: n, the number of operating periods (periods when the machine must operate); $C_{p,i}$, maintenance cost of the current equipment in the period I ($i = 1,2,\ldots n$); $S_{p,i}$, sales value of the current equipment at the end of the time period; A, purchase cost for the new, better equipment; $C_{t,j}$, maintenance cost of the new machine in the dj period after installation ($j = 1,2,\ldots n$); $S_{t,j}$, sale value of the new equipment at the end of the operating period j; r – update factor.

The method aims at determining the value T when replacement should be made with the new, better machine (Figure 7), $T = 0, 1,2,\ldots,n$.

Figure 7. Graphical calculation of T value

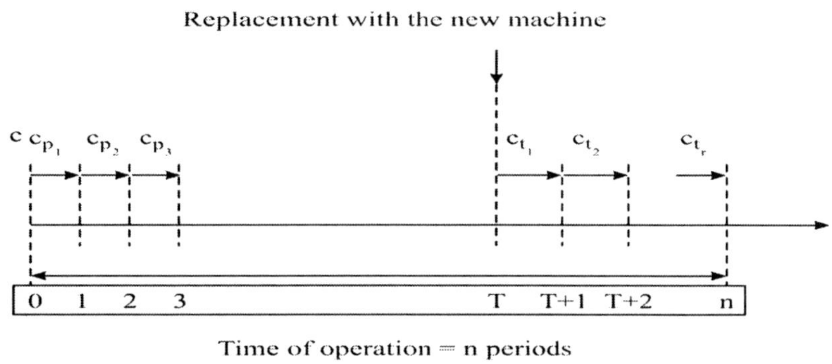

The total updated cost for the *n* periods when replacement occurs at the end of the *T* period is: *C(T)*, updated cost for the maintenance of the current machine in the period *(0,T)*, plus the updated maintenance cost of the new machine in the period *(T,n)*, plus the updated purchase cost of the new machine, minus the updated sale value of the current equipment at the end of the *T* time period, minus the updated sale value of the new equipment at the end of period *n*:

$$C(T) = \left(C_{p,1} \cdot r^1 + C_{p,2} \cdot r^2 + \ldots + C_{p,T} \cdot r^T\right) + \left(C_{t,1} \cdot r^{T+1} + C_{t,2} \cdot r^{T+2} + \ldots + C_{t,n-T} \cdot r^n\right) + A \cdot r^T - \left(S_{p,T} \cdot r^T + C_{t,n-T} \cdot r^n\right) \quad (10)$$

Hence,

$$C(T) = \sum_{i=1}^{T} C_{pi} \cdot r^i + \sum_{j=1}^{n-T} C_{t,j} \cdot r^{T+j} + A \cdot r^T - \left(S_{p,T} \cdot r^T + C_{t,n-T} \cdot r^n\right) \quad (11)$$

As the sole unknown variable is *T*, the minimisation of *C(T)* does not raise further related issues.

DISCUSSION AND CONCLUSION

Because of the fact that above presented methods are more or less applicable based on specific theoretical preconditions, it can be handy to view the given problem from a wider user base. Prior to analysing decisions about equipment replacement in flexible manufacturing cells from practical point of view it is useful recognize two different approaches: deterministic or probabilistic. The probabilistic decisions to replace machines referred to as preventive actions are those decisions where the risk is given by the impossibility to exactly determine the moment when such machine falls or the transition moment from proper operating state to non-operating state. Another source of risk is given by the impossibility to determine the state of the equipment when no inspection or other maintenance activity occurs. Let us consider that there are only two states of the equipment that are always known: a proper operating state or a non-operating state. Then, in order to avoid equipment stoppages in flexible manufacturing cells, the positive replacement decision should come during a proper operating state and accordingly should have a preventive character. In such a way understood preventive replacement for fixed assets implies two conditions:

- *the total* replacement *cost shall be higher after the fall itself at the moment when the preventive replacement is made;*
- the replacement of the machine before the fall itself does not affect the chance that

the equipment may fall at the following moment.

Therefore, preventive replacement is only justified when the rate of replacement grows. In case the machine is damaged, specialists in the department should increase the preventive replacement activities. This may lead to a mistake, as the preventive replacement of machines or their parts is not always justified.

Another above presented approach to equipment replacement decisions is based on considering technological shifts. Even though it is often assumed that an acceleration in technological improvement should result in a more rapid introduction of new technology, according to Cheevaprawatdomrong and Smith (2003), rapid technological improvement may not and indeed should not necessarily lead to more rapid replacement of old technology. In addition, as regards to methods of replacing the equipments based on technological improvement, it has to be 'calculated' with known difficulties and problems such as:

- workers' resistance, as they are used to the old machine;
- lack of will to change the work style;
- fear of the unknown, i.e. workers are scared that they will lose their jobs pursuant to the introduction of new technologies or that they won't be able to adjust to the new working requirements;
- lack of support regarding specialized documentation;
- difference of opinions regarding the operation of the equipment.

Accordingly, a management attitude toward new manufacturing technology will play a major role in determining whether a firm will acquire such technology (Dorf and Kusiak,1994).

REFERENCES

Baker, R. D., & Scarf, P. A. (1995). Can models fitted to small data samples lead to maintenance policies with near-optimum cost? *I.M.A. Journal of Mathematics Applied in Business and Industry*, *6*, 3–12.

Bean, J. C., Lohmann, J. R., & Smith, R. L. (1994). Equipment replacement under technological change. *Nav Res Log*, *41*, 117–128. doi:10.1002/1520-6750(199402)41:1<117::AID-NAV3220410108>3.0.CO;2-U

Cheevaprawatdomrong, T., & Smith, R. L. (2003). A paradox in equipment replacement under technological improvement. *Operations Research Letters*, *31*, 77–82. doi:10.1016/S0167-6377(02)00153-0

Christer, A. H., & Scarf, P. A. (1994). A robust replacement model with application to medical equipment. *The Journal of the Operational Research Society*, *45*, 261–275.

Dehayem Nodem, F. I., Kenné, J. P., & Gharbi, A. (2009). Hierarchical decision making in production and repair/replacement planning with imperfect repairs under uncertainties. *European Journal of Operational Research*, *198*(1), 173–189. doi:10.1016/j.ejor.2008.09.001

Dorf, R. C., & Kusiak, A. (1994). *Handbook of design manufacturing and automation*. John Willey & Sons Inc.doi:10.1002/9780470172452

Fine, C., & Freund, R. (1990). Optimal investment in product-flexible manufacturing capacity. *Management Science*, *36*, 449–466. doi:10.1287/mnsc.36.4.449

Flynn, J., & Chung, C. (2004). A heuristic algorithm for determining replacement policies in consecutive k-out-of-n systems. *Computers & Operations Research*, *31*(8), 1335–1348. doi:10.1016/S0305-0548(03)00094-7

Freeman, J. (1996). Analyzing equipment failure rates. *International Journal of Quality & Reliability Management*, *13*(4), 39–49. doi:10.1108/02656719610114380

Groover, M. P. (2001). *Automation, production systems, and computer integrated manufacturing*. New Jersey: Prentice Hall.

Hartman, J. C., & Ban, J. (2002). The series-parallel replacement problem. *Robotics and Computer-integrated Manufacturing*, *18*, 215–221. doi:10.1016/S0736-5845(02)00012-1

Hritonenko, N., & Yatsenko, Y. (2007). Optimal equipment replacement without paradoxes: A continuous analysis. *Operations Research Letters*, *35*(2), 245–250. doi:10.1016/j.orl.2006.03.001

Jianhua, S., & Fujimoto, Y. (2003). Dynamic scheduling of large-scale flow shops based on relative priority approach. *IEE J Trans. EIS*, *123*(10), 1858–1864. doi:10.1541/ieejeiss.123.1858

Korbaa, O., Camus, H., & Gentina, J.-C. (1997). Cyclic scheduling with overlapping production cycles. *Proceedings of 18th International Conference on Application and Theory of Petri Nets*, Toulouse, France, June 1997.

Leung, L. C., & Tanchoco, J. M. A. (1990). Multiple machine replacement analysis. *Engineering Costs and Production Economics*, *20*, 265–275. doi:10.1016/0167-188X(90)90074-R

Lotfi, V., & Suresh, N. C. (1994). Flexible automation investments: A problem formulation and solution procedure. *European Journal of Operational Research*, *79*(3), 383–404. doi:10.1016/0377-2217(94)90054-X

Love, C., Zhang, Z., Zitron, M., & Guo, R. (2000). Theory and methodology: A discrete semi-Markov decision model to determine the optimal repair/replacement policy under general repairs. *European Journal of Operational Research*, *125*(2), 398–409. doi:10.1016/S0377-2217(99)00009-0

Sarache Castro, W. A., Castrillón, O. D., Gonzales, G., & Viveros Folleco, A. (2009). A multi-criteria application for an equipment replacement decision. *Ing. Desarro., 25,* 80-98. ISSN 0122-3461

Scarf, P. A. (1997). On the application of mathematical models in maintenance. *European Journal of Operational Research*, *99*, 493–506. doi:10.1016/S0377-2217(96)00316-5

Scarf, P. A., & Bouamra, O. (1995). On the application of a capital replacement model for a mixed fleet. *I.M.A. Journal of Mathematics Applied in Business and Industry*, *6*, 39–52.

Sears, S. K., Sears, G. A., & Clough, R. C. (2008). *Construction project management: A practical guide to field construction management*. New York, NY: John Wiley.

Stinson, J. P., & Khumawala, B. M. (1987). The replacement of machines in a serially dependent multi-machine production system. *International Journal of Production Research*, *25*, 677–688. doi:10.1080/00207548708919870

Suresh, N. C. (1991). An extended multi-objective replacement model for flexible automation investments. *International Journal of Production Research*, *29*, 1823–1844. doi:10.1080/00207549108948052

Suresh, N. C. (1992). A generalized multimachine replacement model for flexible automation investments. *IIE Transactions*, *24*, 131–143. doi:10.1080/07408179208964210

Tanchoco, J. M. A., & Leung, L. C. (1987). An input–output model for equipment replacement decisions. *Engineering Costs and Production Economics*, *11*, 69–78.

Chapter 20
Multi-Modal Assembly-Support System for Cellular Manufacturing

Feng Duan
Nankai University, China

Jeffrey Too Chuan Tan
The University of Tokyo, Japan

Ryu Kato
The University of Electro-Communications, Japan

Chi Zhu
Maebashi Institute of Technology, Japan

Tamio Arai
The University of Tokyo, Japan

ABSTRACT

Cellular manufacturing meets the diversified production and quantity requirements flexibly. However, its efficiency mainly depends on the operators' working performance. In order to improve its efficiency, an effective assembly-support system should be developed to assist operators during the assembly process. In this chapter, a multi-modal assembly-support system (MASS) was proposed, which aims to support operators from both information and physical aspects. To protect operators in MASS system, five main safety designs as both hardware and control levels were also discussed. With the information and physical support from the MASS system, the assembly complexity and burden to the assembly operators are reduced. To evaluate the effect of MASS, a group of operators were required to execute a cable harness task. From the experimental results, it can be concluded that by using this system, the operators' assembly performance is improved and their mental work load is reduced. Consequently the efficiency of the cellular manufacturing is improved.

DOI: 10.4018/978-1-61350-047-7.ch020

INTRODUCTION

Traditionally, when the mass production was major in industry production, various assembly systems had been designed as automated manufacturing lines, which are aimed to produce a single specific product without much flexibility. Nowadays, the tastes of consumers change from time to time; therefore, traditional automated manufacturing lines cannot meet the flexibility and efficiency at the same time. To solve this problem, cellular manufacturing system, also called cell production system, has been introduced. In this system, an operator manually assembles each product from start to finish (Isa & Tsuru, 2002; Wemmerlov & Johnson, 1997). The operator enables a cellular manufacturing system to meet the diversified production and quantity requirements flexibly. However, due to the negative growth of the population in Japan, it will become difficult to maintain the cellular manufacturing system with enough skilled operators in the near future. How to improve the assembly performance of the operators and how to reduce their assembly burden are two important factors, which limit the efficiency of the cellular manufacturing system.

Without an effective supporting system, it is difficult to maintain the cellular manufacturing system in Japan. Taking the advantages of the operators and robots, but avoiding their disadvantages at the same time, a new cellular manufacturing system was proposed, namely, the human-robot collaboration assembly system (Duan, 2008). In this human-robot collaboration assembly system, the operators are only required to execute the complicated and flexible assembly tasks that need human assembly skills; while the robots are employed to execute the monotonous and repeated tasks, such as the repetitions of parts feeding during assembly process (Arai, 2009). To make this system has the applicability to assemble a variety of products in different manufacturing circumstances, the following assembly sequence is assumed: each assembly part is collected from the tray shelf by manipulators; all the parts are automatically fed to the operator on a tray as a kit of parts; an operator grasps the individual part respectively and assembles it to form a final product; the assembled product is transferred out to the next station, and so on.

In the following part, a multi-modal assembly-support system (MASS) is introduced, which aims to support an assembly operator in a cellular manufacturing system from both information side and physical side while satisfying the actual manufacturing requirements. MASS system utilizes robots to support the operator and several information devices to monitor and guide the operator during the assembly process. Since it is a human-robot collaboration assembly system, safety strategy must be designed to protect the operator with a reasonable cost benefit balance in the real production line.

The remainder of the chapter is organized as follows: Firstly, the background information and the related studies are introduced. Then, the entire MASS system and its subsystems are briefly described. After that, a description of two manipulators and a mobile base are introduced in physical support part, which are used to feed assembly parts to the assembly operator. Assembly information support part contains a discussion of a multimedia-based assembly table and corresponding devices. Safety standard and safety design are presented in safety strategy part. Taking a cable harness task as an example, the effect of MASS system was evaluated. Finally, the conclusion and the future work are given.

PREVIOUS RELATED STUDIES

To improve the efficiency of the cellular manufacturing system, various cellular manufacturing systems have been designed to improve the assembly performance of the operators and reduce their assembly burden.

Seki (2003) invented a production cell called "Digital Yatai" which monitors the assembly progress and presents information about the next assembly process. Using a semi-transparent head mount display, Reinhart (2003) developed an augmented reality (AR) system to supply information to the operator. These studies support the operator from information aspect. To reduce the operator's physical burden and improve the assembly precision, Hayakawa (1998) employed a manipulator to grasp the assembly parts during the assembly process. This improved the assembly cell in physical support aspect. Sugi (2005) aimed to support the operators from both information side and physical side, and developed an attentive workbench (AWB) system. In this system, a projector was employed to provide assembly information to the operator; a camera was used to detect the direction of an operator's pointing finger; and several self-moving trays were used to deliver parts to the operator. Although AWB achieved its goal of supporting operators from both information aspect and physical aspect, the direct supporting devices are just a projector and several self-moving trays, which are general purpose instruments that cannot meet the actual manufacturing requirements.

In the coming aging society, it will be impossible to maintain the working efficiency if everything is done manually by the operator in the current cellular manufacturing system. In order to increase working efficiency, many researchers have used robot technologies to provide supports to the operator (Kosuge, 1994; Bauer, 2008; Oborski, 2004). According to these studies, human-robot collaboration has potential advantages to improve the operator's working efficiency. However, before implementing this proposal, the most fundamental issue will be the safety strategy, which allows the operators and the robots to execute the collaboration work in their close proximity.

Human-robot collaboration has been studied in many aspects but has not been utilized in the real manufacturing systems. This is mainly because safety codes on industrial robots (ISO 12100, ISO 10218-1, 2006) prohibit the coexistence of an operator in the same space of a robot. According to the current industrial standards and regulations, in a human-robot collaboration system, a physical barrier must be installed to separate the operator and the assisting robot. Under this condition, the greatest limitation is that the close range assisting collaboration is impossible. Based on the definition of Helms (2002), there are four types of human-robot collaboration: *Independent Operation*, *Synchronized Cooperation*, *Simultaneous Cooperation*, and *Assisted Cooperation*. The assisted cooperation is the closest type of collaboration, which involves the same work piece being processed by the operator and the robot together. In this kind of human-robot collaboration, the operator is working close to the working envelope of the assisting robot without physical separation, so that both of them can work on the same work piece in the same process. The most distinguished concept of this study is that the assisting robot in this work is active and is able to work independently as robot manipulator. The advantage of this collaboration is to provide a human-like assistance to the operator, which is similar with the cooperation between two operators. This kind of assistance can improve the working efficiency by automating portion of the work and enable the operator to focus only on the other portion of work which requires human skill and flexibility. However, since the active robot is involved, this kind of collaboration is extremely dangerous and any mistake can be fatal (Beauchamp & Stobbe, 1995).

The challenge of this research work is to design an effective assembly supporting system, which can support the operator in both physical and information aspects. During the assembly process, employing of the assisting robot is an effective method to reduce the operator's assembly burden while improving the working efficiency. This involves the safety issue in this kind of close range active human-robot collaboration. However, there are no industrial safety standards and regulations. Besides the design of the assembly supporting system, the scope of this work also covers both safety design study and development of prototype production cell in cellular manufacturing.

MULTI-MODAL ASSEMBLY-SUPPORT SYSTEM

Structure of the Entire System

Following the fundamental idea that robots and operators share the assembly tasks can maximize their corresponding advantages, the MASS system was designed and its subsystems are shown in *Figure 1* as structure view and in *Figure 2* as system configuration.

The entire MASS system is divided into physical support part and assembly information support part, as shown in *Figure 1*.

Figure 1. Structure of the entire MASS system

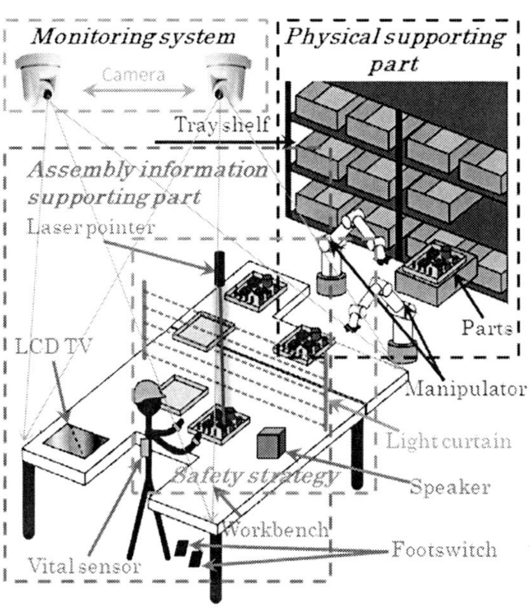

Figure 2. Configuration of MASS system

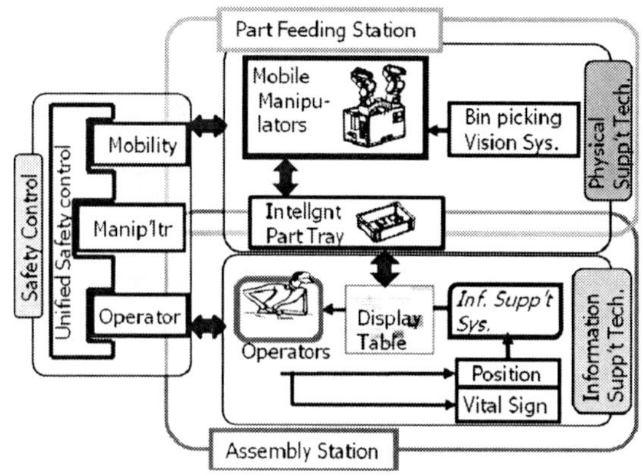

1. **Physical Supporting Part**: The physical supporting part is aimed to support operators from physical aspect, and it is composed of two manipulators with six degrees of freedom and a mobile base, which have two functions: one is to deliver assembly parts from a tray shelf to an assembly table; and the other is to grasp the assembly parts and prevent any wobbling during the assembly process.
2. **Information Supporting Part**: The assembly information supporting part is designed to aid operators in assembly information aspect. An LCD TV, a speaker, and a laser pointer are employed to provide assembly information to guide the operator.
3. **Safety Control Part**: To guarantee the operator's safety during the assembly process, vital sensors are used to monitor the operator's physical conditions during the assembly process, and a serial of safety strategies is used to protect the operator from injury by the manipulators. It controls the collaboration between a robot and an operator (also referred to *Figure* 2).

In the developed MASS system, there are two stations connected through an intelligent part tray as shown in *Figure 2*, on which all the necessary parts are fed into the assembly station and the assembled products are transferred out through a shipment from the assembly station.

1. **Part Feeding Station**: Only robots work here. It is mainly in charge of part handling, such as bin picking, part feeding, kitting and part transferring.
2. **Assembly Station**: An operator executes the assembly tasks with some aid of the robots. Supporting **information** from the MASS system is implemented to accelerate the operator assembly efficiency.

Figure 1 illustrates the setup of the MASS system, in which an operator assembles a product on the workbench in the area of assembly station. The operator is supported with the assembly information and with physical holding of parts for assembly. In this study, the sample product to assemble is a cable harness with several connectors and faster plates. Even experienced operators maybe spend about 15 minutes finishing this assembly task.

Simulator of the Entire System

To reduce the design period, in this study, a simulator of the entire system was developed based on ROBOGUIDE (FANUC ROBOGUIDE) and OpenGL (Neider, 1993), as shown in *Figure 3*. This simulator can not only reproduce the actual motion of the manipulators but also predict collisions in the work space.

Since MASS system is a human-robot cooperation assembly system, considering the operator's safety, the distance between the manipulators and the operator should be optimized to prevent the collisions between them. Furthermore, the moving trajectories of the manipulators should also be optimized to prevent the collisions between themselves. In order to accelerate the development period, all of the optimization assignments are done in this simulator first, and then evaluated in the actual MASS system. With the aid of this simulator, the distance between manipulators and the operator can be adjusted easily, and the moving trajectories of the manipulators' end points can also be reproduced conveniently during the manipulators' moving process. Therefore, based on the simulation results, the actual system could be conveniently constructed.

Figure 3. Simulator of the entire MASS system

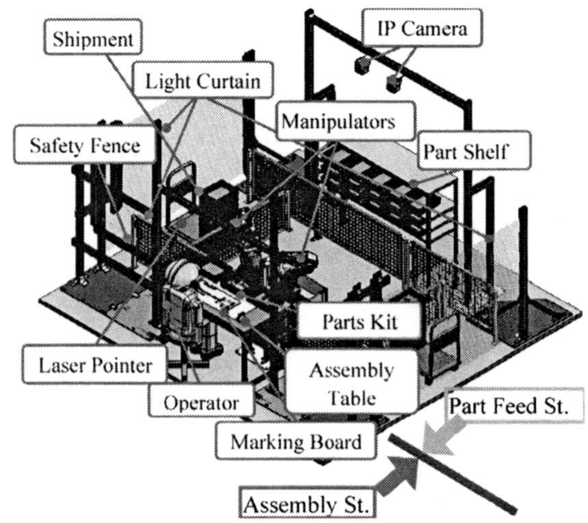

Physical Support

To increase the physical support provided by the MASS system, two manipulators with six degrees of freedom are installed on a mobile base and used to deliver assembly parts to the operator, as shown in *Figures 1-4*. A CCD camera with an LED light is equipped to each manipulator respectively for recognition of picking target from a part bin in scramble.

The manipulators are utilized in part feeding station to

Figure 4. Assembly operations with the aid of manipulators

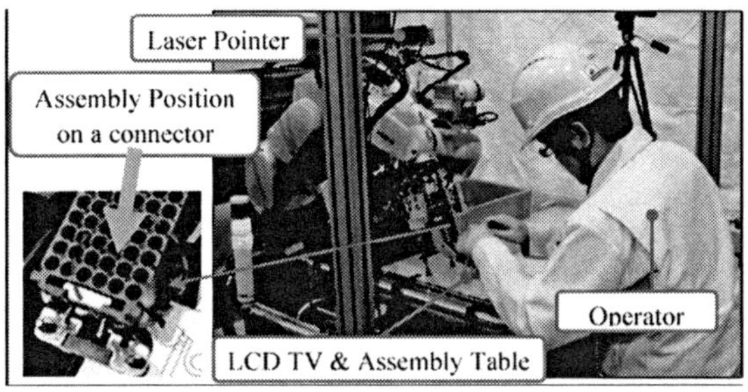

1. Draw a part bin from part shelves;
2. Pick a part from the bin one by one;
3. Kit parts onto a tray;
4. Check visually the parts in a tray.

The parts are efficiently fed by the manipulators, because one manipulator hangs a bin up and the other one grasps a part out like an operator does. Since the bin picking system by manipulators can work 24 hours a day, it enables high productivity. The base carries a few trays and moves to the assembly station, where the base docks in the electric charge connector. In the assembly station, an operator continuously assembles parts one by one, which are transferred by one of the mobile twin manipulators. To increase the precision of assembly and reduce the operator's burden, one manipulator can grasp an assembly part to prevent wobbling during assembly, and the operator executes the assembly task on the basis of the manipulator's assistance, as shown in *Figure 4*.

Obviously, the assistant manipulators move near to the operator during the assembly process. To achieve this collaboration, the manipulators have to penetrate the operator's area. Since the penetration is prohibited by the regulations of the industrial robots (ISO 12100), a new countermeasure must be developed. After finishing an assembly step, the operator pushes a footswitch to send a control command to the manipulators, and the manipulators provide the next assembly part to the operator and the assembly information of the next assembly step is given. Without this control command, the manipulators cannot move to the next step. Furthermore, the operator can stop the manipulators with an emergency button when an accident occurs. These strategies enable the manipulators to support human operators in physical aspect effectively and safely.

Assembly Information Support

Previous studies, Szeauch as Digital Yatai (Seki, 2003), have already testified that providing assembly information to the operator during his assembly process can not only improve his assembly efficiency, but also reduce his assembly errors. Taking the advantages of the previous studies, and also considering the characteristics of human cognition, an assembly information supporting system is designed to guide operators by means of indicating the next assembly sequence and/or an appropriate way of operation.

The developed system has three major advantages:

1. Each assembly sequence is instructed step by step;
2. Considering the characteristics of human cognition, the assembly information can be provided as easily understandable formats for humans, including text, voice, movie, animation and flashing emphasis marks;
3. The assembly information can be selected and provided to the operator according to his assembly skill level.

The total software system of MASS system in *Figure 5* has been developed. It consists of three subsystems as

1. Multi-modal Assembly Skill Transfer (MASTER);
2. Multi-modal Assembly Information SupportER (MAISER);

3. Multi-modal Assembly FOSTER (MAFORSTER).

MASS is designed to extract the skill information from skilled operators by MASTER and to transfer it to novice operators by MAISER as illustrated in *Figure 5*. Here, a human assembly skill model was proposed (Duan, 2009), which extracts and transfers human assembly skills as the cognition skill part and the motor skill part. In the cognition skill part, depending on questionnaire, MASTER obtains the different cognition skills between the skilled operators and the novice operators. In the motor skill part, MASTER mainly utilizes motion capture system to obtain the different motor skills between the skilled operators and the novice operators, especially in the assembly pose aspect and the assembly motion aspect (Duan, Tan, Kato, & Arai, 2009). MAISER provides understandable instructions to novice operators by displaying multi-modal information about assembly operations. MAFOSTER controls interface devices to organize comfortable environment for operators to execute the assembly task like a foster does. MAISER works mainly off-line at a data-preparation phase, and watches on-line the state of an operator to avoid bad motion and dangerous states (Duan, Tan, Kato, & Arai, 2009). MAISER takes the role of an instruction phase.

Interface devices are installed as shown in *Figure1* and *Figure 4* again:

1. **LCD TV:** The horizontal assembly table with built-in 37 inch LCD TV as shown in *Figure 4* may be the first application for assembly. Since it enables operators to read the instructions without moving his/her gaze in different direction, assembly errors can be decreased. The entire assembly scheme is divided into several simple assembly steps, and the corresponding assembly information is written in PowerPoint slides (Zhang, 2008). During the assembly process, these PowerPoint slides are inputted into the LCD TV and switched by footswitch.
2. **Laser Pointer:** Showing the assembly position to the operator is an effective way to reduce assembly mistakes. To this end, a Laser pointer, which is fixed on the environment, is projected onto a task to indicate the accurate position of assembly as shown in the left photo of *Figure 4*. The position can

Figure 5. Software system of MASS system

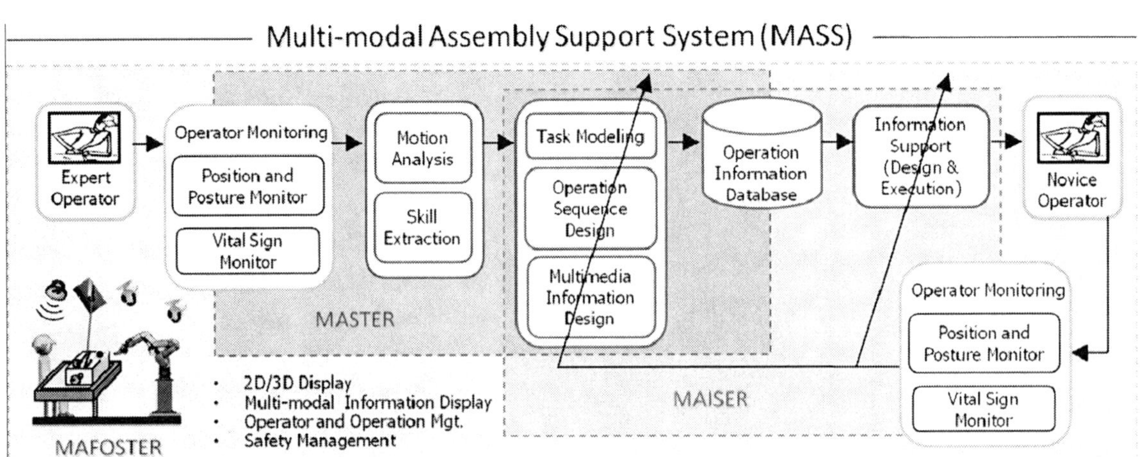

change by the motion of the manipulator. The operator can insert a wire into the instructed assembly position with the aid of the Laser spot.

3. **Audio Speakers:** To easily permit the operator to understand the assembly information, a speaker and a wireless Bluetooth earphone are used to assist the operator with voice information.
4. **Footswitch:** During the assembly process, it is difficult for the operator to switch the PowerPoint slides with his hands. Therefore, a footswitch is used, as shown in *Figure 1*. There are two kinds of footswitches: footswitch A has three buttons, and footswitch B has one button. Just stepping the different buttons on footswitch A, the operator can move the PowerPoint slides forward or backward. Stepping the button on footswitch B, the operator controls the manipulators to supply the necessary assembly parts to the operator, or makes manipulators change the position and orientation of the assembly part during the assembly process.
5. **Assembly Information:** The assembly support information is provided to the operators to improve the productivity by means of good understanding in assembly tasks and of skill transfer with audio-visual aids. As the software structure for the assembly task description is not discussed in this study, please refer to our papers (Duan, 2008; Tan, 2008). Applying Hierarchical Task Analysis (HTA) one assembly task is divided into several simpler assembly steps, whose corresponding information is stored in multimedia. Then appropriate level of information is displayed on LCD panel as shown in *Figure 6*. In each PowerPoint slide, the assembly parts and assembly tools are illustrated with pictures. The assembly positions are noted with color marks. Following the assembly flow chart, videos showing the assembly motions of the experienced operators will appear to guide the novices to execute the assembly tasks. To facilitate the operator's understanding of the assembly process, the colors of the words in the slides are the same as the actual colors used for the assembly parts. For example, there are "blue cable" and "grey cable" in *Figure 6*. In each slide, several design principals of data presentation are introduced such as multimedia principle, coherence principle and spatial contiguity principle (Mayer, 2001). In *Figure 6*, three types of information are displayed as (a) text instruction, (b) pictorial information, (c) movie, and the sequence of assembly is also illustrated. During the assembly process, the PowerPoint slides are output to an LCD TV and switched by the operator's foot with footswitch during the assembly process.
6. **Assembly Information Database:** In this multimedia based assembly supporting system, the assembly information is classified into paper, audio, and video files. The assembly guidance is concisely written in paper files. Guidance of each assembly step is recorded in audio files. After the standard motions of the experienced operators are recorded and analyzed into primitive assembly motions, they are saved into video files. Tan (2008) set up an assembly information database to preserve all of these assembly information files and provide them to the operator depending on the situation. This database contains training data and assembly data: training data are designed for novices, and the assembly information files contain assembly details. Assembly data are used to assist experienced operators by indicating the assembly sequence but not assembly details. As a consequence, this system may promote both novice and experienced operators to enter the workforce.

All the operators who used the assembly table with LCD evaluated positively that the instruction on LCD can be read easily and understood smoothly.

Safety Strategy

MASS system is a kind of human-robot cooperation system. Although employing the assistant robots to support the operator can increase the assembly efficiency and reduce the assembly burden, this collaboration can be extremely dangerous because the active robot is involved and any mistake can be fatal. To protect the operator during the assembly process, several safety designs are proposed and developed in this manufacturing system, which cover both hardware and software to achieve good robot-human collaboration. Fundamental concepts are:

1. Risk assessment by ISO regulation;
2. Area division by safety light curtains as illustrated in Figure 7;
3. Speed/Force limiter by serve controller;
4. Collision protector by physical devices;
5. Collision detector by IP cameras;
6. Inherent safety theory.

Risk Assessment by ISO Regulation

Since no direct industrial safety standards and regulations that govern this type of close range active human-robot collaboration, the safety design in this work is formulated by collective reference to related safety standards and regulations to verify component systems' safety first (non-collaboration safety) and then assess system safety as a whole (collaboration safety). Table 1 summarizes the referred industrial safety standards and regulations in mobile robot manipulators system development and total system development.

This chapter mainly focuses on the discussion on human-robot collaboration safety; therefore the non-collaboration safety of component systems is omitted. However, it is important to bear in mind that the following safety designs for collaboration are built in accordance with the referred standards and regulations in the component level. EU standard permits the collaboration of robots with the operator when the total output of robots is less than 150 (N) at the tip of the end-effecter. Japanese standard defines that each actuator has the power less than 80 (W). The collaboration safety design is presented in hardware design and control design in the following.

Figure 6. Multimedia based assembly supporting information

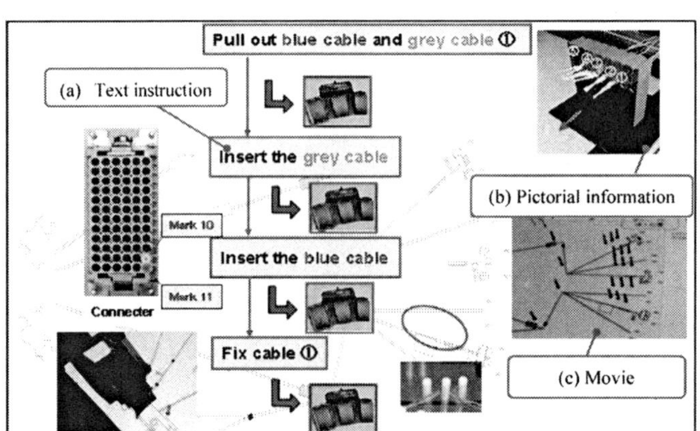

Table 1. Related safety standards and regulations

Standards and Regulations	Descriptions
Related to mobile robot manipulators system development	
IEC 60364-4-41 (JIS C0364-4-41)	Low-voltage electrical installations – Part 4-41: Protection for safety – Protection against electric shock
IEC 60364-7-717	Electrical installations of buildings – Part 7-717: Requirements for special installations or locations – Mobile or transportable units
IEC 61140 (JIS C0365)	Protection against electric shock – Common aspects for installation and equipment
BS EN 1175-1	Safety of industrial trucks – Electrical requirements – Part 1: General requirements for battery powered trucks
ISO 10218-1 (JIS B8433-1)	Robots for industrial environments – Safety requirements – Part 1: Robot
Related to total system development	
ISO 12100-1 (JIS B9700-1)	Safety of machinery – Basic concepts, general principles for design – Part 1: Basic terminology, methodology
ISO 12100-2 (JIS B9700-2)	Safety of machinery – Basic concepts, general principles for design – Part 2: Technical principles
ISO 14121-1 (JIS B9702)	Safety of machinery – Risk assessment – Part 1: Principles
ISO 14121-2	Safety of machinery – Risk assessment – Part 2: Practical guidance and examples of methods
ISO 13849-1 (JIS B9705-1)	Safety of machinery – Safety-related parts of control systems – Part 1: General principles for design
BS EN 954-1	Safety of machinery. Safety related parts of control systems. General principles for design
ANSI/RIA R15.06	Industrial Robots and Robot Systems - Safety Requirements
ISO 13852 (JIS B9707)	Safety of machinery – Safety distances to prevent danger zones being reached by the upper limbs
ISO 14119 (JIS B9710)	Safety of machinery – Interlocking devices associated with guards – Principles for design and selection
ISO 13854 (JIS B9711)	Safety of machinery – Minimum gaps to avoid crushing of parts of the human body
ISO 14118 (JIS B9714)	Safety of machinery – Prevention of unexpected start-up
ISO 13855 (JIS B9715)	Safety of machinery – Positioning of protective equipment with respect to the approach speeds of parts of the human body
ISO 14120 (JIS B9716)	Safety of machinery – Guards – General requirements for the design and construction of fixed and movable guards

Area Division by Safety Light Curtains

The software systems in robot controller and other computers are prepared as Dual Check Safety (DCS), which checks speed and position data of motors with two independent CPUs in the robot controller. In risk assessment, we listed up to 168 risks and take its countermeasure respectively so as to satisfy the required performance level. Whatsoever definition industrial robots are, it is strongly prohibited that robots exist with the operator in the same space. Thus a cage is required to separate the operator from the robots. For the area division, the whole cell in *Figure 7* is divided into human area (**H**), robot area (**R**), and buffer area (**B**) by safety fences, photoelectric sensors and light curtains in order to obtain safe working areas and to monitor border crossing for safety. Robots are allowed to operate in high speed motion in area **R** but low speed movement in area **B**. In area **H**, the strong restrictions are applied to robot motions. When the manipulators move too close to the operators and cross the light curtain 2, the power of the manipulator is cut down by the light curtain. Consequently, the manipulators stop.

Speed/Force Limiter by Serve Controller

As shown in *Figure 8*, by the servo controller, the speed of the mobile manipulators is limited, and the force/torque at the end-effecter is also limited by software. The controller also has a function of abnormal force limiter in case of unexpected collision of the manipulator against the environment. Based on the recommendation from safety standards and risk assessment, during collaboration process, the speed of the mobile manipulators is limited to below 150 (mm/s) and the working area of the robot is restricted within the pink region in *Figure 8*. The minimum distance between the robot gripper and surface of the workbench is 120 (mm) according to ISO 13854.

Figure 7. Three robot working zones for safety

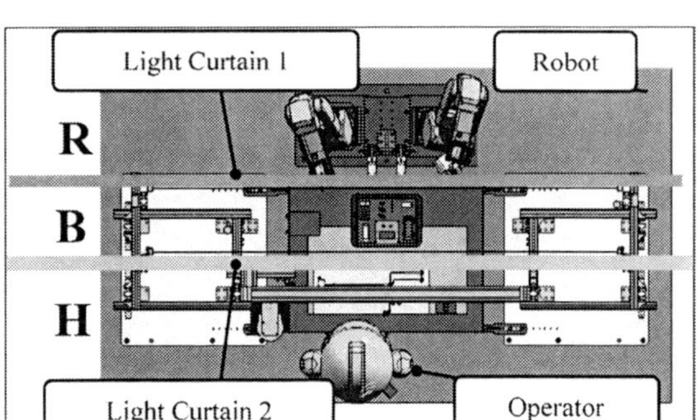

Collision Protectors by Physical Devices

During the assembly process, several collision protectors by physical devices have been designed for accident avoidance and the protection of the operator.

1. **Mobile Base:** To prevent the operator from being hurt by the manipulators, the localization accuracy of the mobile base should be maintained. With vision system to detect marks on the floor, the system has a localization accuracy of 5 (mm) and 0.1°. The base is equipped with bumper switch for object collision detection and wheel guard to prevent foreign object being tangled with the wheels, as illustrated in *Figure 8*.
2. **Footswitch:** In the MASS system, the twin mobile manipulators are used to assist the operator to execute the assembly tasks during the assembly process. Without a safety strategy, the operator could be injured by the manipulators. An effective working sequence is one of the effective ways to reduce the probability of collision between an operator and a manipulator. The manipulators are prevented from moving in the direction of the operator as he performs an assembly task. The probability for collisions is reduced with the introduction of the working sequence. To realize the proposed working sequence, a footswitch is used to control the manipulators, as illustrated in *Figure 9*. When the operator finishes an assembly step, he steps on footswitch, which signals the manipulators to provide the assembly parts to the operator for the next step.
3. **Emergency Button:** When an accident occurs, the operator can just push the emergency button on the right-hand side of the assembly workbench to stop the entire system, as shown in *Figure 9*. After any problem has

Figure 8. Robot speed, force, and area restrictions

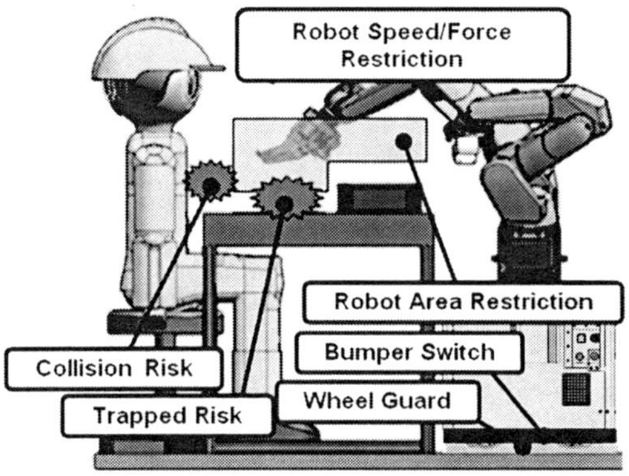

been solved, the operator pushes the reset button to restart the assembly process.

4. **Safe Bar:** In addition, steel safe bar is installed in front of the assembly workbench (referred to *Figure 9*). If other strategies failed to stop the manipulator to collide the operator, this safe bar can protect the operator.

Collision Detector by IP Cameras

The developed system installs a robot with higher ability than both EU and Japan Standard. Even though various countermeasures are introduced, the risk assessment shows residual risks. For the intelligent compensation of safety, two IP cameras are utilized to monitor the operator's safety (referred to *Figure 10*); that is, the cameras track the color marks on the head and shoulders of the operator to measure the body posture and position to estimate the human operation conditions (Duan, 2009). The vision monitoring system has positioning accuracy of 30 (mm) and process delay of 0.6 (s).

Inherent Safety Theory

Although several safety strategies are adopted, there is no guarantee that a collision between a manipulator and an operator will never occur. Therefore, the manipulators should be ameliorated according to inherent safety theory (Ikeda & Saito, 2005) to reduce the injury of the operator. The sharp edges of the manipulators are softened into obtuse-angled brims. The force and speed of the manipulators are reduced as much as possible while still meeting the assembly requirement. In addition, the overall mobile robot manipulators system is built with low center of gravity design to prevent tipping.

Figure 9. Collision protectors by physical devices

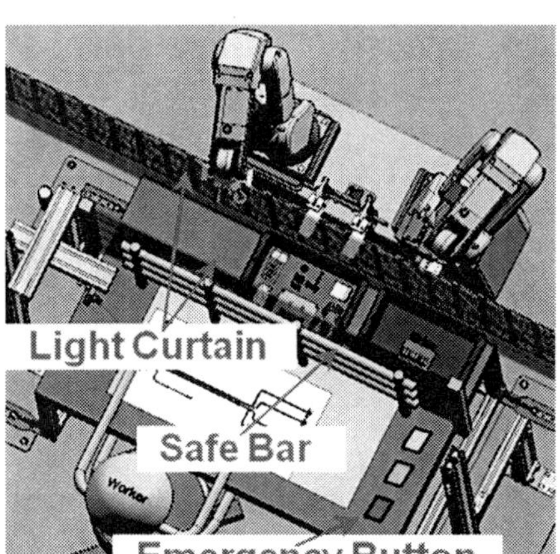

Figure 10. Operator safety monitoring system

Evaluation of MASS System

To evaluate the effect of the MASS system, a group of operators were required to execute an assembly task of cable harness, as illustrated in *Figure 11*. In this task, operators must insert correct cables into corresponding correct holes in the connector. After that, following the cable routes, operators must fix the cables to the jigs on the assembly table. The operators executed the cable harness task in two cases: (1) all of the assembly information, including the cable type, the position of the hole in the connector and the assembly step, was only provided by the assembly manual (Exp I); (2) operators executed the cable harness task under the support of MASS (Exp II). Two parameters were measured in the experiments: assembly time and assembly error. The assembly time is compared between conventional manual assembly setup (Exp I) and the new setup (Exp II). Five novice operators and five expert ones performed three assembly trails respectively for both the setups. From *Figure 12*, it is proved that the overall performance is better (shorter in assembly time) in the new setup (Exp II). Novices and experts show almost the same assembly time from the first trial in the case of the new setup as the dotted lines.

It means that the assembly can be executed at the minimum time even by the unskilled operators. Comparing to the assembly time of the conventional setup (Exp I), the novice operators need only 50% of the time in the MASS system (Exp II), which indicates double productivity. Note that the assembly time at the third trial converges to the minimum by all the cases. This implies that the assembly operation is easy to achieve and the human ability of learning is high. In other words, this system may be beneficial for very frequent change of products. In terms of assembly quality, 10% to 20% of assembly error (insertion error) is observed in conventional setup (Exp I), while in new cell production setup (Exp II) the error is totally being prevented by the robot assistance, especially by guidance of laser pointer and by the instruction of the assembly sequences.

According to the experimental results, it can be concluded that the developed MASS system can accelerate the operator's assembly process as well as prevent assembly errors.

According to Zhang (2008), this cable harness task is a kind of cognitive assembly task (Norman, 1993); therefore, the mental work load of the operators cannot be ignored. To evaluate the mental work load of the operators in (Exp I) and (Exp II), NASA-TLX (Task Load Index) method (NASA-TLX for Windows U.S.) was used. After

Figure 11. Cable harness task

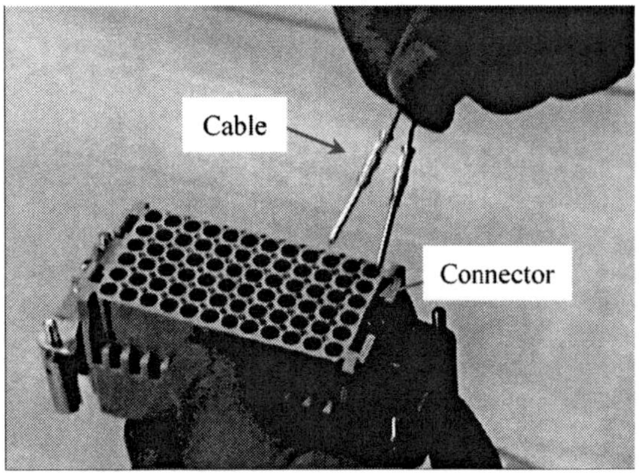

Figure 12. Difference of assembly time by experts and novices

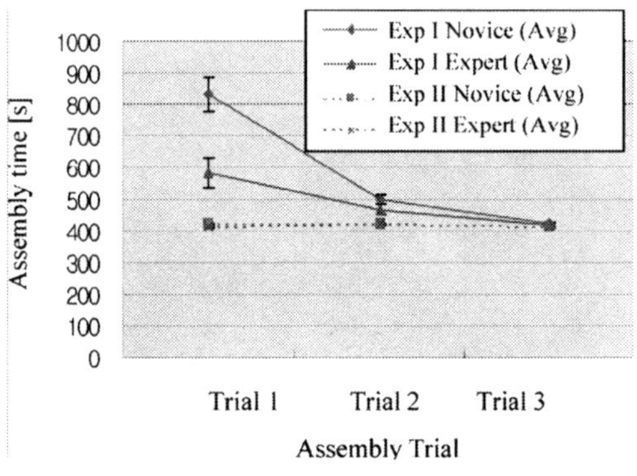

the operators finished the cable harness task, they were required to answer the questionnaires. Based on the NASA-TLX method, the mental work load of the operators can be computed. The mental work load of (Exp I) is 62, which is much higher than that of (Exp II), which is 38. This means that based on the support of MASS, the mental work load of the operators can be reduced significantly.

CONCLUSION

This work aims to realize a new cellular manufacturing system for frequent changes of products. In this chapter, a multi-modal assembly-support system (MASS) was developed for a cellular manufacturing system. In the MASS, two manipulators are used to replace the operators to execute the laborious tasks. Based on the assembly information database and assembly information supporting system, this system is capable of meeting the assembly and training requirements of the experienced and the novice operators. Besides developing the actual system, a simulator for an entire assembly system was created to reduce the time and costs required for development. To protect the operator from harm, several safety strategies and equipments were presented. According to inherent safety theory, two manipulators are ameliorated, which could reduce the injury of the operators even when they were collided by the manipulators.

To evaluate the effect of MASS, a group of experienced operators and novice operators were required to execute a cable harness task. According to the experimental results, basing on the support of MASS, not only the assembly time and the error ratios are reduced, but also the mental work load of the operators is reduced. Therefore, the MASS allows an operator to receive physical and informational support while working in the actual manufacturing assembly process.

Future studies should be directed at identifying and monitoring the conditions that contribute to the operator's fatigue and intention during the assembly process; these efforts will lead to improvements in comfort for the operators and assembly efficiency.

ACKNOWLEDGMENT

This study is supported by NEDO (New Energy and Industrial Technology Development Organization) as one of "Strategic Projects of Element Technology for Advanced Robots". The author Feng DUAN is supported by the Fundamental Research Funds for the Central Universities (No.65010221). We appreciate NEDO and MSTC for accelerating the development of this practical system. In particular, we would like to acknowledge FANUC Company for their excellent cooperation and technical support.

REFERENCES

Arai, T., Duan, F., Kato, R., Tan, J. T. C., Fujita, M., Morioka, M., & Sakakibara, S. (2009). A new cell production assembly system with twin manipulators on mobile base. *Proceeding of 2009 International Symposium on Assembly and Manufacturing* (pp. 149-154). Suwon, Korea.

Bauer, A., Wollherr, D., & Buss, M. (2008). Human-robot collaboration: A survey. *International Journal of Humanoid Robotics*, *5*(1), 47–66. doi:10.1142/S0219843608001303

Beauchamp, Y., & Stobbe, T. J. (1995). A review of experimental studies on human-robot system situations and their design implications. *The International Journal of Human Factors in Manufacturing*, *5*(3), 283–302. doi:10.1002/hfm.4530050306

Colgate, J. E., Wannasuphoprasit, W., & Peshkin, M. A. (1996). Cobots: Robots for collaboration with human operators. *Proceedings of the International Mechanical Engineering Congress and Exhibition*, *58*, 433–439.

Duan, F. (2009). *Assembly skill transfer system for cell production*. Unpublished doctoral dissertation, The University of Tokyo, Japan.

Duan, F., Morioka, M., Tan, J. T. C., & Arai, T. (2008). Multi-modal assembly-support system for cell production. *International Journal of Automation Technology*, *2*(5), 384–389.

Duan, F., Tan, J. T. C., Kato, R., & Arai, T. (2009). Operator monitoring system for cell production. *The International Journal of the Robotics Society of Japan-Advanced Robotics*, *23*, 1373–1391.

Hayakawa, Y., Kitagishi, I., & Sugano, S. (1998). Human intention based physical support robot in assembling work. *Proceedings of the 1998 IEEE/RSJ International Conference on Intelligent Robots and Systems* (pp. 930-935). Victoria, B.C., Canada.

Helms, E., Schraft, R. D., & Hagele, M. (2002). Rob@work: Robot assistant in industrial environments. *Proceedings of 11th IEEE International Workshop on Robot and Human Interactive Communication* (pp. 399-404). Berlin, Germany.

Ikeda, H., & Saito, T. (2005). *Proposal of inherently safe design method and safe design indexes for human-collaborative robots*. (. *Specific Research Reports of the NIIS-SRR-NO.*, *33*, 5–13.

ISO10218-1. (2006). *Robots for industrial environments- Safety requirements - Part 1: Robot*.

Isa, K., & Tsuru, T. (2002). Cell production and workplace innovation in Japan: Towards a new model for Japanese manufacturing? *Industrial Relations*, *41*(4), 548–578. doi:10.1111/1468-232X.00264

ISO 12100. (2010). *Safety of machinery*. Retrieved from http://www.iso.org/iso/iso_catalogue/ catalogue_ics/catalogue_detail_ics.htm?csnumber=27239

ISO 14121-1. (2007). *Safety of machinery - Risk assessment-Part 1: Principles*.

Kosuge, K., Yoshida, H., Taguchi, D., Fukuda, T., Hariki, K., Kanitani, K., & Sakai, M. (1994). Robot-human collaboration for new robotic applications. *Proceeding of the 20th International Conference on Industrial Electronics, Control, and Instrumentation* (pp. 713-718).

Mayer, R. E. (2001). *Multi-media learning*. New York, NY: Cambridge University Press.

NASA. (n.d.). *NASA-TLX for Windows*. US Naval Research Laboratory. Retrieved from http://www.nrl.navy.mil/aic /ide/NASATLX.php

Neider, J., Davis, T., & Woo, M. (1993). *OpenGL programming guide: The official guide to learning OpenGL*. Addison-Wesley Publishing Company.

Norman, D. A. (1993). *Things that make us smart: Defending human attributes in the age of the machine*. Addison Wesley Publish Company.

Oborski, P. (2004). Man-machine interactions in advanced manufacturing systems. *International Journal of Advanced Manufacturing Technology*, *23*(3-4), 227–232. doi:10.1007/s00170-003-1574-5

Reinhart, C., & Patron, C. (2003). Integrating augmented reality in the assembly domain — Fundamentals, benefits and applications. *Annals of the CIRP*, *52*(1), 5–8. doi:10.1016/S0007-8506(07)60517-4

Roboguide, F. A. N. U. C. (n.d.). *Robot system animation tool*. Retrieved from http://www.fanuc.co.jp/en/ product/robot/ro boguide/index.html

Seki, S. (2003). One by one production in the "Digital Yatai" —Practical use of 3D-CAD data in the fabrication. *Journal of the Japan Society of Mechanical Engineering*, *106*(1013), 32–36.

Sugi, M., Nikaido, M., Tamura, Y., Ota, J., Arai, T., & Kotani, K. … Sato, Y. (2005). Motion control of self-moving trays for human supporting production cell "attentive workbench". *Proceedings of the 2005 IEEE International Conference of Robotics and Automation* (pp. 4080-4085). Barcelona, Spain.

Tan, J. T. C., Duan, F., Zhang, Y., Watanabe, K., Pongthanya, N., & Sugi, M. … Arai, T. (2008). Assembly information system for operational support in cell production. *The 41st CIRP Conference on Manufacturing Systems* (pp. 209-212).

Wemmerlov, U., & Johnson, D. J. (1997). Cellular manufacturing at 46 user plants: Implementation experiences and performance improvements. *International Journal of Production Research*, *35*(1), 29–49. doi:10.1080/002075497195966

Zhang, Y., Duan, F., Tan, J. T. C., Watanabe, K., Pongthanya, N., & Sugi, M. … Arai, T. (2008). A study of design factors for information supporting system in cell production. *The 41st CIRP Conference on Manufacturing Systems* (pp. 319-322).

About the Contributors

Vladimír Modrák is Professor of Manufacturing Technology and head of Department of Manufacturing Management at the Technical University of Košice. He obtained a PhD degree in Mechanical Engineering at the same University in 1989. His research interests include manufacturing logistics, cellular manufacturing, and other related disciplines. Prof. Modrák is Vice-Editor in Chief of *Slovak Journal on Manufacturing Engineering* and editorial board member of several international journals. He also served as session chair and chairman at international conferences. He lectured as Visiting Professor at University of Perugia (Italy), University of Applied Sciences Wildau (Germany), University of Czestochowa (Poland), and held seminars at the Keyworth Institute at the University of Leeds (UK), University of Salerno (Italy), and University of Perugia. Presently, he is also chairmen of the Commission for Doctoral Study in the field of Industrial Engineering. Prof. Modrák is co-author and editor of several books on manufacturing logistics, manufacturing technology, and other topics.

R. Suhakara Pandian earned his PhD in Mechanical Engineering under TEQIP fellowship from National Institute of Technology Rourkela, India in 2008. He has done post doctoral work at Technical University of Kosice, Slovak republic in 2009-10. Now he is working as an Associate Professor in the Department of Mechanical Engineering at Kalasalingam University, India where he joined during December 1998 as a Lecturer. He has around twelve years of teaching and research experiences. He is actively supervising four research scholars doing PhD work under him. He is acting as a co-investigator of a DST-India sponsored technical project in the area of cellular manufacturing system. In May 2010, Institution of Engineers (India) honored him with Young Engineers Award in the field of Production Engineering. His research interests include cellular manufacturing system, optimization techniques, operations management, and logistics management.

* * *

Riccardo Accorsi is a PhD student at the Industrial Engineering Doctorate School of Padua University, Italy. His academic research deals with planning, design, and optimization of manufacturing systems, and in particular, storage systems and distribution logistic networks.

Omar Alhawari received his Bachelor of Science degree in Mechanical Engineering in 1996 from Mutah University, Jordan. In 2008, he received his Master's of Sciences degree in Industrial and Systems Engineering from Ohio University, USA. He is currently working as a supervisor engineer in the Hashemite Royal Court, Jordan.

Michele Ambrico is a PhD student at Department of Environmental Engineering and Physics in the Engineering Faculty of Basilicata University (Italy). He took his degree in Industrial Engineering at Basilicata University. His academic studies are focused on cellular manufacturing systems and the development of simulation models in dynamic environment.

Tamio Arai is a Professor at the Department of Precision Engineering, The University of Tokyo. He was an Assistant Professor of the University of Tokyo in 1979. He became a Professor at the Department of Precision Engineering, The University of Tokyo in 1987. He was the Director of Research into Artifacts, Center for Engineering, The University of Tokyo, from 2000 to 2005. He research interests are cellular manufacturing and service engineering.

Jos Bokhorst is an Assistant Professor in the Department of Operations at the Faculty of Economics and Business, University of Groningen, The Netherlands. His main research areas are cellular manufacturing, lean manufacturing, dual resource constrained systems, cross-training, labor assignment, and workforce agility.

Marco Bortolini is a PhD student at the Industrial Engineering Doctorate School of Padua University, Italy. His academic research deals with mechatronics and industrial systems, e.g. the design and control of photovoltaic systems, and the optimization of production and logistic systems.

J. Hector Carlo is a Professor in the Industrial Engineering Department at University of Puerto Rico - Mayaguez. He received his PhD degree in Industrial and Operations Engineering from University of Michigan in 2007. His research interests include design and optimization of facilities planning, distribution and logistics, lean production systems, et cetera.

Karuppasamy Chandrasekar was born at Sivalarkulam, Tamilnadu, India in 1976. He earned his graduate degree in Mechanical Engineering from Madurai Kamaraj University, India in 1999 and his Master's in Computer Aided Design and Computer Aided Manufacturing from Madurai Kamaraj University, India in 2002. Now he is working as an Assistant Professor in the Department of Mechanical Engineering at Sardar Raja College of Engineering, Alangulam, Tamilnadu, India. He is also engaged in research work in the Department of Mechanical Engineering, Kalasalingam University, Krishnankoil, Tamilnadu, India. He has ten years of experience both in teaching and industry. He has published two research papers in journals and nine technical papers in conferences. His research interests include cellular manufacturing systems, operations management, and soft computing applications.

Hedi Chtourou is an Associate Professor in the preparatory institute for engineering studies of Sfax in Tunisia. He received his Master of Science and PhD degrees in mechanical engineering from the Laval University in 1991 and 1996. Dr. Chtourou's research fields are principally resource constrained project scheduling, manufacturing systems sizing, and manufacturing systems layout optimization.

Ioan Constantin Dima has finished his PhD in economics at University of Craiova in 1988. Presently he is in position of full Professor of Economics at Valahia University of Târgovişte. His research is oriented on economic aspects of logistics and decision making models of investment.

Feng Duan is an Associate Professor at the Department of Automation, College of Information Technical Science, Nankai University, Tianjin, P. R. China. He received the BS Degree and MS Degree in Mechanical Engineering from Tianjin University, P. R. China, in 2002 and 2004, respectively. He received the MS Degree and PhD degree in Precision Engineering from The University of Tokyo, Japan in 2006 and 2009, respectively. His main research interests are cellular manufacturing, human skill analysis, image processing, and dynamic simulation.

M. Bulent Durmuşoglu is a Professor of Industrial Engineering at Istanbul Technical University. He obtained his Ph.D. in Industrial Engineering from the same university. His research interests are design and implementation of cellular/lean production and management systems. He is also the author numerous technical articles in these areas.

Vahidreza Ghezavati is currently with the Faculty of Industrial Engineering at Islamic Azad University, south Tehran branch. He received his Ph.D. degree from the Iran University of Science and Technology for mathematical modeling for cellular manufacturing systems under uncertainty 2010. His past and present research interests lie in the fields of operations research, cellular manufacturing, facilities location, metaheuristics, and decision making.

Januzs Grabara is currently Vice-Dean and Director of Institute for Econometrics and Informatics at Faculty of Management, University of Czestochowa, Poland. He was honored as Associated Professor in the field of informatics at Faculty of Electronics and Informatics of Technical University of Kosice in 2008. His research area is focused on information logistics systems and reverse logistics.

Saber Ibrahim is 30 years old. He is a PhD at the Faculty of Economic Sciences and Management of Sfax, the south of Tunisia. He is working on the topic of evolutionary algorithms for solving cellular manufacturing problems. He completed his Master's degree, which consisted of solving machine cell formation problem with Tabou search method. He was a co-author in some international journals like *Computers & Operations Research*. He has also participated in several national and international conferences by some applications in cellular manufacturing systems.

Aleksey Imaev received his PhD degree in Electrical Engineering from Ohio University - Athens in 2009. Currently he is working a Systems Engineer at a consulting company His research interests span the areas of discrete event systems, max-plus algebra, graph theory, and control systems.

Bassem Jarboui is a PhD at the Faculty of Economic sciences and Management of Sfax, the south of Tunisia. He is a graduate Assistant. He obtained his Master's degree with high mention in 2006. His main research area concerns the adaptation of particle swarm optimization to the combinatorial optimization problems. Several publications in international journals and several participations in international conferences have been released on this topic. In addition, he is working on computational intelligence, evolutionary algorithms, and fuzzy logic for many applications.

Mohammad Saeed Jabal-Ameli is an Associate Professor of Industrial Engineering in the Department of Industrial Engineering, Iran University of Science and Technology, Tehran, Iran. He received his Ph.D. in Industrial Engineering from Tarbiat Modarres Univeristy1996. His research interests include facilities location, facilities design, cellular manufacturing, risk management, feasibility study, and operations research.

Abdessalem Jerbi is an Assistant Professor in the Superior Institute for Applied Sciences and Technology of Gabes in Tunisia. He received his Master of Science and PhD degrees in Mechanical Engineering from the National School of Engineers of Sfax in 2004 and 2010. His principal research field is manufacturing systems layout optimization.

P. Robert Judd is a Professor in the Industrial and Systems Engineering Department at Ohio University - Athens. He received his PhD degree in Systems Engineering from Oakland University in 1981. His research interests include discrete event systems, simulation, control of manufacturing systems, et cetera.

Ryu Kato received his Ph.D. degree in Engineering from the University of Tokyo in 2008. He was a project Assistant Professor at the Department of Precision Engineering, the University of Tokyo from 2008 to 2009. Currently, he is an Assistant Professor at Department of Mechanical Engineering and Intelligent Systems, Graduate School of Informatics and Engineering, the University of Electro-Communications, Japan. He is interested in brain machine interface and human-robot collaboration.

Goksu Kaya does research based on the engineering discipline. She is an Industrial Engineering PhD student at Istanbul Technical University. Her research interests are design and implementation of cellular/lean production and knowledge management systems.

Peter Knuth is Assistant Professor at Faculty of Manufacturing Technologies of Technical University of Kosice with a seat in Presov, Slovakia. He obtained PhD degree in Mechanical Engineering Technologies and Materials. His academic research is focused on radio frequency identification, production planning and scheduling, and simulation of cellular manufacturing systems.

Royston Lobo graduated from Manipal Institute of Technology in India in 2002 with a BE in Mechanical Engineering and earned his Master's degree from Ohio University in Industrial & Systems Engineering in 2006. He currently works as an Industrial Engineer at S. S. White Technologies with medical and aerospace products in the CNC department.

Ahmad Makui is an Assistant Professor of Industrial Engineering in the Department of Industrial Engineering, Iran University of Science and Technology, Tehran, Iran. He received his Ph.D. in Industrial Engineering from Iran University of Science and Technology in 1999. His research interests include production planning, supply chain management, MCDM, and mathematical programming.

Riccardo Manzini is an Associate Professor of Logistics and Reliability & Maintenance in the Department of Industrial and Mechanical Engineering at the University of Bologna, Italy. His academic research deals with planning, design, and control of production systems with particular reference to reliability modelling and maintenance system management & optimisation, logistics, storage systems, and supply chain.

Emre Mese completed his Bachelor of Science in Industrial Engineering from Istanbul Technical University, Istanbul, Turkey in June 2006. He earned his Master of Science in Industrial and Systems Engineering in June 2009 from Ohio University. Currently, he lives in Cincinnati, Ohio and works at d.e. Foxx & Associates, Inc.

N. Arun Nambiar is a Professor in the Industrial Technology Department at California State University - Fresno. He received his PhD degree in Integrated Engineering from Ohio University - Athens in 2007. His research interests include production planning, scheduling, lean production systems, RFID, et cetera.

Maria Nowicka-Skowron accomplished her PhD study in 1981 in the field of Economics at Academy of Economics in Katowice. Since 2005 she became Dean of the Management Faculty of Czestochowa University. Dr. Nowicka-Skowron was honored as full Professor in 2002 in the field of logistics. Her research field is econometrics and manufacturing logistics.

Abdelwaheb Rebai is a Professor at the University of Sfax. He occupied various administrative positions in several academic institutions in Tunisia. His main research interests are multi-criteria decision aid, fuzzy logics, statistics, mathematical programming, and so on. He published several papers in international journals. He is the director of more than 30 theses. He organized several training programs for Master's and Doctoral degrees. He is a member in several Tunisian and foreign associations. Also, he was the responsible of sessions in the national and international conferences and referee for national and international journals.

Paolo Renna is an Assistant Professor at Department of Environmental Engineering and Physics in the Engineering Faculty of Basilicata University (Italy). He took Ph.D. degree at Polytechnic of Bari in Advanced Production Systems. His academic researches principally deal with the development of innovative negotiation and production planning in distributed environments and manufacturing scheduling in dynamic environment.

Jose Ribeiro is a Professor of the University of Sao Paulo. He obtained a degree in Mechanical Engineering (Brazil), a MSc in Electrical Engineering (Brazil) and Industrial Engineering (France), and a Ph.D. in Operations Research (France). His research area is mathematical programming applied to manufacturing cell design and scheduling.

Seyed Jafar Sadjadi is a Professor of Industrial Engineering in the Department of Industrial Engineering, Iran University of Science and Technology, Tehran, Iran. He received his Ph.D. in Industrial Engineering from University of Waterloo 1997. His research interests include nonlinear programming, applications of operations research in planning and production, optimization, maintenance, and DEA.

Mohammad Saidi-Mehrabad is a Professor of Industrial Engineering in the Department of Industrial Engineering, Iran University of science and Technology, Tehran, Iran. He received his PhD in Industrial Engineering 1992. He has B.S degree in Mechanical Engineering. His research interests include manufacturing systems analysis and network analysis, automation, metaheuristics, CNC, and CAD/CAM.

Pavol Semanco is currently a PhD student at Faculty of Manufacturing Technologies of Technical University of Kosice with a seat in Presov, Slovakia. His academic research is focused on production planning and scheduling, and simulation of cellular manufacturing systems taking into consideration dynamic environment.

Jannes Slomp is an Associate Professor at the Faculty of Economics and Business of the University of Groningen, The Netherlands. His research interests are in flexible manufacturing systems, cellular manufacturing, lean manufacturing, and modular product development. He has published in academic journals such as *International Journal of Production Research, International Journal of Production Economics, Production Planning & Control, European Journal of Operational Research, IIE Transactions,* and *Interfaces.* He is Director of the Lean Operations Research Center (see www.rug.nl/feb/lo-rc) of the University of Groningen.

Gürsel Süer is a Professor in the Industrial and Systems Engineering Department at Ohio University. Currently he serves as the manufacturing area editor of the *Computers and Industrial Engineering Journal.* He has consulted various companies and organized/co-organized Group Technology/Cellular Manufacturing Conferences held in 2000-Puerto Rico, 2003-Ohio, 2006-Netherlands, 2009-Japan.

Jeffrey Too Chuan Tan received BS Degree and MS Degree from Universiti Tenaga Nasional, Malaysia, in 2003 and 2007, respectively. He received PhD Degree from the Department of Precision Engineering, The University of Tokyo, Japan in 2010. He is now a project Assistant Professor at the Department of Precision Engineering, the University of Tokyo. His main research interests are cell production and human task analysis.

Ponnusamy Venkumar was born at Srivilliputtur, Tamilnadu, India in 1967. He received his B.E (Mechanical Engineering) in 1994 and ME (Industrial Engineering) degree in 1997 from Madurai Kamaraj University, India. He received his PhD (Mechanical Engineering) in 2006 from Manonmanium Sundaranar University, India. Now he is the Professor in the Department of Mechanical Engineering at Kalasalingam University, India from 1997 onwards. He has around fifteen years of experience in teaching and research. He has published eleven papers in various international / national journals and presented twenty five papers in the conferences. He has organized national / international conferences and workshops. His research interests are in the field of cellular manufacturing system, optimization, and supply chain management.

Gen'ichi Yasuda joined the Faculty of Engineering of Nagasaki Institute of Applied Science, Japan, in 1983. He was a Professor of Department of Mechanical Engineering during 1996-2005, a Professor of Department of Human and Computer Intelligence, Faculty of Informatics during 2005-2010, and also a Professor of Department of Electronics and Information Technology, Graduate School of Engineering during 1996-2010. His teaching and research interests include industrial intelligent robotics, advanced intelligent mechatronics, and distributed autonomous robotic systems.

Linda Zhang is currently an Assistant Professor in Department of Operations at University of Groningen, The Netherlands. She obtained her BEng and PhD degrees in Industrial Engineering in 1998 and 2007, respectively. Her research interests lie in mass customization, production reconfiguration, and product family development. On these areas, she has published a number of articles in journals, such as *IIE Transactions, IEEE Transactions on Engineering Management, Decision Support Systems, International Journal of Production Research*, et cetera.

Chi Zhu is an associate professor at the Department of Systems Life Engineering, Maebashi Institute of Technology, Japan. He received his PhD from the University of Tokyo, Japan in 1999. He was a Researcher at National Institute for Resource and Environment, Japan and Yokohama National University in 1999 and 2000, respectively. Then he was a Post-doctoral Associate at Massachusetts Institute of Technology during 2001– 2002. His research interests include dynamic robotic manipulation, biped walking and jumping robots, humanoid robots, rehabilitation robots and power assist devices.

Index

A

Adaptive Resonance Theory (ART1) 11, 20, 70, 74, 79, 114-117, 119
affecting parameters 322, 325
Analytic Hierarchy Process (AHP) 21-23, 30-31, 33-36, 357, 360, 364
ANOVA 189, 194, 200-202, 219
Applied Mathematics (AM) 3, 291, 315
Artificial Immune System (AIS) 99, 101-102, 120-121, 124-129, 279, 291
Artificial Intelligence (AI) 10-11, 14, 128, 165-166, 178-181, 224, 311, 334
Artificial Neural Networks (ANN) 70, 99, 101-102, 113, 126
assembly burden 413-414, 420
assistant robots 420
assisted cooperation 414
attentive workbench (AWB) 414, 427
Augmented Reality (AR) 414, 426
Average Linkage (AL) 9-10, 100, 147, 166
Average Link Clustering (ALC) 100, 166
Axiomatic Design (AD) 342, 344-347, 357, 359, 363-365

B

base model 190
Batch Size (BS) 46, 61, 69, 191, 193-196, 200-203, 205, 343
bayesian optimization algorithm (BOA) 167, 170, 181
big picture mapping 352-353
Bill of Materials (BOM) 143, 332
bivariate marginal distribution algorithm 169, 181
Block-Diagonal Structure (BDS) 9, 68, 75
Bond Energy Algorithm (BEA) 9, 100, 165
branch-and-bound (BB) 11, 166, 210, 228, 230-231, 252, 279, 292
bumping phenomenon 389
business activities 6

C

Cell Formation (CF) 2, 7-12, 14-20, 57, 63-70, 74-76, 78-81, 99-101, 109, 121, 123, 126-129, 140-142, 145, 148-150, 153, 159, 161-167, 171-172, 174, 177-180, 182, 200-201, 205, 260, 298-303, 311-316, 342-343, 345, 360-361, 369, 383-384
Cell Formation Problem (CFP) 7-8, 10-11, 14-20, 64-66, 68-69, 74, 78-79, 81, 99-101, 109, 121, 126-127, 129, 140-141, 153, 159, 161, 163-165, 171-172, 174, 177-180, 303, 315-316, 383-384
cell loading 208-211, 215, 224-225, 259, 261, 267-268, 276
Cellular Equipment Effectiveness (CEE) 361
Cellular Layout (CL) 10, 21, 26, 39-40, 189-198, 201, 204-206, 356, 369
cellular manufacturing (CM) 1-2, 7-10, 15-20, 22, 24, 26, 28-29, 33-38, 40, 50-51, 62, 64-66, 68-69, 80, 99-100, 102, 127-129, 141-143, 149-150, 160, 162-166, 175, 177-178, 180-182, 189-190, 205-206, 208-210, 223-225, 246, 248, 258, 260-261, 275-276, 278, 293, 298-300, 302-303, 305, 307, 309, 312, 314-316, 318, 333, 343-345, 356, 363-367, 369-370, 375-376, 381, 383-384, 412-415, 425, 427
Cellular Manufacturing Layout (CML) 24, 40
Cellular Manufacturing System (CMS) 7, 10-11, 16-17, 19, 35-38, 40, 51, 64-66, 68-69, 99-100, 102, 126-129, 160, 162-165, 181, 206, 208-209, 224, 246, 248, 258, 261, 275, 278, 293, 298-301, 303-305, 307, 314-316, 343, 345-346, 356, 363-373, 375-376, 378-379, 381, 383-384, 413-414, 425
chance-constrained programming 303
Closeness Gap (CG) 24

Cluster Analysis (CA) 8-11, 14-15, 66, 140-143, 153, 160, 162-163, 165-166, 181
cluster analysis methods 8-11
Cluster Identification Algorithm (CIA) 166
collision protectors 420, 422-423
Complete Linkage Method (CLINK) 147
connected cells 37-42, 44, 46-47, 50, 224
Constraint Satisfaction Problem (CSP) 320
Control Factor Interactions (CFI) 200-203
Control Factors (CF) 7-8, 11, 14-15, 69-70, 75-76, 141-142, 149-150, 166-167, 198, 200-203, 205, 298, 300-301, 312
control loops 26, 32
crossover 70-74, 78, 103-105, 107, 109-110, 123-125, 167, 218-219, 286, 288

D

data analysis 44, 54
Data Envelopment Analysis (DEA) 23, 34, 36
Decoupling Point (DP) 40, 52, 345-346, 349, 351, 353-354
defining process parameters 352
defining relations 352
Degree of Decomposability (DD) 191, 193
Demand Rate (DEMAND) 191-194
descriptive methods 8, 11, 165
Design Parameters (DPs) 40, 320-322, 325, 329, 332, 344-346, 349, 351, 353-354, 369
digital yatai 414, 417, 427
Direct Clustering Algorithm (DCA) 9, 17, 100, 127, 165, 179
disconnected cells 37-43, 46, 48, 50
discrete event robotic manufacturing cells 385
disruptive innovations 6
distributed control 385, 387, 389-392, 394, 396-397, 400
Dual Check Safety (DCS) 421
Dynamic Cell Manufacturing System (DCMS) 368-369

E

economic globalization 4
economies of scale 343, 401-402
equipment costs 25-26, 404
equipment replacement decisions rules 401
Estimation Distribution Algorithm (EDA) 172, 177
Estimation of Bayesian Network Algorithm (EBNA) 170
Evolutionary Algorithm (EA) 17, 64, 128, 164, 166, 175, 179, 181
Evolutionary Strategy (ES) 163, 166, 182
Exceptional Elements (EE) 9, 68, 70, 72, 76, 79, 101, 109, 122-123, 125-126, 142, 150-152, 159-161, 164-166, 171-173, 175, 177, 301, 368

F

fabrication cells 41-42, 44-45, 48
Facilities Layout Flexibility (FLF) 24
family scheduling 208-209, 211, 215
firability 390, 392-393
Flexible Manufacturing Cells (FMC) 386-387, 401-402, 405, 409
Flow Direction (FLOW) 191-195, 279, 281
flow shop scheduling 249, 253, 256-257, 277-279, 291-293
Flow Shop Scheduling Problem (FSP) 257, 277-279
flow shop sequencing 254, 257, 278-279, 281, 291, 293
flow times 22, 26, 192, 205, 210, 224, 255-257, 262, 268, 278-279, 369-370
Fractal Cellular Manufacturing System (FCMS) 366, 368, 371-378, 381, 383
Functional Layout (FL) 21-22, 24, 26, 30, 40, 189-191, 193-197, 200-201, 204-206, 343
Functional Requirements (FRs) 344-358
fuzzy logic 10, 34, 54, 369

G

gate arc 388, 394
generic models 317, 319, 321, 327-329, 333
generic process structure 320, 333
Genetic Algorithm (GA) 15-20, 63-68, 70-74, 78-81, 99, 101-108, 122, 124-126, 166, 169, 175-176, 178-182, 208, 215, 217, 219-223, 228, 231, 254-255, 257, 261, 276-277, 279, 282, 285, 287-289, 292, 314, 316, 319, 334, 363-365, 370, 383
Genetic Algorithms Software Application (GASA) 215
Global Growth Companies (GGCs) 4-5, 18
graph partitioning methods 10-11
graph theory 10, 53, 55, 63, 65, 68-69, 80, 100
Grouping Genetic Algorithm (GGA) 17, 66, 81, 175, 178-179
group scheduling (GS) 60, 146, 210, 215, 224, 278, 299
Group Technology (GT) 7, 17-20, 33-36, 38-39, 51-52, 54, 63-67, 69-70, 79-81, 99-100, 126-129, 141, 152, 163-165, 172, 178-182, 189, 205-206, 209-210, 224-225, 275-276, 278, 292, 299-300, 333, 364-365, 367, 384

H

heuristic methods 10-11, 15, 60, 279, 293
Highly Dynamic Cellular Environment 258, 275
human-robot collaboration 413-414, 420, 425
hybrid algorithm (HA) 68, 74, 77-79, 164
Hybrid Grouping Genetic Algorithm (HGGA) 17, 175, 179
hybrid manufacturing systems (HMS) 342-346, 358, 363-364
hybrid methods 11

I

independent operations 414
Industrial Engineering (IE) 2-3, 16-20, 33-35, 51, 63-66, 80-81, 127, 129, 163, 179-182, 205-206, 224-225, 251-256, 275-276, 291-292, 314-315, 334-335, 364, 384
industrial robots 395, 401, 414, 417, 421
inherent safety theory 420, 423, 425
interchangeability of equipment 27
inventory costs 25-26
IP cameras 420, 423

J

Jacson network 309
job shop systems 53, 141, 242

K

kanban cards 358, 361
knowledge processes 351-352

L

lean manufacturing 342-346, 349, 351-352, 358, 361, 364-365
lean production 343, 364
lean thinking based investment planning 342, 344, 346, 358, 363
Learning Factorised Distribution Algorithm (LFDA) 170
Linear Cell Clustering (LCC) 66, 166
linearization approaches 313
local transitions 390, 392-394, 396

M

machine controllers 387, 390, 392-396, 398-399
Machine Group Identification (MGI) 20, 141
Machine-Part Cell Formation (MPCF) 10, 16, 20, 69, 79, 172, 174, 178
Machine-Part Incidence Matrix (MPIM) 9, 68-69, 74, 121, 144, 153-154
Machine Utilization Index (MUI) 172
Machine Utilization rate (MUR) 192-194
Make-to-Order (MTO) 37-38, 40, 50-52
Make-to-Stock (MTS) 40, 51
Management of Technology (MoT) 5-6, 16-17
manufacturing cells 7-8, 11, 16-20, 28, 30-31, 34-35, 38, 41, 52-57, 62-67, 80, 101, 121, 123, 127, 140-144, 147, 149, 154-156, 162-164, 171-172, 179-180, 182, 191, 195, 208-212, 224-225, 232, 246, 258-261, 275-277, 291-292, 303-305, 315-316, 343, 368, 370-371, 383, 385-387, 389-390, 393, 395, 397-399, 401-403, 409
manufacturing process knowledge 319
Manufacturing System (MS) 28-29, 34-35, 38, 40, 51-52, 68-69, 99-100, 116, 127, 162-164, 189-193, 195-198, 200-201, 204-205, 208-209, 211, 227, 232-233, 248, 261, 263, 267, 298-299, 303, 305, 307, 310, 314-316, 343-346, 349, 356, 358, 360, 363-365, 367-371, 374-379, 381, 383, 385-386, 388, 393-394, 398, 402, 413-414, 420, 425
market signals 6
Markov chain 302, 314
material costs 25-26
mathematical modeling 208-209, 255
mathematical programming (MP) 2, 10-11, 14-16, 54, 66, 69, 79, 101, 142, 165, 230, 253, 298-300, 319
maximum transportation 307
max-min strategy 258, 262-263, 266-267, 271
Max-Plus Algebra 227-228, 231-232, 251, 253-254, 257
max-probability problems 303
max strategy 262-263, 266, 271
Mean Earliness (ME) 192, 194
Mean Flow Time (MFT) 192-194, 200-201, 210, 224, 257, 279, 369
Mean Tardiness (MT) 192, 194, 369
mean value model 302, 314
meta-heuristics 10-11, 14-16, 54, 58, 62, 69-70, 80, 279, 282, 289-290, 292, 301, 314-315
meta model 317, 321-322, 324-327
middle managers 26
modern operations management 1-2, 4-5, 15
Modified Grouping Efficiency (MGE) 69-70, 76, 79
Modified Rank Order Clustering (MODROC) 9, 17, 69, 79, 165, 179
Multi-Modal Assembly-Support System (MASS) 412-413, 415-418, 420, 422, 424-426

Multiple-Attribute Decision Making (MADM) 24, 36
mutation 20, 70-74, 78, 103-104, 106-107, 120-125, 166-168, 218-219, 286, 288

N

neural networks (NN) 10-11, 17, 19, 54, 64-65, 68-70, 79-80, 99, 101-102, 113-114, 116, 124, 126, 128, 147, 166, 180, 230, 251, 253, 256-257
Noise Factors (NF) 198, 200

O

Objective Comparison Methodology (OCM) 190, 198-200, 202-204
one-piece flow 41, 44-46, 353, 356, 364
Operating Costs (OC) 126, 227, 260, 404, 406-407
Operating Equipment Cost (OEC) 404
operations engineers 2
Operations Management (OM) 1-6, 15-17, 20, 34-36, 52, 163, 205
Operations Research (OR) 2-3, 16-18, 63-65, 118, 128, 179-180, 254-257, 291-293, 315, 346, 363, 365, 383, 403, 410-411
operations strategy 6, 19
operator assignment decisions 258
Operator Utilization Rate (OPUR) 192-193
Order Crossover (OX) 218
organizational changes 6
overlapping mode 26, 28

P

packaging cells 38, 41-42, 44-48, 50
Part Family Identification (PFI) 141
Particle Swarm Optimization (PSO) 15, 167, 178
Permutation Flow Shop Scheduling Problem (PFSP) 277-280
personnel costs 25-26
Petri net 385-390, 392-400
Petri net based controller 385, 392, 394, 400
Position-Based Crossover (P-B) 218
probability distribution 108, 167-170, 298, 301, 303
process families 317-323, 325, 327-329, 332-333, 336-337
process family planning 317-323, 325, 327-329, 332-333, 336-337
Processing Time (PT) 73, 126, 149, 151, 192-196, 200-203, 205-206, 230, 239, 241, 243, 250, 252, 262, 264, 267-268, 279-281, 283-285, 299-300, 303, 312, 314-315, 371, 373, 402
product flexibility 28, 343
product flow analysis (PFA) 8, 16, 18, 30, 34, 68-69, 79-80, 100, 127-128, 162, 165, 178
Production and Operations Management (POM) 2-3, 15-16, 34, 36, 52
Production Area Utilisation (PAU) 24
production control 28, 31-32, 206, 358, 360
production management 1-2, 16, 19, 35-36, 162, 206, 364
production planning 34, 65, 103, 224, 227-228, 252, 256, 275, 286, 300-301, 303, 314, 316, 347, 369, 383-384, 402
production speed 26
product oriented similarity index 143
product sequencing 209-210, 224-225, 259, 261, 267-268, 276
product structure 143, 314, 320, 333
product variants 318, 320-322, 324-328, 333, 336
Programmable Logic Controllers (PLCs) 386, 394, 399-400
PROgrammed Graph Rewriting System (PRO-GRES) 317, 319, 321-322, 333-334

Q

Quadratic Assignment Problem (QAP) 101, 166
queuing theory 300, 302-304, 307, 311, 314

R

Rank Order Clustering (ROC) 9, 17-18, 69, 79-80, 100, 165, 179
RCMS 366, 370-375, 377, 379, 381-383
Reciprocal Exchange (R-E) 219
Replacement Equipment Cost (REC) 151, 404
representation 8, 10, 70-71, 78, 100, 102, 116, 121, 147, 172, 178, 213, 218, 232-233, 241, 253, 287, 318-319, 321-322, 329, 333, 352, 386-387, 389-390, 394-397, 405
reproduction 72-74, 103-104, 106-107, 167, 218-219, 286, 288
research utility 1-2
robotic manufacturing cells 385, 390, 398
robust optimization 301-303, 314
Rotary Molding Machine Minicell (RMMM) 212-213

S

scenario planning 301-302, 305, 315
Scheduling Rules (RULE) 191, 193-195, 200-203, 205, 262, 370

Self-Organizing Neural Network (SONN) 11, 17, 64
sequential heuristics 53, 58, 61-62
Set up Time (ST) 26, 28, 44, 53-54, 61, 80, 192-194, 196, 200-203, 205, 209-211, 215, 217, 219, 222-224, 276, 278-279, 291-292, 313-314, 316, 329, 352, 357, 360, 362, 371, 378, 400, 408, 427
Sheep Flock Heredity Algorithm (SFHA) 99, 101-102, 125-126
Shortest Processing Time Technique (SPT) 191, 197, 262, 268
Signal Factors (SF) 198-200
Similarity Coefficient Methods (SCF) 19, 69, 163
similarity coefficient (SC) 10, 19, 57, 64, 68-69, 145-146, 153, 155-157, 162-163, 166, 233, 236-237, 246-248, 301
Simulated Annealing (SA) 11, 15, 19-20, 53-54, 58-59, 61-64, 67, 70, 80-81, 99, 101-102, 108-110, 124-126, 166, 181, 228, 230-231, 254-255, 279, 282, 292
simultaneous cooperation 414
Single Linkage Method (SLINK) 147
Single Minute Exchange of Dye (SMED) 357, 360
Single Model Assembly Line Balancing (SALBP) 230
slope index 279, 282, 285-287, 289
stochastic optimization 300, 302-304, 314
synchronized cooperation 414

t

Tabu Search (TS) 15-16, 20, 53-54, 58-62, 64, 66, 70, 80-81, 99, 101-102, 110-113, 124, 126, 166, 180, 279, 282, 291
Taguchi Method 190, 198, 205
time bucket 367
TOPSIS 16, 24
total expected cost 302, 307
total productive maintenance (TPM) 349, 357, 360, 405
Toyota Production system (TPS) 255, 343, 364
Transfer Mode 191
Transfer Time (TT) 191-195, 200-203, 310, 369
transport equipment 26
transport operations 26

U

Unweighted Pair-Group Method using Arithmetic Average (UPGMA) 147
Unweighted Pair-Group Method using Centroid (UPGMC) 147

V

Value Stream Mapping (VSM) 342-343, 345-347, 351-354, 363-365
Variable Neighborhood Search (VNS) 164-165, 173-175, 178, 180
Virtual Cell Manufacturing System (VCMS) 368, 370
Virtual Manufacturing Cells (VMCs) 34, 63, 66, 370, 383

W

Water Flow-Like Algorithm (WFA) 15, 20
Work-In-Process Inventory (WIP) 26, 37-41, 48, 50-51, 70, 192-194, 209, 367-368, 370-371, 375, 378

Z

ZODIAC 10, 17, 69, 79, 175-176, 179

CPSIA information can be obtained at www.ICGtesting.com
Printed in the USA
BVOW020602221011

273893BV00007B/1/P

9 781613 500477